PLATO

IN TWELVE VOLUMES

IX

TIMAEUS · CRITIAS
CLEITOPHON · MENEXENUS
EPISTLES

WITH AN ENGLISH TRANSLATION BY

R. G. BURY

HARVARD UNIVERSITY PRESS
CAMBRIDGE, MASSACHUSETTS
LONDON, ENGLAND

First published 1929
Reprinted 1942, 1952, 1961, 1966, 1975, 1981, 1989

ISBN 0-674-99257-1

*Printed in Great Britain by St. Edmundsbury Press Ltd,
Bury St. Edmunds, Suffolk, on wood-free paper.
Bound by Hunter & Foulis Ltd, Edinburgh, Scotland.*

CONTENTS

	PAGE
LIST OF PLATO'S WORKS	vii
TIMAEUS	1
CRITIAS	255
CLEITOPHON	309
MENEXENUS	329
EPISTLES	383
INDEX	629

LIST OF PLATO'S WORKS showing their division into volumes in this edition and their place in the edition of H. Stephanus (vols. I–III, Paris, 1578).

VOLUME		PAGES
I. Euthyphro	I.	2A–16
Apology	I.	17A–42
Crito	I.	43A–54E
Phaedo	I.	57A–118
Phaedrus	III.	227A–279C
II. Laches	II	178A–201C
Protagoras	I.	309A–362
Meno	II.	70A–100B
Euthydemus	I.	271A–307C
III. Lysis	II.	203A–223B
Symposium	III.	172A–223D
Gorgias	I.	447A–527E
IV. Cratylus	I.	383A–440E
Parmenides	III.	126A–166C
Greater Hippias	III.	281A–304E
Lesser Hippias	I.	363A–376C
V. Republic I: Books I-V	II.	327A–480
VI. Republic II: Books VI-X	II.	484A–621D

vii

LIST OF PLATO'S WORKS

VOLUME		PAGES
VII.	Theaetetus	I. 142A–210D
	Sophist	I. 216A–268B
VIII.	The Statesman	II. 257A–311C
	Philebus	II. 11A–67B
	Ion	I. 530A–542B
IX.	Timaeus	III. 17A–92C
	Critias	III. 106A–121C
	Cleitophon	III. 406A–410E
	Menexenus	II. 234A–249E
	Epistles	III. 309A–363C
X.	The Laws I : Books I-VI	II. 624A–785B
XI.	The Laws II : Books VII-XII	II. 788A–969D
XII.	Charmides	II. 153A–176D
	Alcibiades I and II	II. 103A–151C
	Hipparchus	II. 225A–232C
	The Lovers	I. 132A–139
	Theages	I. 121A–131
	Minos	II. 313A–321D
	Epinomis	II. 973A–992E

TIMAEUS

INTRODUCTION TO THE *TIMAEUS*

From the time of Aristotle downwards the *Timaeus* has been regarded as one of the most significant writings of Plato. It has arrested the attention of philosophic inquirers because of the scope of its speculations; while the literary genius displayed in the style and diction of its central Myth has compelled the admiration of the artistic reader. The theme of that central Myth is nothing less than the Creation of the Universe—" in the beginning How the Heaven and Earth rose out of Chaos "; and the oracular obscurity with which this abstruse theme is expounded has served to excite rather than repel the interest of the curious student of every age.

The *Timaeus* is professedly a sequel to the *Republic*, portions of which are recapitulated in the prefatory chapter, and it is also the first section of a projected trilogy, which was intended to contain also a *Critias* (of which only a fragment remains) and a *Hermocrates*. The interlocutors of the dialogue are Socrates, Timaeus, Hermocrates, and Critias. Of Timaeus of Locri nothing authentic is known beyond what we learn from the dialogue itself—that he was a man of high social position and wide culture, especially skilled in mathematics and astronomy: later tradition makes him out a leader of the Pythagorean school. Hermo-

crates is generally identified with the famous Syracusan general of that name, who is said to have spent his later years in exile at Sparta and in Asia Minor. Critias, a connexion of Plato's, poet and orator, chief of the 30 tyrants of 404 B.C., has already figured in the earlier dialogues, *Charmides* and *Protagoras*. Of him it was said, according to the Scholiast, that he was "an amateur among philosophers, a philosopher among amateurs."

The dialogue may conveniently be divided into three main sections:

(1) The Introduction, including the Atlantis legend as related by Solon (19 A-27 C).

(2) The making of the Soul of the World; the doctrine of the Elements; and the theory of matter and of sense-objects (27 C-69 A).

(3) The making of Man's Soul and Body; physiology and pathology (69 A-end).

Of the introductory section little need be said. The Atlantis legend serves as a connecting link between the *Timaeus* and the *Critias*, but how far the legend as here given is really based on earlier tradition, and whether there ever existed any island of the kind described, seems to be quite uncertain. The fact that it would lie somewhere near that volcanic belt of the earth's crust which stretches from Etna and Vesuvius to Teneriffe may seem to lend some plausibility to the story. In any case the account of the ancient Athenians and the islanders of Atlantis whom they routed is a fine piece of literary fiction; and in its eulogy of Athens it has many points of resemblance to such appreciations as we find in the *Menexenus* and the third book of the *Laws*: "the men of Marathon" had their proto-

INTRODUCTION TO THE *TIMAEUS*

types in the saviours of Greece from Atlantis 9000 years before.

The central portion of the *Timaeus* is that which has chiefly won it fame. In it is handled the most tremendous of subjects, the construction of the Universe. We are transported in imagination to a point " before the beginning of years," when time was not, and " the earth was without form and void." There we follow, step by step, the process whereby the World was built up into a harmonious structure, and Cosmos evolved out of Chaos. We need not repeat here all the details of that process; nor can we deal at length with the many obscure problems and points of controversy which are raised by the mythical language in which the exposition is clothed. But it may be helpful to the reader unversed in Plato's philosophy if we add some observations, mostly of a general character, calling attention to the most notable points of his doctrine as here set forth.

In a well-known passage of the *Phaedo* (96 A ff.) Plato had criticized the earlier philosophers for their failure to indicate the Cause of the physical processes by which they explained the World: even Anaxagoras, when he introduces his World-ordering " Reason," makes very little use of it as a causative principle. A thorough-going teleological explanation of the Universe is what Plato desiderated in the *Phaedo*, and what he attempts to provide in the *Timaeus*. But in this connexion it is important to notice that a distinction is drawn between Cause proper, or final cause, and auxiliary Cause, which is the sum of necessary physical conditions. Related to this is the distinction between the operations of Intelligence and the works of Necessity in the formation of the

PLATO

World; we have to recognize that the Divine Reason which in all things designs the best cannot always and completely realize its designs because of an intrinsic and incorrigible element which subsists always in the World.

Another distinction, fundamental in Plato's philosophy, is that which meets us at the outset of this section—the distinction between Being and Becoming. Being is changeless, eternal, self-existent, apprehensible by thought only; Becoming is the opposite—ever-changing, never truly existent, and the object of irrational sensation. The one is the Ideal sphere, the other the Phenomenal; and it is evident that the visible Universe belongs to the latter. Consequently it cannot be an object of pure thought and does not admit of investigation by " science " in the strict sense of the term. And thus we find Plato emphasizing repeatedly the provisional character of his exposition: the *true* must be distinguished from the " *likely* " or probable, and it is only the latter which we can hope to attain in dealing with matters of this kind. In making this distinction Plato was not innovating: long before him Parmenides had divided his exposition into two sections, " the Way of Truth " and " the Way of Opinion," while Democritus had drawn a sharp line of division between the " Dark Knowledge " we have of sensibles and the " Genuine Knowledge " which apprehends the only realities, the Atoms and the Void. But while Plato is thus careful to mark the hypothetical and uncertain character which attaches to all physical investigation as such, he is obviously serious with the explanations he gives, and regards them as a summary of the best knowledge available *de rerum natura*.

INTRODUCTION TO THE *TIMAEUS*

The rôle of the World-Artificer, the "Demiurge," is similar to that of the Anaxagorean "Nous" (Reason). He is not a Creator, in the strict sense; that is to say, he does not make things *ex nihilo* but only imposes order and system on pre-existing Chaos. Nor does he continue to act directly, *in propria persona*, throughout the process of world-building, but, at a certain stage, hands over his task to the created star-gods and retires into his primal solitude. This, at least, is what the mythical narrative tells us; but later on we hear of "God" (in the singular) or "the Divine" Power, as well as "the Gods" (in the plural), taking an active part in operations marked by rational purpose. The chief characteristic of the Demiurge is his *goodness*: he aims always at *the best*: he is, in fact, the embodiment of "the Good" regarded as efficient Cause. This inevitably reminds us of the famous passage in the *Republic* which describes "the Self-Good," or supreme Idea of Good, as the source of all knowledge and all existence, the Sun of the intelligible sphere; and many interpreters of Platonism have identified the Demiurge with this Idea. But if the Demiurge is the Idea of the Good, what then is the "Model" or Pattern (*paradeigma*) in view of which he operates, the "Self-Animal"? Surely it must be an Idea, and a final cause, *i.e.* the Idea of the Good; in which case the Demiurge cannot be this (or any) Idea. And, in general, we may say that the Ideas are described as objects of intelligence, not intelligences, whereas the Demiurge obviously is an intelligence. Consequently there would seem to be no way of maintaining this identification of the Demiurge with the Idea of the Good except by supposing that there is no real

PLATO

distinction between the Demiurge and his Model, he being at once subject and object, intellect and intelligible, equivalent, in fact, to Aristotle's Deity—"Thought thinking upon Thought." But there is no hint in the text of the *Timaeus* that this is Plato's meaning. A more plausible identification finds the Demiurge in the "Cause of the Mixture" as expounded in the *Philebus* (23 D ff.), many phrases of which echo the language of the *Timaeus*, describing the efficient Cause as a "Demiurgic" power (27 B), "veritable and divine Reason" (22 C, 28 D, 30 C).

The primary operation of the Demiurge is to construct the World-Soul. In his description of this process Plato mixes myth with mathematics in a peculiarly baffling way. The World-Soul is said to be composed of three elements, Sameness, Otherness, and Being; or of "the indivisible and changeless essence," "the divisible," and a third essence which is a mean proportional between these two. Moreover, it is constructed so as to combine within itself one "Circle of the Same" and seven "Circles of the Other." This fact, together with the mathematical details of the construction, suggests at once that the main purpose of the doctrine of the World-Soul is to supply an explanation of the heavenly bodies, their revolutions and periods, the "Circle of the Same" being that of the fixed stars, and the "Circles of the Other" those of the seven planets. The details of Plato's astronomy (a subject in which he shows great interest both here and elsewhere) need not concern us, but his views about the nature of Soul are of such importance for his philosophy in general that further explanation may be desirable.

Although Timaeus mentions Motion and its kinds

INTRODUCTION TO THE *TIMAEUS*

before he treats of Soul, this order is really illogical inasmuch as there can be no motion apart from Soul. For already in the *Phaedrus*, and again in the *Laws* (896 A), Plato defines Soul as " the self-moving," the first principle and originator of all movement. It is at once the cause of its own subjective affections (desire, emotion, opinion, etc., see *Laws* 897 A), and of all motion and change in the corporeal world, which give rise to the sense-qualities of material things. In the *Timaeus* (34 A) we have seven kinds or modes of Motion distinguished, and in the *Laws* ten kinds; but both dialogues agree in making one special kind correspond to " the revolutions and reckonings of Reason," namely the uniform revolution of a sphere revolving on its own axis and in the same spot. That is the highest and best type of motion, the type proper to that prime embodiment of rational activity, the World-All; and in the outer Circle of the Cosmic Sphere, the heaven of the fixed stars, this type of motion is seen at its purest. Moreover, Plato seems to attach a special importance to this " revolution of Reason " because he sees in it an example of the union of opposites, Motion reconciled with Rest, " the Same " with " the Other."

The precise relation of Reason to Soul is frequently left obscure. If the Demiurge is Reason personified, as one seems bound to suppose, and also the Constructor of Soul, it appears to follow that Reason is separate from Soul and prior thereto. On the other hand, if Soul is the first cause of Motion and Reason possesses Motion, Soul must be prior to Reason. To resolve this apparent contradiction it seems best to regard Reason as a species of Soul, or a part of Soul (as in the *Republic*, 435 E ff.); or, to use the language

of the *Theaetetus* (185 E), Reason is Soul functioning "itself by itself." This view is supported not only by the assigning of "the Motion of Reason" to the (seven) Motions of the Soul, but also by the position ascribed to human reason as one of the three parts of the embodied soul. Further confirmation of this partial identity of Soul and Reason is afforded by the statement (in 37 B-C) that Soul apprehends intelligible objects through the action of "the Circle of the Same," which is a part of herself, on the principle "like is known by like." Taking all this into consideration we seem forced to the conclusion that the Demiurge is no separate Power or independent Divinity, but merely a part or faculty of the World-Soul, his apparent independence being due solely to the mythical form of the exposition.

Turning next to the account given of the Body of the Universe, there is no point more obscure or more vehemently disputed than the nature of the so-called "Platonic matter." The passage which deals with this subject (48 E ff.) begins by stating that besides the Model and the Copy a third principle is necessary to the formation of the World. The Forms which pass in and out of the world of changing elements require a permanent substrate upon which they may imprint themselves. Such a substrate must itself be wholly formless, but capable of receiving forms of every kind; it is essentially "the all-receptive," "that *in* which" the forms appear, or "place." The view taken by Zeller, and other important authorities, identifies this with Space. No doubt there is a good deal in Plato's account which seems to favour this view; but it does not seem quite satisfactory. For one thing, if Plato means simply "empty space," or

INTRODUCTION TO THE *TIMAEUS*

"the void," why does he not say so plainly? Why so much mystification about it? Again, irregular motion (52 D) and weight (53 A) are ascribed to the substrate, but are hardly attributes of empty space. The supporters of the identification with Space try to avoid some of the difficulties it involves by distinguishing between a "primary matter," which is Space, and a "secondary matter" of a mythical character; but this device is neither legitimate nor successful. The fact that the assumption of this substrate is closely connected with Plato's doctrine of the Elements (53 c ff.) leads us rather to the belief that it corresponds to the Democritean conception of a primitive ground-stuff, void of quality, from which the infinity of atoms are split off. So it may help us to understand the doctrine of the material substrate, regarded in itself, if we consider briefly the doctrine of the elements.

In 31 B ff. we are told that the Body of the Universe is compounded of the four elements—fire, air, water, earth—of which the second and third are mean proportionals between the first and last. In this Plato is merely adopting the doctrine of Empedocles, who affirmed these elements to be "the four roots of all things"; and he gives no hint that they are not basic substances. But when we come to 53 c ff. we are told that these so-called elements are not, strictly speaking, "elements" at all (in the sense of simple basic substances) but compounds; and they are analysed for us into their ultimate constituents. By the aid of the latest researches in Stereometry, Plato explains the formation of the five regular solids (the so-called "Platonic bodies"), four of which he assigns to the elements. Thus it is shown how each of these

is an aggregate of basic triangles. For the details of this analysis we must refer to the text itself; but one of the questions it raises deserves attention, since it bears on the problem of the material substrate. If the basic triangles are merely mathematical figures existent in pure space, how can they form solid bodies endowed with corporeal qualities? Is it not more likely that Plato conceived his triangles to be a kind of prisms, corresponding to the solid atoms of Democritus; and if so, must not the substrate in which they are located be something more substantial than pure space?

It is in connexion with the " dim and baffling form " of the Substrate and with the doctrine of the Elements that Plato makes his most positive statement about the Ideas (51 B ff.). Already, in 27 D ff., he had distinguished sharply between eternal Being, the object of thought, and that which is ceaselessly in process of Becoming and Perishing, the object of opinion and sense. This distinction was required in order to make clear the relation between the visible Universe and its Model, the " Self-Animal "; and this Idea of the Universe as a Living Whole is the only Idea referred to in the earlier passage. In the later passage it is the Idea of Fire which is specially mentioned, and other Ideas are implied, though not specified. The terms used in describing Ideas, or " intelligible Forms," are very similar to those employed in such earlier dialogues as the *Symposium*, *Phaedo*, and *Republic*: the Ideas are eternal, immutable, self-subsistent essences, cognizable only by pure thought. And, what is most remarkable, the proof of the reality of the Ideas is made to lie in the difference between reason and opinion: since reason and

INTRODUCTION TO THE *TIMAEUS*

opinion are two distinct faculties, they must deal with two different classes of objects, and the objects of Reason can only be intelligible Forms or Ideas. Plato, as it appears from this, approaches the doctrine of the Ideas from the side of Epistemology rather than of Ontology: the reality of the Ideas is an assumption necessary to provide an objective basis for any valid science or system of knowledge. But it cannot be said that the *Timaeus* adds anything new to our knowledge of the Ideal Theory, and no unprejudiced reader would venture to claim that "the *Timaeus*, and the *Timaeus* alone, enables us to recognize Platonism as a complete and coherent system of monistic idealism," or to assert that "the *Timaeus* furnishes us with a master-key, whereby alone we may enter into Plato's secret chambers."

In truth, there is but little of metaphysics in the *Timaeus*; it is mainly occupied with the attempt to give a "probable" account of matters which belong to the sciences of physics and physiology. And the departments of "natural" science in which Plato displays most interest are those which admit of treatment by applied mathematics. Indeed we may fairly suppose that one of the main purposes of the *Timaeus* is to provide a permanent record of the discoveries of Plato's friends Theaetetus and Eudoxus in the field of mathematics and astronomy, by enshrining them in a general treatise for which no fitter title could be found than the words "God geometrizes." Nor was it only for its own sake that Plato valued this branch of learning: he valued it also as a help to the knowledge of law and order, measure and symmetry, uniformity and regularity, harmony and rhythm; and to the application of these to the art of life. By

discovering the " numbers and forms " of the divinely ordered Cosmos, and the laws of the motions of the stellar deities, we are determining a standard and pattern for our own souls and their emotions ; since our well-being lies in conforming to the Universal Order, keeping step with the rhythm of the Cosmos and in tune with the celestial harmonies. So true is it that " in the image of God made He man," and that the chief end of Man and his final felicity is " assimilation to God "—Man the visible Microcosm imaging God the invisible Macrocosm. Thus Cosmology is for the sake of Ethics and Politics : the Cosmic Goodness and Beauty are worth study if they teach us to make ourselves more beautiful and good " in the inner man " : for *virtue* is the first theme of Plato's teaching and the last.

In Plato's view the " natural sciences " with which this dialogue is mainly occupied—physics, biology, and pathology—are precisely those branches of study which are least " scientific " in the proper sense of the word, inasmuch as they deal with objects of sense. Nevertheless they admit of degrees of probability ; and in the *Timaeus* Plato, as we must assume, is giving us what he believes to be a summary of the best available knowledge about these subjects—a summary which might serve as an official hand-book for the members of the Academy. As regards his sources, he draws freely, no doubt, on the writings of earlier physicists, such as Empedocles and Alcmaeon, and of the medical Schools of Cos and Cnidos ; and many friends and disciples of his own may have contributed to his knowledge by their researches and speculations ; nor is it likely that there is much that is original in his treatment, beyond the strictly

INTRODUCTION TO THE *TIMAEUS*

teleological standpoint. And, unfortunately, it is just this standpoint which tends most to hamper the student of "nature" by luring him to look for "design" in the wrong place, and by fixing his gaze on what "ought" to be rather than what is. Plato, in fact, was too much of an idealist and too much of a mathematician to be a good naturalist; and yet we must give him the credit of making a brave effort, in the *Timaeus*, to master and set down the best that was then known about the world of Nature and of Man.

The text here printed is based on that of the Zurich edition (Zur.), the main deviations from which are indicated in the footnotes.

For help in the interpretation of the mathematical passage (35 A-36 D), with the relevant annotations, I am specially indebted to the kindness of Dr. A. L. Peck, of Christ's College, Cambridge.

Besides the well-known edition by Mr. R. D. Archer-Hind, with its stylish translation, there is a recent English Commentary on the *Timaeus*, as well as a separate translation of the *Timaeus* and *Critias*, by Prof. A. E. Taylor.

R. G. B.

ΤΙΜΑΙΟΣ

ΤΑ ΤΟΥ ΔΙΑΛΟΓΟΥ ΠΡΟΣΩΠΑ

ΣΩΚΡΑΤΗΣ, ΚΡΙΤΙΑΣ, ΤΙΜΑΙΟΣ, ΕΡΜΟΚΡΑΤΗΣ

17 Εἷς, δύο, τρεῖς· ὁ δὲ δὴ τέταρτος ἡμῖν, ὦ φίλε Τίμαιε, ποῦ, τῶν χθὲς μὲν δαιτυμόνων, τὰ νῦν δ' ἑστιατόρων;

ΤΙ. Ἀσθένειά τις αὐτῷ ξυνέπεσεν, ὦ Σώκρατες· οὐ γὰρ ἂν ἑκὼν τῆσδε ἀπελείπετο τῆς ξυνουσίας.

ΣΩ. Οὐκοῦν σὸν τῶνδέ τε ἔργον καὶ τὸ ὑπὲρ τοῦ ἀπόντος ἀναπληροῦν μέρος;

ΤΙ. Πάνυ μὲν οὖν, καὶ κατὰ δύναμίν γε οὐδὲν B ἐλλείψομεν· οὐδὲ γὰρ ἂν εἴη δίκαιον χθὲς ὑπὸ σοῦ ξενισθέντας οἷς ἦν πρέπον ξενίοις μὴ οὐ προθύμως σὲ τοὺς λοιποὺς ἡμῶν ἀντεφεστιᾶν.

ΣΩ. Ἆρ' οὖν μέμνησθε ὅσα ὑμῖν καὶ περὶ ὧν ἐπέταξα εἰπεῖν;

ΤΙ. Τὰ μὲν μεμνήμεθα, ὅσα δὲ μή, σὺ παρὼν ὑπομνήσεις. μᾶλλον δέ, εἰ μή τί σοι χαλεπόν, ἐξ ἀρχῆς διὰ βραχέων πάλιν ἐπάνελθε αὐτά, ἵνα βεβαιωθῇ μᾶλλον παρ' ἡμῖν.

C ΣΩ. Ταῦτ' ἔσται. χθές που τῶν ὑπ' ἐμοῦ ῥηθέν-

---

[1] This fourth guest cannot be identified. Some have supposed that Plato himself is intended.

TIMAEUS

CHARACTERS

SOCRATES, TIMAEUS, HERMOCRATES, CRITIAS

soc. One, two, three,—but where, my dear Timaeus, is the fourth [1] of our guests of yesterday, our hosts of to-day?

TIM. Some sickness has befallen him, Socrates; for he would never have stayed away from our gathering of his own free will.

soc. Then the task of filling the place of the absent one falls upon you and our friends here, does it not?

TIM. Undoubtedly, and we shall do our best not to come short; for indeed it would not be at all right, after the splendid hospitality we received from you yesterday, if we—that is, those who are left of us—failed to entertain you cordially in return.

soc. Well, then, do you remember the extent and character of the subjects which I proposed for your discussion?

TIM. In part we do remember them; and of what we have forgotten you are present to remind us. Or rather, if it is not a trouble, recount them again briefly from the beginning, so as to fix them more firmly in our minds.

soc. It shall be done. The main part of the dis-

17 τῶν λόγων περὶ πολιτείας ἦν τὸ κεφάλαιον, οἷά τε καὶ ἐξ οἵων ἀνδρῶν ἀρίστη κατεφαίνετ' ἄν μοι γενέσθαι.

ΤΙ. Καὶ μάλα γε ἡμῖν, ὦ Σώκρατες, ῥηθεῖσα πᾶσι κατὰ νοῦν.

ΣΩ. Ἆρ' οὖν οὐ τὸ τῶν γεωργῶν, ὅσαι τε ἄλλαι τέχναι, πρῶτον ἐν αὐτῇ χωρὶς διειλόμεθα ἀπὸ τοῦ γένους τοῦ τῶν προπολεμησόντων;

ΤΙ. Ναί.

ΣΩ. Καὶ κατὰ φύσιν δὴ δόντες τὸ καθ' αὑτὸν
D ἑκάστῳ πρόσφορον ἓν μόνον ἐπιτήδευμα [καὶ μίαν ἑκάστῳ τέχνην],[1] τούτους οὓς πρὸ πάντων ἔδει πολεμεῖν εἴπομεν, ὡς ἄρα αὐτοὺς δέοι φύλακας εἶναι μόνον τῆς πόλεως, εἴ τέ τις ἔξωθεν ἢ καὶ τῶν ἔνδον ἴοι κακουργήσων, δικάζοντας μὲν πράως τοῖς ἀρχομένοις ὑπ' αὐτῶν ἅτε φύσει φίλοις οὖσι,
18 χαλεποὺς δὲ ἐν ταῖς μάχαις τοῖς ἐντυγχάνουσι τῶν ἐχθρῶν γιγνομένους.

ΤΙ. Παντάπασι μὲν οὖν.

ΣΩ. Φύσιν γὰρ οἶμαί τινα τῶν φυλάκων τῆς ψυχῆς ἐλέγομεν ἅμα μὲν θυμοειδῆ, ἅμα δὲ φιλόσοφον δεῖν εἶναι διαφερόντως, ἵνα πρὸς ἑκατέρους δύναιντο ὀρθῶς πρᾶοι καὶ χαλεποὶ γίγνεσθαι.

ΤΙ. Ναί.

ΣΩ. Τί δὲ τροφήν; ἆρ' οὐ γυμναστικῇ καὶ μουσικῇ μαθήμασί τε, ὅσα προσήκει τούτοις, ἐν ἅπασι τεθράφθαι;

ΤΙ. Πάνυ μὲν οὖν.

[1] καὶ . . . τέχνην bracketed by Hermann : καὶ is omitted by best MSS.

TIMAEUS

course I delivered yesterday [1] was concerned with the kind of constitution which seemed to me likely to prove the best, and the character of its citizens.

TIM. And in truth, Socrates, the polity you described was highly approved by us all.

SOC. Did we not begin by dividing off the class of land-workers in it, and all other crafts, from the class of its defenders? [2]

TIM. Yes.

SOC. And when, in accordance with Nature, we had assigned to each citizen his one proper and peculiar occupation, we declared that those whose duty it is to fight in defence of all must act solely as guardians of the State, in case anyone from without or any of those within should go about to molest it; and that they should judge leniently such as are under their authority and their natural friends, but show themselves stern in battle towards all the enemies they encounter.[3]

TIM. Very true.

SOC. For we said, as I think, that the soul of the Guardians ought to be of a nature at once spirited and philosophic in a superlative degree, so that they might be able to treat their friends rightly with leniency and their foes with sternness.

TIM. Yes.

SOC. And what of their training? Did we not say that they were trained in gymnastic, in music, and in all the studies proper for such men? [4]

TIM. Certainly.

---

[1] *i.e.* the *Republic*, of which the political part (books ii.-v.) is here briefly recapitulated.
[2] See *Rep.* 369 E ff., 374 E ff.
[3] *Cf. Rep.* 375 B ff.
[4] *Cf. Rep.* 376 D ff.

PLATO

ΣΩ. Τοὺς δέ γε οὕτω τραφέντας ἐλέχθη που μήτε χρυσὸν μήτε ἄργυρον μήτε ἄλλο ποτὲ μηδὲν κτῆμα ἑαυτῶν ἴδιον νομίζειν δεῖν, ἀλλ' ὡς ἐπικούρους μισθὸν λαμβάνοντας τῆς φυλακῆς παρὰ τῶν σωζομένων ὑπ' αὐτῶν, ὅσος σώφροσι μέτριος, ἀναλίσκειν τε δὴ κοινῇ καὶ ξυνδιαιτωμένους μετ' ἀλλήλων ζῆν, ἐπιμέλειαν ἔχοντας ἀρετῆς διὰ παντός, τῶν ἄλλων ἐπιτηδευμάτων ἄγοντας σχολήν.

ΤΙ. Ἐλέχθη καὶ ταῦτα ταύτῃ.

ΣΩ. Καὶ μὲν δὴ καὶ περὶ γυναικῶν ἐπεμνήσθημεν, ὡς τὰς φύσεις τοῖς ἀνδράσι παραπλησίας εἴη ξυναρμοστέον καὶ τὰ ἐπιτηδεύματα πάντα κοινὰ κατά τε πόλεμον καὶ κατὰ τὴν ἄλλην δίαιταν δοτέον πάσαις.

ΤΙ. Ταύτῃ καὶ ταῦτα ἐλέγετο.

ΣΩ. Τί δὲ δὴ τὸ περὶ τῆς παιδοποιίας; ἢ τοῦτο μὲν διὰ τὴν ἀήθειαν τῶν λεχθέντων εὐμνημόνευτον, ὅτι κοινὰ τὰ τῶν γάμων καὶ τὰ τῶν παίδων πᾶσιν ἁπάντων ἐτίθεμεν, μηχανώμενοι ὅπως μηδείς ποτε τὸ γεγενημένον αὐτῷ ἰδίᾳ γνώσοιτο, νομιοῦσι δὲ πάντες πάντας αὐτοὺς ὁμογενεῖς, ἀδελφὰς μὲν καὶ ἀδελφοὺς ὅσοιπερ ἂν τῆς πρεπούσης ἐντὸς ἡλικίας γίγνωνται, τοὺς δ' ἔμπροσθεν καὶ ἄνωθεν γονέας τε καὶ γονέων προγόνους, τοὺς δ' εἰς τὸ κάτωθεν ἐκγόνους παῖδάς τ' ἐκγόνων;

ΤΙ. Ναί, καὶ ταῦτα εὐμνημόνευτα ᾗ λέγεις.

ΣΩ. Ὅπως δὲ δὴ κατὰ δύναμιν εὐθὺς γίγνοιντο ὡς ἄριστοι τὰς φύσεις, ἆρ' οὐ μεμνήμεθα ὡς τοὺς ἄρχοντας ἔφαμεν καὶ τὰς ἀρχούσας δεῖν εἰς τὴν τῶν γάμων σύνερξιν λάθρᾳ μηχανᾶσθαι κλήροις

---

[1] *Cf. Rep.* 416 D ff.

TIMAEUS

soc. And it was said, I believe, that the men thus trained should never regard silver or gold or anything else as their own private property; but as auxiliaries, who in return for their guard-work receive from those whom they protect such a moderate wage as suffices temperate men, they should spend their wage in common and live together in fellowship one with another, devoting themselves unceasingly to virtue, but keeping free from all other pursuits.[1]

TIM. That too was stated as you say.

soc. Moreover, we went on to say about women [2] that their natures must be attuned into accord with the men, and that the occupations assigned to them, both in war and in all other activities of life, should in every case be the same for all alike.

TIM. This matter also was stated exactly so.

soc. And what about the matter of child-production? Or was this a thing easy to recollect because of the strangeness of our proposals? For we ordained that as regards marriages and children all should have all in common, so that no one should ever recognize his own particular offspring, but all should regard all as their actual kinsmen—as brothers and sisters, if of a suitable age; as parents and grandparents, if more advanced in age; and as children and children's children, if junior in age.[3]

TIM. Yes, this also, as you say, is easy to recollect.

soc. And in order that, to the best of our power, they might at once become as good as possible in their natural characters, do we not recollect how we said that the rulers, male and female, in dealing with marriage-unions must contrive to secure, by

[2] *Cf. Rep.* 451 c ff.
[3] *Cf. Rep.* 457 ff., 461 D.

PLATO

τισὶν ὅπως οἱ κακοὶ χωρὶς οἵ τ' ἀγαθοὶ ταῖς ὁμοίαις ἑκάτεροι ξυλλήξονται, καὶ μή τις αὐτοῖς ἔχθρα διὰ ταῦτα γίγνηται, τύχην ἡγουμένοις αἰτίαν τῆς ξυλλήξεως;

ΤΙ. Μεμνήμεθα.

ΣΩ. Καὶ μὴν ὅτι γε τὰ μὲν τῶν ἀγαθῶν θρεπτέον ἔφαμεν εἶναι, τὰ δὲ τῶν κακῶν εἰς τὴν ἄλλην λάθρᾳ διαδοτέον πόλιν· ἐπαυξανομένων δὲ σκοποῦντας ἀεὶ τοὺς ἀξίους πάλιν ἀνάγειν δεῖν, τοὺς δὲ παρὰ σφίσιν ἀναξίους εἰς τὴν τῶν ἐπανιόντων χώραν μεταλλάττειν;

ΤΙ. Οὕτως.

ΣΩ. Ἆρ' οὖν δὴ διεληλύθαμεν ἤδη καθάπερ χθές, ὡς ἐν κεφαλαίοις πάλιν ἐπανελθεῖν; ἢ ποθοῦμεν ἔτι τι τῶν ῥηθέντων, ὦ φίλε Τίμαιε, ὡς[1] ἀπολειπόμενον;

ΤΙ. Οὐδαμῶς, ἀλλὰ αὐτὰ[2] ταῦτα ἦν τὰ λεχθέντα, ὦ Σώκρατες.

ΣΩ. Ἀκούοιτ' ἂν ἤδη τὰ μετὰ ταῦτα περὶ τῆς πολιτείας ἣν διήλθομεν, οἷόν τι πρὸς αὐτὴν πεπονθὼς τυγχάνω. προσέοικε δὲ δή τινί μοι τοιῷδε τὸ πάθος, οἷον εἴ τις ζῷα καλά που θεασάμενος, εἴτε ὑπὸ γραφῆς εἰργασμένα εἴτε καὶ ζῶντα ἀληθινῶς, ἡσυχίαν δὲ ἄγοντα, εἰς ἐπιθυμίαν ἀφίκοιτο θεάσασθαι κινούμενά τε αὐτὰ καί τι τῶν τοῖς σώμασι δοκούντων προσήκειν κατὰ τὴν ἀγωνίαν ἀθλοῦντα. ταὐτὸν καὶ ἐγὼ πέπονθα πρὸς τὴν πόλιν ἣν διήλθομεν· ἡδέως γὰρ ἄν του λόγῳ διεξιόντος ἀκούσαιμ' ἄν, ἄθλους οὓς πόλις ἀθλεῖ,

---

[1] ὡς is omitted by most MSS. and Zur.
[2] αὐτὰ Stephens: ταὐτὰ Zur. with best MS.

---

[1] *Cf. Rep.* 458 ff.

TIMAEUS

some secret method of allotment, that the two classes of bad men and good shall each be mated by lot with women of a like nature, and that no enmity shall occur amongst them because of this, seeing that they will ascribe the allotment to chance?[1]

TIM. We recollect.

SOC. And do you recollect further how we said that the offspring of the good were to be reared, but those of the bad were to be sent privily to various other parts of the State; and as these grew up the rulers should keep constantly on the watch for the deserving amongst them and bring them back again, and into the place of those thus restored transplant the undeserving among themselves?[2]

TIM. So we said.

SOC. May we say then that we have now gone through our discourse of yesterday, so far as is requisite in a summary review; or is there any point omitted, my dear Timaeus, which we should like to see added?

TIM. Certainly not: this is precisely what was said, Socrates.

SOC. And now, in the next place, listen to what my feeling is with regard to the polity we have described. I may compare my feeling to something of this kind: suppose, for instance, that on seeing beautiful creatures, whether works of art or actually alive but in repose, a man should be moved with desire to behold them in motion and vigorously engaged in some such exercise as seemed suitable to their physique; well, that is the very feeling I have regarding the State we have described. Gladly would I listen to anyone who should depict in words our State contending

[2] *Cf. Rep.* 415 B, C, 459 D ff.

19 τούτους αὐτὴν ἀγωνιζομένην πρὸς πόλεις ἄλλας, πρεπόντως εἴς τε πόλεμον ἀφικομένην καὶ ἐν τῷ πολεμεῖν τὰ προσήκοντα ἀποδιδοῦσαν τῇ παιδείᾳ καὶ τροφῇ κατά τε τὰς ἐν τοῖς ἔργοις πράξεις καὶ κατὰ τὰς ἐν τοῖς λόγοις διερμηνεύσεις πρὸς ἑκάστας τῶν πόλεων. ταῦτ᾽ οὖν, ὦ Κριτία καὶ
D Ἑρμόκρατες, ἐμαυτοῦ μὲν αὐτὸς κατέγνωκα μή ποτ᾽ ἂν δυνατὸς γενέσθαι τοὺς ἄνδρας καὶ τὴν πόλιν ἱκανῶς ἐγκωμιάσαι. καὶ τὸ μὲν ἐμὸν οὐδὲν θαυμαστόν· ἀλλὰ τὴν αὐτὴν δόξαν εἴληφα καὶ περὶ τῶν πάλαι γεγονότων καὶ τῶν νῦν ὄντων ποιητῶν, οὔ τι τὸ ποιητικὸν ἀτιμάζων γένος, ἀλλὰ παντὶ δῆλον ὡς τὸ μιμητικὸν ἔθνος, οἷς ἂν ἐντραφῇ, ταῦτα μιμήσεται ῥᾷστα καὶ ἄριστα, τὸ δ᾽ ἐκτὸς τῆς τροφῆς ἑκάστοις γιγνόμενον χαλεπὸν μὲν
E ἔργοις, ἔτι δὲ χαλεπώτερον λόγοις εὖ μιμεῖσθαι. τὸ δὲ τῶν σοφιστῶν γένος αὖ πολλῶν μὲν λόγων καὶ καλῶν ἄλλων μάλα ἔμπειρον ἥγημαι, φοβοῦμαι δὲ μή πως, ἅτε πλανητὸν ὂν κατὰ πόλεις οἰκήσεις τε ἰδίας οὐδαμῇ διῳκηκός, ἄστοχον ἅμα φιλοσόφων ἀνδρῶν ᾖ καὶ πολιτικῶν, ὅσ᾽ ἂν οἷά τε ἐν πολέμῳ καὶ μάχαις πράττοντες ἔργῳ καὶ λόγῳ προσομιλοῦντες᾽ ἑκάστοις πράττοιεν καὶ λέγοιεν. καταλέλειπται δὴ τὸ τῆς ὑμετέρας ἕξεως
20 γένος, ἅμα ἀμφοτέρων φύσει καὶ τροφῇ μετέχον. Τίμαιός τε γὰρ ὅδε, εὐνομωτάτης ὢν πόλεως τῆς ἐν Ἰταλίᾳ Λοκρίδος, οὐσίᾳ καὶ γένει οὐδενὸς ὕστερος ὢν τῶν ἐκεῖ, τὰς μεγίστας μὲν ἀρχάς τε καὶ τιμὰς τῶν ἐν τῇ πόλει μετακεχείρισται, φιλο-

---

[1] For poetry as an "imitative" art cf. *Rep.* 392 D, 597 E ff.
[2] Cf. *Laws* 638 B. The laws of Epizephyrian Locri were ascribed to Zaleucus (*circa* 650 B.C.).

TIMAEUS

against others in those struggles which States wage; in how proper a spirit it enters upon war, and how in its warring it exhibits qualities such as befit its education and training in its dealings with each several State whether in respect of military actions or in respect of verbal negotiations. And herein, Critias and Hermocrates, I am conscious of my own inability ever to magnify sufficiently our citizens and our State. Now in this inability of mine there is nothing surprising; but I have formed the same opinion about the poets also, those of the present as well as those of the past; not that I disparage in any way the poetic clan, but it is plain to all that the imitative [1] tribe will imitate with most ease and success the things amidst which it has been reared, whereas it is hard for any man to imitate well in action what lies outside the range of his rearing, and still harder in speech. Again, as to the class of Sophists, although I esteem them highly versed in many fine discourses of other kinds, yet I fear lest haply, seeing they are a class which roams from city to city and has no settled habitations of its own, they may go wide of the mark in regard to men who are at once philosophers and statesmen, and what they would be likely to do and say, in their several dealings with foemen in war and battle, both by word and deed. Thus there remains only that class which is of your complexion—a class which, alike by nature and nurture, shares the qualities of both the others. For our friend Timaeus is a native of a most well-governed State, Italian Locris,[2] and inferior to none of its citizens either in property or in rank; and not only has he occupied the highest offices and posts of honour in his State, but he has

PLATO

20 σοφίας δ' αὖ κατ' ἐμὴν δόξαν ἐπ' ἄκρον ἁπάσης ἐλήλυθε· Κριτίαν δέ που πάντες οἱ τῇδ' ἴσμεν οὐδενὸς ἰδιώτην ὄντα ὧν λέγομεν· τῆς δ' Ἑρμοκράτους αὖ περὶ φύσεως καὶ τροφῆς, πρὸς
B ἅπαντα ταῦτ' εἶναι ἱκανῆς πολλῶν μαρτυρούντων πιστευτέον. διὸ καὶ χθὲς ἐγὼ διανοούμενος ὑμῶν δεομένων τὰ περὶ τῆς πολιτείας διελθεῖν προθύμως ἐχαριζόμην, εἰδὼς ὅτι τὸν ἑξῆς λόγον οὐδένες ἂν ὑμῶν ἐθελόντων ἱκανώτερον ἀποδοῖεν· εἰς γὰρ πόλεμον πρέποντα καταστήσαντες τὴν πόλιν ἅπαντ' αὐτῇ τὰ προσήκοντα ἀποδοῖτ' ἂν μόνοι τῶν νῦν. εἰπὼν δὴ τἀπιταχθέντα ἀντεπέταξα ὑμῖν ἃ καὶ νῦν λέγω. ξυνωμολογήσατ' οὖν κοινῇ σκεψάμενοι
C πρὸς ὑμᾶς αὐτοὺς εἰς νῦν ἀνταποδώσειν μοι τὰ τῶν λόγων ξένια, πάρειμί τε οὖν δὴ κεκοσμημένος ἐπ' αὐτὰ καὶ πάντων ἑτοιμότατος ὢν δέχεσθαι.

ΕΡ. Καὶ μὲν δή, καθάπερ εἶπε Τίμαιος ὅδε, ὦ Σώκρατες, οὔτε ἐλλείψομεν προθυμίας οὐδὲν οὔτ' ἔστιν οὐδεμία πρόφασις ἡμῖν τοῦ μὴ δρᾶν ταῦτα· ὥστε καὶ χθὲς εὐθὺς ἐνθένδε, ἐπειδὴ παρὰ Κριτίαν πρὸς τὸν ξενῶνα, οὗ καὶ καταλύομεν, ἀφικόμεθα, καὶ ἔτι πρότερον καθ' ὁδὸν αὖ ταῦτα ἐσκοποῦμεν.
D ὅδε[1] οὖν ἡμῖν λόγον εἰσηγήσατο ἐκ παλαιᾶς ἀκοῆς· ὃν καὶ νῦν λέγε, ὦ Κριτία, τῷδε, ἵνα ξυνδοκιμάσῃ πρὸς τὴν ἐπίταξιν εἴτ' ἐπιτήδειος εἴτ' ἀνεπιτήδειός ἐστιν.

ΚΡ. Ταῦτα χρὴ δρᾶν, εἰ καὶ τῷ τρίτῳ κοινωνῷ Τιμαίῳ ξυνδοκεῖ.

[1] ὅδε best MS.: ὁ δ' other MSS., Zur.

TIMAEUS

also attained, in my opinion, the very summit of eminence in all branches of philosophy. As to Critias, all of us here know that he is no novice in any of the subjects we are discussing. As regards Hermocrates, we must believe the many witnesses who assert that both by nature and by nurture he is competent for all these inquiries. So, with this in my mind, when you requested me yesterday to expound my views of the polity I gratified you most willingly, since I knew that none could deal more adequately than you (if you were willing) with the next subject of discourse; for you alone, of men now living, could show our State engaged in a suitable war and exhibiting all the qualities which belong to it. Accordingly, when I had spoken upon my prescribed theme, I in turn prescribed for you this theme which I am now explaining. And you, after consulting together among yourselves, agreed to pay me back to-day with a feast of words; so here I am, ready for that feast in festal garb, and eager above all men to begin.

HERM. Of a truth, Socrates, as our friend Timaeus has said, we will show no lack of zeal, nor have we any excuse for refusing to do as you say. Yesterday, in fact, immediately after our return from you to the guest-chamber at Critias's where we are lodging— aye, and earlier still, on our way there—we were considering these very subjects. Critias here mentioned to us a story derived from ancient tradition; and now, Critias, pray tell it again to our friend here, so that he may help us to decide whether or not it is pertinent to our prescribed theme.

CRIT. That I must certainly do, if our third partner, Timaeus, also approves.

PLATO

ΤΙ. Δοκεῖ μήν.

ΚΡ. Ἄκουε δή, ὦ Σώκρατες, λόγου μάλα μὲν ἀτόπου, παντάπασί γε μὴν ἀληθοῦς, ὡς ὁ τῶν ἑπτὰ σοφώτατος Σόλων ποτ' ἔφη. ἦν μὲν οὖν οἰκεῖος καὶ σφόδρα φίλος ἡμῖν Δρωπίδου τοῦ προπάππου, καθάπερ λέγει πολλαχοῦ καὶ αὐτὸς ἐν τῇ ποιήσει· πρὸς δὲ Κριτίαν τὸν ἡμέτερον πάππον εἶπεν, ὡς ἀπεμνημόνευεν αὖ πρὸς ἡμᾶς ὁ γέρων, ὅτι μεγάλα καὶ θαυμαστὰ τῆσδ' εἴη παλαιὰ ἔργα τῆς πόλεως ὑπὸ χρόνου καὶ φθορᾶς ἀνθρώπων ἠφανισμένα, πάντων δὲ ἓν μέγιστον, οὗ νῦν ἐπιμνησθεῖσι πρέπον ἂν ἡμῖν εἴη σοί τε ἀποδοῦναι χάριν καὶ τὴν θεὸν ἅμα ἐν τῇ πανηγύρει δικαίως τε καὶ ἀληθῶς οἷόνπερ ὑμνοῦντας ἐγκωμιάζειν.

ΣΩ. Εὖ λέγεις. ἀλλὰ δὴ ποῖον ἔργον τοῦτο Κριτίας οὐ λεγόμενον μέν, ὡς δὲ πραχθὲν ὄντως ὑπὸ τῆσδε τῆς πόλεως ἀρχαῖον διηγεῖτο κατὰ τὴν Σόλωνος ἀκοήν;

ΚΡ. Ἐγὼ φράσω παλαιὸν ἀκηκοὼς λόγον οὐ νέου ἀνδρός. ἦν μὲν γὰρ δὴ τότε Κριτίας, ὡς ἔφη, σχεδὸν ἐγγὺς ἤδη τῶν ἐνενήκοντα ἐτῶν, ἐγὼ δέ πῃ μάλιστα δεκέτης· ἡ δὲ Κουρεῶτις ἡμῖν οὖσα ἐτύγχανεν Ἀπατουρίων. τὸ δὴ τῆς ἑορτῆς· σύνηθες ἑκάστοτε καὶ τότε ξυνέβη τοῖς παισίν· ἆθλα γὰρ ἡμῖν οἱ πατέρες ἔθεσαν ῥαψῳδίας. πολλῶν μὲν οὖν δὴ καὶ πολλὰ ἐλέχθη ποιητῶν ποιήματα, ἅτε δὲ νέα κατ' ἐκεῖνον τὸν χρόνον ὄντα τὰ Σόλωνος πολλοὶ τῶν παίδων ᾔσαμεν. εἶπεν οὖν δή τις τῶν

---

[1] i.e. the lesser Panathenaea, held early in June, just after the Bendideia.

[2] The Apaturia was a feast held in October in honour of Dionysus. On the third day of the feast the children

TIMAEUS

TIM. Assuredly I approve.

CRIT. Listen then, Socrates, to a tale which, though passing strange, is yet wholly true, as Solon, the wisest of the Seven, once upon a time declared. Now Solon—as indeed he often says himself in his poems—was a relative and very dear friend of our great-grandfather Dropides; and Dropides told our grandfather Critias—as the old man himself, in turn, related to us—that the exploits of this city in olden days, the record of which had perished through time and the destruction of its inhabitants, were great and marvellous, the greatest of all being one which it would be proper for us now to relate both as a payment of our debt of thanks to you and also as a tribute of praise, chanted as it were duly and truly, in honour of the Goddess on this her day of Festival.[1]

SOC. Excellent! But come now, what was this exploit described by Critias, following Solon's report, as a thing not verbally recorded, although actually performed by this city long ago?

CRIT. I will tell you : it is an old tale, and I heard it from a man not young. For indeed at that time, as he said himself, Critias was already close upon ninety years of age, while I was somewhere about ten; and it chanced to be that day of the Apaturia which is called " Cureotis." [2] The ceremony for boys which was always customary at the feast was held also on that occasion, our fathers arranging contests in recitation. So while many poems of many poets were declaimed, since the poems of Solon were at that time new, many of us children chanted them. And one of our fellow-tribesmen—whether he really

born during the year were registered (hence the name *Cureotis* : κοῦροι = youths).

21

φρατέρων, εἴτε δὴ δοκοῦν αὐτῷ τότε εἴτε καὶ
χάριν τινὰ τῷ Κριτίᾳ φέρων, δοκεῖν οἱ τά τε ἄλλα
σοφώτατον γεγονέναι Σόλωνα καὶ κατὰ τὴν ποίησιν
αὖ τῶν ποιητῶν πάντων ἐλευθεριώτατον. ὁ δὴ
γέρων, σφόδρα γὰρ οὖν μέμνημαι, μάλα τε ἤσθη
καὶ διαμειδιάσας εἶπεν· Εἴ γε, ὦ Ἀμύνανδρε, μὴ
παρέργῳ τῇ ποιήσει κατεχρήσατο, ἀλλ' ἐσπου-
δάκει καθάπερ ἄλλοι, τόν τε λόγον ὃν ἀπ' Αἰγύπτου
δεῦρο ἠνέγκατο ἀπετέλεσε, καὶ μὴ διὰ τὰς στάσεις
ὑπὸ κακῶν τε ἄλλων, ὅσα εὗρεν ἐνθάδε ἥκων,
ἠναγκάσθη καταμελῆσαι, κατά γ' ἐμὴν δόξαν οὔτε
Ἡσίοδος οὔτε Ὅμηρος οὔτε ἄλλος οὐδεὶς ποιητὴς
εὐδοκιμώτερος ἐγένετο ἄν ποτε αὐτοῦ. Τίς δ' ἦν
ὁ λόγος, ἦ δ' ὅς, ὦ Κριτία; Ἡ περὶ μεγίστης,
ἔφη, καὶ ὀνομαστοτάτης πασῶν δικαιότατ' ἂν
πράξεως οὔσης, ἣν ἥδε ἡ πόλις ἔπραξε μέν, διὰ δὲ
χρόνον καὶ φθορὰν τῶν ἐργασαμένων οὐ διήρκεσε
δεῦρο ὁ λόγος. Λέγε ἐξ ἀρχῆς, ἦ δ' ὅς, τί τε καὶ
πῶς καὶ παρὰ τίνων ὡς ἀληθῆ διακηκοὼς ἔλεγεν
ὁ Σόλων.

Ἔστι τις κατ' Αἴγυπτον, ἦ δ' ὅς, ἐν τῷ Δέλτα,
περὶ ὃ κατὰ κορυφὴν σχίζεται τὸ τοῦ Νείλου
ῥεῦμα, Σαϊτικὸς ἐπικαλούμενος νομός, τούτου δὲ
τοῦ νομοῦ μεγίστη πόλις Σάϊς, ὅθεν δὴ καὶ Ἄμασις
ἦν ὁ βασιλεύς· οἷς τῆς πόλεως θεὸς ἀρχηγός τίς
ἐστιν, Αἰγυπτιστὶ μὲν τοὔνομα Νηΐθ, Ἑλληνιστὶ
δέ, ὡς ὁ ἐκείνων λόγος, Ἀθηνᾶ· μάλα δὲ φιλ-
αθήναιοι καί τινα τρόπον οἰκεῖοι τῶνδ' εἶναί φασιν.
οἱ δὴ Σόλων ἔφη πορευθεὶς σφόδρα τε γενέσθαι
παρ' αὐτοῖς ἔντιμος, καὶ δὴ καὶ τὰ παλαιὰ

---

[1] Amasis (Aahmes) was king of Egypt 569-525 B.C., and a phil-Hellene; cf. Hdt. ii. 162 ff.

TIMAEUS

thought so at the time or whether he was paying a compliment to Critias—declared that in his opinion Solon was not only the wisest of men in all else, but in poetry also he was of all poets the noblest. Whereat the old man (I remember the scene well) was highly pleased and said with a smile, " If only, Amynander, he had not taken up poetry as a by-play but had worked hard at it like others, and if he had completed the story he brought here from Egypt, instead of being forced to lay it aside owing to the seditions and all the other evils he found here on his return,—why then, I say, neither Hesiod nor Homer nor any other poet would ever have proved more famous than he." " And what was the story, Critias? " said the other. " Its subject," replied Critias, " was a very great exploit, worthy indeed to be accounted the most notable of all exploits, which was performed by this city, although the record of it has not endured until now owing to lapse of time and the destruction of those who wrought it." " Tell us from the beginning," said Amynander, " what Solon related and how, and who were the informants who vouched for its truth."

" In the Delta of Egypt," said Critias, " where, at its head, the stream of the Nile parts in two, there is a certain district called the Saitic. The chief city in this district is Sais—the home of King Amasis,[1]—the founder of which, they say, is a goddess whose Egyptian name is Neïth,[2] and in Greek, as they assert, Athena. These people profess to be great lovers of Athens and in a measure akin to our people here. And Solon said that when he travelled there he was held in great esteem amongst them; moreover, when he was questioning

[2] Neïth is identified by Plutarch with Isis; *cf.* Hdt. ii. 28.

22 ἀνερωτῶν ποτε[1] τοὺς μάλιστα περὶ ταῦτα τῶν ἱερέων ἐμπείρους σχεδὸν οὔτε αὑτὸν οὔτε ἄλλον Ἕλληνα οὐδένα οὐδέν, ὡς ἔπος εἰπεῖν, εἰδότα περὶ τῶν τοιούτων ἀνευρεῖν. καί ποτε προαγαγεῖν βουληθεὶς αὐτοὺς περὶ τῶν ἀρχαίων εἰς λόγους, τῶν τῇδε τὰ ἀρχαιότατα λέγειν ἐπιχειρεῖν, περὶ Φορωνέως τε τοῦ πρώτου λεχθέντος καὶ Νιόβης, καὶ μετὰ τὸν κατακλυσμὸν αὖ περὶ Δευκαλίωνος καὶ Πύρρας ὡς διεγένοντο μυθολογεῖν, καὶ τοὺς **B** ἐξ αὐτῶν γενεαλογεῖν, καὶ τὰ τῶν ἐτῶν ὅσα ἦν οἷς ἔλεγε πειρᾶσθαι διαμνημονεύων τοὺς χρόνους ἀριθμεῖν· καί τινα εἰπεῖν τῶν ἱερέων εὖ μάλα παλαιόν· Ὦ Σόλων, Σόλων, Ἕλληνες ἀεὶ παῖδές ἐστε, γέρων δὲ Ἕλλην οὐκ ἔστιν. Ἀκούσας οὖν, Πῶς τί τοῦτο λέγεις; φάναι. Νέοι ἐστέ, εἰπεῖν, τὰς ψυχὰς πάντες· οὐδεμίαν γὰρ ἐν αὐταῖς ἔχετε δι' ἀρχαίαν ἀκοὴν παλαιὰν δόξαν οὐδὲ μάθημα χρόνῳ **C** πολιὸν οὐδέν. τὸ δὲ τούτων αἴτιον τόδε. πολλαὶ καὶ κατὰ πολλὰ φθοραὶ γεγόνασιν ἀνθρώπων καὶ ἔσονται, πυρὶ μὲν καὶ ὕδατι μέγισται, μυρίοις δὲ ἄλλοις ἕτεραι βραχύτεραι. τὸ γὰρ οὖν καὶ παρ' ὑμῖν λεγόμενον, ὥς ποτε Φαέθων Ἡλίου παῖς τοῦ πατρὸς ἅρμα ζεύξας διὰ τὸ μὴ δυνατὸς εἶναι κατὰ τὴν τοῦ πατρὸς ὁδὸν ἐλαύνειν τά τ' ἐπὶ γῆς ξυνέκαυσε καὶ αὐτὸς κεραυνωθεὶς διεφθάρη, τοῦτο μύθου μὲν σχῆμα ἔχον λέγεται, τὸ δ' ἀληθές ἐστι **D** τῶν περὶ γῆν καὶ κατ' οὐρανὸν ἰόντων παράλλαξις

[1] ποτε is omitted by some MSS. and Zur.

---

[1] *Cf. Laws* 676 ff.
[2] For the legend of Phaethon see Ovid, *Met.* i. 751 ff.

TIMAEUS

such of their priests as were most versed in ancient lore about their early history, he discovered that neither he himself nor any other Greek knew anything at all, one might say, about such matters. And on one occasion, when he wished to draw them on to discourse on ancient history, he attempted to tell them the most ancient of our traditions, concerning Phoroneus, who was said to be the first man, and Niobe; and he went on to tell the legend about Deucalion and Pyrrha after the Flood, and how they survived it, and to give the genealogy of their descendants; and by recounting the number of years occupied by the events mentioned he tried to calculate the periods of time. Whereupon one of the priests, a prodigiously old man, said, "O Solon, Solon, you Greeks are always children: there is not such a thing as an old Greek." And on hearing this he asked, "What mean you by this saying?" And the priest replied, "You are young in soul, every one of you. For therein you possess not a single belief that is ancient and derived from old tradition, nor yet one science that is hoary with age. And this is the cause thereof: There have been and there will be many and divers destructions of mankind,[1] of which the greatest are by fire and water, and lesser ones by countless other means. For in truth the story that is told in your country as well as ours, how once upon a time Phaethon, son of Helios,[2] yoked his father's chariot, and, because he was unable to drive it along the course taken by his father, burnt up all that was upon the earth and himself perished by a thunderbolt,—that story, as it is told, has the fashion of a legend, but the truth of it lies in the occurrence of a shifting of the bodies in the heavens which move

33

PLATO

22 καὶ διὰ μακρῶν χρόνων γιγνομένη τῶν ἐπὶ γῆς
πυρὶ πολλῷ φθορά. τότε οὖν ὅσοι κατ' ὄρη καὶ
ἐν ὑψηλοῖς τόποις καὶ ἐν ξηροῖς οἰκοῦσι, μᾶλλον
διόλλυνται τῶν ποταμοῖς καὶ θαλάττῃ προσοικούν-
των· ἡμῖν δὲ ὁ Νεῖλος εἴς τε τὰ ἄλλα σωτὴρ καὶ
τότε ἐκ ταύτης τῆς ἀπορίας σώζει αὐξόμενος.[1]
ὅταν δ' αὖ οἱ θεοὶ τὴν γῆν ὕδασι καθαίροντες
κατακλύζωσιν, οἱ μὲν ἐν τοῖς ὄρεσι διασώζονται
E βουκόλοι νομεῖς τε, οἱ δ' ἐν ταῖς παρ' ὑμῖν πόλεσιν
εἰς τὴν θάλατταν ὑπὸ τῶν ποταμῶν φέρονται· κατὰ
δὲ τήνδε τὴν χώραν οὔτε τότε οὔτε ἄλλοτε ἄνωθεν
ἐπὶ τὰς ἀρούρας ὕδωρ ἐπιρρεῖ, τὸ δ' ἐναντίον
κάτωθεν πᾶν[2] ἐπανιέναι πέφυκεν. ὅθεν καὶ δι' ἃς
αἰτίας τἀνθάδε σωζόμενα λέγεται παλαιότατα. τὸ
δὲ ἀληθές, ἐν πᾶσι τοῖς τόποις ὅπου μὴ χειμὼν
ἐξαίσιος ἢ καῦμα ἀπείργει, πλέον, τοτὲ δὲ ἔλαττον
23 ἀεὶ γένος ἐστὶν ἀνθρώπων· ὅσα δὲ ἢ παρ' ὑμῖν
ἢ τῇδε ἢ καὶ κατ' ἄλλον τόπον ὧν ἀκοῇ ἴσμεν, εἴ
πού τι καλὸν ἢ μέγα γέγονεν ἢ καί τινα διαφορὰν
ἄλλην ἔχον, πάντα γεγραμμένα ἐκ παλαιοῦ τῇδ'
ἐστὶν ἐν τοῖς ἱεροῖς καὶ σεσωσμένα. τὰ δὲ παρ'
ὑμῖν καὶ τοῖς ἄλλοις ἄρτι κατεσκευασμένα ἑκάστοτε
τυγχάνει γράμμασι καὶ ἅπασιν ὁπόσων πόλεις
δέονται, καὶ πάλιν δι' εἰωθότων ἐτῶν ὥσπερ
νόσημα ἥκει φερόμενον αὐτοῖς ῥεῦμα οὐράνιον
B καὶ τοὺς ἀγραμμάτους τε καὶ ἀμούσους ἔλιπεν
ὑμῶν, ὥστε πάλιν ἐξ ἀρχῆς οἷον νέοι γίγνεσθε,
οὐδὲν εἰδότες οὔτε τῶν τῇδε οὔτε τῶν παρ' ὑμῖν,

---

[1] αὐξόμενος J. Cook Wilson: λυόμενος mss., Zur.
[2] πᾶν is omitted by some mss. and Zur.

TIMAEUS

round the earth, and a destruction of the things on the earth by fierce fire, which recurs at long intervals. At such times all they that dwell on the mountains and in high and dry places suffer destruction more than those who dwell near to rivers or the sea ; and in our case the Nile, our Saviour in other ways, saves us also at such times from this calamity by rising high. And when, on the other hand, the Gods purge the earth with a flood of waters, all the herdsmen and shepherds that are in the mountains are saved,[1] but those in the cities of your land are swept into the sea by the streams ; whereas in our country neither then nor at any other time does the water pour down over our fields from above, on the contrary it all tends naturally to well up from below. Hence it is, for these reasons, that what is here preserved is reckoned to be most ancient ; the truth being that in every place where there is no excessive heat or cold to prevent it there always exists some human stock, now more, now less in number. And if any event has occurred that is noble or great or in any way conspicuous, whether it be in your country or in ours or in some other place of which we know by report, all such events are recorded from of old and preserved here in our temples ; whereas your people and the others are but newly equipped, every time, with letters and all such arts as civilized States require ; and when, after the usual interval of years, like a plague, the flood from heaven comes sweeping down afresh upon your people, it leaves none of you but the unlettered and uncultured, so that you become young as ever, with no knowledge of all that happened in old times in this

[1] *Cf. Laws* 677 B.

PLATO

23 ὅσα ἦν ἐν τοῖς παλαιοῖς χρόνοις. τὰ γοῦν νῦν δὴ γενεαλογηθέντα, ὦ Σόλων, περὶ τῶν παρ' ὑμῖν ἃ διῆλθες, παίδων βραχύ τι διαφέρει μύθων, οἳ πρῶτον μὲν ἕνα γῆς κατακλυσμὸν μέμνησθε πολλῶν ἔμπροσθεν γεγονότων, ἔτι δὲ τὸ κάλλιστον καὶ ἄριστον γένος ἐπ' ἀνθρώπους ἐν τῇ χώρᾳ τῇ παρ' ὑμῖν οὐκ ἴστε γεγονός, ἐξ ὧν σύ τε καὶ πᾶσα ἡ
C πόλις ἔστι τὰ νῦν ὑμῶν περιλειφθέντος ποτὲ σπέρματος βραχέος, ἀλλ' ὑμᾶς λέληθε διὰ τὸ τοὺς περιγενομένους ἐπὶ πολλὰς γενεὰς γράμμασι τελευτᾶν ἀφώνους. ἦν γὰρ δή ποτε, ὦ Σόλων, ὑπὲρ τὴν μεγίστην φθορὰν ὕδασιν ἡ νῦν Ἀθηναίων οὖσα πόλις ἀρίστη πρός τε τὸν πόλεμον καὶ κατὰ πάντα εὐνομωτάτη διαφερόντως· ᾗ κάλλιστα ἔργα καὶ πολιτεῖαι γενέσθαι λέγονται κάλλισται πασῶν, ὁπόσων νῦν ὑπὸ τὸν οὐρανὸν ἡμεῖς ἀκοὴν παρεδεξάμεθα.

D Ἀκούσας οὖν ὁ Σόλων ἔφη θαυμάσαι καὶ πᾶσαν προθυμίαν σχεῖν δεόμενος τῶν ἱερέων πάντα δι' ἀκριβείας οἱ τὰ περὶ τῶν πάλαι πολιτῶν ἑξῆς διελθεῖν. τὸν οὖν ἱερέα φάναι· Φθόνος οὐδείς, ὦ Σόλων, ἀλλὰ σοῦ τε ἕνεκα ἐρῶ καὶ τῆς πόλεως ὑμῶν, μάλιστα δὲ τῆς θεοῦ χάριν, ἣ τήν τε ὑμετέραν καὶ τήνδ' ἔλαχε καὶ ἔθρεψε καὶ ἐπαίδευσε, προτέραν μὲν τὴν παρ' ὑμῖν ἔτεσι χιλίοις, ἐκ Γῆς
E τε καὶ Ἡφαίστου τὸ σπέρμα παραλαβοῦσα ὑμῶν, τήνδε δὲ ὑστέραν. τῆς δὲ ἐνθάδε διακοσμήσεως παρ' ἡμῖν ἐν τοῖς ἱεροῖς γράμμασιν ὀκτακισχιλίων ἐτῶν ἀριθμὸς γέγραπται. περὶ δὴ τῶν ἐνακισχίλια γεγονότων ἔτη πολιτῶν σοι δηλώσω διὰ βραχέων

---

[1] *i.e.* from the elements *earth* and *fire*, *cf.* 3i B.

For the legend of Erechtheus, son of Gē and Hephaestus, and king of Athens (Hom. *Il.* ii. 547), see Eurip. *Ion.*

TIMAEUS

land or in your own. Certainly the genealogies which you related just now, Solon, concerning the people of your country, are little better than children's tales; for, in the first place, you remember but one deluge, though many had occurred previously; and next, you are ignorant of the fact that the noblest and most perfect race amongst men were born in the land where you now dwell, and from them both you yourself are sprung and the whole of your existing city, out of some little seed that chanced to be left over; but this has escaped your notice because for many generations the survivors died with no power to express themselves in writing. For verily at one time, Solon, before the greatest destruction by water, what is now the Athenian State was the bravest in war and supremely well organized also in all other respects. It is said that it possessed the most splendid works of art and the noblest polity of any nation under heaven of which we have heard tell."

Upon hearing this, Solon said that he marvelled, and with the utmost eagerness requested the priest to recount for him in order and exactly all the facts about those citizens of old. The priest then said: "I begrudge you not the story, Solon; nay, I will tell it, both for your own sake and that of your city, and most of all for the sake of the Goddess who has adopted for her own both your land and this of ours, and has nurtured and trained them,—yours first by the space of a thousand years, when she had received the seed of you from Gê and Hephaestus,[1] and after that ours. And the duration of our civilization as set down in our sacred writings is 8000 years. Of the citizens, then, who lived 9000 years ago, I will

PLATO

νόμους τε καὶ τῶν ἔργων αὐτοῖς ὃ κάλλιστον ἐπράχθη· τὸ δ' ἀκριβὲς περὶ πάντων ἐφεξῆς εἰσαῦθις κατὰ σχολήν, αὐτὰ τὰ γράμματα λαβόντες, διέξιμεν. τοὺς μὲν οὖν νόμους σκόπει πρὸς τοὺς τῇδε· πολλὰ γὰρ παραδείγματα τῶν τότε παρ' ὑμῖν ὄντων ἐνθάδε νῦν ἀνευρήσεις, πρῶτον μὲν τὸ τῶν ἱερέων γένος ἀπὸ τῶν ἄλλων χωρὶς ἀφωρισμένον, μετὰ δὲ τοῦτο τὸ τῶν δημιουργῶν, ὅτι καθ' αὑτὸ ἕκαστον ἄλλῳ δὲ οὐκ ἐπιμιγνύμενον δημιουργεῖ, τό τε τῶν νομέων καὶ τὸ τῶν θηρευτῶν τό τε τῶν γεωργῶν. καὶ δὴ καὶ τὸ μάχιμον γένος ᾔσθησαί που τῇδε ἀπὸ πάντων τῶν γενῶν κεχωρισμένον, οἷς οὐδὲν ἄλλο πλὴν τὰ περὶ τὸν πόλεμον ὑπὸ τοῦ νόμου προσετάχθη μέλειν. ἔτι δὲ ἡ τῆς ὁπλίσεως αὐτῶν σχέσις ἀσπίδων καὶ δοράτων, οἷς ἡμεῖς πρῶτοι τῶν περὶ τὴν Ἀσίαν ὡπλίσμεθα, τῆς θεοῦ, καθάπερ ἐν ἐκείνοις τοῖς τόποις, παρ' ὑμῖν πρώτοις ἐνδειξαμένης. τὸ δ' αὖ περὶ τῆς φρονήσεως, ὁρᾷς που τὸν νόμον τῇδε ὅσην ἐπιμέλειαν ἐποιήσατο εὐθὺς κατ' ἀρχὰς περί τε τὸν κόσμον ἅπαντα μέχρι μαντικῆς καὶ ἰατρικῆς πρὸς ὑγίειαν ἐκ τούτων θείων ὄντων εἰς τὰ ἀνθρώπινα ἀνευρών, ὅσα τε ἄλλα τούτοις ἕπεται μαθήματα πάντα κτησάμενος. ταύτην οὖν δὴ τότε ξύμπασαν τὴν διακόσμησιν καὶ σύνταξιν ἡ θεὸς προτέρους ὑμᾶς διακοσμήσασα κατῴκισεν, ἐκλεξαμένη τὸν τόπον ἐν ᾧ γεγένησθε, τὴν εὐκρασίαν τῶν ὡρῶν ἐν αὐτῷ κατιδοῦσα, ὅτι φρονιμωτάτους ἄνδρας οἴσοι. ἅτ' οὖν φιλοπόλεμός τε καὶ φιλόσοφος ἡ

---

[1] Egypt being reckoned as part of Asia.

TIMAEUS

declare to you briefly certain of their laws and the noblest of the deeds they performed: the full account in precise order and detail we shall go through later at our leisure, taking the actual writings. To get a view of their laws, look at the laws here; for you will find existing here at the present time many examples of the laws which then existed in your city. You see, first, how the priestly class is separated off from the rest; next, the class of craftsmen, of which each sort works by itself without mixing with any other; then the classes of shepherds, hunters, and farmers, each distinct and separate. Moreover, the military class here, as no doubt you have noticed, is kept apart from all the other classes, being enjoined by the law to devote itself solely to the work of training for war. A further feature is the character of their equipment with shields and spears; for we were the first of the peoples of Asia [1] to adopt these weapons, it being the Goddess who instructed us, even as she instructed you first of all the dwellers in yonder lands. Again, with regard to wisdom, you perceive, no doubt, the law here,—how much attention it has devoted from the very beginning to the Cosmic Order, by discovering all the effects which the divine causes produce upon human life, down to divination and the art of medicine which aims at health, and by its mastery also of all the other subsidiary studies. So when, at that time, the Goddess had furnished you, before all others, with all this orderly and regular system, she established your State, choosing the spot wherein you were born since she perceived therein a climate duly blended, and how that it would bring forth men of supreme wisdom. So it was that the Goddess, being

PLATO

24 θεὸς οὖσα τὸν προσφερεστάτους αὐτῇ μέλλοντα οἴσειν τόπον ἄνδρας, τοῦτον ἐκλεξαμένη πρῶτον κατῴκισεν. ᾠκεῖτε δὴ οὖν νόμοις τε τοιούτοις χρώμενοι καὶ ἔτι μᾶλλον εὐνομούμενοι πάσῃ τε πάντας ἀνθρώπους ὑπερβεβηκότες ἀρετῇ, καθάπερ εἰκὸς γεννήματα καὶ παιδεύματα θεῶν ὄντας. πολλὰ μὲν οὖν ὑμῶν καὶ μεγάλα ἔργα τῆς πόλεως τῆδε γεγραμμένα θαυμάζεται, πάντων μὴν ἓν **E** ὑπερέχει μεγέθει καὶ ἀρετῇ· λέγει γὰρ τὰ γεγραμμένα, ὅσην ἡ πόλις ὑμῶν ἔπαυσέ ποτε δύναμιν ὕβρει πορευομένην ἅμα ἐπὶ πᾶσαν Εὐρώπην καὶ Ἀσίαν, ἔξωθεν ὁρμηθεῖσαν ἐκ τοῦ Ἀτλαντικοῦ πελάγους. τότε γὰρ πορεύσιμον ἦν τὸ ἐκεῖ πέλαγος· νῆσον γὰρ πρὸ τοῦ στόματος εἶχεν, ὃ καλεῖτε, ὥς φατε, ὑμεῖς Ἡρακλέους στήλας·[1] ἡ δὲ νῆσος ἅμα Λιβύης ἦν καὶ Ἀσίας μείζων, ἐξ ἧς ἐπιβατὸν ἐπὶ τὰς ἄλλας νήσους τοῖς τότ᾽ ἐγίγνετο πορευομένοις, ἐκ δὲ τῶν νήσων ἐπὶ τὴν καταντι-
25 κρὺ πᾶσαν ἤπειρον τὴν περὶ τὸν ἀληθινὸν ἐκεῖνον πόντον. τάδε μὲν γάρ, ὅσα ἐντὸς τοῦ στόματος οὗ λέγομεν, φαίνεται λιμὴν στενόν τινα ἔχων εἴσπλουν· ἐκεῖνο δὲ πέλαγος ὄντως ἥ τε περιέχουσα αὐτὸ γῆ παντελῶς [ἀληθῶς][1] ὀρθότατ᾽ ἂν λέγοιτο ἤπειρος. ἐν δὲ δὴ τῇ Ἀτλαντίδι νήσῳ ταύτῃ μεγάλη συνέστη καὶ θαυμαστὴ δύναμις βασιλέων, κρατοῦσα μὲν ἁπάσης τῆς νήσου, πολλῶν δὲ ἄλλων νήσων καὶ μερῶν τῆς ἠπείρου· πρὸς δὲ
B τούτοις ἔτι τῶν ἐντὸς τῇδε Λιβύης μὲν ἦρχον μέχρι

---

[1] ἀληθῶς erased in best MS.

---

[1] *i.e.* the Straits of Gibraltar.   [2] *i.e.* Africa.
[3] *i.e.* the Mediterranean Sea, contrasted with the Atlantic Ocean.

TIMAEUS

herself both a lover of war and a lover of wisdom, chose the spot which was likely to bring forth men most like unto herself, and this first she established. Wherefore you lived under the rule of such laws as these,—yea, and laws still better,—and you surpassed all men in every virtue, as became those who were the offspring and nurslings of gods. Many, in truth, and great are the achievements of your State, which are a marvel to men as they are here recorded; but there is one which stands out above all both for magnitude and for nobleness. For it is related in our records how once upon a time your State stayed the course of a mighty host, which, starting from a distant point in the Atlantic ocean, was insolently advancing to attack the whole of Europe, and Asia to boot. For the ocean there was at that time navigable; for in front of the mouth which you Greeks call, as you say, 'the pillars of Heracles,'[1] there lay an island which was larger than Libya[2] and Asia together; and it was possible for the travellers of that time to cross from it to the other islands, and from the islands to the whole of the continent over against them which encompasses that veritable ocean. For all that we have here, lying within the mouth of which we speak,[3] is evidently a haven having a narrow entrance; but that yonder is a real ocean, and the land surrounding it may most rightly be called, in the fullest and truest sense, a continent. Now in this island of Atlantis there existed a confederation of kings, of great and marvellous power, which held sway over all the island, and over many other islands also and parts of the continent; and, moreover, of the lands here within the Straits they ruled over Libya as far as Egypt,

25 πρὸς Αἴγυπτον, τῆς δὲ Εὐρώπης μέχρι Τυρρηνίας. αὕτη δὴ πᾶσα ξυναθροισθεῖσα εἰς ἓν ἡ δύναμις τόν τε παρ' ὑμῖν καὶ τὸν παρ' ἡμῖν καὶ τὸν ἐντὸς τοῦ στόματος πάντα τόπον μιᾷ ποτ' ἐπεχείρησεν ὁρμῇ δουλοῦσθαι. τότε οὖν ὑμῶν, ὦ Σόλων, τῆς πόλεως ἡ δύναμις εἰς ἅπαντας ἀνθρώπους διαφανὴς ἀρετῇ τε καὶ ῥώμῃ ἐγένετο· πάντων γὰρ προστᾶσα
C εὐψυχίᾳ καὶ τέχναις ὅσαι κατὰ πόλεμον, τὰ μὲν τῶν Ἑλλήνων ἡγουμένη, τὰ δ' αὐτὴ μονωθεῖσα ἐξ ἀνάγκης τῶν ἄλλων ἀποστάντων, ἐπὶ τοὺς ἐσχάτους ἀφικομένη κινδύνους, κρατήσασα μὲν τῶν ἐπιόντων τρόπαια ἔστησε, τοὺς δὲ μήπω δεδουλωμένους διεκώλυσε δουλωθῆναι, τοὺς δ' ἄλλους, ὅσοι κατοικοῦμεν ἐντὸς ὅρων Ἡρακλείων, ἀφθόνως ἅπαντας ἠλευθέρωσεν. ὑστέρῳ δὲ χρόνῳ σεισμῶν ἐξαισίων καὶ κατακλυσμῶν γενομένων,
D μιᾶς ἡμέρας καὶ νυκτὸς χαλεπῆς ἐλθούσης, τό τε παρ' ὑμῶν μάχιμον πᾶν ἀθρόον ἔδυ κατὰ γῆς, ἥ τε Ἀτλαντὶς νῆσος ὡσαύτως κατὰ τῆς θαλάττης δῦσα ἠφανίσθη· διὸ καὶ νῦν ἄπορον καὶ ἀδιερεύνητον γέγονε τὸ ἐκεῖ πέλαγος, πηλοῦ καταβραχέος[1] ἐμποδὼν ὄντος, ὃν ἡ νῆσος ἱζομένη παρέσχετο.

Τὰ μὲν δὴ ῥηθέντα, ὦ Σώκρατες, ὑπὸ τοῦ παλαιοῦ Κριτίου κατ' ἀκοὴν τὴν Σόλωνος, ὡς
E συντόμως εἰπεῖν, ἀκήκοας· λέγοντος δὲ δὴ χθὲς σοῦ περὶ πολιτείας καὶ τῶν ἀνδρῶν οὓς ἔλεγες, ἐθαύμαζον ἀναμιμνησκόμενος αὐτὰ ἃ νῦν λέγω, κατανοῶν ὡς δαιμονίως ἔκ τινος τύχης οὐκ ἄπο σκοποῦ ξυνηνέχθης τὰ πολλὰ οἷς Σόλων εἶπεν. οὐ
26 μὴν ἐβουλήθην παραχρῆμα εἰπεῖν· διὰ χρόνου γὰρ

[1] καταβραχέος] κάρτα βαθέος best MS. and Zur.

TIMAEUS

and over Europe as far as Tuscany. So this host, being all gathered together, made an attempt one time to enslave by one single onslaught both your country and ours and the whole of the territory within the Straits. And then it was, Solon, that the manhood of your State showed itself conspicuous for valour and might in the sight of all the world. For it stood pre-eminent above all in gallantry and all warlike arts, and acting partly as leader of the Greeks, and partly standing alone by itself when deserted by all others, after encountering the deadliest perils, it defeated the invaders and reared a trophy ; whereby it saved from slavery such as were not as yet enslaved, and all the rest of us who dwell within the bounds of Heracles it ungrudgingly set free. But at a later time there occurred portentous earthquakes and floods, and one grievous day and night befell them, when the whole body of your warriors was swallowed up by the earth, and the island of Atlantis in like manner was swallowed up by the sea and vanished ; wherefore also the ocean at that spot has now become impassable and unsearchable, being blocked up by the shoal mud which the island created as it settled down."

You have now heard, Socrates, in brief outline, the account given by the elder Critias of what he heard from Solon ; and when you were speaking yesterday about the State and the citizens you were describing, I marvelled as I called to mind the facts I am now relating, reflecting what a strange piece of fortune it was that your description coincided so exactly for the most part with Solon's account. I was loth, however, to speak on the instant ; for owing to lapse of time my recollection of his account was not

οὐχ ἱκανῶς ἐμεμνήμην. ἐνενόησα οὖν ὅτι χρεὼν
εἴη με πρὸς ἐμαυτὸν πρῶτον ἱκανῶς πάντα ἀνα-
λαβόντα λέγειν οὕτως. ὅθεν ταχὺ ξυνωμολόγησά
σοι τἀπιταχθέντα χθές, ἡγούμενος, ὅπερ ἐν ἅπασι
τοῖς τοιοῖσδε μέγιστον ἔργον, λόγον τινὰ πρέποντα
τοῖς βουλήμασιν ὑποθέσθαι, τούτου μετρίως ἡμᾶς
εὐπορήσειν. οὕτω δή, καθάπερ ὅδ' εἶπε, χθές τε
εὐθὺς ἐνθένδε ἀπιὼν πρὸς τούσδε ἀνέφερον αὐτὰ
B ἀναμιμνησκόμενος, ἀπελθών τε σχεδόν τι πάντα
ἐπισκοπῶν τῆς νυκτὸς ἀνέλαβον. ὡς δή τοι, τὸ
λεγόμενον, τὰ παίδων μαθήματα θαυμαστὸν ἔχει
τι μνημεῖον! ἐγὼ γὰρ ἃ μὲν χθὲς ἤκουσα, οὐκ ἂν
οἶδα εἰ δυναίμην ἅπαντα ἐν μνήμῃ πάλιν λαβεῖν·
ταῦτα δὲ ἃ πάμπολυν χρόνον διακήκοα, παντάπασι
θαυμάσαιμ' ἂν εἴ τί με αὐτῶν διαπέφευγεν. ἦν
μὲν οὖν μετὰ πολλῆς ἡδονῆς καὶ παιδικῆς τότε
C ἀκουόμενα, καὶ τοῦ πρεσβύτου προθύμως με διδά-
σκοντος, ἅτ' ἐμοῦ πολλάκις ἐπανερωτῶντος, ὥστε
οἷον ἐγκαύματα ἀνεκπλύτου γραφῆς ἔμμονά μοι
γέγονε. καὶ δὴ καὶ τοῖσδε εὐθὺς ἔλεγον ἕωθεν
αὐτὰ ταῦτα, ἵνα εὐποροῖεν λόγων μετ' ἐμοῦ.

Νῦν οὖν, οὗπερ ἕνεκα πάντα ταῦτα εἴρηται, λέγειν
εἰμὶ ἕτοιμος, ὦ Σώκρατες, μὴ μόνον ἐν κεφαλαίοις
ἀλλ' ὥσπερ ἤκουσα καθ' ἕκαστον. τοὺς δὲ πολίτας
καὶ τὴν πόλιν ἣν χθὲς ἡμῖν ὡς ἐν μύθῳ διῄεισθα σύ,
D νῦν[1] μετενεγκόντες ἐπὶ τἀληθὲς δεῦρο θήσομεν ὡς
ἐκείνην τήνδε οὖσαν, καὶ τοὺς πολίτας οὓς διενοοῦ

---

[1] νῦν is omitted by the best MS. and Zur.

TIMAEUS

sufficiently clear. So I decided that I ought not to relate it until I had first gone over it all carefully in my own mind. Consequently, I readily consented to the theme you proposed yesterday, since I thought that we should be reasonably well provided for the task of furnishing a satisfactory discourse—which in all such cases is the greatest task. So it was that, as Hermocrates has said, the moment I left your place yesterday I began to relate to them the story as I recollected it, and after I parted from them I pondered it over during the night and recovered, as I may say, the whole story. Marvellous, indeed, is the way in which the lessons of one's childhood " grip the mind," as the saying is. For myself, I know not whether I could recall to mind all that I heard yesterday; but as to the account I heard such a great time ago, I should be immensely surprised if a single detail of it has escaped me. I had then the greatest pleasure and amusement in hearing it, and the old man was eager to tell me, since I kept questioning him repeatedly, so that the story is stamped firmly on my mind like the encaustic designs of an indelible painting. Moreover, immediately after daybreak I related this same story to our friends here, so that they might share in my rich provision of discourse.

Now, therefore,—and this is the purpose of all that I have been saying,—I am ready to tell my tale, not in summary outline only but in full detail just as I heard it. And the city with its citizens which you described to us yesterday, as it were in a fable, we will now transport hither into the realm of fact; for we will assume that the city is that ancient city of ours, and declare that the citizens you conceived are in

26 φήσομεν ἐκείνους τοὺς ἀληθινοὺς εἶναι προγόνους
ἡμῶν οὓς ἔλεγεν ὁ ἱερεύς· πάντως ἁρμόσουσι, καὶ
οὐκ ἀπᾳσόμεθα λέγοντες αὐτοὺς εἶναι τοὺς ἐν
τῷ τότε ὄντας χρόνῳ. κοινῇ δὲ διαλαμβάνοντες
ἅπαντες πειρασόμεθα τὸ πρέπον εἰς δύναμιν οἷς
ἐπέταξας ἀποδοῦναι. σκοπεῖν οὖν δὴ χρή, ὦ
Σώκρατες, εἰ κατὰ νοῦν ὁ λόγος ἡμῖν οὗτος, ἤ
Ε τινα ἔτ' ἄλλον ἀντ' αὐτοῦ ζητητέον.

ΣΩ. Καὶ τίν' ἄν, ὦ Κριτία, μᾶλλον ἀντὶ τούτου
μεταλάβοιμεν, ὃς τῇ τε παρούσῃ τῆς θεοῦ θυσίᾳ
διὰ τὴν οἰκειότητα ἂν πρέποι μάλιστα, τό τε μὴ
πλασθέντα μῦθον ἀλλ' ἀληθινὸν λόγον εἶναι πάμ-
μεγά που. πῶς γὰρ καὶ πόθεν ἄλλους ἀνευρήσομεν
ἀφέμενοι τούτων; οὐκ ἔστιν, ἀλλ' ἀγαθῇ τύχῃ χρὴ
λέγειν μὲν ὑμᾶς, ἐμὲ δὲ ἀντὶ τῶν χθὲς λόγων νῦν
27 ἡσυχίαν ἄγοντα ἀντακούειν.

ΚΡ. Σκόπει δὴ τὴν τῶν ξενίων σοι διάθεσιν, ὦ
Σώκρατες, ᾗ διέθεμεν. ἔδοξε γὰρ ἡμῖν Τίμαιον
μέν, ἅτε ὄντα ἀστρονομικώτατον ἡμῶν καὶ περὶ
φύσεως τοῦ παντὸς εἰδέναι μάλιστα ἔργον πε-
ποιημένον, πρῶτον λέγειν ἀρχόμενον ἀπὸ τῆς τοῦ
κόσμου γενέσεως, τελευτᾶν δὲ εἰς ἀνθρώπων φύσιν·
ἐμὲ δὲ μετὰ τοῦτον, ὡς παρὰ μὲν τούτου δεδεγ-
μένον ἀνθρώπους τῷ λόγῳ γεγονότας, παρὰ σοῦ δὲ
Β πεπαιδευμένους διαφερόντως αὐτῶν τινάς, κατὰ δὴ[1]
τὸν Σόλωνος λόγον τε καὶ νόμον εἰσαγαγόντα
αὐτοὺς ὡς εἰς δικαστὰς ὑμᾶς ποιῆσαι πολίτας τῆς

[1] δὴ Stallbaum: δὲ mss., Zur.

TIMAEUS

truth those actual progenitors of ours, of whom the priest told. In all ways they will correspond, nor shall we be out of tune if we affirm that those citizens of yours are the very men who lived in that age. Thus, with united effort, each taking his part, we will endeavour to the best of our powers to do justice to the theme you have prescribed. Wherefore, Socrates, we must consider whether this story is to our mind, or we have still to look for some other to take its place.

soc. What story should we adopt, Critias, in preference to this? For this story will be admirably suited to the festival of the Goddess which is now being held, because of its connexion with her; and the fact that it is no invented fable but genuine history is all-important. How, indeed, and where shall we discover other stories if we let these slip? Nay, it is impossible. You, therefore, must now deliver your discourse (and may Good Fortune attend you!), while I, in requital for my speech of yesterday, must now keep silence in my turn and hearken.

crit. Consider now, Socrates, the order of the feast as we have arranged it. Seeing that Timaeus is our best astronomer and has made it his special task to learn about the nature of the Universe, it seemed good to us that he should speak first, beginning with the origin of the Cosmos and ending with the generation of mankind. After him I am to follow, taking over from him mankind, already as it were created by his speech, and taking over from you a select number of men superlatively well trained. Then, in accordance with the word and law of Solon, I am to bring these before ourselves, as before a court of judges, and make them citizens of this State of ours,

πόλεως τῆσδε ὡς ὄντας τοὺς τότε Ἀθηναίους, οὓς ἐμήνυσεν ἀφανεῖς ὄντας ἡ τῶν ἱερῶν γραμμάτων φήμη, τὰ λοιπὰ δὲ ὡς περὶ πολιτῶν καὶ Ἀθηναίων ὄντων ἤδη ποιεῖσθαι τοὺς λόγους.

ΣΩ. Τελέως τε καὶ λαμπρῶς ἔοικα ἀνταπολήψεσθαι τὴν τῶν λόγων ἑστίασιν. σὸν οὖν ἔργον λέγειν ἄν, ὦ Τίμαιε, εἴη τὸ μετὰ τοῦτο, ὡς ἔοικεν, καλέσαντα κατὰ νόμον θεούς.

C ΤΙ. Ἀλλ', ὦ Σώκρατες, τοῦτό γε δὴ πάντες ὅσοι καὶ κατὰ βραχὺ σωφροσύνης μετέχουσιν ἐπὶ παντὸς ὁρμῇ καὶ σμικροῦ καὶ μεγάλου πράγματος θεὸν ἀεί που καλοῦσιν· ἡμᾶς δὲ τοὺς περὶ τοῦ παντὸς λόγους ποιεῖσθαί πῃ μέλλοντας, ᾗ γέγονεν ἢ καὶ ἀγενές ἐστιν, εἰ μὴ παντάπασι παραλλάττομεν, ἀνάγκη θεούς τε καὶ θεὰς ἐπικαλουμένους εὔχεσθαι πάντα κατὰ νοῦν ἐκείνοις μὲν μάλιστα, ἑπομένως δὲ ἡμῖν εἰπεῖν. καὶ τὰ μὲν περὶ θεῶν D ταύτῃ παρακεκλήσθω· τὸ δ' ἡμέτερον παρακλητέον, ᾗ ῥᾷστ' ἂν ὑμεῖς μὲν μάθοιτε, ἐγὼ δὲ ᾗ διανοοῦμαι μάλιστ' ἂν περὶ τῶν προκειμένων ἐνδειξαίμην.

Ἔστιν οὖν δὴ κατ' ἐμὴν δόξαν πρῶτον διαιρετέον τάδε· τί τὸ ὂν ἀεί, γένεσιν δὲ οὐκ ἔχον, καὶ τί τὸ γιγνόμενον μὲν ἀεί, ὂν δὲ οὐδέποτε; τὸ μὲν δὴ νοήσει μετὰ λόγου περιληπτὸν ἀεὶ κατὰ ταὐτὰ ὄν, τὸ δ' αὖ δόξῃ μετ' αἰσθήσεως ἀλόγου δοξαστὸν γιγνόμενον καὶ ἀπολλύμενον, ὄντως δὲ οὐδέποτε ὄν. πᾶν δὲ αὖ τὸ γιγνόμενον ὑπ' αἰτίου τινὸς ἐξ ἀνάγκης γίγνεσθαι· παντὶ γὰρ ἀδύνατον χωρὶς

TIMAEUS

regarding them as Athenians of that bygone age whose existence, so long forgotten, has been revealed to us by the record of the sacred writings; and thenceforward I am to proceed with my discourse as if I were speaking of men who already are citizens and men of Athens.

soc. Bounteous and magnificent, methinks, is the feast of speech with which I am to be requited. So then, Timaeus, it will be your task, it seems, to speak next, when you have duly invoked the gods.

TIM. Nay, as to that, Socrates, all men who possess even a small share of good sense call upon God always at the outset of every undertaking, be it small or great; we therefore who are purposing to deliver a discourse concerning the Universe, how it was created or haply is uncreate, must needs invoke Gods and Goddesses (if so be that we are not utterly demented), praying that all we say may be approved by them in the first place, and secondly by ourselves. Grant, then, that we have thus duly invoked the deities; ourselves we must also invoke so to proceed, that you may most easily learn and I may most clearly expound my views regarding the subject before us.

Now first of all we must, in my judgement, make the following distinction. What is that which is Existent always and has no Becoming? And what is that which is Becoming always and never is Existent? Now the one of these is apprehensible by thought with the aid of reasoning, since it is ever uniformly existent; whereas the other is an object of opinion with the aid of unreasoning sensation, since it becomes and perishes and is never really existent. Again, everything which becomes must of necessity become owing to some Cause; for without

28 αἰτίου γένεσιν σχεῖν. ὅτου μὲν οὖν ἂν ὁ δημιουργὸς πρὸς τὸ κατὰ ταὐτὰ ἔχον βλέπων ἀεί, τοιούτῳ τινὶ προσχρώμενος παραδείγματι, τὴν ἰδέαν καὶ δύναμιν αὐτοῦ ἀπεργάζηται, καλὸν ἐξ
B ἀνάγκης οὕτως ἀποτελεῖσθαι πᾶν· οὗ δ᾽ ἂν εἰς τὸ γεγονός, γεννητῷ παραδείγματι προσχρώμενος, οὐ καλόν. ὁ δὴ πᾶς οὐρανὸς ἢ κόσμος ἢ καὶ ἄλλο ὅ τί ποτε ὀνομαζόμενος μάλιστ᾽ ἂν δέχοιτο, τοῦθ᾽ ἡμῖν ὠνομάσθω—, σκεπτέον δ᾽ οὖν περὶ αὐτοῦ πρῶτον, ὅπερ ὑπόκειται περὶ παντὸς ἐν ἀρχῇ δεῖν σκοπεῖν, πότερον ἦν ἀεί, γενέσεως ἀρχὴν ἔχων οὐδεμίαν, ἢ γέγονεν, ἀπ᾽ ἀρχῆς τινὸς ἀρξάμενος. γέγονεν· ὁρατὸς γὰρ ἁπτός τέ ἐστι καὶ σῶμα ἔχων, πάντα δὲ
C τὰ τοιαῦτα αἰσθητά, τὰ δὲ αἰσθητά, δόξῃ περιληπτὰ μετὰ αἰσθήσεως, γιγνόμενα καὶ γεννητὰ ἐφάνη. τῷ δ᾽ αὖ γενομένῳ φαμὲν ὑπ᾽ αἰτίου τινὸς ἀνάγκην εἶναι γενέσθαι. τὸν μὲν οὖν ποιητὴν καὶ πατέρα τοῦδε τοῦ παντὸς εὑρεῖν τε ἔργον καὶ εὑρόντα εἰς πάντας ἀδύνατον λέγειν· τόδε δ᾽ οὖν πάλιν ἐπισκεπτέον περὶ αὐτοῦ, πρὸς πότερον τῶν παραδειγμάτων ὁ τεκταινόμενος αὐτὸν ἀπειργάζετο,
29 πότερον πρὸς τὸ κατὰ ταὐτὰ καὶ ὡσαύτως ἔχον ἢ πρὸς τὸ γεγονός. εἰ μὲν δὴ καλός ἐστιν ὅδε ὁ κόσμος ὅ τε δημιουργὸς ἀγαθός, δῆλον ὡς πρὸς τὸ ἀΐδιον ἔβλεπεν· εἰ δέ, ὃ μηδ᾽ εἰπεῖν τινι θέμις, πρὸς

---

[1] *Cf.* 28 A.

a cause it is impossible for anything to attain becoming. But when the artificer of any object, in forming its shape and quality, keeps his gaze fixed on that which is uniform, using a model of this kind, that object, executed in this way, must of necessity be beautiful; but whenever he gazes at that which has come into existence and uses a created model, the object thus executed is not beautiful. Now the whole Heaven, or Cosmos, or if there is any other name which it specially prefers, by that let us call it,— so, be its name what it may, we must first investigate concerning it that primary question which has to be investigated at the outset in every case,—namely, whether it has existed always, having no beginning of generation, or whether it has come into existence, having begun from some beginning. It has come into existence; for it is visible and tangible and possessed of a body; and all such things are sensible, and things sensible, being apprehensible by opinion with the aid of sensation, come into existence, as we saw,[1] and are generated. And that which has come into existence must necessarily, as we say, have come into existence by reason of some Cause. Now to discover the Maker and Father of this Universe were a task indeed; and having discovered Him, to declare Him unto all men were a thing impossible. However, let us return and inquire further concerning the Cosmos, —after which of the Models did its Architect construct it? Was it after that which is self-identical and uniform, or after that which has come into existence? Now if so be that this Cosmos is beautiful and its Constructor good, it is plain that he fixed his gaze on the Eternal; but if otherwise (which is an impious supposition), his gaze was on that which has come

τὸ γεγονός. παντὶ δὴ σαφὲς ὅτι πρὸς τὸ ἀίδιον· ὁ μὲν γὰρ κάλλιστος τῶν γεγονότων, ὁ δ' ἄριστος τῶν αἰτίων. οὕτω δὴ γεγενημένος πρὸς τὸ λόγῳ καὶ φρονήσει περιληπτὸν καὶ κατὰ ταὐτὰ ἔχον δεδημιούργηται.

B Τούτων δὲ ὑπαρχόντων αὖ πᾶσα ἀνάγκη τόνδε τὸν κόσμον εἰκόνα τινὸς εἶναι. μέγιστον δὴ παντὸς ἄρξασθαι κατὰ φύσιν ἀρχήν· ὧδε οὖν περί τε εἰκόνος καὶ περὶ τοῦ παραδείγματος αὐτῆς διοριστέον, ὡς ἄρα τοὺς λόγους, ὧνπέρ εἰσιν ἐξηγηταί, τούτων αὐτῶν καὶ ξυγγενεῖς ὄντας. τοῦ μὲν οὖν μονίμου καὶ βεβαίου καὶ μετὰ τοῦ καταφανοῦς μονίμους καὶ ἀμεταπτώτους, καθ' ὅσον οἷόν τε ἀνελέγκτοις προσήκει λόγοις εἶναι καὶ C ἀνικήτοις, τούτου δεῖ μηδὲν ἐλλείπειν· τοὺς δὲ τοῦ πρὸς μὲν ἐκεῖνο ἀπεικασθέντος, ὄντος δὲ εἰκόνος εἰκότας ἀνὰ λόγον τε ἐκείνων ὄντας· ὅ τί περ πρὸς γένεσιν οὐσία, τοῦτο πρὸς πίστιν ἀλήθεια. ἐὰν οὖν, ὦ Σώκρατες, πολλὰ πολλῶν [εἰπόντων] πέρι,[1] θεῶν καὶ τῆς τοῦ παντὸς γενέσεως μὴ δυνατοὶ γιγνώμεθα πάντῃ πάντως αὐτοὺς αὑτοῖς ὁμολογουμένους λόγους καὶ ἀπηκριβωμένους ἀποδοῦναι, μὴ θαυμάσῃς, ἀλλ' ἐὰν ἄρα μηδενὸς ἧττον παρεχώμεθα εἰκότας, ἀγαπᾶν χρή, μεμνημένους[2] ὡς ὁ λέγων D ἐγὼ ὑμεῖς τε οἱ κριταὶ φύσιν ἀνθρωπίνην ἔχομεν, ὥστε περὶ τούτων τὸν εἰκότα μῦθον ἀποδεχομένους πρέπει τούτου μηδὲν ἔτι πέρα ζητεῖν.

ΣΩ. Ἄριστα, ὦ Τίμαιε, παντάπασί τε ὡς

[1] εἰπόντων is absent from best mss.: πέρι (for περὶ of mss.) Diehl.
[2] μεμνημένους] μεμνημένον best ms. and Zur.

TIMAEUS

into existence. But it is clear to everyone that his gaze was on the Eternal; for the Cosmos is the fairest of all that has come into existence, and He the best of all the Causes. So having in this wise come into existence, it has been constructed after the pattern of that which is apprehensible by reason and thought and is self-identical.

Again, if these premisses be granted, it is wholly necessary that this Cosmos should be a Copy of something. Now in regard to every matter it is most important to begin at the natural beginning. Accordingly, in dealing with a copy and its model, we must affirm that the accounts given will themselves be akin to the diverse objects which they serve to explain; those which deal with what is abiding and firm and discernible by the aid of thought will be abiding and unshakable; and in so far as it is possible and fitting for statements to be irrefutable and invincible, they must in no wise fall short thereof; whereas the accounts of that which is copied after the likeness of that Model, and is itself a likeness, will be analogous thereto and possess likelihood; for as Being is to Becoming, so is Truth to Belief. Wherefore, Socrates, if in our treatment of a great host of matters regarding the Gods and the generation of the Universe we prove unable to give accounts that are always in all respects self-consistent and perfectly exact, be not thou surprised; rather we should be content if we can furnish accounts that are inferior to none in likelihood, remembering that both I who speak and you who judge are but human creatures, so that it becomes us to accept the likely account of these matters and forbear to search beyond it.

soc. Excellent, Timaeus! We must by all means

κελεύεις ἀποδεκτέον· τὸ μὲν οὖν προοίμιον θαυμασίως ἀπεδεξάμεθά σου, τὸν δὲ δὴ νόμον[1] ἡμῖν ἐφεξῆς πέραινε.

ΤΙ. Λέγωμεν δὴ δι' ἥν τινα αἰτίαν γένεσιν καὶ τὸ πᾶν τόδε ὁ ξυνιστὰς ξυνέστησεν. ἀγαθὸς ἦν, ἀγαθῷ δὲ οὐδεὶς περὶ οὐδενὸς οὐδέποτε ἐγγίγνεται φθόνος· τούτου δ' ἐκτὸς ὢν πάντα ὅ τι μάλιστα γενέσθαι ἐβουλήθη παραπλήσια ἑαυτῷ. ταύτην δὲ γενέσεως καὶ κόσμου μάλιστ' ἄν τις ἀρχὴν κυριωτάτην παρ' ἀνδρῶν φρονίμων ἀποδεχόμενος ὀρθότατα ἀποδέχοιτ' ἄν. βουληθεὶς γὰρ ὁ θεὸς ἀγαθὰ μὲν πάντα, φλαῦρον δὲ μηδὲν εἶναι κατὰ δύναμιν, οὕτω δὴ πᾶν ὅσον ἦν ὁρατὸν παραλαβὼν οὐχ ἡσυχίαν ἄγον ἀλλὰ κινούμενον πλημμελῶς καὶ ἀτάκτως, εἰς τάξιν αὐτὸ ἤγαγεν ἐκ τῆς ἀταξίας, ἡγησάμενος ἐκεῖνο τούτου πάντως ἄμεινον. θέμις δὲ οὔτ' ἦν οὔτ' ἔστι τῷ ἀρίστῳ δρᾶν ἄλλο πλὴν τὸ κάλλιστον· λογισάμενος οὖν εὕρισκεν ἐκ τῶν κατὰ φύσιν ὁρατῶν οὐδὲν ἀνόητον τοῦ νοῦν ἔχοντος ὅλον ὅλου κάλλιον ἔσεσθαί ποτ' ἔργον, νοῦν δ' αὖ χωρὶς ψυχῆς ἀδύνατον παραγενέσθαι τῳ. διὰ δὴ τὸν λογισμὸν τόνδε νοῦν μὲν ἐν ψυχῇ, ψυχὴν δὲ ἐν σώματι ξυνιστὰς τὸ πᾶν ξυνετεκταίνετο, ὅπως ὅ τι κάλλιστον εἴη κατὰ φύσιν ἄριστόν τε ἔργον ἀπειργασμένος. οὕτως οὖν δὴ κατὰ λόγον τὸν εἰκότα δεῖ λέγειν τόνδε τὸν κόσμον ζῷον ἔμψυχον ἔννουν τε τῇ ἀληθείᾳ διὰ τὴν τοῦ θεοῦ γενέσθαι πρόνοιαν.

Τούτου δ' ὑπάρχοντος αὖ τὰ τούτοις ἐφεξῆς ἡμῖν

---

[1] νόμον] λόγον best MS. and Zur.

TIMAEUS

accept it, as you suggest; and certainly we have most cordially accepted your prelude; so now, we beg of you, proceed straight on with the main theme.

TIM. Let us now state the Cause wherefor He that constructed it constructed Becoming and the All. He was good, and in him that is good no envy ariseth ever concerning anything; and being devoid of envy He desired that all should be, so far as possible, like unto Himself. This principle, then, we shall be wholly right in accepting from men of wisdom as being above all the supreme originating principle of Becoming and the Cosmos. For God desired that, so far as possible, all things should be good and nothing evil; wherefore, when He took over all that was visible, seeing that it was not in a state of rest but in a state of discordant and disorderly motion, He brought it into order out of disorder, deeming that the former state is in all ways better than the latter. For Him who is most good it neither was nor is permissible to perform any action save what is most fair. As He reflected, therefore, He perceived that of such creatures as are by nature visible, none that is irrational will be fairer, comparing wholes with wholes, than the rational; and further, that reason cannot possibly belong to any apart from Soul. So because of this reflexion He constructed reason within soul and soul within body as He fashioned the All, that so the work He was executing might be of its nature most fair and most good. Thus, then, in accordance with the likely account, we must declare that this Cosmos has verily come into existence as a Living Creature endowed with soul and reason owing to the providence of God.

This being established, we must declare that which

30 λεκτέον, τίνι τῶν ζώων αὐτὸν εἰς ὁμοιότητα ὁ ξυνιστὰς ξυνέστησε. τῶν μὲν οὖν ἐν μέρους εἴδει πεφυκότων μηδενὶ καταξιώσωμεν· ἀτελεῖ γὰρ ἐοικὸς οὐδέν ποτ' ἂν γένοιτο καλόν· οὗ δ' ἔστι τἆλλα ζῷα καθ' ἓν καὶ κατὰ γένη μόρια, τούτῳ πάντων ὁμοιότατον αὐτὸν εἶναι τιθῶμεν. τὰ γὰρ δὴ νοητὰ ζῷα πάντα ἐκεῖνο ἐν ἑαυτῷ περιλαβὸν ἔχει, καθάπερ ὅδε ὁ κόσμος ἡμᾶς ὅσα τε ἄλλα
D θρέμματα ξυνέστηκεν ὁρατά. τῷ γὰρ τῶν νοουμένων καλλίστῳ καὶ κατὰ πάντα τελέῳ μάλιστ' αὐτὸν ὁ θεὸς ὁμοιῶσαι βουληθεὶς ζῷον ἓν ὁρατόν, πάνθ' ὅσα αὐτοῦ κατὰ φύσιν ξυγγενῆ ζῷα ἐντὸς ἔχον ἑαυτοῦ, ξυνέστησε.

31 Πότερον οὖν ὀρθῶς ἕνα οὐρανὸν προσειρήκαμεν, ἢ πολλοὺς καὶ ἀπείρους λέγειν ἦν ὀρθότερον; ἕνα, εἴπερ κατὰ τὸ παράδειγμα δεδημιουργημένος ἔσται. τὸ γὰρ περιέχον πάντα ὁπόσα νοητὰ ζῷα μεθ' ἑτέρου δεύτερον οὐκ ἄν ποτ' εἴη· πάλιν γὰρ ἂν ἕτερον εἶναι τὸ περὶ ἐκείνω δέοι ζῶον, οὗ μέρος ἂν εἴτην ἐκείνω, καὶ οὐκ ἂν ἔτι ἐκείνοιν ἀλλ' ἐκείνῳ τῷ περιέχοντι τόδ' ἂν ἀφωμοιωμένον
B λέγοιτο ὀρθότερον. ἵνα οὖν τόδε κατὰ τὴν μόνωσιν ὅμοιον ᾖ τῷ παντελεῖ ζῴῳ, διὰ ταῦτα οὔτε δύο οὔτ'

[1] Cf. 55 c ff. The Atomists held that there is an infinite number of worlds.

TIMAEUS

comes next in order. In the semblance of which of the living Creatures did the Constructor of the Cosmos construct it? We shall not deign to accept any of those which belong by nature to the category of " parts "; for nothing that resembles the imperfect would ever become fair. But we shall affirm that the Cosmos, more than aught else, resembles most closely that Living Creature of which all other living creatures, severally and generically, are portions. For that Living Creature embraces and contains within itself all the intelligible Living Creatures, just as this Universe contains us and all the other visible living creatures that have been fashioned. For since God desired to make it resemble most closely that intelligible Creature which is fairest of all and in all ways most perfect, He constructed it as a Living Creature, one and visible, containing within itself all the living creatures which are by nature akin to itself.

Are we right, then, in describing the Heaven as one, or would it be more correct to speak of heavens as many or infinite [1] in number? One it must be termed, if it is to be framed after its Pattern. For that which embraces all intelligible Living Creatures could never be second, with another beside it; for if so, there must needs exist yet another Living Creature, which should embrace them both, and of which they two would each be a part; in which case this Universe could no longer be rightly described as modelled on these two, but rather on that third Creature which contains them both. Wherefore, in order that this Creature might resemble the all-perfect Living Creature in respect of its uniqueness, for this reason its Maker made neither two Universes

PLATO

ἀπείρους ἐποίησεν ὁ ποιῶν κόσμους, ἀλλ' εἷς ὅδε μονογενὴς οὐρανὸς γεγονὼς ἔστι τε καὶ ἔτ' ἔσται.

Σωματοειδὲς δὲ δὴ καὶ ὁρατὸν ἁπτόν τε δεῖ τὸ γενόμενον εἶναι. χωρισθὲν δὲ πυρὸς οὐδὲν ἄν ποτε ὁρατὸν γένοιτο, οὐδὲ ἁπτὸν ἄνευ τινὸς στερεοῦ, στερεὸν δὲ οὐκ ἄνευ γῆς· ὅθεν ἐκ πυρὸς καὶ γῆς τὸ τοῦ παντὸς ἀρχόμενος ξυνιστάναι σῶμα ὁ θεὸς ἐποίει. δύο δὲ μόνω καλῶς ξυνίστασθαι τρίτου C χωρὶς οὐ δυνατόν· δεσμὸν γὰρ ἐν μέσῳ δεῖ τινὰ ἀμφοῖν ξυναγωγὸν γίγνεσθαι. δεσμῶν δὲ κάλλιστος ὃς ἂν αὑτὸν καὶ τὰ ξυνδούμενα ὅ τι μάλιστα ἓν ποιῇ. τοῦτο δὲ πέφυκεν ἀναλογία κάλλιστα ἀποτελεῖν· ὁπόταν γὰρ ἀριθμῶν τριῶν εἴτε ὄγκων εἴτε 32 δυνάμεων ὡντινωνοῦν ᾖ τὸ μέσον, ὅ τί περ τὸ πρῶτον πρὸς αὐτό, τοῦτο αὐτὸ πρὸς τὸ ἔσχατον, καὶ πάλιν αὖθις ὅ τι τὸ ἔσχατον πρὸς τὸ μέσον, τοῦτο τὸ μέσον πρὸς τὸ πρῶτον, τότε τὸ μέσον μὲν πρῶτον καὶ ἔσχατον γιγνόμενον, τὸ δὲ ἔσχατον καὶ τὸ πρῶτον αὖ μέσα ἀμφότερα, πάνθ' οὕτως ἐξ ἀνάγκης τὰ αὐτὰ εἶναι ξυμβήσεται, τὰ αὐτὰ δὲ γενόμενα ἀλλήλοις ἓν πάντα ἔσται. εἰ μὲν οὖν ἐπίπεδον μέν, βάθος δὲ μηδὲν ἔχον ἔδει γίγνεσθαι B τὸ τοῦ παντὸς σῶμα, μία μεσότης ἂν ἐξήρκει τά τε μεθ' αὑτῆς ξυνδεῖν καὶ ἑαυτήν· νῦν δέ—στερεοειδῆ γὰρ αὐτὸν προσῆκεν εἶναι, τὰ δὲ στερεὰ μία μὲν οὐδέποτε, δύο δὲ ἀεὶ μεσότητες συναρμόττουσιν· οὕτω δὴ πυρός τε καὶ γῆς ὕδωρ ἀέρα τε ὁ θεὸς ἐν μέσῳ θείς, καὶ πρὸς ἄλληλα καθ' ὅσον ἦν

---

[1] *Cf.* 92 c.
[2] Dealing first with "square" numbers, the proportion here indicated is—$a^2 : ab :: ab : b^2$; conversely, $b^2 : ab :: ab : a^2$; alternately, $ab : a^2 :: b^2 : ab$.

TIMAEUS

nor an infinite number, but there is and will continue to be this one generated Heaven, unique of its kind.[1]

Now that which has come into existence must needs be of bodily form, visible and tangible; yet without fire nothing could ever become visible, nor tangible without some solidity, nor solid without earth. Hence, in beginning to construct the body of the All, God was making it of fire and earth. But it is not possible that two things alone should be conjoined without a third; for there must needs be some intermediary bond to connect the two. And the fairest of bonds is that which most perfectly unites into one both itself and the things which it binds together; and to effect this in the fairest manner is the natural property of proportion. For whenever the middle term of any three numbers, cubic or square,[2] is such that as the first term is to it, so is it to the last term, —and again, conversely, as the last term is to the middle, so is the middle to the first,—then the middle term becomes in turn the first and the last, while the first and last become in turn middle terms, and the necessary consequence will be that all the terms are interchangeable, and being interchangeable they all form a unity. Now if the body of the All had had to come into existence as a plane surface, having no depth, one middle term would have sufficed to bind together both itself and its fellow-terms; but now it is otherwise: for it behoved it to be solid of shape, and what brings solids into unison is never one middle term alone but always two.[3] Thus it was that in the midst between fire and earth God set water and air, and having bestowed upon them so far as possible

---

[3] Two mean terms are required for a continuous proportion of "solid" (or cubic) numbers, e.g. $a^3 : a^2b :: a^2b : ab^2 :: ab^2 : b^3$.

32 δυνατὸν ἀνὰ τὸν αὐτὸν λόγον ἀπεργασάμενος, ὅ τί περ πῦρ πρὸς ἀέρα, τοῦτο ἀέρα πρὸς ὕδωρ, καὶ ὅ τι ἀὴρ πρὸς ὕδωρ, ὕδωρ πρὸς γῆν, ξυνέδησε καὶ ξυνεστήσατο οὐρανὸν ὁρατὸν καὶ ἁπτόν. καὶ διὰ C ταῦτα ἔκ τε δὴ τούτων τοιούτων καὶ τὸν ἀριθμὸν τεττάρων τὸ τοῦ κόσμου σῶμα ἐγεννήθη δι' ἀναλογίας ὁμολογῆσαν, φιλίαν τε ἔσχεν ἐκ τούτων, ὥστ' εἰς ταὐτὸν αὑτῷ ξυνελθὸν ἄλυτον ὑπό του ἄλλου πλὴν ὑπὸ τοῦ ξυνδήσαντος γενέσθαι.

Τῶν δὲ δὴ τεττάρων ἓν ὅλον ἕκαστον εἴληφεν ἡ τοῦ κόσμου ξύστασις· ἐκ γὰρ πυρὸς παντὸς ὕδατός τε καὶ ἀέρος καὶ γῆς ξυνέστησεν αὐτὸν ὁ ξυνιστάς, μέρος οὐδὲν οὐδενὸς οὐδὲ δύναμιν ἔξωθεν ὑπολιπών, D τάδε διανοηθείς, πρῶτον μὲν ἵνα ὅλον ὅ τι μάλιστα ζῷον τέλεον ἐκ τελέων τῶν μερῶν εἴη, πρὸς δὲ 33 τούτοις ἕν, ἅτε οὐχ ὑπολελειμμένων ἐξ ὧν ἄλλο τοιοῦτον γένοιτ' ἄν, ἔτι δὲ ἵνα ἀγήρων καὶ ἄνοσον ᾖ, κατανοῶν ὡς συστάτῳ σώματι[1] θερμὰ καὶ ψυχρὰ καὶ πάνθ' ὅσα δυνάμεις ἰσχυρὰς ἔχει περιιστάμενα ἔξωθεν καὶ προσπίπτοντα ἀκαίρως λύει καὶ νόσους γῆράς τε ἐπάγοντα φθίνειν ποιεῖ. διὰ δὴ τὴν αἰτίαν καὶ τὸν λογισμὸν τόνδε ἓν ὅλον ὅλων ἐξ ἁπάντων τέλεον καὶ ἀγήρων καὶ ἄνοσον αὐτὸν B ἐτεκτήνατο. σχῆμα δὲ ἔδωκεν αὐτῷ τὸ πρέπον καὶ τὸ ξυγγενές. τῷ δὲ τὰ πάντ' ἐν αὑτῷ ζῷα περιέχειν μέλλοντι ζῴῳ πρέπον ἂν εἴη σχῆμα τὸ

[1] συστάτῳ σώματι Proclus: ἃ ξυνιστᾷ τὰ σώματα Zur.

TIMAEUS

a like ratio one towards another—air being to water as fire to air, and water being to earth as air to water,—he joined together and constructed a Heaven visible and tangible. For these reasons and out of these materials, such in kind and four in number, the body of the Cosmos was harmonized by proportion and brought into existence. These conditions secured for it Amity, so that being united in identity with itself it became indissoluble by any agent other than Him who had bound it together.

Now of the four elements the construction of the Cosmos had taken up the whole of every one. For its Constructor had constructed it of all the fire and water and air and earth that existed, leaving over, outside it, no single particle or potency of any one of these elements. And these were his intentions: first, that it might be, so far as possible, a Living Creature, perfect and whole, with all its parts perfect; and next, that it might be One, inasmuch as there was nothing left over out of which another like Creature might come into existence; and further, that it might be secure from age and ailment, since He perceived that when heat and cold, and all things which have violent potencies, surround a composite body from without and collide with it they dissolve it unduly and make it to waste away by bringing upon it ailments and age. Wherefore, because of this reasoning, He fashioned it to be One single Whole, compounded of all wholes, perfect and ageless and unailing. And he bestowed on it the shape which was befitting and akin. Now for that Living Creature which is designed to embrace within itself all living creatures the fitting shape will be that which comprises within itself all

PLATO

33 περιειληφὸς ἐν αὑτῷ πάντα ὁπόσα σχήματα. διὸ
καὶ σφαιροειδές, ἐκ μέσου πάντη πρὸς τὰς τελευτὰς
ἴσον ἀπέχον, κυκλοτερὲς αὐτὸ ἐτορνεύσατο, πάντων
τελεώτατον ὁμοιότατόν τε αὐτὸ ἑαυτῷ σχημάτων,
νομίσας μυρίῳ κάλλιον ὅμοιον ἀνομοίου. λεῖον δὲ
δὴ κύκλῳ πᾶν ἔξωθεν αὐτὸ ἀπηκριβοῦτο, πολλῶν
C χάριν. ὀμμάτων τε γὰρ ἐπεδεῖτο οὐδέν, ὁρατὸν γὰρ
οὐδὲν ὑπελείπετο ἔξωθεν, οὐδ' ἀκοῆς, οὐδὲ γὰρ
ἀκουστόν· πνεῦμά τε οὐκ ἦν περιεστὸς δεόμενον
ἀναπνοῆς. οὐδ' αὖ τινὸς ἐπιδεὲς ἦν ὀργάνου σχεῖν,
ᾧ τὴν μὲν εἰς ἑαυτὸ τροφὴν δέξοιτο, τὴν δὲ πρότερον ἐξικμασμένην ἀποπέμψοι πάλιν· ἀπῄει τε γὰρ
οὐδὲν οὐδὲ προσῄειν αὐτῷ ποθέν· οὐδὲ γὰρ ἦν.
αὐτὸ γὰρ ἑαυτῷ τροφὴν τὴν ἑαυτοῦ φθίσιν παρέχον
D καὶ πάντα ἐν ἑαυτῷ καὶ ὑφ' ἑαυτοῦ πάσχον καὶ
δρῶν ἐκ τέχνης γέγονεν· ἡγήσατο γὰρ αὐτὸ ὁ
ξυνθεὶς αὔταρκες ὂν ἄμεινον ἔσεσθαι μᾶλλον ἢ
προσδεὲς ἄλλων. χειρῶν δέ, αἷς οὔτε λαβεῖν οὔτε
αὖ τινα ἀμύνασθαι χρεία τις ἦν, μάτην οὐκ ᾤετο
δεῖν αὐτῷ προσάπτειν, οὐδὲ ποδῶν οὐδὲ ὅλως τῆς
34 περὶ τὴν βάσιν ὑπηρεσίας. κίνησιν γὰρ ἀπένειμεν
αὐτῷ τὴν τοῦ σώματος οἰκείαν, τῶν ἑπτὰ τὴν περὶ
νοῦν καὶ φρόνησιν μάλιστα οὖσαν. διὸ δὴ κατὰ
ταὐτὰ ἐν τῷ αὐτῷ καὶ ἐν ἑαυτῷ περιαγαγὼν αὐτὸ
ἐποίησε κύκλῳ κινεῖσθαι στρεφόμενον, τὰς δὲ ἓξ
ἁπάσας κινήσεις ἀφεῖλε καὶ ἀπλανὲς ἀπειργάσατο
ἐκείνων. ἐπὶ δὲ τὴν περίοδον ταύτην ἅτ' οὐδὲν
ποδῶν δέον ἀσκελὲς καὶ ἄπουν αὐτὸ ἐγέννησεν.

---

[1] For "the seven motions" see 43 B; and for the (rotatory) "motion of reason" *cf. Laws* 898 A. *Cf.* also 37 A ff., 42 C, 47 D, 77 B.

TIMAEUS

the shapes there are; wherefore He wrought it into a round, in the shape of a sphere, equidistant in all directions from the centre to the extremities, which of all shapes is the most perfect and the most self-similar, since He deemed that the similar is infinitely fairer than the dissimilar. And on the outside round about, it was all made smooth with great exactness, and that for many reasons. For of eyes it had no need, since outside of it there was nothing visible left over; nor yet of hearing, since neither was there anything audible; nor was there any air surrounding it which called for respiration; nor, again, did it need any organ whereby it might receive the food that entered and evacuate what remained undigested. For nothing went out from it or came into it from any side, since nothing existed; for it was so designed as to supply its own wastage as food for itself, and to experience by its own agency and within itself all actions and passions, since He that had constructed it deemed that it would be better if it were self-sufficing rather than in need of other things. Hands, too, He thought He ought not to attach unto it uselessly, seeing they were not required either for grasping or for repelling anyone; nor yet feet, nor any instruments of locomotion whatsoever. For movement He assigned unto it that which is proper to its body, namely, that one of the seven motions [1] which specially belongs to reason and intelligence; wherefore He spun it round uniformly in the same spot and within itself and made it move revolving in a circle; and all the other six motions He took away and fashioned it free from their aberrations. And seeing that for this revolving motion it had no need of feet, He begat it legless and footless.

PLATO

34

Οὗτος δὴ πᾶς ὄντος ἀεὶ λογισμὸς θεοῦ περὶ τὸν
B ποτὲ ἐσόμενον θεὸν λογισθεὶς λεῖον καὶ ὁμαλὸν
πανταχῇ τε ἐκ μέσου ἴσον καὶ ὅλον καὶ τέλεον ἐκ
τελέων σωμάτων σῶμα ἐποίησε. ψυχὴν δὲ εἰς τὸ
μέσον αὐτοῦ θεὶς διὰ παντός τε ἔτεινε καὶ ἔτι
ἔξωθεν τὸ σῶμα αὐτῇ περιεκάλυψε ταύτῃ, καὶ
κύκλῳ δὴ κύκλον στρεφόμενον οὐρανὸν ἕνα μόνον
ἔρημον κατέστησε, δι' ἀρετὴν δὲ αὐτὸν αὑτῷ δυνάμενον ξυγγίγνεσθαι καὶ οὐδενὸς ἑτέρου προσδεό-
μενον, γνώριμον δὲ καὶ φίλον ἱκανῶς αὐτὸν αὑτῷ.
διὰ πάντα δὴ ταῦτα εὐδαίμονα θεὸν αὐτὸν ἐγεν-
νήσατο.

Τὴν δὲ δὴ ψυχὴν οὐχ ὡς νῦν ὑστέραν ἐπιχειροῦ-
C μεν λέγειν, οὕτως ἐμηχανήσατο καὶ ὁ θεὸς νεω-
τέραν· οὐ γὰρ ἂν ἄρχεσθαι πρεσβύτερον ὑπὸ
νεωτέρου ξυνέρξας εἴασεν· ἀλλά πως ἡμεῖς πολὺ
μετέχοντες τοῦ προστυχόντος τε καὶ εἰκῇ ταύτῃ
πῃ καὶ λέγομεν, ὁ δὲ καὶ γενέσει καὶ ἀρετῇ προτέραν καὶ πρεσβυτέραν ψυχὴν σώματος, ὡς δε-
σπότιν καὶ ἄρξουσαν ἀρξομένου, συνεστήσατο ἐκ
35 τῶνδέ τε καὶ τοιῷδε τρόπῳ.

Τῆς ἀμερίστου καὶ ἀεὶ κατὰ ταὐτὰ ἐχούσης
οὐσίας καὶ τῆς αὖ περὶ τὰ σώματα γιγνομένης
μεριστῆς, τρίτον ἐξ ἀμφοῖν ἐν μέσῳ συνεκεράσατο
οὐσίας εἶδος, τῆς τε ταὐτοῦ φύσεως [αὖ πέρι]¹
καὶ τῆς θατέρου, καὶ κατὰ ταὐτὰ ξυνέστησεν
ἐν μέσῳ τοῦ τε ἀμεροῦς αὐτῶν καὶ τοῦ κατὰ
τὰ σώματα μεριστοῦ. καὶ τρία λαβὼν αὐτὰ ὄντα

¹ αὖ πέρι bracketed by Hermann, after Sextus Empir.

---

[1] For the priority of Soul *cf. Laws* 892 A, B, 896 C ff.; and for the right of the elder to rule *cf. Laws* 714 E.

TIMAEUS

Such, then, was the sum of the reasoning of the ever-existing God concerning the god which was one day to be existent, whereby He made it smooth and even and equal on all sides from the centre, a whole and perfect body compounded of perfect bodies. And in the midst thereof He set Soul, which He stretched throughout the whole of it, and therewith He enveloped also the exterior of its body; and as a Circle revolving in a circle He stablished one sole and solitary Heaven, able of itself because of its excellence to company with itself and needing none other beside, sufficing unto itself as acquaintance and friend. And because of all this He generated it to be a blessed God.

Now as regards the Soul, although we are essaying to describe it after the body, God did not likewise plan it to be younger than the body [1]; for, when uniting them, He would not have permitted the elder to be ruled by the younger; but as for us men, even as we ourselves partake largely of the accidental and casual, so also do our words. God, however, constructed Soul to be older than Body and prior in birth and excellence, since she was to be the mistress and ruler and it the ruled; and He made her of the materials and in the fashion which I shall now describe.

Midway between the Being which is indivisible and remains always the same and the Being which is transient and divisible in bodies, He blended a third form of Being compounded out of the twain, that is to say, out of the Same and the Other; and in like manner He compounded it midway between that one of them which is indivisible and that one which is divisible in bodies. And He took the three of

PLATO

35

συνεκεράσατο εἰς μίαν πάντα ἰδέαν, τὴν θατέρου
φύσιν δύσμικτον οὖσαν εἰς ταὐτὸν ξυναρμόττων
B βίᾳ. μιγνὺς δὲ μετὰ τῆς οὐσίας καὶ ἐκ τριῶν
ποιησάμενος ἓν πάλιν ὅλον τοῦτο μοίρας ὅσας
προσῆκε διένειμεν, ἑκάστην δὲ ἔκ τε ταὐτοῦ καὶ
θατέρου καὶ τῆς οὐσίας μεμιγμένην. ἤρχετο δὲ
διαιρεῖν ὧδε.[1] μίαν ἀφεῖλε τὸ πρῶτον ἀπὸ παντὸς
μοῖραν, μετὰ δὲ ταύτην ἀφῄρει διπλασίαν ταύτης,
τὴν δ' αὖ τρίτην ἡμιολίαν μὲν τῆς δευτέρας,
τριπλασίαν δὲ τῆς πρώτης, τετάρτην δὲ τῆς
δευτέρας διπλῆν, πέμπτην δὲ τριπλῆν τῆς τρίτης,
C τὴν δ' ἕκτην τῆς πρώτης ὀκταπλασίαν, ἑβδόμην
δὲ ἑπτακαιεικοσαπλασίαν τῆς πρώτης. μετὰ δὲ
36 ταῦτα ξυνεπληροῦτο τά τε διπλάσια καὶ τρι-
πλάσια διαστήματα, μοίρας ἔτι ἐκεῖθεν ἀποτέμνων
καὶ τιθεὶς εἰς τὸ μεταξὺ τούτων, ὥστε ἐν ἑκάστῳ
διαστήματι δύο εἶναι μεσότητας, τὴν μὲν ταὐτῷ
μέρει τῶν ἄκρων αὐτῶν ὑπερέχουσαν καὶ ὑπερ-

---

[1] The choice of these three as constituents of the Soul is explained by the use of the same terms in the *Sophist* (244-245) to denote certain "Greatest Kinds" or main categories. As Professor Paul Shorey has aptly observed (*Amer. Journ. Philol.* ix. p. 298), "It is necessary that the Soul should recognize everywhere . . . the *same*, the *other* and *essence*, those three μέγιστα γένη of the . . . *Sophist*. Hence, on the Greek principle that like is known by like, Plato makes real substances out of these three abstractions and puts them as plastic material into the hands of the Demiurgus for the formation of the Soul."

[2] These seven numbers may be arranged in two branches, in order to show the two series of which Timaeus immediately goes on to speak:

```
              1 (the 1st)
    2 (the 2nd)        3 (the 3rd)
    4 (the 4th)        9 (the 5th)
    8 (the 6th)       27 (the 7th)
```

TIMAEUS

them, and blent them all together into one form, by forcing the Other into union with the Same, in spite of its being naturally difficult to mix. And when with the aid of Being He had mixed them, and had made of them one out of three, straightway He began to distribute the whole thereof into so many portions as was meet; and each portion was a mixture of the Same, of the Other, and of Being.[1] And He began making the division thus:

First He took one portion from the whole;
then He took a portion double of this;
then a third portion, half as much again as the second portion, that is, three times as much as the first;
the fourth portion He took was twice as much as the second;
the fifth three times as much as the third;
the sixth eight times as much as the first; and
the seventh twenty-seven times as much as the first.[2]

After that He went on to fill up the intervals in the series of the powers of 2 and the intervals in the series of powers of 3 in the following manner[3]:

He cut off yet further portions from the original mixture, and set them in between the portions above rehearsed, so as to place two Means in each interval, —one a Mean which exceeded its Extremes and was by them exceeded by the same *proportional part* or *fraction* of each of the Extremes respectively[4]; the

---

The left-hand branch contains the " double intervals," *i.e.* the powers of 2; the right-hand one the " triple intervals," *i.e.* the powers of 3.

[3] *Lit.* " the double intervals and the triple intervals." See the preceding note.

[4] The " harmonic Mean."

67

PLATO

ἐχομένην, τὴν δ' ἴσῳ μὲν κατ' ἀριθμὸν ὑπερέχουσαν, ἴσῳ δὲ ὑπερεχομένην. ἡμιολίων δὲ διαστάσεων καὶ ἐπιτρίτων καὶ ἐπογδόων γενομένων ἐκ τούτων τῶν δεσμῶν ἐν ταῖς πρόσθεν διαστάσεσι, τῷ τοῦ ἐπογδόου διαστήματι τὰ ἐπίτριτα πάντα ξυνεπληροῦτο, λείπων αὐτῶν ἑκάστου μόριον, τῆς τοῦ μορίου ταύτης διαστάσεως λειφθείσης ἀριθμοῦ πρὸς ἀριθμὸν ἐχούσης τοὺς ὅρους ἓξ καὶ πεντήκοντα καὶ διακοσίων πρὸς τρία καὶ τετταράκοντα

---

[1] The "arithmetical Mean."

The Means are inserted as follows:

In the " double intervals "
1. h.m. a.m. 2. h.m. a.m. 4. h.m. a.m. 8.

In the " triple intervals "
1. h.m. a.m. 3. h.m. a.m. 9. h.m. a.m. 27.

Evaluated, these Means are

In the " double intervals "
1. 1⅓. 1½. 2. 2⅔. 3. 4. 5⅓. 6. 8. OR 1. 4/3. 3/2. 2. 8/3. 3. 4. 16/3. 6. 8.

In the " triple intervals "
1. 1½. 2. 3. 4½. 6. 9. 13½. 18. 27. OR 1. 3/2. 2. 3. 9/2. 6. 9. 27/2. 18. 27.

[2] The fresh intervals formed are:

	In the first series.	In the second series.
Between the 1st and 2nd terms	4 : 3	3 : 2
" " 2nd " 3rd "	9 : 8	4 : 3
" " 3rd " 4th "	4 : 3	3 : 2
" " 4th " 5th "	4 : 3	3 : 2
" " 5th " 6th "	9 : 8	4 : 3
" " 6th " 7th "	4 : 3	3 : 2
" " 7th " 8th "	4 : 3	3 : 2
" " 8th " 9th "	9 : 8	4 : 3
" " 9th " 10th "	4 : 3	3 : 2

Wherever there is an interval of 4 : 3, he fills it up with

TIMAEUS

other a Mean which exceeded one Extreme by the same *number* or *integer* as it was exceeded by its other Extreme.[1]

And whereas the insertion of these links formed fresh intervals in the former intervals, that is to say, intervals of 3 : 2 and 4 : 3 and 9 : 8, He went on to fill up the 4 : 3 intervals with 9 : 8 intervals. This still left over in each case a fraction, which is represented by the terms of the numerical ratio 256 : 243.[2]

intervals of 9 : 8. There is room for two 9 : 8 intervals in every 4 : 3 interval; but there is also an odd fraction left over, which turns out to be an interval of 256 : 243. For example, between the terms 1 and $1\frac{1}{3}$ in the first series, which is an interval of 4 : 3, he inserts the terms $\frac{9}{8}$, $\frac{81}{64}$, thus:

$$1 \cdot \tfrac{9}{8} \cdot \tfrac{81}{64} \cdot \tfrac{4}{3}.$$

The interval between the first and second and between the second and third is 9 : 8; between the last two 256 : 243. This occurs every time there is an interval of 4 : 3.

By thus filling up the intervals in the "double interval" series, we get a fresh series containing twenty-two terms, and therefore twenty-one intervals. These intervals come in the following succession:

9 : 8, 9 : 8, 256 : 243, 9 : 8, 9 : 8, 9 : 8, 256 : 243, the whole three times over.

Timaeus does not say how the intervals in the series of the "triple intervals" are to be filled up; but it can now be seen that the "double interval" series and the "triple interval" series are in reality but one series. For even when we have inserted no more than the harmonic and arithmetical Means (as shown above, note 1), out of all the terms in the "triple" series, so far as it covers the ground of the "double" series, only one is not also to be found in the "double" series, viz. $\frac{9}{2}$; and even this makes its appearance in the "double" series at the next step, when we fill up the 4 : 3 intervals with 9 : 8 intervals.

Hence, when we come to the end of the "double interval" series, with its harmonic and arithmetical Means inserted, and with the 9 : 8 intervals also inserted—that is, when we

69

PLATO

καὶ διακόσια. καὶ δὴ τὸ μιχθὲν ἐξ οὗ ταῦτα κατέτεμνεν οὕτως ἤδη πᾶν καταναλώκει. ταύτην οὖν τὴν ξύστασιν πᾶσαν διπλῆν κατὰ μῆκος σχίσας, μέσην πρὸς μέσην ἑκατέραν ἀλλήλαις οἷον χῖ προσβαλὼν κατέκαμψεν εἰς κύκλον, ξυνάψας αὐταῖς τε καὶ ἀλλήλαις ἐν τῷ καταντικρὺ τῆς προσβολῆς, καὶ τῇ κατὰ ταὐτὰ καὶ ἐν ταὐτῷ περιαγομένῃ κινήσει πέριξ αὐτὰς ἔλαβε, καὶ τὸν μὲν ἔξω, τὸν δ' ἐντὸς ἐποιεῖτο τῶν κύκλων. τὴν μὲν οὖν ἔξω φορὰν ἐπεφήμισεν εἶναι τῆς ταὐτοῦ φύσεως, τὴν δ' ἐντὸς τῆς θατέρου. τὴν μὲν δὴ ταὐτοῦ κατὰ πλευρὰν ἐπὶ δεξιὰ περιήγαγε, τὴν δὲ

---

have got the series of 22 terms ending with the term 8 (as described above)—we continue the series till we reach as far as the remaining terms of the original "triple interval" series will take us, that is, as far as the term 27, building up the series by inserting terms in the same way as before: intervals of 9 : 8, 9 : 8, 256 : 243, 9 : 8, 9 : 8, 9 : 8, 256 : 243, and so on *da capo*. This gives us twelve more terms (among which are, of course, the 9, $\frac{27}{2}$, 18, and 27 which appeared in the second version of the "triple interval" series), the last of which is 27.

The complete series of terms, from 1 to 27, making thirty-four terms in all, is intended to correspond with the notes of a musical scale, having a compass of four octaves and a major "sixth."

For the sake of convenience, I give here a complete list of the terms of the series. An *a* indicates the relationship 9 : 8 between a term and its predecessor, a letter *b* the relationship 256 : 243.

a	*a*	*b*	*a*		*a*	*a*	*b*	*a*	*a*	*b*	*a*	*a*
1	$\frac{9}{8}$	$\frac{81}{64}$	$\frac{4}{3}$		$\frac{3}{2}$	$\frac{27}{16}$	$\frac{243}{128}$	2	$\frac{9}{4}$	$\frac{81}{32}$	$\frac{8}{3}$	3

TIMAEUS

And thus the mixture, from which He had been cutting these portions off, was now all spent.

Next, He split all this that He had put together into two parts lengthwise; and then He laid the twain one against the other, the middle of one to the middle of the other, like a great cross $+$; and bent either of them into a circle, and joined them, each to itself and also to the other, at a point opposite to where they had first been laid together.[1] And He compassed them about with the motion that revolves in the same spot continually, and He made the one circle outer and the other inner. And the outer motion He ordained to be the Motion of the Same, and the inner motion the Motion of the Other. And He made the Motion of the Same to be toward the right along the side, and the Motion of the Other to be toward the left along the

a	b	a	a	b	a	a	a	b	a	a
$\frac{27}{8}$	$\frac{243}{64}$	4	$\frac{9}{2}$	$\frac{81}{16}$	$\frac{16}{3}$	6	$\frac{27}{4}$	$\frac{243}{32}$	8	9

b	a	a	a	b	a	a	b	a	a
$\frac{81}{8}$	$\frac{32}{3}$	12	$\frac{27}{2}$	$\frac{243}{16}$	16	18	$\frac{81}{4}$	$\frac{64}{3}$	24	27

The octave-terms are naturally 1, 2, 4, 8, 16.

[1] The accompanying figure indicates how the two strips were applied to each other. The place where they were originally laid together across each other is, in the diagram, on the further side, and is marked by a dot; the place where the two ends of each band are joined together, and where the two bands are themselves again joined together is, in the diagram, on the near side, and is indicated by a line on the outer band. The second place of meeting is, as the dotted line indicates, immediately opposite to the first.

The outer band, as Timaeus goes on to say, is the Revolution of the Same, and the inner the Revolution of the Other.

36 θατέρου κατὰ διάμετρον ἐπ' ἀριστερά. κράτος δ'
D ἔδωκε τῇ ταὐτοῦ καὶ ὁμοίου περιφορᾷ· μίαν γὰρ
αὐτὴν ἄσχιστον εἴασε, τὴν δ' ἐντὸς σχίσας ἑξαχῇ
ἑπτὰ κύκλους ἀνίσους κατὰ τὴν τοῦ διπλασίου
καὶ τριπλασίου διάστασιν ἑκάστην, οὐσῶν ἑκατέρων τριῶν, κατὰ τἀναντία μὲν ἀλλήλοις προσέταξεν ἰέναι τοὺς κύκλους, τάχει δὲ τρεῖς μὲν
ὁμοίως, τοὺς δὲ τέτταρας ἀλλήλοις καὶ τοῖς τρισὶν
ἀνομοίως, ἐν λόγῳ δὲ φερομένους.

Ἐπεὶ δὲ κατὰ νοῦν τῷ ξυνιστάντι πᾶσα ἡ τῆς
ψυχῆς ξύστασις ἐγεγένητο, μετὰ τοῦτο πᾶν τὸ
E σωματοειδὲς ἐντὸς αὐτῆς ἐτεκταίνετο καὶ μέσον
μέσῃ ξυναγαγὼν προσήρμοττεν. ἡ δ' ἐκ μέσου
πρὸς τὸν ἔσχατον οὐρανὸν πάντῃ διαπλακεῖσα
κύκλῳ τε αὐτὸν ἔξωθεν περικαλύψασα, αὐτή τε
ἐν αὑτῇ στρεφομένη, θείαν ἀρχὴν ἤρξατο ἀπαύστου
καὶ ἔμφρονος βίου πρὸς τὸν ξύμπαντα χρόνον.
καὶ τὸ μὲν δὴ σῶμα ὁρατὸν οὐρανοῦ γέγονεν, αὐτὴ
δὲ ἀόρατος μέν, λογισμοῦ δὲ μετέχουσα καὶ
37 ἁρμονίας [ψυχή],¹ τῶν νοητῶν ἀεί τε ὄντων ὑπὸ
τοῦ ἀρίστου ἀρίστη γενομένη τῶν γεννηθέντων.
ἅτε οὖν ἐκ τῆς ταὐτοῦ καὶ τῆς θατέρου φύσεως
ἔκ τε οὐσίας τριῶν τούτων συγκραθεῖσα μοιρῶν,

¹ ψυχὴ I bracket.

---

¹ He now tilts the inner band, so that it makes an oblique angle with the outer, which is set at the horizontal; from which we see that the Revolution of the Same represents the celestial Equator, moving " horizontally to the right " (from East to West), and the Revolution of the Other represents the Ecliptic, which moves in a contrary direction to the Equator (from West to East), and at an angle to it. The Ecliptic He divides into seven, to represent the seven planets.

² *Viz.* 2, 3, 4, 8, 9, 27.

TIMAEUS

diagonal[1]; and He gave the sovranty to the Revolution of the Same and of the Uniform. For this alone He suffered to remain uncloven, whereas He split the inner Revolution in six places into seven unequal circles, according to each of the intervals of the double and triple intervals,[2] three double and three triple.[3] These two circles then He appointed to go in contrary directions; and of the seven circles into which He split the inner circle, He appointed three to revolve at an equal speed, the other four[4] to go at speeds equal neither with each other nor with the speed of the aforesaid three, yet moving at speeds the ratios of which one to another are those of natural integers.

And when the construction of the Soul had all been completed to the satisfaction of its Constructor, then He fabricated within it all the Corporeal, and uniting them centre to centre He made them fit together. And the Soul, being woven throughout the Heaven every way from the centre to the extremity, and enveloping it in a circle from without, and herself revolving within herself, began a divine beginning of unceasing and intelligent life lasting throughout all time. And whereas the body of the Heaven is visible, the Soul is herself invisible but partakes in reasoning and in harmony, having come into existence by the agency of the best of things intelligible and ever-existing as the best of things generated. Inasmuch, then, as she is a compound, blended of the natures of the Same and the Other and Being, these three portions, and is proportionately divided

[3] *Viz.* 2, 4, 8 double; 3, 9, 27 triple.
[4] The three are Sun, Venus, Mercury; the four Moon, Mars, Jupiter, Saturn.

PLATO

37 καὶ ἀνὰ λόγον μερισθεῖσα καὶ ξυνδεθεῖσα, αὐτή τε ἀνακυκλουμένη πρὸς αὑτήν, ὅταν οὐσίαν σκεδαστὴν ἔχοντός τινος ἐφάπτηται καὶ ὅταν ἀμέριστον, λέγει κινουμένη διὰ πάσης ἑαυτῆς, ὅτῳ τ' ἄν τι
B ταὐτὸν ᾖ καὶ ὅτου ἂν ἕτερον, πρὸς ὅ τί τε μάλιστα καὶ ὅπῃ καὶ ὅπως καὶ ὁπότε ξυμβαίνει κατὰ τὰ γιγνόμενά τε πρὸς ἕκαστον ἕκαστα εἶναι καὶ πάσχειν καὶ πρὸς τὰ κατὰ ταὐτὰ ἔχοντα ἀεί. λόγος δὲ ὁ κατὰ ταὐτὸν ἀληθὴς γιγνόμενος, περί τε θατέρου ὢν καὶ περὶ τὸ ταὐτόν, ἐν τῷ κινουμένῳ ὑφ' αὑτοῦ φερόμενος ἄνευ φθόγγου καὶ ἠχῆς, ὅταν μὲν περὶ τὸ αἰσθητὸν γίγνηται καὶ ὁ τοῦ θατέρου κύκλος ὀρθὸς ἰὼν[1] εἰς πᾶσαν αὐτὰ[2] τὴν ψυχὴν διαγγείλῃ, δόξαι καὶ πίστεις γίγνονται βέβαιοι καὶ ἀληθεῖς· ὅταν δὲ αὖ περὶ τὸ λογι-
C στικὸν ᾖ καὶ ὁ τοῦ ταὐτοῦ κύκλος εὔτροχος ὢν αὐτὰ μηνύσῃ, νοῦς ἐπιστήμη τε ἐξ ἀνάγκης ἀποτελεῖται. τούτω δὲ ἐν ᾧ τῶν ὄντων ἐγγίγνεσθον, ἄν ποτέ τις αὐτὸ ἄλλο πλὴν ψυχὴν εἴπῃ, πᾶν μᾶλλον ἢ τἀληθὲς ἐρεῖ.

Ὡς δὲ κινηθὲν αὐτὸ καὶ ζῶν ἐνόησε τῶν ἀϊδίων θεῶν γεγονὸς ἄγαλμα ὁ γεννήσας πατήρ, ἠγάσθη τε καὶ εὐφρανθεὶς ἔτι δὴ μᾶλλον ὅμοιον πρὸς
D τὸ παράδειγμα ἐπενόησεν ἀπεργάσασθαι. καθάπερ οὖν αὐτὸ τυγχάνει ζῶον ἀΐδιον ὄν, καὶ τόδε τὸ πᾶν οὕτως εἰς δύναμιν ἐπεχείρησε τοιοῦτον ἀποτελεῖν. ἡ μὲν οὖν τοῦ ζῴου φύσις ἐτύγχανεν οὖσα αἰώνιος. καὶ τοῦτο μὲν δὴ τῷ γεννητῷ παντελῶς προσάπτειν οὐκ ἦν δυνατόν· εἰκὼ δ' ἐπινοεῖ κινητόν

---
[1] ἰὼν some mss.: ὢν other mss., Zur.
[2] αὐτὰ Hoffmann: αὐτοῦ mss., Zur.

TIMAEUS

and bound together, and revolves back upon herself, whenever she touches anything which has its substance dispersed or anything which has its substance undivided she is moved throughout her whole being and announces what the object is identical with and from what it is different, and in what relation, where and how and when, it comes about that each thing exists and is acted upon by others both in the sphere of the Becoming and in that of the ever-uniform. And her announcement, being identically true concerning both the Other and the Same, is borne through the self-moved without speech or sound; and whenever it is concerned with the sensible, and the circle of the Other moving in straight course proclaims it to the whole of its Soul, opinions and beliefs arise which are firm and true; and again, when it is concerned with the rational, and the circle of the Same, spinning truly, declares the facts, reason and knowledge of necessity result. But should anyone assert that the substance in which these two states arise is something other than Soul, his assertion will be anything rather than the truth.

And when the Father that engendered it perceived it in motion and alive, a thing of joy to the eternal gods, He too rejoiced[1]; and being well-pleased He designed to make it resemble its Model still more closely. Accordingly, seeing that that Model is an eternal Living Creature, He set about making this Universe, so far as He could, of a like kind. But inasmuch as the nature of the Living Creature was eternal, this quality it was impossible to attach in its entirety to what is generated; wherefore He

[1] Note the play on ἄγαλμα (" thing of joy " or " statue ") and ἠγάσθη (" rejoiced ").

PLATO

37 τινα αἰῶνος ποιῆσαι, καὶ διακοσμῶν ἅμα οὐρανὸν ποιεῖ μένοντος αἰῶνος ἐν ἑνὶ κατ' ἀριθμὸν ἰοῦσαν αἰώνιον εἰκόνα, τοῦτον ὃν δὴ χρόνον ὠνομάκαμεν. Ε ἡμέρας γὰρ καὶ νύκτας καὶ μῆνας καὶ ἐνιαυτοὺς οὐκ ὄντας πρὶν οὐρανὸν γενέσθαι, τότε ἅμα ἐκείνῳ ξυνισταμένῳ τὴν γένεσιν αὐτῶν μηχανᾶται. ταῦτα δὲ πάντα μέρη χρόνου, καὶ τό τ' ἦν τό τ' ἔσται, χρόνου γεγονότα εἴδη, ἃ δὴ φέροντες λανθάνομεν ἐπὶ τὴν ἀΐδιον οὐσίαν οὐκ ὀρθῶς. λέγομεν γὰρ δὴ ὡς ἦν ἔστι τε καὶ ἔσται, τῇ δὲ τὸ ἔστι μόνον
38 κατὰ τὸν ἀληθῆ λόγον προσήκει, τὸ δὲ ἦν τό τ' ἔσται περὶ τὴν ἐν χρόνῳ γένεσιν ἰοῦσαν πρέπει λέγεσθαι· κινήσεις γάρ ἐστον, τὸ δὲ ἀεὶ κατὰ ταὐτὰ ἔχον ἀκινήτως οὔτε πρεσβύτερον οὔτε νεώτερον προσήκει γίγνεσθαι διὰ χρόνου οὐδὲ γενέσθαι ποτὲ οὐδὲ γεγονέναι νῦν οὐδ' εἰσαῦθις ἔσεσθαι, τὸ παράπαν τε οὐδὲν ὅσα γένεσις τοῖς ἐν αἰσθήσει φερομένοις προσῆψεν, ἀλλὰ χρόνου ταῦτα αἰῶνά τε μιμουμένου καὶ κατ' ἀριθμὸν κυκλουμένου γέγονεν εἴδη. καὶ πρὸς τούτοις ἔτι τὰ
Β τοιάδε, τό τε γεγονὸς εἶναι γεγονὸς καὶ τὸ γιγνόμενον εἶναι γιγνόμενον, ἔτι δὲ τὸ γενησόμενον εἶναι γενησόμενον καὶ τὸ μὴ ὂν μὴ ὂν εἶναι, ὧν οὐδὲν ἀκριβὲς λέγομεν. περὶ μὲν οὖν τούτων τάχ' ἂν οὐκ εἴη καιρὸς πρέπων ἐν τῷ παρόντι διακριβολογεῖσθαι.

Χρόνος δ' οὖν μετ' οὐρανοῦ γέγονεν, ἵνα ἅμα

---

[1] *i.e.* it is incorrect to use the term " is " (ἐστί) both as a mere copula and in the sense of " exists."

TIMAEUS

planned to make a movable image of Eternity, and, as He set in order the Heaven, of that Eternity which abides in unity He made an eternal image, moving according to number, even that which we have named Time. For simultaneously with the construction of the Heaven He contrived the production of days and nights and months and years, which existed not before the Heaven came into being. And these are all portions of Time; even as "Was" and "Shall be" are generated forms of Time, although we apply them wrongly, without noticing, to Eternal Being. For we say that it "is" or "was" or "will be," whereas, in truth of speech, "is" alone is the appropriate term; "was" and "will be," on the other hand, are terms properly applicable to the Becoming which proceeds in Time, since both of these are motions; but it belongs not to that which is ever changeless in its uniformity to become either older or younger through time, nor ever to have become so, nor to be so now, nor to be about to be so hereafter, nor in general to be subject to any of the conditions which Becoming has attached to the things which move in the world of Sense, these being generated forms of Time, which imitates Eternity and circles round according to number. And besides these we make use of the following expressions,—that what is become *is* become, and what is becoming *is* becoming, and what is about to become *is* about to become, and what is non-existent *is* non-existent; but none of these expressions is accurate.[1] But the present is not, perhaps, a fitting occasion for an exact discussion of these matters.

Time, then, came into existence along with the

PLATO

γεννηθέντες ἅμα καὶ λυθῶσιν, ἄν ποτε λύσις τις αὐτῶν γίγνηται, καὶ κατὰ τὸ παράδειγμα τῆς διαιωνίας φύσεως, ἵν' ὡς ὁμοιότατος αὐτῷ κατὰ δύναμιν ᾖ· τὸ μὲν γὰρ δὴ παράδειγμα πάντα αἰῶνά ἐστιν ὄν, ὁ δ' αὖ διὰ τέλους τὸν ἅπαντα χρόνον γεγονώς τε καὶ ὢν καὶ ἐσόμενος. ἐξ οὖν λόγου καὶ διανοίας θεοῦ τοιαύτης πρὸς χρόνου γένεσιν, ἵνα γεννηθῇ χρόνος, ἥλιος καὶ σελήνη καὶ πέντε ἄλλα ἄστρα, ἐπίκλην ἔχοντα πλανητά, εἰς διορισμὸν καὶ φυλακὴν ἀριθμῶν χρόνου γέγονε. σώματα δὲ αὐτῶν ἑκάστων ποιήσας ὁ θεὸς ἔθηκεν εἰς τὰς περιφορὰς ἃς ἡ θατέρου περίοδος ᾔειν, ἑπτὰ οὔσας ὄντα ἑπτά, σελήνην μὲν εἰς τὸν περὶ γῆν πρῶτον, ἥλιον δ' εἰς τὸν δεύτερον ὑπὲρ γῆς, ἑωσφόρον δὲ καὶ τὸν ἱερὸν Ἑρμοῦ λεγόμενον εἰς τοὺς[1] τάχει μὲν ἰσόδρομον ἡλίῳ κύκλον ἰόντας, τὴν δ' ἐναντίαν εἰληχότας αὐτῷ δύναμιν· ὅθεν καταλαμβάνουσί τε καὶ καταλαμβάνονται κατὰ ταὐτὰ ὑπ' ἀλλήλων ἥλιός τε καὶ ὁ τοῦ Ἑρμοῦ καὶ ἑωσφόρος. τὰ δ' ἄλλα οἷ δὴ καὶ δι' ἃς αἰτίας ἱδρύσατο, εἴ τις ἐπεξίοι πάσας, ὁ λόγος πάρεργος ὢν πλέον ἂν ἔργον ὧν ἕνεκα λέγεται παράσχοι. ταῦτα μὲν οὖν ἴσως τάχ' ἂν κατὰ σχολὴν ὕστερον τῆς ἀξίας τύχοι διηγήσεως.

Ἐπειδὴ δὲ οὖν εἰς τὴν ἑαυτῷ πρέπουσαν ἕκαστον ἀφίκετο φορὰν τῶν ὅσα ἔδει ξυναπεργάζεσθαι

---

[1] τοὺς some MSS. : τὸν best MSS., Zur.

---

[1] Cf. 36 D. [2] i.e. Venus.
[3] i.e. a tendency as to direction.

TIMAEUS

Heaven, to the end that having been generated together they might also be dissolved together, if ever a dissolution of them should take place; and it was made after the pattern of the Eternal Nature, to the end that it might be as like thereto as possible; for whereas the pattern is existent through all eternity, the copy, on the other hand, is through all time, continually having existed, existing, and being about to exist. Wherefore, as a consequence of this reasoning and design on the part of God, with a view to the generation of Time, the sun and moon and five other stars, which bear the appellation of "planets," came into existence for the determining and preserving of the numbers of Time. And when God had made the bodies of each of them He placed them in the orbits along which the revolution of the Other was moving, seven orbits for the seven bodies.[1] The Moon He placed in the first circle around the Earth, the Sun in the second above the Earth; and the Morning Star[2] and the Star called Sacred to Hermes He placed in those circles which move in an orbit equal to the Sun in velocity, but endowed with a power[3] contrary thereto; whence it is that the Sun and the Star of Hermes and the Morning Star regularly overtake and are overtaken by one another. As to the rest of the stars, were one to describe in detail the positions in which He set them, and all the reasons therefor, the description, though but subsidiary, would prove a heavier task than the main argument which it subserves. Later on, perhaps, at our leisure these points may receive the attention they merit.

So when each of the bodies whose co-operation was required for the making of Time had arrived in its

PLATO

38 χρόνον, δεσμοῖς τε ἐμψύχοις σώματα δεθέντα
39 ζῷα ἐγεννήθη τό τε προσταχθὲν ἔμαθε, κατὰ
δὴ τὴν θατέρου φορὰν πλαγίαν οὖσαν διὰ τῆς
ταὐτοῦ φορᾶς ἰοῦσαν τε καὶ κρατουμένην,[1] τὸ
μὲν μείζονα αὐτῶν, τὸ δὲ ἐλάττω κύκλον ἰόν,
θᾶττον μὲν τὰ τὸν ἐλάττω, τὰ δὲ τὸν μείζω
βραδύτερον περιῄειν. τῇ δὴ ταὐτοῦ φορᾷ τὰ
τάχιστα περιόντα ὑπὸ τῶν βραδύτερον ἰόντων
ἐφαίνετο καταλαμβάνοντα καταλαμβάνεσθαι· πάντας γὰρ τοὺς κύκλους αὐτῶν στρέφουσα ἕλικα,
B διὰ τὸ διχῇ κατὰ τὰ ἐναντία ἅμα προϊέναι, τὸ
βραδύτατα ἀπιὸν ἀφ' αὑτῆς οὔσης ταχίστης ἐγγύτατα ἀπέφαινεν. ἵνα δὲ εἴη μέτρον ἐναργές τι
πρὸς ἄλληλα βραδυτῆτι καὶ τάχει καθ' ἃ[2] περὶ τὰς
ὀκτὼ φορὰς πορεύοιτο, φῶς ὁ θεὸς ἀνῆψεν ἐν τῇ
πρὸς γῆν δευτέρᾳ τῶν περιόδων, ὃ δὴ νῦν κεκλή-
καμεν ἥλιον, ἵνα ὅ τι μάλιστα εἰς ἅπαντα φαίνοι
τὸν οὐρανὸν μετάσχοι τε ἀριθμοῦ τὰ ζῷα, ὅσοις
ἦν προσῆκον, μαθόντα παρὰ τῆς ταὐτοῦ καὶ ὁμοίου
C περιφορᾶς. νὺξ μὲν οὖν ἡμέρα τε γέγονεν οὕτω
καὶ διὰ ταῦτα, ἡ τῆς μιᾶς καὶ φρονιμωτάτης
κυκλήσεως περίοδος· μεὶς δὲ ἐπειδὰν σελήνη περι-
ελθοῦσα τὸν ἑαυτῆς κύκλον ἥλιον ἐπικαταλάβῃ,
ἐνιαυτὸς δὲ ὁπόταν ἥλιος τὸν ἑαυτοῦ περιέλθῃ
κύκλον. τῶν δ' ἄλλων τὰς περιόδους οὐκ ἐννενοη-

[1] ἰοῦσαν ... κρατουμένην MS. corr.: ἰούσης ... κρατουμένης MSS., Zur.

[2] καθ' ἃ Archer-Hind: καὶ τὰ MSS., Zur.

[1] *i.e.* a planet moving along the Ecliptic from W. to E. is at the same time drawn from E. to W. (in the plane of the Equator) by the regular motion of the sphere of the fixed stars (the circle of " the Same " which moves at a higher velocity than that of " the Other ").

TIMAEUS

proper orbit; and when they had been generated as living creatures, having their bodies bound with living bonds, and had learnt their appointed duties; then they kept revolving around the circuit of the Other, which is transverse and passes through the circuit of the Same and is dominated thereby; and part of them moved in a greater, part in a smaller circle, those in the smaller moving more quickly and those in the greater more slowly. And because of the motion of the Same, the stars which revolved most quickly appeared to be overtaken by those which moved most slowly, although in truth they overtook them; for, because of their simultaneous progress in two opposite directions,[1] the motion of the Same, which is the swiftest of all motions, twisted all their circles into spirals and thus caused the body which moves away from it most slowly to appear the nearest.[2] And in order that there might be a clear measure of the relative speeds, slow and quick, with which they travelled round their eight orbits, in that circle which is second from the earth God kindled a light which now we call the Sun, to the end that it might shine, so far as possible, throughout the whole Heaven, and that all the living creatures entitled thereto might participate in number, learning it from the revolution of the Same and Similar. In this wise and for these reasons were generated Night and Day, which are the revolution of the one and most intelligent circuit; and Month, every time that the Moon having completed her own orbit overtakes the Sun; and Year, as often as the Sun has completed his own orbit. Of the other stars the revolutions

[2] *i.e.* Saturn appears to be nearest to the sphere of the fixed stars in point of velocity. *Cf. Laws* 822 A ff.

PLATO

39 κότες ἄνθρωποι, πλὴν ὀλίγοι τῶν πολλῶν, οὔτε ὀνομάζουσιν οὔτε πρὸς ἄλληλα ξυμμετροῦνται σκοποῦντες ἀριθμοῖς, ὥστε ὡς ἔπος εἰπεῖν οὐκ
D ἴσασι χρόνον ὄντα τὰς τούτων πλάνας, πλήθει μὲν ἀμηχάνῳ χρωμένας, πεποικιλμένας δὲ θαυμαστῶς. ἔστι δ' ὅμως οὐδὲν ἧττον κατανοῆσαι δυνατὸν ὡς ὅ γε τέλεος ἀριθμὸς χρόνου τὸν τέλεον ἐνιαυτὸν πληροῖ τότε, ὅταν ἁπασῶν τῶν ὀκτὼ περιόδων τὰ πρὸς ἄλληλα ξυμπερανθέντα τάχη σχῇ κεφαλὴν τῷ τοῦ ταὐτοῦ καὶ ὁμοίως ἰόντος ἀναμετρηθέντα κύκλῳ. κατὰ ταῦτα δὴ καὶ τούτων ἕνεκα ἐγεννήθη τῶν ἄστρων ὅσα δι' οὐρανοῦ πορευόμενα ἔσχε τροπάς, ἵνα τόδ' ὡς ὁμοιότατον ᾖ τῷ τελέῳ καὶ
E νοητῷ ζῴῳ πρὸς τὴν τῆς διαιωνίας μίμησιν φύσεως.

Καὶ τὰ μὲν ἄλλα ἤδη μέχρι χρόνου γενέσεως ἀπείργαστο εἰς ὁμοιότητα ᾧπερ ἀπεικάζετο, τῷ δὲ μήπω τὰ πάντα ζῷα ἐντὸς αὐτοῦ γεγενημένα περιειληφέναι, ταύτῃ ἔτι εἶχεν ἀνομοίως. τοῦτο δὴ τὸ κατάλοιπον ἀπειργάζετο αὐτοῦ πρὸς τὴν τοῦ παραδείγματος ἀποτυπούμενος φύσιν. ᾗπερ οὖν νοῦς ἐνούσας ἰδέας τῷ ὃ ἔστι ζῷον, οἷαί τε ἔνεισι καὶ ὅσαι, καθορᾷ, τοιαύτας καὶ τοσαύτας διενοήθη δεῖν καὶ τόδε σχεῖν. εἰσὶ δὴ τέτταρες, μία μὲν οὐράνιον
40 θεῶν γένος, ἄλλη δὲ πτηνὸν καὶ ἀεροπόρον, τρίτη δὲ ἔνυδρον εἶδος, πεζὸν δὲ καὶ χερσαῖον τέταρτον.

---

[1] An allusion to the name "planets," *i.e.* "wanderers"; *cf.* 38 c.

[2] *i.e.* the Great World-Year, which is completed when all the planets return simultaneously to their original starting-points. Its length was variously computed: Plato seems to have put it at 36,000 years (*cf. Rep.* 546 B ff.).

TIMAEUS

have not been discovered by men (save for a few out of the many); wherefore they have no names for them, nor do they compute and compare their relative measurements, so that they are not aware, as a rule, that the "wanderings"[1] of these bodies, which are hard to calculate and of wondrous complexity, constitute Time. Nevertheless, it is still quite possible to perceive that the complete number of Time fulfils the Complete Year[2] when all the eight circuits, with their relative speeds, finish together and come to a head, when measured by the revolution of the Same and Similarly-moving. In this wise and for these reasons were generated all those stars which turn themselves about as they travel through Heaven, to the end that this Universe might be as similar as possible to the perfect and intelligible Living Creature in respect of its imitation of the Eternal Nature thereof.

Now in all other respects this World had already, with the birth of Time, been wrought in the similitude of that whereunto it was being likened, but inasmuch as it did not as yet contain generated within it the whole range of living creatures, therein it was still dissimilar. So this part of the work which was still undone He completed by moulding it after the nature of the Model. According, then, as Reason perceives Forms existing in the Absolute Living Creature, such and so many as exist therein did He deem that this World also should possess. And these Forms are four,—one the heavenly kind of gods[3]; another the winged kind which traverses the air; thirdly, the class which inhabits the waters; and fourthly, that

[3] *i.e.* the stars.

PLATO

τοῦ μὲν οὖν θείου τὴν πλείστην ἰδέαν ἐκ πυρὸς ἀπειργάζετο,[1] ὅπως ὅ τι λαμπρότατον ἰδεῖν τε κάλλιστον εἴη, τῷ δὲ παντὶ προσεικάζων εὔκυκλον ἐποίει, τίθησί τε εἰς τὴν τοῦ κρατίστου φρόνησιν ἐκείνῳ ξυνεπόμενον, νείμας περὶ πάντα κύκλῳ τὸν οὐρανόν, κόσμον ἀληθινὸν αὐτῷ πεποικιλμένον εἶναι καθ᾽ ὅλον. κινήσεις δὲ δύο προσῆψεν ἑκάστῳ, τὴν μὲν ἐν ταὐτῷ κατὰ ταὐτὰ περὶ τῶν αὐτῶν ἀεὶ τὰ αὐτὰ ἑαυτῷ διανοουμένῳ, τὴν δὲ εἰς τὸ πρόσθεν ὑπὸ τῆς ταὐτοῦ καὶ ὁμοίου περιφορᾶς κρατουμένῳ· τὰς δὲ πέντε κινήσεις ἀκίνητον καὶ ἑστός, ἵν᾽ ὅ τι μάλιστα αὐτῶν ἕκαστον γένοιτο ὡς ἄριστον. ἐξ ἧς δὴ τῆς αἰτίας γέγονεν ὅσ᾽ ἀπλανῆ τῶν ἄστρων ζῷα θεῖα ὄντα καὶ ἀΐδια καὶ κατὰ ταὐτὰ ἐν ταὐτῷ στρεφόμενα ἀεὶ μένει· τὰ δὲ τρεπόμενα καὶ πλάνην τοιαύτην ἴσχοντα, καθάπερ ἐν τοῖς πρόσθεν ἐρρήθη, κατ᾽ ἐκεῖνα γέγονε. γῆν δὲ τροφὸν μὲν ἡμετέραν, εἰλλομένην δὲ περὶ τὸν διὰ παντὸς πόλον τεταμένον φύλακα καὶ δημιουργὸν νυκτός τε καὶ ἡμέρας ἐμηχανήσατο, πρώτην καὶ πρεσβυτάτην θεῶν ὅσοι ἐντὸς οὐρανοῦ γεγόνασι. χορείας δὲ τούτων αὐτῶν καὶ παραβολὰς ἀλλήλων, καὶ [περὶ][2] τὰς τῶν

[1] ἀπειργάζετο some MSS.: ἀπήρξατο best MS., Zur.
[2] περὶ bracketed by Ast.

[1] *i.e.* the fixed stars, and their sphere which moves with the daily rotation of the spherical Cosmos (the motion proper to "intelligence," *cf.* 36 c, *Cratyl.* 411 D).

[2] *i.e.* the "intelligent" outermost sphere of "the Same" (*cf.* the derivation of φρόνησις from φορά in *Cratyl.* 411 D).

[3] There is a play here on the word κόσμος, as meaning (1) "adornment," (2) "universe."

[4] *i.e.* (1) the rotation of the star on its own axis; (2) the diurnal revolution of the sphere of fixed stars.

[5] *Cf.* 34 A, 43 B.

TIMAEUS

which goes on foot on dry land. The form of the divine class [1] He wrought for the most part out of fire, that this kind might be as bright as possible to behold and as fair; and likening it to the All He made it truly spherical; and He placed it in the intelligence [2] of the Supreme to follow therewith, distributing it round about over all the Heaven, to be unto it a veritable adornment [3] cunningly traced over the whole. And each member of this class He endowed with two motions,[4] whereof the one is uniform motion in the same spot, whereby it conceives always identical thoughts about the same objects, and the other is a forward motion due to its being dominated by the revolution of the Same and Similar; but in respect of the other five motions [5] they are at rest and move not, so that each of them may attain the greatest possible perfection. From this cause, then, came into existence all those unwandering stars which are living creatures divine and eternal and abide for ever revolving uniformly in the same spot; and those which keep swerving and wandering have been generated in the fashion previously described. And Earth, our nurse, which is globed around [6] the pole that stretches through all, He framed to be the wardress and fashioner of night and day, she being the first and eldest of all the gods which have come into existence within the Heaven. But the choric dances of these same stars and their crossings one of

[6] The word εἴλλεσθαι (or ἴλλεσθαι) is taken by some to imply "oscillation" or "rotation" (*cf.* Aristot. *De caelo* ii. 293 b 30); but it seems best to suppose that Plato is here regarding the Earth as stationary. Her potential motion (we may assume) is equal and opposite to that of the Universe, of which she is the centre, and by thus neutralizing it she remains at rest.

κύκλων πρὸς ἑαυτοὺς ἐπανακυκλήσεις καὶ προχωρήσεις, ἔν τε ταῖς ξυνάψεσιν ὁποῖοι τῶν θεῶν κατ' ἀλλήλους γιγνόμενοι καὶ ὅσοι καταντικρύ, μεθ' οὕστινάς τε ἐπίπροσθεν ἀλλήλοις ἡμῖν τε κατὰ χρόνους οὕστινας ἕκαστοι κατακαλύπτονται καὶ πάλιν ἀναφαινόμενοι φόβους καὶ σημεῖα τῶν μετὰ ταῦτα γενησομένων τοῖς οὐ[1] δυναμένοις λογίζεσθαι πέμπουσι, τὸ λέγειν ἄνευ διόψεως τούτων αὐτῶν μιμημάτων μάταιος ἂν εἴη πόνος· ἀλλὰ ταῦτά τε ἱκανῶς ἡμῖν ταύτῃ καὶ τὰ περὶ θεῶν ὁρατῶν καὶ γεννητῶν εἰρημένα φύσεως ἐχέτω τέλος.

Περὶ δὲ τῶν ἄλλων δαιμόνων εἰπεῖν καὶ γνῶναι τὴν γένεσιν μεῖζον ἢ καθ' ἡμᾶς, πειστέον δὲ τοῖς εἰρηκόσιν ἔμπροσθεν, ἐκγόνοις μὲν θεῶν οὖσιν, ὡς ἔφασαν, σαφῶς δέ που τούς γε αὑτῶν προγόνους εἰδόσιν· ἀδύνατον οὖν θεῶν παισὶν ἀπιστεῖν, καίπερ ἄνευ τε εἰκότων καὶ ἀναγκαίων ἀποδείξεων λέγουσιν, ἀλλ' ὡς οἰκεῖα φάσκουσιν ἀπαγγέλλειν ἑπομένους τῷ νόμῳ πιστευτέον. οὕτως οὖν κατ' ἐκείνους ἡμῖν ἡ γένεσις περὶ τούτων τῶν θεῶν ἐχέτω καὶ λεγέσθω. Γῆς τε καὶ Οὐρανοῦ παῖδες Ὠκεανός τε καὶ Τηθὺς ἐγενέσθην, τούτων δὲ Φόρκυς Κρόνος τε καὶ Ῥέα καὶ ὅσοι μετὰ τούτων, ἐκ δὲ Κρόνου καὶ Ῥέας Ζεὺς Ἥρα τε καὶ πάντες ὅσους ἴσμεν ἀδελφοὺς λεγομένους αὐτῶν, ἔτι τε τούτων ἄλλους ἐκγόνους.

Ἐπεὶ δ' οὖν πάντες ὅσοι τε περιπολοῦσι φανερῶς καὶ ὅσοι φαίνονται καθ' ὅσον ἂν ἐθέλωσι θεοὶ

---

[1] οὐ omitted by most MSS. and Zur.

---

[1] *i.e.* such instruments as a celestial globe or planetarium.

TIMAEUS

another, and the relative reversals and progressions of their orbits, and which of the gods meet in their conjunctions, and how many are in opposition, and behind which and at what times they severally pass before one another and are hidden from our view, and again re-appearing send upon men unable to calculate alarming portents of the things which shall come to pass hereafter,—to describe all this without an inspection of models [1] of these movements would be labour in vain. Wherefore, let this account suffice us, and let our discourse concerning the nature of the visible and generated gods have an end.

Concerning the other divinities, to discover and declare their origin is too great a task for us, and we must trust to those who have declared it aforetime, they being, as they affirmed, descendants of gods and knowing well, no doubt, their own forefathers.[2] It is, as I say, impossible to disbelieve the children of gods, even though their statements lack either probable or necessary demonstration; and inasmuch as they profess to speak of family matters, we must follow custom and believe them. Therefore let the generation of these gods be stated by us, following their account, in this wise: Of Gê and Uranus were born the children Oceanus and Tethys; and of these, Phorkys, Cronos, Rhea, and all that go with them; and of Cronos and Rhea were born Zeus and Hera and all those who are, as we know, called their brethren; and of these again, other descendants.

Now when all the gods, both those who revolve manifestly [3] and those who manifest themselves so

---

[2] This is, obviously, ironical; cf. *Cratyl.* 402 B, *Phileb.* 66 c.
[3] *i.e.* the Stars; the others are the deities of popular belief (such as Homer depicts).

PLATO

41 γένεσιν ἔσχον, λέγει πρὸς αὐτοὺς ὁ τόδε τὸ πᾶν γεννήσας τάδε· Θεοὶ θεῶν, ὧν ἐγὼ δημιουργὸς πατήρ τε ἔργων, [ἃ δι' ἐμοῦ γενόμενα]¹ ἄλυτα ἐμοῦ γε μὴ² ἐθέλοντος. τὸ μὲν οὖν δὴ δεθὲν πᾶν λυτόν, B τό γε μὴν καλῶς ἁρμοσθὲν καὶ ἔχον εὖ λύειν ἐθέλειν κακοῦ. δι' ἃ καὶ ἐπείπερ γεγένησθε, ἀθάνατοι μὲν οὐκ ἐστὲ οὐδ' ἄλυτοι τὸ πάμπαν, οὔ τι μὲν δὴ λυθήσεσθέ γε οὐδὲ τεύξεσθε θανάτου μοίρας, τῆς ἐμῆς βουλήσεως μείζονος ἔτι δεσμοῦ καὶ κυριωτέρου λαχόντες ἐκείνων οἷς ὅτ' ἐγίγνεσθε ξυνεδεῖσθε. νῦν οὖν ὃ λέγω πρὸς ὑμᾶς ἐνδεικνύμενος, μάθετε. θνητὰ ἔτι γένη λοιπὰ τρί' ἀγέννητα. τούτων δὲ μὴ γενομένων οὐρανὸς ἀτελὴς ἔσται· τὰ γὰρ ἅπαντα ἐν αὑτῷ γένη ζῴων οὐχ ἕξει, δεῖ δέ, εἰ C μέλλει τέλεος ἱκανῶς εἶναι. δι' ἐμοῦ δὲ ταῦτα γενόμενα καὶ βίου μετασχόντα θεοῖς ἰσάζοιτ' ἄν· ἵν' οὖν θνητά τε ᾖ τό τε πᾶν τόδε ὄντως ἅπαν ᾖ, τρέπεσθε κατὰ φύσιν ὑμεῖς ἐπὶ τὴν τῶν ζῴων δημιουργίαν, μιμούμενοι τὴν ἐμὴν δύναμιν περὶ τὴν ὑμετέραν γένεσιν. καὶ καθ' ὅσον μὲν αὐτῶν ἀθανάτοις ὁμώνυμον εἶναι προσήκει, θεῖον λεγόμενον ἡγεμονοῦν τ' ἐν αὐτοῖς τῶν ἀεὶ δίκῃ καὶ ὑμῖν ἐθελόντων ἕπεσθαι, σπείρας καὶ ὑπαρξάμενος ἐγὼ D παραδώσω· τὸ δὲ λοιπὸν ὑμεῖς, ἀθανάτῳ θνητὸν προσυφαίνοντες, ἀπεργάζεσθε ζῷα καὶ γεννᾶτε

¹ ἃ ... γενόμενα bracketed by Rawack, after Philo, al.
² γε μὴ] γ' Zur., with some MSS.

---

[1] An intensive form of expression, like the Biblical "King of kings and Lord of lords."

TIMAEUS

far as they choose, had come to birth, He that generated this All addressed them thus:

"Gods of gods,[1] those works whereof I am framer and father are indissoluble save by my will. For though all that is bound may be dissolved, yet to will to dissolve that which is fairly joined together and in good case were the deed of a wicked one. Wherefore ye also, seeing that ye were generated, are not wholly immortal or indissoluble, yet in no wise shall ye be dissolved nor incur the doom of death, seeing that in my will ye possess a bond greater and more sovereign than the bonds wherewith, at your birth, ye were bound together. Now, therefore, what I manifest and declare unto you do ye learn. Three mortal kinds[2] still remain ungenerated; but if these come not into being the Heaven will be imperfect; for it will not contain within itself the whole sum of the kinds of living creatures, yet contain them it must if it is to be fully perfect. But if by my doing these creatures came into existence and partook of life, they would be made equal unto gods; in order, therefore, that they may be mortal and that this World-all may be truly All, do ye turn yourselves, as Nature directs, to the work of fashioning these living creatures, imitating the power showed by me in my generating of you. Now so much of them as it is proper to designate 'immortal,' the part we call divine which rules supreme in those who are fain to follow justice always and yourselves, that part I will deliver unto you when I have sown it and given it origin. For the rest, do ye weave together the mortal with the immortal, and thereby fashion and generate living

[2] *Viz.* the inhabitants of air, of water, and of earth.

41 τροφήν τε διδόντες αὐξάνετε καὶ φθίνοντα πάλιν δέχεσθε.

Ταῦτ' εἶπε, καὶ πάλιν ἐπὶ τὸν πρότερον κρατῆρα, ἐν ᾧ τὴν τοῦ παντὸς ψυχὴν κεραννὺς ἔμισγε, τὰ τῶν πρόσθεν ὑπόλοιπα κατεχεῖτο μίσγων τρόπον μέν τινα τὸν αὐτόν, ἀκήρατα δ' οὐκέτι κατὰ ταὐτὰ ὡσαύτως, ἀλλὰ δεύτερα καὶ τρίτα. ξυστήσας δὲ τὸ πᾶν διεῖλε ψυχὰς ἰσαρίθμους τοῖς ἄστροις ἔνειμέ θ' E ἑκάστην πρὸς ἕκαστον, καὶ ἐμβιβάσας ὡς ἐς ὄχημα τὴν τοῦ παντὸς φύσιν ἔδειξε, νόμους τε τοὺς εἱμαρμένους εἶπεν αὐταῖς, ὅτι γένεσις πρώτη μὲν ἔσοιτο τεταγμένη μία πᾶσιν, ἵνα μήτις ἐλαττοῖτο ὑπ' αὐτοῦ, δέοι δὲ σπαρείσας αὐτὰς εἰς τὰ προσήκοντα ἑκάσταις ἕκαστα ὄργανα χρόνων φῦναι ζῴων τὸ 42 θεοσεβέστατον, διπλῆς δὲ οὔσης τῆς ἀνθρωπίνης φύσεως τὸ κρεῖττον τοιοῦτον εἴη γένος ὃ καὶ ἔπειτα κεκλήσοιτο ἀνήρ. ὁπότε δὴ σώμασιν ἐμφυτευθεῖεν ἐξ ἀνάγκης, καὶ τὸ μὲν προσίοι, τὸ δ' ἀπίοι τοῦ σώματος αὐτῶν, πρῶτον μὲν αἴσθησιν ἀναγκαῖον εἴη μίαν πᾶσιν ἐκ βιαίων παθημάτων ξύμφυτον γίγνεσθαι, δεύτερον δὲ ἡδονῇ καὶ λύπῃ μεμιγμένον ἔρωτα, πρὸς δὲ τούτοις φόβον καὶ θυμὸν B ὅσα τε ἑπόμενα αὐτοῖς καὶ ὁπόσα ἐναντίως πέφυκε διεστηκότα· ὧν εἰ μὲν κρατήσοιεν, δίκῃ βιώσοιντο, κρατηθέντες δὲ ἀδικίᾳ. καὶ ὁ μὲν εὖ τὸν προσήκοντα χρόνον βιούς, πάλιν εἰς τὴν τοῦ ξυννόμου

---

[1] *Cf. Laws* 899 A.   [2] *i.e.* star.

TIMAEUS

creatures, and give them food that they may grow, and when they waste away receive them to yourselves again."

Thus He spake, and once more into the former bowl, wherein He had blended and mixed the Soul of the Universe, He poured the residue of the previous material, mixing it in somewhat the same manner, yet no longer with a uniform and invariable purity, but second and third in degree of purity. And when He had compounded the whole He divided it into souls equal in number to the stars, and each several soul He assigned to one star, and setting them each as it were in a chariot [1] He showed them the nature of the Universe, and declared unto them the laws of destiny,—namely, how that the first birth should be one and the same ordained for all, in order that none might be slighted by Him; and how it was needful that they, when sown each into his own proper organ of time,[2] should grow into the most god-fearing of living creatures; and that, since human nature is two-fold, the superior sex is that which hereafter should be designated "man." And when, by virtue of Necessity, they should be implanted in bodies, and their bodies are subject to influx and efflux, these results would necessarily follow,—firstly, sensation that is innate and common to all proceeding from violent affections; secondly, desire mingled with pleasure and pain; and besides these, fear and anger and all such emotions as are naturally allied thereto, and all such as are of a different and opposite character. And if they shall master these they will live justly, but if they are mastered, unjustly. And he that has lived his appointed time well shall return again to his abode

πορευθεὶς οἴκησιν ἄστρου, βίον εὐδαίμονα καὶ
συνήθη ἕξοι· σφαλεὶς δὲ τούτων εἰς γυναικὸς φύσιν
ἐν τῇ δευτέρᾳ γενέσει μεταβαλοῖ· μὴ παυόμενός τε
ἐν τούτοις ἔτι κακίας, τρόπον ὃν κακύνοιτο, κατὰ
τὴν ὁμοιότητα τῆς τοῦ τρόπου γενέσεως εἴς τινα
τοιαύτην ἀεὶ μεταβαλοῖ θήρειον φύσιν, ἀλλάττων τε
οὐ πρότερον πόνων λήξοι, πρὶν τῇ ταὐτοῦ καὶ ὁμοίου
περιόδῳ τῇ ἐν αὑτῷ ξυνεπισπώμενος τὸν πολὺν
ὄχλον καὶ ὕστερον προσφύντα ἐκ πυρὸς καὶ ὕδατος
καὶ ἀέρος καὶ γῆς, θορυβώδη καὶ ἄλογον ὄντα λόγῳ
κρατήσας εἰς τὸ τῆς πρώτης καὶ ἀρίστης ἀφίκοιτο
εἶδος ἕξεως.

Διαθεσμοθετήσας δὲ πάντα αὐτοῖς ταῦτα, ἵνα
τῆς ἔπειτα εἴη κακίας ἑκάστων ἀναίτιος, ἔσπειρε
τοὺς μὲν εἰς γῆν, τοὺς δ' εἰς σελήνην, τοὺς δ' εἰς
τἆλλα ὅσα ὄργανα χρόνου. τὸ δὲ μετὰ τὸν σπόρον
τοῖς νέοις παρέδωκε θεοῖς σώματα πλάττειν θνητά,
τό τε ἐπίλοιπον ὅσον ἔτι ἦν ψυχῆς ἀνθρωπίνης
δέον προσγενέσθαι, τοῦτο καὶ πάνθ' ὅσα ἀκόλουθα
ἐκείνοις ἀπεργασαμένους ἄρχειν, καὶ κατὰ δύναμιν
ὅ τι κάλλιστα καὶ ἄριστα τὸ θνητὸν διακυβερνᾶν
ζῷον, ὅ τι μὴ κακῶν αὐτὸ ἑαυτῷ γίγνοιτο αἴτιον.

Καὶ ὁ μὲν δὴ ἅπαντα ταῦτα διατάξας ἔμενεν ἐν
τῷ ἑαυτοῦ κατὰ τρόπον ἤθει· μένοντος δὲ νοήσαντες
οἱ παῖδες τὴν τοῦ πατρὸς πρόσταξιν[1] ἐπείθοντο αὐτῇ,
καὶ λαβόντες ἀθάνατον ἀρχὴν θνητοῦ ζῴου, μιμούμενοι τὸν σφέτερον δημιουργόν, πυρὸς καὶ γῆς
ὕδατός τε καὶ ἀέρος ἀπὸ τοῦ κόσμου δανειζόμενοι

---

[1] πρόσταξιν one ms.: διάταξιν Zur.

[1] Cf. 90 E.

TIMAEUS

in his native star, and shall gain a life that is blessed and congenial; but whoso has failed therein shall be changed into woman's nature at the second birth;[1] and if, in that shape, he still refraineth not from wickedness he shall be changed every time, according to the nature of his wickedness, into some bestial form after the similitude of his own nature; nor in his changings shall he cease from woes until he yields himself to the revolution of the Same and Similar that is within him, and dominating by force of reason that burdensome mass which afterwards adhered to him of fire and water and earth and air, a mass tumultuous and irrational, returns again to the semblance of his first and best state.

When He had fully declared unto them all these ordinances, to the end that He might be blameless in respect of the future wickedness of any one of them, He proceeded to sow them, some in the Earth, some in the Moon, others in the rest of the organs of Time. Following upon this sowing, He delivered over to the young gods the task of moulding mortal bodies, and of framing and controlling all the rest of the human soul which it was still necessary to add, together with all that belonged thereto, and of governing this mortal creature in the fairest and best way possible, to the utmost of their power, except in so far as it might itself become the cause of its own evils.

So He, then, having given all these commands, was abiding in His own proper and wonted state. And as He thus abode, His children gave heed to their Father's command and obeyed it. They took the immortal principle of the mortal living creature, and imitating their own Maker, they borrowed from the

43 μόρια ὡς ἀποδοθησόμενα πάλιν, εἰς ταὐτὸν τὰ λαμβανόμενα ξυνεκόλλων, οὐ τοῖς ἀλύτοις οἷς αὐτοὶ ξυνείχοντο δεσμοῖς, ἀλλὰ διὰ σμικρότητα ἀοράτοις, πυκνοῖς γόμφοις ξυντήκοντες, ἓν ἐξ ἁπάντων ἀπεργαζόμενοι σῶμα ἕκαστον, τὰς τῆς ἀθανάτου ψυχῆς περιόδους ἐνέδουν εἰς ἐπίρρυτον σῶμα καὶ ἀπόρρυτον. αἱ δὲ εἰς ποταμὸν ἐνδεθεῖσαι πολὺν οὔτ' ἐκράτουν οὔτ' ἐκρατοῦντο, βίᾳ δ'
B ἐφέροντο καὶ ἔφερον, ὥστε τὸ μὲν ὅλον κινεῖσθαι ζῷον, ἀτάκτως μὴν ὅπῃ τύχοι προϊέναι καὶ ἀλόγως, τὰς ἓξ ἁπάσας κινήσεις ἔχον· εἴς τε γὰρ τὸ πρόσθε καὶ ὄπισθεν καὶ πάλιν εἰς δεξιὰ καὶ ἀριστερὰ κάτω τε καὶ ἄνω καὶ πάντῃ κατὰ τοὺς ἓξ τόπους πλανώμενα προῄειν. πολλοῦ γὰρ ὄντος τοῦ κατακλύζοντος καὶ ἀπορρέοντος κύματος, ὃ τὴν τροφὴν παρεῖχεν, ἔτι μείζω θόρυβον ἀπειργάζετο τὰ
C τῶν προσπιπτόντων παθήματα ἑκάστοις, ὅτε πυρὶ προσκρούσειε τὸ σῶμά τινος ἔξωθεν ἀλλοτρίῳ περιτυχὸν ἢ καὶ στερεῷ γῆς ὑγροῖς τε ὀλισθήμασιν ὑδάτων, εἴτε ζάλῃ πνευμάτων ὑπ' ἀέρος φερομένων καταληφθείη, καὶ ὑπὸ πάντων τούτων διὰ τοῦ σώματος αἱ κινήσεις ἐπὶ τὴν ψυχὴν φερόμεναι προσπίπτοιεν· αἳ δὴ καὶ ἔπειτα διὰ ταῦτα ἐκλήθησάν τε καὶ νῦν ἔτι αἰσθήσεις ξυνάπασαι κέκληνται. καὶ δὴ καὶ τότε ἐν τῷ παρόντι πλείστην καὶ μεγίστην παρεχόμεναι κίνησιν, μετὰ τοῦ ῥέοντος
D ἐνδελεχῶς ὀχετοῦ κινοῦσαι καὶ σφοδρῶς σείουσαι

---

[1] *i.e.* omitting the seventh motion ("rotation"), *cf.* 34 A.
[2] *i.e.* αἴσθησις ("sensation") is here derived from ἀΐσσω ("dart," "rush").

TIMAEUS

Cosmos portions of fire and earth and water and air, as if meaning to pay them back, and the portions so taken they cemented together; but it was not with those indissoluble bonds wherewith they themselves were joined that they fastened together the portions but with numerous pegs, invisible for smallness; and thus they constructed out of them all each several body, and within bodies subject to inflow and outflow they bound the revolutions of the immortal Soul. The souls, then, being thus bound within a mighty river neither mastered it nor were mastered, but with violence they rolled along and were rolled along themselves, so that the whole of the living creature was moved, but in such a random way that its progress was disorderly and irrational, since it partook of all the six motions [1]: for it progressed forwards and backwards, and again to right and to left, and upwards and downwards, wandering every way in all the six directions. For while the flood which foamed in and streamed out, as it supplied the food, was immense, still greater was the tumult produced within each creature as a result of the colliding bodies, when the body of a creature happened to meet and collide with alien fire from without, or with a solid lump of earth or liquid glidings of waters, or when it was overtaken by a tempest of winds driven by air, and when the motions due to all these causes rushing through the body impinged upon the Soul. And for these reasons all such motions were then termed "Sensations," [2] and are still so termed to-day. Moreover, since at that time they were causing, for the moment, constant and widespread motion, joining with the perpetually flowing stream in moving and violently shaking the revolutions of the Soul, they

PLATO

43 τὰς τῆς ψυχῆς περιόδους, τὴν μὲν ταὐτοῦ παντάπασιν ἐπέδησαν ἐναντία αὐτῇ ῥέουσαι, καὶ ἐπέσχον ἄρχουσαν καὶ ἰοῦσαν, τὴν δ' αὖ θατέρου διέσεισαν, ὥστε τὰς τοῦ διπλασίου καὶ τριπλασίου τρεῖς ἑκατέρας ἀποστάσεις καὶ τὰς τῶν ἡμιολίων καὶ ἐπιτρίτων καὶ ἐπογδόων μεσότητας καὶ ξυνδέσεις, ἐπειδὴ παντελῶς λυταὶ οὐκ ἦσαν πλὴν ὑπὸ τοῦ ξυνδήσαντος, πάσας μὲν στρέψαι στροφάς, πάσας
E δὲ κλάσεις καὶ διαφθορὰς τῶν κύκλων ἐμποιεῖν, ὁσαχῇπερ ἦν δυνατόν, ὥστε μετ' ἀλλήλων μόγις ξυνεχομένας φέρεσθαι μέν, ἀλόγως δὲ φέρεσθαι, τοτὲ μὲν ἀντίας, ἄλλοτε δὲ πλαγίας, τοτὲ δ' ὑπτίας· οἷον ὅταν τις ὕπτιος ἐρείσας τὴν κεφαλὴν μὲν ἐπὶ γῆς, τοὺς δὲ πόδας ἄνω προσβαλὼν ἔχῃ πρός τινι, τότε ἐν τούτῳ τῷ πάθει τοῦ τε πάσχοντος καὶ τῶν ὁρώντων τά τε δεξιὰ ἀριστερὰ καὶ τὰ ἀριστερὰ δεξιὰ ἑκατέροις τὰ ἑκατέρων φαντάζεται. ταὐτὸν δὴ τοῦτο καὶ τοιαῦτα ἕτερα αἱ περιφοραὶ πάσχουσαι
44 σφοδρῶς, ὅταν γέ τῳ τῶν ἔξωθεν τοῦ ταὐτοῦ γένους ἢ τοῦ θατέρου περιτύχωσι, τότε ταὐτόν τῳ καὶ θάτερόν του τἀναντία τῶν ἀληθῶν προσαγορεύουσαι ψευδεῖς καὶ ἀνόητοι γεγόνασιν, οὐδεμία τε ἐν αὐταῖς τότε περίοδος ἄρχουσα οὐδ' ἡγεμών ἐστιν· αἷς δ' ἂν ἔξωθεν αἰσθήσεις τινὲς φερόμεναι καὶ προσπεσοῦσαι ξυνεπισπάσωνται καὶ τὸ τῆς ψυχῆς ἅπαν κύτος, τόθ' αὗται κρατούμεναι κρατεῖν δοκοῦσι. καὶ διὰ δὴ ταῦτα πάντα τὰ παθήματα νῦν
B κατ' ἀρχάς τε ἄνους ψυχὴ γίγνεται τὸ πρῶτον, ὅταν

---

[1] Cf. 35 B.   [2] Cf. 37 A.

TIMAEUS

totally blocked the course of the Same by flowing contrary thereto, and hindered it thereby in its ruling and its going; while, on the other hand, they so shook up the course of the Other that in the three several intervals of the double and the triple,[1] and in the mean terms and binding links of the $\frac{3}{2}$, $\frac{4}{3}$, and $\frac{9}{8}$,— these being not wholly dissoluble save by Him who had bound them together,—they produced all manner of twistings, and caused in their circles fractures and disruptions of every possible kind, with the result that, as they barely held together one with another, they moved indeed but moved irrationally, being at one time reversed, at another oblique, and again upside down. Suppose, for example, that a man is in an upside down position, with his head resting on the earth and his feet touching something above, then, in this position of the man relative to that of the onlookers, his right will appear left to them, and his left right, and so will theirs to him. This, and such like, are just what the revolutions of the Soul experience with intensity; and every time they happen upon any external object, whether it be of the class of the Same or of the Other,[2] they proclaim it to be the same as something or other than something contrary to the truth, and thereby prove themselves false and foolish, and devoid, at such times, of any revolution that rules and guides. And whenever external sensations in their movement collide with these revolutions and sweep along with them also the whole vessel of the Soul, then the revolutions, though actually mastered, appear to have the mastery. Hence it comes about that, because of all these affections, now as in the beginning, so often as the Soul is bound within a mortal body it

εἰς σῶμα ἐνδεθῇ θνητόν· ὅταν δὲ τὸ τῆς αὔξης καὶ
τροφῆς ἔλαττον ἐπίῃ ῥεῦμα, πάλιν δὲ αἱ περίοδοι
λαμβανόμεναι γαλήνης τὴν ἑαυτῶν ὁδὸν ἴωσι καὶ
καθιστῶνται μᾶλλον ἐπιόντος τοῦ χρόνου, τότε ἤδη
πρὸς τὸ κατὰ φύσιν ἰόντων σχῆμα ἑκάστων τῶν
κύκλων αἱ περιφοραὶ κατευθυνόμεναι, τό τε θάτερον
καὶ τὸ ταὐτὸν προσαγορεύουσαι κατ' ὀρθόν, ἔμφρονα τὸν ἔχοντα αὐτὰς γιγνόμενον ἀποτελοῦσιν. ἂν
C μὲν οὖν δὴ καὶ ξυνεπιλαμβάνηταί τις ὀρθὴ τροφὴ
παιδεύσεως, ὁλόκληρος ὑγιής τε παντελῶς, τὴν
μεγίστην ἀποφυγὼν νόσον, γίγνεται· καταμελήσας
δέ, χωλὴν τοῦ βίου διαπορευθεὶς ζωήν, ἀτελὴς καὶ
ἀνόητος εἰς Ἅιδου πάλιν ἔρχεται. ταῦτα μὲν οὖν
ὕστερά ποτε γίγνεται· περὶ δὲ τῶν νῦν προτεθέντων
δεῖ διελθεῖν ἀκριβέστερον. τὰ δὲ πρὸ τούτων, περὶ
σωμάτων κατὰ μέρη τῆς γενέσεως καὶ περὶ ψυχῆς,
δι' ἅς τε αἰτίας καὶ προνοίας γέγονε θεῶν, τοῦ
D μάλιστα εἰκότος ἀντεχομένοις, οὕτω καὶ κατὰ
ταῦτα πορευομένοις διεξιτέον.

Τὰς μὲν δὴ θείας περιόδους δύο οὔσας, τὸ τοῦ
παντὸς σχῆμα ἀπομιμησάμενοι περιφερὲς ὄν, εἰς
σφαιροειδὲς σῶμα ἐνέδησαν, τοῦτο ὃ νῦν κεφαλὴν
ἐπονομάζομεν, ὃ θειότατόν τ' ἐστὶ καὶ τῶν ἐν ἡμῖν
πάντων δεσποτοῦν. ᾧ καὶ πᾶν τὸ σῶμα παρέδοσαν
ὑπηρεσίαν αὐτῷ ξυναθροίσαντες θεοί, κατανοή-
σαντες ὅτι πασῶν ὅσαι κινήσεις ἔσοιντο μετέχοι·
E ἵν' οὖν μὴ κυλινδούμενον ἐπὶ γῆς ὕψη τε καὶ βάθη
παντοδαπὰ ἐχούσης ἀποροῖ τὰ μὲν ὑπερβαίνειν,

---

[1] *Cf.* 86 E; *Phaedo* 81 c, 83 D.
[2] *i.e.* ignorance; *cf.* 86 B ff., *Laws* 863 c ff.
[3] *Cf.* 86 B ff.   [4] *Cf.* 29 c, D.
[5] *Cf.* 73 c, 81 D.

TIMAEUS

becomes at the first irrational.[1] But as soon as the stream of increase and nutriment enters in less volume, and the revolutions calm down and pursue their own path, becoming more stable as time proceeds, then at length, as the several circles move each according to its natural track, their revolutions are straightened out and they announce the Same and the Other aright, and thereby they render their possessor intelligent. And if so be that this state of his soul be reinforced by right educational training, the man becomes wholly sound and faultless, having escaped the worst of maladies [2]; but if he has been wholly negligent therein, after passing a lame existence in life he returns again unperfected and unreasoning to Hades. These results, however, come about at a later time.[3] Regarding the subjects now before us, we must give a more exact exposition; and also regarding the subjects anterior to these, namely, the generation of bodies in their several parts, and the causes and divine counsels whereby the Soul has come into existence, we must hold fast to the most probable [4] account, and proceed accordingly, in the exposition now to be given.

The divine revolutions, which are two, they bound within a sphere-shaped body, in imitation of the spherical form [5] of the All, which body we now call the "head," it being the most divine part and reigning over all the parts within us. To it the gods delivered over the whole of the body they had assembled to be its servant, having formed the notion that it should partake in all the motions which were to be. In order, then, that it should not go rolling upon the earth, which has all manner of heights and hollows, and be at a loss how to climb over the one and climb out of the

PLATO

44 ἔνθεν δὲ ἐκβαίνειν, ὄχημ' αὐτῷ τοῦτο καὶ εὐπορίαν ἔδοσαν. ὅθεν δὴ μῆκος τὸ σῶμα ἔσχεν, ἐκτατά τε κῶλα καὶ καμπτὰ ἔφυσε τέτταρα θεοῦ μηχανησαμένου πορεῖα, οἷς ἀντιλαμβανόμενον καὶ ἀπ-
45 ερειδόμενον διὰ πάντων τόπων πορεύεσθαι δυνατὸν γέγονε, τὴν τοῦ θειοτάτου καὶ ἱερωτάτου φέρον οἴκησιν ἐπάνωθεν ἡμῶν. σκέλη μὲν οὖν χεῖρές τε ταύτῃ καὶ διὰ ταῦτα προσέφυ πᾶσι· τοῦ δ' ὄπισθεν τὸ πρόσθεν τιμιώτερον καὶ ἀρχικώτερον νομίζοντες θεοὶ ταύτῃ τὸ πολὺ τῆς πορείας ἡμῖν ἔδοσαν. ἔδει δὴ διωρισμένον ἔχειν καὶ ἀνόμοιον τοῦ σώματος τὸ πρόσθεν ἄνθρωπον. διὸ πρῶτον μὲν περὶ τὸ τῆς κεφαλῆς κύτος, ὑποθέντες αὐτόσε τὸ πρόσωπον,
B ὄργανα ἐνέδησαν τούτῳ πάσῃ τῇ τῆς ψυχῆς προνοίᾳ, καὶ διετάξαντο τὸ μετέχον ἡγεμονίας τοῦτ' εἶναι τὸ κατὰ φύσιν πρόσθεν. τῶν δὲ ὀργάνων πρῶτον μὲν φωσφόρα ξυνετεκτήναντο ὄμματα, τοιᾷδε ἐνδήσαντες αἰτίᾳ. τοῦ πυρὸς ὅσον τὸ μὲν κάειν οὐκ ἔσχε, τὸ δὲ παρέχειν φῶς ἥμερον, οἰκεῖον ἑκάστης ἡμέρας, σῶμα ἐμηχανήσαντο γίγνεσθαι. τὸ γὰρ ἐντὸς ἡμῶν ἀδελφὸν ὂν τούτου πῦρ εἰλικρινὲς ἐποίησαν διὰ τῶν ὀμμάτων ῥεῖν λεῖον καὶ πυκνὸν
C ὅλον μέν, μάλιστα δὲ τὸ μέσον ξυμπιλήσαντες τῶν ὀμμάτων, ὥστε τὸ μὲν ἄλλο ὅσον παχύτερον στέγειν πᾶν, τὸ τοιοῦτον δὲ μόνον αὐτὸ καθαρὸν διηθεῖν. ὅταν οὖν μεθημερινὸν ᾖ φῶς περὶ τὸ τῆς ὄψεως ῥεῦμα, τότ' ἐκπῖπτον ὅμοιον πρὸς ὅμοιον, ξυμπαγὲς γενόμενον, ἓν σῶμα· οἰκειωθὲν ξυνέστη

---

[1] There is a play here on the words ἥμερον ("mild")... ἡμέρας ("day"); cf. Cratyl. 418 c.

[2] Vision is explained on the principle that "like is known by like": a fire-stream issuing from the eye meets a fire-

TIMAEUS

other, they bestowed upon it the body as a vehicle and means of transport. And for this reason the body acquired length, and, by God's contriving, shot forth four limbs, extensible and flexible, to serve as instruments of transport, so that grasping with these and supported thereon it was enabled to travel through all places, bearing aloft the chamber of our most divine and holy part. In this wise and for these reasons were legs and hands attached to all men; and inasmuch as they demand the forepart superior to the hinder part in honour and dignity, the Gods gave us the most part of our going in this direction. Thus it was necessary that man should have the forepart of his body distinct and dissimilar. Wherefore, dealing first with the vessel of the head, they set the face in the front thereof and bound within it organs for all the forethought of the Soul; and they ordained that this, which is the natural front, should be the leading part. And of the organs they constructed first light-bearing eyes, and these they fixed in the face for the reason following. They contrived that all such fire as had the property not of burning but of giving a mild light should form a body akin to the light of every day.[1] For they caused the pure fire within us, which is akin to that of day, to flow through the eyes in a smooth and dense stream; and they compressed the whole substance, and especially the centre, of the eyes, so that they occluded all other fire that was coarser and allowed only this pure kind of fire to filter through. So whenever the stream of vision is surrounded by mid-day light, it flows out like unto like,[2] and coalescing therewith it forms one

stream coming from the object of vision (*cf.* the view of Empedocles).

κατὰ τὴν τῶν ὀμμάτων εὐθυωρίαν, ὅπηπερ ἂν ἀντερείδῃ τὸ προσπῖπτοι ἔνδοθεν πρὸς ὃ τῶν ἔξω ξυνέπεσεν. ὁμοιοπαθὲς δὴ δι' ὁμοιότητα πᾶν γενόμενον, ὅτου τε ἂν αὐτό ποτε ἐφάπτηται καὶ ὃ ἂν ἄλλο ἐκείνου, τούτων τὰς κινήσεις διαδιδὸν εἰς ἅπαν τὸ σῶμα μέχρι τῆς ψυχῆς αἴσθησιν παρέσχετο ταύτην, ᾗ δὴ ὁρᾶν φαμέν. ἀπελθόντος δὲ εἰς νύκτα τοῦ ξυγγενοῦς πυρὸς ἀποτέτμηται· πρὸς γὰρ ἀνόμοιον ἐξιὸν ἀλλοιοῦταί τε αὐτὸ καὶ κατασβέννυται, ξυμφυὲς οὐκέτι τῷ πλησίον ἀέρι γιγνόμενον, ἅτε πῦρ οὐκ ἔχοντι. παύεταί τε οὖν ὁρῶν, ἔτι τε ἐπαγωγὸν ὕπνου γίγνεται· σωτηρίαν γὰρ ἣν οἱ θεοὶ τῆς ὄψεως ἐμηχανήσαντο, τὴν τῶν βλεφάρων φύσιν, ὅταν ταῦτα ξυμμύσῃ, καθείργνυσι τὴν τοῦ πυρὸς ἐντὸς δύναμιν, ἡ δὲ διαχεῖ τε καὶ ὁμαλύνει τὰς ἐντὸς κινήσεις, ὁμαλυνθεισῶν δὲ ἡσυχία γίγνεται, γενομένης δὲ πολλῆς μὲν ἡσυχίας βραχυόνειρος ὕπνος ἐμπίπτει, καταλειφθεισῶν δέ τινων κινήσεων μειζόνων, οἷαι καὶ ἐν οἵοις ἂν τόποις λείπωνται, τοιαῦτα καὶ τοσαῦτα παρέσχοντο ἀφομοιωθέντα ἐντός, ἔξω τε ἐγερθεῖσιν ἀπομνημονευόμενα, φαντάσματα.

Τὸ δὲ περὶ τὴν τῶν κατόπτρων εἰδωλοποιίαν, καὶ πάντα ὅσα ἐμφανῆ καὶ λεῖα, κατιδεῖν οὐδὲν ἔτι χαλεπόν· ἐκ γὰρ τῆς ἐντὸς ἐκτός τε τοῦ πυρὸς ἑκατέρου κοινωνίας ἀλλήλοις, ἑνός τε αὖ περὶ τὴν λειότητα ἑκάστοτε γενομένου καὶ πολλαχῇ μεταρρυθμισθέντος, πάντα τὰ τοιαῦτα ἐξ ἀνάγκης ἐμφαίνεται, τοῦ περὶ τὸ πρόσωπον πυρὸς τῷ περὶ τὴν

TIMAEUS

kindred substance along the path of the eyes' vision, wheresoever the fire which streams from within collides with an obstructing object without. And this substance, having all become similar in its properties because of its similar nature, distributes the motions of every object it touches, or whereby it is touched, throughout all the body even unto the Soul, and brings about that sensation which we now term "seeing." But when the kindred fire vanishes into night, the inner fire is cut off; for when it issues forth into what is dissimilar it becomes altered in itself and is quenched, seeing that it is no longer of like nature with the adjoining air, since that air is devoid of fire. Wherefore it leaves off seeing, and becomes also an inducement to sleep. For the eyelids —whose structure the Gods devised as a safeguard for the vision,—when they are shut close, curb the power of the inner fire; which power dissipates and allays the inward motions, and upon their allaying quiet ensues; and when this quiet has become intense there falls upon us a sleep that is well-nigh dreamless; but when some greater motions are still left behind, according to their nature and the positions they occupy such and so great are the images they produce, which images are copied within and are remembered by the sleepers when they awake out of the dream.

And it is no longer difficult to perceive the truth about the formation of images in mirrors and in bright and smooth surfaces of every kind. It is from the combination with each other of the inner and the outer fires, every time that they unite on the smooth surface and are variously deflected, that all such reflections necessarily result, owing to the fire of the reflected face coalescing with the fire of the vision

PLATO

ὄψιν πυρὶ περὶ τὸ λεῖον καὶ λαμπρὸν ξυμπαγοῦς γιγνομένου. δεξιὰ δὲ φαντάζεται τὰ ἀριστερά, ὅτι τοῖς ἐναντίοις μέρεσι τῆς ὄψεως περὶ τἀναντία μέρη γίγνεται ἐπαφὴ παρὰ τὸ καθεστὸς ἔθος τῆς προσβολῆς· δεξιὰ δὲ τὰ δεξιὰ καὶ τὰ ἀριστερὰ ἀριστερὰ τοὐναντίον, ὅταν μεταπέσῃ ξυμπηγνύμενον ᾧ ξυμπήγνυται φῶς· τοῦτο δέ, ὅταν ἡ τῶν κατόπτρων λειότης, ἔνθεν καὶ ἔνθεν ὕψη λαβοῦσα, τὸ δεξιὸν εἰς τὸ ἀριστερὸν μέρος ἀπώσῃ τῆς ὄψεως καὶ θάτερον ἐπὶ θάτερον. κατὰ δὲ τὸ μῆκος στραφὲν τοῦ προσώπου ταὐτὸν τοῦτο ὕπτιον ἐποίησε πᾶν φαίνεσθαι, τὸ κάτω πρὸς τὸ ἄνω τῆς αὐγῆς τό τ' ἄνω πρὸς τὸ κάτω πάλιν ἀπῶσαν.

Ταῦτ' οὖν πάντ' ἔστι τῶν ξυναιτίων, οἷς θεὸς ὑπηρετοῦσι χρῆται τὴν τοῦ ἀρίστου κατὰ τὸ δυνατὸν ἰδέαν ἀποτελῶν· δοξάζεται δὲ ὑπὸ τῶν πλείστων οὐ ξυναίτια ἀλλ' αἴτια εἶναι τῶν πάντων, ψύχοντα καὶ θερμαίνοντα πηγνύντα τε καὶ διαχέοντα καὶ ὅσα τοιαῦτα ἀπεργαζόμενα. λόγον δὲ οὐδένα οὐδὲ νοῦν εἰς οὐδὲν δυνατὰ ἔχειν ἐστί. τῶν γὰρ ὄντων ᾧ νοῦν μόνῳ κτᾶσθαι προσήκει, λεκτέον ψυχήν· τοῦτο δὲ ἀόρατον, πῦρ δὲ καὶ ὕδωρ καὶ γῆ καὶ ἀὴρ σώματα πάντα ὁρατὰ γέγονε. τὸν δὲ νοῦ καὶ ἐπιστήμης ἐραστὴν ἀνάγκη τὰς τῆς ἔμφρονος φύσεως αἰτίας πρώτας μεταδιώκειν, ὅσαι δὲ ὑπ' ἄλλων μὲν κινουμένων, ἕτερα δ' ἐξ ἀνάγκης κινούντων γίγνονται, δευτέρας. ποιητέον δὴ κατὰ ταῦτα

---

[1] *e.g.* when a man looks at his own face reflected in a mirror. *Cf. Soph.* 266 c.

[2] *i.e.* concave (and hemi-cylindrical).

[3] These causes are "secondary," as contrasted with the "primary" or First Cause (which is also the "final Cause"), "the Good"; *cf.* 29 E, 68 E, *Phaedo* 99 B.

TIMAEUS

on the smooth and bright surface.[1] And left appears as right, because contact takes place between opposite portions of the visual stream and opposite portions of the object, contrary to the regular mode of collision. Contrariwise, right appears as right and left as left whenever the fire changes sides on coalescing with the object wherewith it coalesces; and this occurs whenever the smooth surface of the mirrors, being elevated on this side and on that,[2] repels the right portion of the visual stream to the left and the left to the right. And when this same mirror is turned lengthwise to the face it makes the whole face appear upside down, since it repels the bottom of the ray to the top, and conversely the top to the bottom.

Now all these are among the auxiliary Causes[3] which God employs as his ministers in perfecting, so far as possible, the Form of the Most Good; but by the most of men[4] they are supposed to be not auxiliary but primary causes of all things—cooling and heating, solidifying and dissolving, and producing all such effects. Yet they are incapable of possessing reason and thought for any purpose. For, as we must affirm, the one and only existing thing which has the property of acquiring thought is Soul; and Soul is invisible, whereas fire and water and earth and air are all visible bodies; and the lover of thought and knowledge must needs pursue first the causes which belong to the Intelligent Nature, and put second all such as are of the class of things which are moved by others, and themselves, in turn, move others because they cannot help it. And we

[4] *e.g.* Anaxagoras and the Atomists.

PLATO

46 καὶ ἡμῖν· λεκτέα μὲν ἀμφότερα τὰ τῶν αἰτιῶν γένη, χωρὶς δὲ ὅσαι μετὰ νοῦ καλῶν καὶ ἀγαθῶν δημιουργοὶ καὶ ὅσαι μονωθεῖσαι φρονήσεως τὸ τυχὸν ἄτακτον ἑκάστοτε ἐξεργάζονται.

Τὰ μὲν οὖν τῶν ὀμμάτων ξυμμεταίτια πρὸς τὸ σχεῖν τὴν δύναμιν ἣν νῦν εἴληχεν εἰρήσθω· τὸ δὲ 47 μέγιστον αὐτῶν εἰς ὠφέλειαν ἔργον, δι' ὃ θεὸς αὔθ' ἡμῖν δεδώρηται, μετὰ τοῦτο ῥητέον. ὄψις δὴ κατὰ τὸν ἐμὸν λόγον αἰτία τῆς μεγίστης ὠφελείας γέγονεν ἡμῖν, ὅτι τῶν νῦν λόγων περὶ τοῦ παντὸς λεγομένων οὐδεὶς ἄν ποτε ἐρρήθη μήτε ἄστρα μήτε ἥλιον μήτ' οὐρανὸν ἰδόντων. νῦν δ' ἡμέρα τε καὶ νὺξ ὀφθεῖσαι μῆνές τε καὶ ἐνιαυτῶν περίοδοι μεμηχάνηνται μὲν ἀριθμόν, χρόνου δὲ ἔννοιαν περί τε τῆς τοῦ παντὸς φύσεως ζήτησιν ἔδοσαν· ἐξ ὧν ἐπορισάμεθα φιλο-
B σοφίας γένος, οὗ μεῖζον ἀγαθὸν οὔτ' ἦλθεν οὔθ' ἥξει ποτὲ τῷ θνητῷ γένει δωρηθὲν ἐκ θεῶν. λέγω δὴ τοῦτο ὀμμάτων μέγιστον ἀγαθόν· τἆλλα δέ, ὅσα ἐλάττω, τί ἂν ὑμνοῖμεν; ὧν ὁ μὴ φιλόσοφος τυφλωθεὶς ὀδυρόμενος ἂν θρηνοῖ μάτην. ἀλλὰ τούτου[1] λεγέσθω παρ' ἡμῶν αὕτη ἐπὶ ταῦτα αἰτία, θεὸν ἡμῖν ἀνευρεῖν δωρήσασθαί τε ὄψιν, ἵνα τὰς ἐν οὐρανῷ κατιδόντες τοῦ νοῦ περιόδους χρησαίμεθα ἐπὶ τὰς περιφορὰς τὰς τῆς παρ' ἡμῖν διανοήσεως,
C ξυγγενεῖς ἐκείναις οὔσας, ἀταράκτοις τεταραγμένας, ἐκμαθόντες δὲ καὶ λογισμῶν κατὰ φύσιν ὀρθότητος μετασχόντες, μιμούμενοι τὰς τοῦ θεοῦ πάντως

---

[1] τούτου some mss.: τοῦτο Zur.

[1] Cf. 37 D ff.
[2] Cf. Phileb. 16 c ff.
[3] An echo of Eurip. Phoenissae 1762 ἀλλὰ γὰρ τί ταῦτα θρηνῶ καὶ μάτην ὀδύρομαι;

TIMAEUS

also must act likewise. We must declare both kinds of Causes, but keep distinct those which, with the aid of thought, are artificers of things fair and good, and all those which are devoid of intelligence and produce always accidental and irregular effects.

Now regarding the auxiliary causes which have helped the eyes to acquire the power which they now possess, let this statement suffice. Next we must declare the most important benefit effected by them, for the sake of which God bestowed them upon us. Vision, in my view, is the cause of the greatest benefit to us, inasmuch as none of the accounts now given concerning the Universe would ever have been given if men had not seen the stars or the sun or the heaven. But as it is, the vision of day and night and of months and circling years has created the art of number and has given us not only the notion of Time [1] but also means of research into the nature of the Universe. From these we have procured Philosophy in all its range, than which no greater boon ever has come or will come, by divine bestowal, unto the race of mortals.[2] This I affirm to be the greatest good of eyesight. As for all the lesser goods, why should we celebrate them? He that is no philosopher when deprived of the sight thereof may utter vain lamentations![3] But the cause and purpose of that best good, as we must maintain, is this,—that God devised and bestowed upon us vision to the end that we might behold the revolutions of Reason in the Heaven and use them for the revolvings of the reasoning that is within us, these being akin to those, the perturbable to the imperturbable ; and that, through learning and sharing in calculations which are correct by their nature, by imitation of the absolutely un-

PLATO

47 ἀπλανεῖς οὔσας τὰς ἐν ἡμῖν πεπλανημένας καταστησαίμεθα.

Φωνῆς τε δὴ καὶ ἀκοῆς πέρι πάλιν ὁ αὐτὸς λόγος, ἐπὶ ταὐτὰ τῶν αὐτῶν ἕνεκα παρὰ θεῶν δεδωρῆσθαι. λόγος τε γὰρ ἐπ᾽ αὐτὰ ταῦτα τέτακται, μεγίστην ξυμβαλλόμενος εἰς αὐτὰ μοῖραν, ὅσον τ᾽ αὖ μουσικῆς φωνῇ[1] χρηστικὸν[2] πρὸς ἀκοὴν
D ἕνεκα ἁρμονίας ἐστὶ δοθέν· ἡ δὲ ἁρμονία, ξυγγενεῖς ἔχουσα φορὰς ταῖς ἐν ἡμῖν τῆς ψυχῆς περιόδοις, τῷ μετὰ νοῦ προσχρωμένῳ Μούσαις οὐκ ἐφ᾽ ἡδονὴν ἄλογον, καθάπερ νῦν, εἶναι δοκεῖ χρήσιμος, ἀλλ᾽ ἐπὶ τὴν γεγονυῖαν ἐν ἡμῖν ἀνάρμοστον ψυχῆς περίοδον εἰς κατακόσμησιν καὶ συμφωνίαν ἑαυτῇ ξύμμαχος ὑπὸ Μουσῶν δέδοται· καὶ ῥυθμὸς αὖ διὰ τὴν
E ἄμετρον ἐν ἡμῖν καὶ χαρίτων ἐπιδεᾶ γιγνομένην ἐν τοῖς πλείστοις ἕξιν ἐπίκουρος ἐπὶ ταῦτα ὑπὸ τῶν αὐτῶν ἐδόθη.

Τὰ μὲν οὖν παρεληλυθότα τῶν εἰρημένων, πλὴν βραχέων, ἐπιδέδεικται τὰ διὰ νοῦ δεδημιουργημένα· δεῖ δὲ καὶ τὰ δι᾽ ἀνάγκης γιγνόμενα τῷ λόγῳ
48 παραθέσθαι. μεμιγμένη γὰρ οὖν ἡ τοῦδε τοῦ κόσμου γένεσις ἐξ ἀνάγκης τε καὶ νοῦ συστάσεως ἐγεννήθη· νοῦ δὲ ἀνάγκης ἄρχοντος τῷ πείθειν αὐτὴν τῶν γιγνομένων τὰ πλεῖστα ἐπὶ τὸ βέλτιστον ἄγειν, ταύτῃ κατὰ ταὐτά τε δι᾽ ἀνάγκης ἡττωμένης ὑπὸ πειθοῦς ἔμφρονος οὕτω κατ᾽ ἀρχὰς ξυνίστατο τόδε

[1] φωνῇ best mss.: φωνῆς Zur.
[2] χρηστικὸν] χρήσιμον mss., Zur.]

---

[1] For the importance of music in education *cf. Rep.* 401 D, *Laws* 666 D ff.: also *Tim.* 80 B.

TIMAEUS

varying revolutions of the God we might stabilize the variable revolutions within ourselves.

Concerning sound also and hearing, once more we make the same declaration, that they were bestowed by the Gods with the same object and for the same reasons; for it was for these same purposes that speech was ordained, and it makes the greatest contribution thereto; music too, in so far as it uses audible sound, was bestowed for the sake of harmony.[1] And harmony, which has motions akin to the revolutions of the Soul within us, was given by the Muses to him who makes intelligent use of the Muses, not as an aid to irrational pleasure, as is now supposed, but as an auxiliary to the inner revolution of the Soul, when it has lost its harmony, to assist in restoring it to order and concord with itself. And because of the unmodulated condition, deficient in grace, which exists in most of us, Rhythm also was bestowed upon us to be our helper by the same deities and for the same ends.

The foregoing part of our discourse, save for a small portion, has been an exposition of the operations of Reason; but we must also furnish an account of what comes into existence through Necessity.[2] For, in truth, this Cosmos in its origin was generated as a compound, from the combination of Necessity and Reason. And inasmuch as Reason was controlling Necessity by persuading her to conduct to the best end the most part of the things coming into existence, thus and thereby it came about, through Necessity yielding to intelligent persuasion, that this Universe of ours was being in this wise constructed at the

[2] *i.e.* the sphere of mechanical causation, physical and physiological processes and results.

PLATO

τὸ πᾶν. εἴ τις οὖν ᾗ γέγονε, κατὰ ταῦτα ὄντως ἐρεῖ, μικτέον καὶ τὸ τῆς πλανωμένης εἶδος αἰτίας, ᾗ φέρειν πέφυκεν. ὧδε οὖν πάλιν ἀναχωρητέον, καὶ λαβοῦσιν αὐτῶν τούτων προσήκουσαν ἑτέραν ἀρχὴν αὖθις αὖ, καθάπερ περὶ τῶν τότε, νῦν οὕτω περὶ τούτων πάλιν ἀρκτέον ἀπ' ἀρχῆς. τὴν δὴ πρὸ τῆς οὐρανοῦ γενέσεως πυρὸς ὕδατός τε καὶ ἀέρος καὶ γῆς φύσιν θεατέον αὐτὴν καὶ τὰ πρὸ τούτου πάθη. νῦν γὰρ οὐδείς πω γένεσιν αὐτῶν μεμήνυκεν, ἀλλ' ὡς εἰδόσι πῦρ ὅ τί ποτε ἔστι καὶ ἕκαστον αὐτῶν λέγομεν ἀρχὰς αὐτὰ τιθέμενοι στοιχεῖα τοῦ παντός, προσῆκον αὐτοῖς οὐδ' ὡς ἐν συλλαβῆς εἴδεσι μόνον εἰκότως ὑπὸ τοῦ καὶ βραχὺ φρονοῦντος ἀπεικασθῆναι. νῦν δὲ οὖν τό γε παρ' ἡμῶν ὧδε ἐχέτω· τὴν μὲν περὶ ἁπάντων εἴτε ἀρχὴν εἴτε ἀρχὰς εἴτε ὅπῃ δοκεῖ τούτων πέρι, τὸ νῦν οὐ ῥητέον, δι' ἄλλο μὲν οὐδέν, διὰ δὲ τὸ χαλεπὸν εἶναι κατὰ τὸν παρόντα τρόπον τῆς διεξόδου δηλῶσαι τὰ δοκοῦντα· μήτ' οὖν ὑμεῖς οἴεσθε δεῖν ἐμὲ λέγειν, οὔτ' αὐτὸς αὖ πείθειν ἐμαυτὸν εἴην ἂν δυνατὸς ὡς ὀρθῶς ἐγχειροῖμ' ἂν τοσοῦτον ἐπιβαλλόμενος ἔργον. τὸ δὲ κατ' ἀρχὰς ῥηθὲν διαφυλάττων, τὴν τῶν εἰκότων λόγων δύναμιν, πειράσομαι μηδενὸς ἧττον εἰκότα, μᾶλλον δέ, ὡς[1] ἔμπροσθεν ἀπ' ἀρχῆς περὶ ἑκάστων καὶ ξυμπάντων λέγειν. θεὸν δὴ καὶ νῦν ἐπ' ἀρχῇ τῶν λεγομένων σωτῆρα ἐξ ἀτόπου καὶ ἀήθους

[1] ὡς] καὶ MSS., Zur.

---

[1] στοιχεῖα, here applied to physical "elements," was the regular term for "letters" of the alphabet; *cf. Theaet.* 203 B ff., *Rep.* 402 A ff.

[2] *i.e.* a method which aims only at "probability" or "likelihood": to attain to "first principles" we should need to employ the "dialectic" method. [3] *Cf.* 27 c.

TIMAEUS

beginning. Wherefore if one is to declare how it actually came into being on this wise, he must include also the form of the Errant Cause, in the way that it really acts. To this point, therefore, we must return, and taking once again a fresh starting-point suitable to the matter we must make a fresh start in dealing therewith, just as we did with our previous subjects. We must gain a view of the real nature of fire and water, air and earth, as it was before the birth of Heaven, and the properties they had before that time; for at present no one has as yet declared their generation, but we assume that men know what fire is, and each of these things, and we call them principles and presume that they are elements[1] of the Universe, although in truth they do not so much as deserve to be likened with any likelihood, by the man who has even a grain of sense, to the class of syllables. For the present, however, let our procedure be as follows. We shall not now expound the principle of all things—or their principles, or whatever term we use concerning them; and that solely for this reason, that it is difficult for us to explain our views while keeping to our present method of exposition.[2] You, therefore, ought not to suppose that I should expound them, while as for me—I should never be able to convince myself that I should be right in attempting to undertake so great a task. Strictly adhering, then, to what we previously affirmed, the import of the "likely" account, I will essay (as I did before) to give as "likely" an exposition as any other (nay, more so), regarding both particular things and the totality of things from the very beginning. And as before,[3] so now, at the commencement of our account, we must call upon God the Saviour to bring us safe

διηγήσεως πρὸς τὸ τῶν εἰκότων δόγμα διασώζειν ἡμᾶς ἐπικαλεσάμενοι πάλιν ἀρχώμεθα λέγειν.

Ἡ δ' οὖν αὖθις ἀρχὴ περὶ τοῦ παντὸς ἔστω μειζόνως τῆς πρόσθεν διῃρημένη. τότε μὲν γὰρ δύο εἴδη διειλόμεθα, νῦν δὲ τρίτον ἄλλο γένος ἡμῖν δηλωτέον. τὰ μὲν γὰρ δύο ἱκανὰ ἦν ἐπὶ τοῖς ἔμπροσθεν λεχθεῖσιν, ἓν μὲν ὡς παραδείγματος εἶδος ὑποτεθέν, νοητὸν καὶ ἀεὶ κατὰ ταὐτὰ ὄν, μίμημα δὲ παραδείγματος δεύτερον, γένεσιν ἔχον καὶ ὁρατόν. τρίτον δὲ τότε μὲν οὐ διειλόμεθα, νομίσαντες τὰ δύο ἕξειν ἱκανῶς· νῦν δὲ ὁ λόγος ἔοικεν εἰσαναγκάζειν χαλεπὸν καὶ ἀμυδρὸν εἶδος ἐπιχειρεῖν λόγοις ἐμφανίσαι. τίνα οὖν ἔχον δύναμιν κατὰ φύσιν αὐτὸ ὑποληπτέον; τοιάνδε μάλιστα, πάσης εἶναι γενέσεως ὑποδοχὴν αὐτό, οἷον τιθήνην. εἴρηται μὲν οὖν τἀληθές, δεῖ δ' ἐναργέστερον εἰπεῖν περὶ αὐτοῦ. χαλεπὸν δὲ ἄλλως τε καὶ διότι προαπορηθῆναι περὶ πυρὸς καὶ τῶν μετὰ πυρὸς ἀναγκαῖον τούτου χάριν· τούτων γὰρ εἰπεῖν ἕκαστον, ὁποῖον ὄντως ὕδωρ χρὴ λέγειν μᾶλλον ἢ πῦρ καὶ ὁποῖον ὁτιοῦν μᾶλλον ἢ καὶ ἅπαντα καθ' ἕκαστόν τε, οὕτως ὥστε τινὶ πιστῷ καὶ βεβαίῳ χρήσασθαι λόγῳ, χαλεπόν. πῶς οὖν δὴ τοῦτ' αὐτὸ καὶ πῇ καὶ τί περὶ αὐτῶν εἰκότως διαπορηθέντες ἂν λέγοιμεν; πρῶτον μὲν ὃ δὴ νῦν ὕδωρ ὠνομάκαμεν, πηγνύμενον, ὡς δοκοῦμεν, λίθους καὶ γῆν γιγνόμενον ὁρῶμεν, τηκόμενον δὲ καὶ διακρινόμενον αὖ ταὐτὸν τοῦτο πνεῦμα καὶ ἀέρα, ξυγκαυθέντα δὲ ἀέρα πῦρ, ἀνάπαλιν δὲ πῦρ συγκριθὲν καὶ κατα-

---

[1] Cf. 28 A.

TIMAEUS

through a novel and unwonted exposition to a conclusion based on likelihood, and thus begin our account once more.

We must, however, in beginning our fresh account of the Universe make more distinctions than we did before; for whereas then we distinguished two Forms,[1] we must now declare another third kind. For our former exposition those two were sufficient, one of them being assumed as a Model Form, intelligible and ever uniformly existent, and the second as the model's Copy, subject to becoming and visible. A third kind we did not at that time distinguish, considering that those two were sufficient; but now the argument seems to compel us to try to reveal by words a Form that is baffling and obscure. What essential property, then, are we to conceive it to possess? This in particular,—that it should be the receptacle, and as it were the nurse, of all Becoming. Yet true though this statement is, we must needs describe it more plainly. That, however, is a difficult task, especially because it is necessary, for its sake, to discuss first the problem of fire and its fellow elements. For in regard to these it is hard to say which particular element we ought really to term water rather than fire, and which we ought to term any one element rather than each and all of them, while still employing a terminology that is reliable and stable. How, then, shall we handle this problem, and what likely solution can we offer? First of all, we see that which we now call "water" becoming by condensation, as we believe, stones and earth; and again, this same substance, by dissolving and dilating, becoming breath and air; and air through combustion becoming fire; and conversely, fire when

PLATO

49 σβεσθὲν εἰς ἰδέαν τε ἀπιὸν αὖθις ἀέρος, καὶ πάλιν ἀέρα ξυνιόντα καὶ πυκνούμενον νέφος καὶ ὁμίχλην, ἐκ δὲ τούτων ἔτι μᾶλλον ξυμπιλουμένων ῥέον ὕδωρ, ἐξ ὕδατος δὲ γῆν καὶ λίθους αὖθις, κύκλον τε οὕτω διαδιδόντα εἰς ἄλληλα, ὡς φαί-
D νεται, τὴν γένεσιν. οὕτω δὴ τούτων οὐδέποτε τῶν αὐτῶν ἑκάστων φανταζομένων, ποῖον αὐτῶν ὡς ὂν ὁτιοῦν τοῦτο καὶ οὐκ ἄλλο παγίως διισχυριζόμενος οὐκ αἰσχυνεῖταί τις ἑαυτόν; οὐκ ἔστιν, ἀλλ᾽ ἀσφαλέστατον μακρῷ περὶ τούτων τιθεμένους ὧδε λέγειν· ἀεὶ ὃ καθορῶμεν ἄλλοτε ἄλλῃ γιγνόμενον, ὡς πῦρ, μὴ τοῦτο ἀλλὰ τὸ τοιοῦτον ἑκάστοτε προσαγορεύειν πῦρ, μηδὲ ὕδωρ τοῦτο ἀλλὰ τὸ τοιοῦτον ἀεί, μηδὲ ἄλλο ποτὲ μηδὲν ὥς τιν᾽ ἔχον
E βεβαιότητα, ὅσα δεικνύντες τῷ ῥήματι τῷ τόδε καὶ τοῦτο προσχρώμενοι δηλοῦν ἡγούμεθά τι· φεύγει γὰρ οὐχ ὑπομένον τὴν τοῦ τόδε καὶ τοῦτο [καὶ τὴν τῷδε]¹ καὶ πᾶσαν ὅση μόνιμα ὡς ὄντα αὐτὰ ἐνδείκνυται φάσις. ἀλλὰ ταῦτα μὲν ἕκαστα μὴ λέγειν, τὸ δὲ τοιοῦτον ἀεὶ περιφερομένων² ὅμοιον ἑκάστου πέρι καὶ ξυμπάντων οὕτω καλεῖν· καὶ δὴ καὶ πῦρ τὸ διὰ παντὸς τοιοῦτον, καὶ ἅπαν ὅσονπερ ἂν ἔχῃ γένεσιν. ἐν ᾧ δὲ ἐγγιγνόμενα ἀεὶ ἕκαστα
50 αὐτῶν φαντάζεται καὶ πάλιν ἐκεῖθεν ἀπόλλυται, μόνον ἐκεῖνο αὖ προσαγορεύειν τῷ τε τοῦτο καὶ τῷ τόδε προσχρωμένους ὀνόματι, τὸ δὲ ὁποιονοῦν τι, θερμὸν ἢ λευκὸν ἢ καὶ ὁτιοῦν τῶν ἐναντίων,

---

¹ καὶ τὴν τῷδε I bracket, after E. Sachs.
² περιφερομένων] περιφερόμενον mss., Zur.

TIMAEUS

contracted and quenched returning back to the form of air; and air once more uniting and condensing into cloud and mist; and issuing from these, when still further compressed, flowing water; and from water earth and stones again: thus we see the elements passing on to one another, as it would seem, in an unbroken circle the gift of birth. Accordingly, since no one of these ever remains identical in appearance, which of them shall a man definitely affirm to be any one particular element and no other without incurring ridicule? None such exists. On the contrary, by far the safest plan in treating of these elements is to proceed thus: Whatsoever object we perceive to be constantly changing from one state to another, like fire, that object, be it fire, we must never describe as "this" but as "suchlike," nor should we ever call water "this" but "suchlike"; nor should we describe any other element, as though it possessed stability, of all those which we indicate by using the terms "this" and "that" and suppose ourselves to refer to a definite object. For such an object shuns and eludes the names "this" and "that" and every name which indicates that they are stable. Thus we must not call the several elements "these," but in regard to each of them and all together we must apply the term "suchlike" to represent what is always circling round: thus we shall call that which is constantly "suchlike" by the name of fire, and so with everything else that is generated. But that "wherein" they are always, in appearance, coming severally into existence, and "wherefrom" in turn they perish, in describing that and that alone should we employ the terms "this" and "that"; whereas, in describing what is "suchlike"—hot, for instance, or

καὶ πάνθ' ὅσα ἐκ τούτων, μηδὲν ἐκεῖνο αὖ τούτων καλεῖν.

Ἔτι δὲ σαφέστερον αὐτοῦ πέρι προθυμητέον αὖθις εἰπεῖν. εἰ γὰρ πάντα τις σχήματα πλάσας ἐκ χρυσοῦ μηδὲν μεταπλάττων παύοιτο ἕκαστα εἰς ἅπαντα, δεικνύντος δή τινος αὐτῶν ἓν καὶ ἐρομένου τί ποτ' ἔστι, μακρῷ πρὸς ἀλήθειαν ἀσφαλέστατον εἰπεῖν ὅτι χρυσός, τὸ δὲ τρίγωνον ὅσα τε ἄλλα σχήματα ἐνεγίγνετο, μηδέποτε λέγειν ταῦτα ὡς ὄντα, ἅ γε μεταξὺ τιθεμένου μεταπίπτει, ἀλλ' ἐὰν ἄρα καὶ τὸ τοιοῦτον μετ' ἀσφαλείας ἐθέλῃ δέχεσθαί τινος, ἀγαπᾶν. ὁ αὐτὸς δὴ λόγος καὶ περὶ τῆς τὰ πάντα δεχομένης σώματα φύσεως· ταὐτὸν αὐτὴν ἀεὶ προσρητέον· ἐκ γὰρ τῆς ἑαυτῆς τὸ παράπαν οὐκ ἐξίσταται δυνάμεως· δέχεταί τε γὰρ ἀεὶ τὰ πάντα, καὶ μορφὴν οὐδεμίαν ποτὲ οὐδενὶ τῶν εἰσιόντων ὁμοίαν εἴληφεν οὐδαμῇ οὐδαμῶς· ἐκμαγεῖον γὰρ φύσει παντὶ κεῖται, κινούμενόν τε καὶ διασχηματιζόμενον ὑπὸ τῶν εἰσιόντων· φαίνεται δὲ δι' ἐκεῖνα ἄλλοτε ἀλλοῖον. τὰ δὲ εἰσιόντα καὶ ἐξιόντα τῶν ὄντων ἀεὶ μιμήματα, τυπωθέντα ἀπ' αὐτῶν τρόπον τινὰ δύσφραστον καὶ θαυμαστόν, ὃν εἰσαῦθις μέτιμεν.

Ἐν δ' οὖν τῷ παρόντι χρὴ γένη διανοηθῆναι τριττά, τὸ μὲν γιγνόμενον, τὸ δ' ἐν ᾧ γίγνεται, τὸ δ' ὅθεν ἀφομοιούμενον φύεται τὸ γιγνόμενον. καὶ δὴ καὶ

---
[1] *Cf.* 53 c.

TIMAEUS

white, or any of the opposite qualities, or any compounds thereof—we ought never to apply to it any of these terms.

But we must bestir ourselves to explain this matter again yet more clearly. Now imagine that a man were to model all possible figures out of gold, and were then to proceed without cessation to remodel each of these into every other,—then, if someone were to point to one of the figures and ask what it *is*, by far the safest reply, in point of truth, would be that it is gold; but as for the triangle and all the other figures which were formed in it, one should never describe them as "being" seeing that they change even while one is mentioning them; rather one should be content if the figure admits of even the title "suchlike" being applied to it with any safety. And of the substance which receives all bodies the same account must be given. It must be called always by the same name; for from its own proper quality it never departs at all; for while it is always receiving all things, nowhere and in no wise does it assume any shape similar to any of the things that enter into it. For it is laid down by nature as a moulding-stuff for everything, being moved and marked by the entering figures, and because of them it appears different at different times. And the figures that enter and depart are copies of those that are always existent, being stamped from them in a fashion marvellous and hard to describe, which we shall investigate hereafter.[1]

For the present, then, we must conceive of three kinds,—the Becoming, that "Wherein" it becomes, and the source "Wherefrom" the Becoming is copied and produced. Moreover, it is proper to liken the

117

προσεικάσαι πρέπει τὸ μὲν δεχόμενον μητρί, τὸ δ' ὅθεν πατρί, τὴν δὲ μεταξὺ τούτων φύσιν ἐκγόνῳ, νοῆσαί τε ὡς οὐκ ἂν ἄλλως ἐκτυπώματος ἔσεσθαι μέλλοντος ἰδεῖν ποικίλου πάσας ποικιλίας τοῦτ' αὐτὸ ἐν ᾧ ἐκτυπούμενον ἐνίσταται γένοιτ' ἂν παρεσκευασμένον εὖ, πλὴν ἄμορφον ὂν ἐκείνων ἁπασῶν τῶν ἰδεῶν ὅσας μέλλοι δέχεσθαί ποθεν. ὅμοιον γὰρ ὂν τῶν ἐπεισιόντων τινὶ τὰ τῆς ἐναντίας τά τε τῆς τὸ παράπαν ἄλλης φύσεως, ὁπότ' ἔλθοι, δεχόμενον κακῶς ἂν ἀφομοιοῖ, τὴν αὑτοῦ παρεμφαῖνον ὄψιν. διὸ καὶ πάντων ἐκτὸς εἰδῶν εἶναι χρεὼν τὸ τὰ πάντα ἐκδεξόμενον ἐν αὑτῷ γένη, καθάπερ περὶ τὰ ἀλείμματα, ὁπόσα εὐώδη, τέχνῃ μηχανῶνται πρῶτον τοῦτ' αὐτὸ ὑπάρχον, ποιοῦσιν ὅ τι μάλιστα ἀώδη τὰ δεξόμενα ὑγρὰ τὰς ὀσμάς· ὅσοι τε ἔν τισι τῶν μαλακῶν σχήματα ἀπομάττειν ἐπιχειροῦσι, τὸ παράπαν σχῆμα οὐδὲν ἔνδηλον ὑπάρχειν ἐῶσι, προομαλύναντες δὲ ὅ τι λειότατον ἀπεργάζονται. ταὐτὸν οὖν καὶ τῷ τὰ τῶν ⟨νοητῶν⟩[1] πάντων ἀεί τε ὄντων κατὰ πᾶν ἑαυτοῦ πολλάκις ἀφομοιώματα καλῶς μέλλοντι δέχεσθαι πάντων ἐκτὸς αὐτῷ προσήκει πεφυκέναι τῶν εἰδῶν. διὸ δὴ τὴν τοῦ γεγονότος ὁρατοῦ καὶ πάντως αἰσθητοῦ μητέρα καὶ ὑποδοχὴν μήτε γῆν μήτε ἀέρα μήτε πῦρ μήτε ὕδωρ λέγωμεν, μήτε ὅσα ἐκ τούτων μήτε ἐξ ὧν ταῦτα γέγονεν· ἀλλ' ἀνόρατον εἶδός τι καὶ ἄμορφον, πανδεχές, μεταλαμβάνον δὲ ἀπορώτατά πῃ τοῦ νοητοῦ καὶ δυσαλωτότατον αὐτὸ λέγοντες οὐ ψευσόμεθα.

Καθ' ὅσον δὲ ἐκ τῶν προειρημένων δυνατὸν ἐφ-

[1] νοητῶν I add, after Cook-Wilson (*cf.* 37 A).

TIMAEUS

Recipient to the Mother, the Source to the Father, and what is engendered between these two to the Offspring; and also to perceive that, if the stamped copy is to assume diverse appearances of all sorts, that substance wherein it is set and stamped could not possibly be suited to its purpose unless it were itself devoid of all those forms which it is about to receive from any quarter. For were it similar to any of the entering forms, on receiving forms of an opposite or wholly different kind, as they arrived, it would copy them badly, through obtruding its own visible shape. Wherefore it is right that the substance which is to receive within itself all the kinds should be void of all forms; just as with all fragrant ointments, men bring about this condition by artistic contrivance and make the liquids which are to receive the odours as odourless as possible; and all who essay to mould figures in any soft material utterly refuse to allow any previous figure to remain visible therein, and begin by making it even and as smooth as possible before they execute the work. So likewise it is right that the substance which is to be fitted to receive frequently over its whole extent the copies of all things intelligible and eternal should itself, of its own nature, be void of all the forms. Wherefore, let us not speak of her that is the Mother and Receptacle of this generated world, which is perceptible by sight and all the senses, by the name of earth or air or fire or water, or any aggregates or constituents thereof: rather, if we describe her as a Kind invisible and unshaped, all-receptive, and in some most perplexing and most baffling way partaking of the intelligible, we shall describe her truly.

In so far as it is possible to arrive at the nature of

51ˇ ἱκνεῖσθαι τῆς φύσεως αὐτοῦ, τῇδ' ἄν τις ὀρθότατα λέγοι, πῦρ μὲν ἑκάστοτε αὐτοῦ τὸ πεπυρωμένον μέρος φαίνεσθαι, τὸ δὲ ὑγρανθὲν ὕδωρ, γῆν δὲ καὶ ἀέρα καθ' ὅσον ἂν μιμήματα τούτων δέχηται. λόγῳ δὲ δὴ μᾶλλον τὸ τοιόνδε διοριζομένους περὶ αὐτῶν διασκεπτέον· ἆρ' ἔστι τι πῦρ αὐτὸ ἐφ' ἑαυτοῦ,
C καὶ πάντα περὶ ὧν ἀεὶ λέγομεν οὕτως αὐτὰ καθ' αὑτὰ ὄντα ἕκαστα, ἢ ταῦτα ἅπερ καὶ βλέπομεν ὅσα τε ἄλλα διὰ τοῦ σώματος αἰσθανόμεθα μόνα ἔστι, τοιαύτην ἔχοντα ἀλήθειαν, ἄλλα δὲ οὐκ ἔστι παρὰ ταῦτα οὐδαμῇ οὐδαμῶς, ἀλλὰ μάτην ἑκάστοτε εἶναί τί φαμεν εἶδος ἑκάστου νοητόν, τὸ δὲ οὐδὲν ἄρ' ἦν πλὴν λόγος; οὔτε οὖν δὴ τὸ παρὸν ἄκριτον καὶ ἀδίκαστον ἀφέντα ἄξιον φάναι διισχυριζόμενον ἔχειν οὕτως, οὔτ' ἐπὶ λόγου μήκει πάρεργον ἄλλο
D μῆκος ἐπεμβλητέον· εἰ δέ τις ὅρος ὁρισθεὶς μέγας διὰ βραχέων φανείη, τοῦτο μάλιστ' ἐγκαιριώτατον γένοιτ' ἄν.

Ὧδε οὖν τήν γ' ἐμὴν αὐτὸς τίθεμαι ψῆφον· εἰ μὲν νοῦς καὶ δόξα ἀληθής ἐστον δύο γένη, παντάπασιν εἶναι καθ' αὑτὰ ταῦτα τὰ ἀναίσθητα ὑφ' ἡμῶν εἴδη, νοούμενα μόνον· εἰ δ' ὥς τισι φαίνεται δόξα ἀληθὴς νοῦ διαφέρει τὸ μηδέν, πάνθ' ὁπόσα αὖ διὰ τοῦ σώματος αἰσθανόμεθα
E θετέον βεβαιότατα. δύο δὴ λεκτέον ἐκείνω, διότι χωρὶς γεγόνατον ἀνομοίως τε ἔχετον. τὸ μὲν γὰρ

TIMAEUS

this kind from the foregoing account, one may state it most correctly in this way. That part of it which is made fiery appears each time as fire, that which has been liquefied as water; and it appears as earth and air in so far as it receives copies of these. But let us investigate the matter by more exact reasoning, and consider this question. Does there exist any self-subsisting fire or any of those other objects which we likewise term " self-subsisting realities " ? Or is it only these things which we see, or otherwise perceive by means of bodily senses, that exist, possessed of sensible reality; beside which no other things exist anywhere or anyhow, and it is merely an idle assertion of ours that there always exists an intelligible Form of every object, whereas it is really nothing more than a verbal phrase? Now, on the one hand, it would be improper to dismiss the question before us without a trial and a verdict, and simply to asseverate that the fact is so; while, on the other hand, we ought not to burden a lengthy discourse with another subsidiary argument. If, however, it were possible to disclose briefly some main determining principle, that would best serve our purpose.

This, then, is the view for which I, for my part, cast my vote. If Reason and True Opinion are two distinct Kinds, most certainly these self-subsisting Forms do exist, imperceptible by our senses, and objects of Reason only; whereas if, as appears to some, True Opinion differs in naught from Reason, then, on the contrary, all the things which we perceive by our bodily senses must be judged to be most stable. Now these two Kinds must be declared to be two, because they have come into existence separately and are unlike in condition. For the one of

PLATO

51 αὑτῶν διὰ διδαχῆς, τὸ δ' ὑπὸ πειθοῦς ἡμῖν ἐγγίγνεται· καὶ τὸ μὲν ἀεὶ μετὰ ἀληθοῦς λόγου, τὸ δὲ ἄλογον· καὶ τὸ μὲν ἀκίνητον πειθοῖ, τὸ δὲ μεταπειστόν· καὶ τοῦ μὲν πάντα ἄνδρα μετέχειν φατέον, νοῦ δὲ θεούς, ἀνθρώπων δὲ γένος βραχύ τι. τούτων δὲ οὕτως ἐχόντων ὁμολογητέον ἓν μὲν εἶναι **52** τὸ κατὰ ταὐτὰ εἶδος ἔχον, ἀγέννητον καὶ ἀνώλεθρον, οὔτε εἰς ἑαυτὸ εἰσδεχόμενον ἄλλο ἄλλοθεν οὔτε αὐτὸ εἰς ἄλλο ποι ἰόν, ἀόρατον δὲ καὶ ἄλλως ἀναίσθητον, τοῦτο ὃ δὴ νόησις εἴληχεν ἐπισκοπεῖν· τὸ δ' ὁμώνυμον ὅμοιόν τε ἐκείνῳ δεύτερον, αἰσθητόν, γεννητόν, πεφορημένον ἀεί, γιγνόμενόν τε ἔν τινι τόπῳ καὶ πάλιν ἐκεῖθεν ἀπολλύμενον, δόξῃ μετ' αἰσθήσεως περιληπτόν· τρίτον δὲ αὖ γένος ὂν τὸ **B** τῆς χώρας ἀεί, φθορὰν οὐ προσδεχόμενον, ἕδραν δὲ παρέχον ὅσα ἔχει γένεσιν πᾶσιν, αὐτὸ δὲ μετ' ἀναισθησίας ἁπτὸν λογισμῷ τινι νόθῳ, μόγις πιστόν, πρὸς ὃ δὴ καὶ ὀνειροπολοῦμεν βλέποντες καὶ φαμεν ἀναγκαῖον εἶναί που τὸ ὂν ἅπαν ἔν τινι τόπῳ καὶ κατέχον χώραν τινά, τὸ δὲ μήτε ἐν γῇ μήτε που κατ' οὐρανὸν οὐδὲν εἶναι. ταῦτα δὴ πάντα καὶ τούτων ἄλλ' ἀδελφὰ καὶ περὶ τὴν ἄϋπνον καὶ ἀληθῶς φύσιν ὑπάρχουσαν ὑπὸ ταύτης **C** τῆς ὀνειρώξεως οὐ δυνατοὶ γιγνόμεθα ἐγερθέντες διοριζόμενοι τἀληθὲς λέγειν, ὡς εἰκόνι μέν, ἐπείπερ οὐδ' αὐτὸ τοῦτο ἐφ' ᾧ γέγονεν ἑαυτῆς ἐστίν, ἑτέρου δέ τινος ἀεὶ φέρεται φάντασμα, διὰ ταῦτα ἐν

TIMAEUS

them arises in us by teaching, the other by persuasion; and the one is always in company with true reasoning, whereas the other is irrational; and the one is immovable by persuasion, whereas the other is alterable by persuasion; and of the one we must assert that every man partakes, but of Reason only the gods and but a small class of men. This being so, we must agree that One Kind is the self-identical Form, ungenerated and indestructible, neither receiving into itself any other from any quarter nor itself passing anywhither into another, invisible and in all ways imperceptible by sense, it being the object which it is the province of Reason to contemplate; and a second Kind is that which is named after the former and similar thereto, an object perceptible by sense, generated, ever carried about, becoming in a place and out of it again perishing, apprehensible by Opinion with the aid of Sensation; and a third Kind is ever-existing Place, which admits not of destruction, and provides room for all things that have birth, itself being apprehensible by a kind of bastard reasoning by the aid of non-sensation, barely an object of belief; for when we regard this we dimly dream and affirm that it is somehow necessary that all that exists should exist *in* some spot and occupying some *place*, and that that which is neither on earth nor anywhere in the Heaven is nothing. So because of all these and other kindred notions, we are unable also on waking up to distinguish clearly the unsleeping and truly subsisting substance, owing to our dreamy condition, or to state the truth—how that it belongs to a copy —seeing that it has not for its own even that substance for which it came into being, but fleets ever as a phantom of something else—to come into exist-

ἑτέρῳ προσήκει τινὶ γίγνεσθαι, οὐσίας ἀμῶς γέ
πως ἀντεχομένην, ἢ μηδὲν τὸ παράπαν αὐτὴν
εἶναι, τῷ δὲ ὄντως ὄντι βοηθὸς ὁ δι' ἀκριβείας
ἀληθὴς λόγος, ὡς ἕως ἄν τι τὸ μὲν ἄλλο ᾖ, τὸ δὲ
ἄλλο, οὐδέτερον ἐν οὐδετέρῳ ποτὲ γεγενημένον ἓν
D ἵμα ταὐτὸν καὶ δύο γενήσεσθον.

Οὗτος μὲν οὖν δὴ παρὰ τῆς ἐμῆς ψήφου λογισθεὶς ἐν κεφαλαίῳ δεδόσθω λόγος, ὅν τε καὶ χώραν
καὶ γένεσιν εἶναι, τρία τριχῇ, καὶ πρὶν οὐρανὸν
γενέσθαι· τὴν δὲ γενέσεως τιθήνην ὑγραινομένην
καὶ πυρουμένην καὶ τὰς γῆς τε καὶ ἀέρος μορφὰς
δεχομένην, καὶ ὅσα ἄλλα τούτοις πάθη ξυνέπεται
E πάσχουσαν, παντοδαπὴν μὲν ἰδεῖν φαίνεσθαι, διὰ
δὲ τὸ μήθ' ὁμοίων δυνάμεων μήτ' ἰσορρόπων
ἐμπίπλασθαι κατ' οὐδὲν αὐτῆς ἰσορροπεῖν, ἀλλ'
ἀνωμάλως πάντη ταλαντουμένην σείεσθαι μὲν ὑπ'
ἐκείνων αὐτήν, κινουμένην δ' αὖ πάλιν ἐκεῖνα
σείειν· τὰ δὲ κινούμενα ἄλλα ἄλλοσε ἀεὶ φέρεσθαι
διακρινόμενα, ὥσπερ τὰ ὑπὸ τῶν πλοκάνων τε καὶ
ὀργάνων τῶν περὶ τὴν τοῦ σίτου κάθαρσιν σειόμενα
καὶ ἀναλικμώμενα τὰ μὲν πυκνὰ καὶ βαρέα ἄλλῃ,
53 τὰ δὲ μανὰ καὶ κοῦφα εἰς ἑτέραν ἵζει φερόμενα
ἕδραν· τότε οὕτω τὰ τέτταρα γένη σειόμενα ὑπὸ
τῆς δεξαμενῆς, κινουμένης αὐτῆς οἷον ὀργάνου
σεισμὸν παρέχοντος, τὰ μὲν ἀνομοιότατα πλεῖστον
αὐτὰ ἀφ' αὑτῶν ὁρίζειν, τὰ δ' ὁμοιότατα μάλιστα
εἰς ταὐτὸν ξυνωθεῖν, διὸ δὴ καὶ χώραν ταῦτα ἄλλα
ἄλλην ἴσχειν, πρὶν καὶ τὸ πᾶν ἐξ αὐτῶν διακοσμη-

ence *in* some other thing, clinging to existence as best it may, on pain of being nothing at all ; whereas to the aid of the really existent there comes the accurately true argument, that so long as one thing is one thing, and another something different, neither of the two will ever come to exist in the other so that the same thing becomes simultaneously both one and two.

Let this, then, be, according to my verdict, a reasoned account of the matter summarily stated,— that Being and Place and Becoming were existing, three distinct things, even before the Heaven came into existence ; and that the Nurse of Becoming, being liquefied and ignified and receiving also the forms of earth and of air, and submitting to all the other affections which accompany these, exhibits every variety of appearance ; but owing to being filled with potencies that are neither similar nor balanced, in no part of herself is she equally balanced, but sways unevenly in every part, and is herself shaken by these forms and shakes them in turn as she is moved. And the forms, as they are moved, fly continually in various directions and are dissipated ; just as the particles that are shaken and winnowed by the sieves and other instruments used for the cleansing of corn fall in one place if they are solid and heavy, but fly off and settle elsewhere if they are spongy and light. So it was also with the Four Kinds when shaken by the Recipient : her motion, like an instrument which causes shaking, was separating farthest from one another the dissimilar, and pushing most closely together the similar ; wherefore also these Kinds occupied different places even before that the Universe was organized and generated out of them.

θὲν γενέσθαι. καὶ τὸ μὲν δὴ πρὸ τούτου πάντα ταῦτ' ἔχειν ἀλόγως καὶ ἀμέτρως· ὅτε δ' ἐπεχειρεῖτο κοσμεῖσθαι τὸ πᾶν, πῦρ πρῶτον καὶ ὕδωρ καὶ γῆν καὶ ἀέρα, ἴχνη μὲν ἔχοντα αὐτῶν ἄττα, παντάπασί γε μὴν διακείμενα ὥσπερ εἰκὸς ἔχειν ἅπαν ὅταν ἀπῇ τινος θεός, οὕτω δὴ τότε πεφυκότα ταῦτα πρῶτον διεσχηματίσατο εἴδεσί τε καὶ ἀριθμοῖς. τὸ δὲ ᾗ δυνατὸν ὡς κάλλιστα ἄριστά τε ἐξ οὐχ οὕτως ἐχόντων τὸν θεὸν αὐτὰ ξυνιστάναι, παρὰ πάντα ἡμῖν ὡς ἀεὶ τοῦτο λεγόμενον ὑπαρχέτω. νῦν δ' οὖν τὴν διάταξιν αὐτῶν ἐπιχειρητέον ἑκάστων καὶ γένεσιν ἀήθει λόγῳ πρὸς ὑμᾶς δηλοῦν· ἀλλὰ γὰρ ἐπεὶ μετέχετε τῶν κατὰ παίδευσιν ὁδῶν, δι' ὧν ἐνδείκνυσθαι τὰ λεγόμενα ἀνάγκη, ξυνέψεσθε.

Πρῶτον μὲν δὴ πῦρ καὶ γῆ καὶ ὕδωρ καὶ ἀὴρ ὅτι σώματά ἐστι, δῆλόν που καὶ παντί. τὸ δὲ τοῦ σώματος εἶδος πᾶν καὶ βάθος ἔχει. τὸ δὲ βάθος αὖ πᾶσα ἀνάγκη τὴν ἐπίπεδον περιειληφέναι φύσιν. ἡ δὲ ὀρθὴ τῆς ἐπιπέδου βάσεως ἐκ τριγώνων συνέστηκε. τὰ δὲ τρίγωνα πάντα ἐκ δυοῖν ἄρχεται τριγώνοιν, μίαν μὲν ὀρθὴν ἔχοντος ἑκατέρου γωνίαν, τὰς δὲ ὀξείας· ὧν τὸ μὲν ἕτερον ἑκατέρωθεν ἔχει μέρος γωνίας ὀρθῆς πλευραῖς ἴσαις διῃρημένης, τὸ δὲ ἕτερον ἀνίσοις ἄνισα μέρη νενεμημένης. ταύτην δὴ πυρὸς ἀρχὴν καὶ τῶν ἄλλων σωμάτων ὑποτιθέμεθα κατὰ τὸν μετ' ἀνάγκης εἰκότα λόγον πορευ-

---

[1] *i.e.* the rectangular isosceles triangle and the rectangular scalene; all other triangles can be built up from these two (*e.g.* see 54 E n.).

TIMAEUS

Before that time, in truth, all these things were in a state devoid of reason or measure, but when the work of setting in order this Universe was being undertaken, fire and water and earth and air, although possessing some traces of their own nature, were yet so disposed as everything is likely to be in the absence of God; and inasmuch as this was then their natural condition, God began by first marking them out into shapes by means of forms and numbers. And that God constructed them, so far as He could, to be as fair and good as possible, whereas they had been otherwise,—this above all else must always be postulated in our account. Now, however, it is the disposition and origin of each of these Kinds which I must endeavour to explain to you in an exposition of an unusual type; yet, inasmuch as you have some acquaintance with the technical method which I must necessarily employ in my exposition, you will follow me.

In the first place, then, it is plain I presume to everyone that fire and earth and water and air are solid bodies; and the form of a body, in every case, possesses depth also. Further, it is absolutely necessary that depth should be bounded by a plane surface; and the rectilinear plane is composed of triangles. Now all triangles derive their origin from two triangles, each having one angle right and the others acute [1]; and the one of these triangles has on each side half a right angle marked off by equal sides, while the other has the right angle divided into unequal parts by unequal sides. These we lay down as the principles of fire and all the other bodies, proceeding according to a method in which the probable is combined with the necessary; but the principles

53

όμενοι· τὰς δ' ἔτι τούτων ἀρχὰς ἄνωθεν θεὸς οἶδε
E καὶ ἀνδρῶν ὃς ἂν ἐκείνῳ φίλος ᾖ. δεῖ δὴ λέγειν
ποῖα κάλλιστα σώματα γένοιτ' ἂν τέτταρα, ἀνόμοια μὲν ἑαυτοῖς, δυνατὰ δὲ ἐξ ἀλλήλων αὐτῶν
ἄττα διαλυόμενα γίγνεσθαι. τούτου γὰρ τυχόντες
ἔχομεν τὴν ἀλήθειαν γενέσεως πέρι γῆς τε καὶ
πυρὸς τῶν τε ἀνὰ λόγον ἐν μέσῳ· τότε γὰρ οὐδενὶ
συγχωρησόμεθα καλλίω τούτων ὁρώμενα σώματα
εἶναί που καθ' ἓν γένος ἕκαστον ὄν. τοῦτ' οὖν
προθυμητέον, τὰ διαφέροντα κάλλει σωμάτων τέτταρα γένη συναρμόσασθαι καὶ φάναι τὴν τούτων
54 ἡμᾶς φύσιν ἱκανῶς εἰληφέναι. τοῖν δὴ δυοῖν
τριγώνοιν τὸ μὲν ἰσοσκελὲς μίαν εἴληχε φύσιν, τὸ
δὲ πρόμηκες ἀπεράντους. προαιρετέον οὖν αὖ
τῶν ἀπείρων τὸ κάλλιστον, εἰ μέλλομεν ἄρξεσθαι
κατὰ τρόπον. ἂν οὖν τις ἔχῃ κάλλιον ἐκλεξάμενος
εἰπεῖν εἰς τὴν τούτων ξύστασιν, ἐκεῖνος οὐκ ἐχθρὸς
ὢν ἀλλὰ φίλος κρατεῖ· τιθέμεθα δ' οὖν τῶν πολλῶν
τριγώνων κάλλιστον ἕν, ὑπερβάντες τἆλλα, ἐξ οὗ
B τὸ ἰσόπλευρον τρίγωνον ἐκ τρίτου συνέστηκε.
διότι δέ, ὁ λόγος πλείων· ἀλλὰ τῷ τοῦτο ἐλέγξαντι
καὶ ἀνευρόντι δὴ μὴ οὕτως ἔχον κεῖται φιλία τὰ
ἆθλα. προῃρήσθω δὴ δύο τρίγωνα, ἐξ ὧν τό τε
τοῦ πυρὸς καὶ τὰ τῶν ἄλλων σώματα μεμηχά-
νηται, τὸ μὲν ἰσοσκελές, τὸ δὲ τριπλῆν κατὰ
δύναμιν ἔχον τῆς ἐλάττονος τὴν μείζω πλευρὰν ἀεί.

---

[1] *i.e.* the half of an equilateral triangle; *e.g.* if the triangle ABC is bisected by the line AD, we have two such triangles in ADB and ADC.

TIMAEUS

which are still higher than these are known only to God and the man who is dear to God. We must now declare what will be the four fairest bodies, dissimilar to one another, but capable in part of being produced out of one another by means of dissolution; for if we succeed herein we shall grasp the truth concerning the generation of earth and fire and the mean proportionals. For to no one will we concede that fairer bodies than these, each distinct of its kind, are anywhere to be seen. Wherefore we must earnestly endeavour to frame together these four kinds of bodies which excel in beauty, and to maintain that we have apprehended their nature adequately. Now of the two triangles, the isosceles possesses one single nature, but the scalene an infinite number; and of these infinite natures we must select the fairest, if we mean to make a suitable beginning. If, then, anyone can claim that he has chosen one that is fairer for the construction of these bodies, he, as friend rather than foe, is the victor. We, however, shall pass over all the rest and postulate as the fairest of the triangles that triangle out of which, when two are conjoined, the equilateral triangle is constructed as a third.[1] The reason why is a longer story; but should anyone refute us and discover that it is not so, we begrudge him not the prize. Accordingly, let these two triangles be selected as those wherefrom are contrived the bodies of fire and of the other elements,—one being the isosceles, and the other that which always has the square on its greater side three times the square on the lesser side.[2]

---

[2] *i.e.* in the triangle ADB (see last note) $AB = 2BD$, and $(AB)^2 = (BD)^2 + (AD)^2$; therefore $4(BD)^2 = (BD)^2 + (AD)^2$, and so $3(BD)^2 = (AD)^2$.

Τὸ δὴ πρόσθεν ἀσαφῶς ῥηθὲν νῦν μᾶλλον διοριστέον. τὰ γὰρ τέτταρα γένη δι' ἀλλήλων εἰς ἄλληλα ἐφαίνετο πάντα γένεσιν ἔχειν, οὐκ ὀρθῶς φανταζόμενα· γίγνεται μὲν γὰρ ἐκ τῶν τριγώνων ὧν προῃρήμεθα γένη τέτταρα, τρία μὲν ἐξ ἑνὸς τοῦ τὰς πλευρὰς ἀνίσους ἔχοντος, τὸ δὲ τέταρτον ἓν μόνον ἐκ τοῦ ἰσοσκελοῦς τριγώνου ξυναρμοσθέν. οὔκουν δυνατὰ πάντα εἰς ἄλληλα διαλυόμενα ἐκ πολλῶν σμικρῶν ὀλίγα μεγάλα καὶ τοὐναντίον γίγνεσθαι, τὰ δὲ τρία οἷόν τε· ἐκ γὰρ ἑνὸς ἅπαντα πεφυκότα, λυθέντων τε τῶν μειζόνων πολλὰ σμικρὰ ἐκ τῶν αὐτῶν ξυστήσεται, δεχόμενα τὰ προσήκοντα ἑαυτοῖς σχήματα, καὶ σμικρὰ ὅταν αὖ πολλὰ κατὰ τὰ τρίγωνα διασπαρῇ, γενόμενος εἷς ἀριθμὸς ἑνὸς ὄγκου μέγα ἀποτελέσειεν ἂν ἄλλο εἶδος ἕν. ταῦτα μὲν οὖν λελέχθω περὶ τῆς εἰς ἄλληλα γενέσεως.

Οἷον δὲ ἕκαστον αὐτῶν γέγονεν εἶδος καὶ ἐξ ὅσων ξυμπεσόντων ἀριθμῶν, λέγειν ἂν ἑπόμενον εἴη. ἄρξει δὴ τό τε πρῶτον εἶδος καὶ σμικρότατον ξυνιστάμενον· στοιχεῖον δ' αὐτοῦ τὸ τὴν ὑποτείνουσαν τῆς ἐλάττονος πλευρᾶς διπλασίαν ἔχον μήκει· ξύνδυο δὲ τοιούτων κατὰ διάμετρον ξυντιθεμένων καὶ τρὶς τούτου γενομένου, τὰς διαμέτρους καὶ τὰς βραχείας πλευρὰς εἰς ταὐτὸν ὡς κέντρον ἐρεισάντων, ἓν ἰσόπλευρον τρίγωνον ἐξ ἓξ τὸν ἀριθμὸν ὄντων γέγονε· τρίγωνα δὲ

TIMAEUS

Moreover, a point about which our previous statement was obscure must now be defined more clearly. It appeared as if the four Kinds, in being generated, all passed through one another into one another, but this appearance was deceptive. For out of the triangles which we have selected four Kinds are generated, three of them out of that one triangle which has its sides unequal, and the fourth Kind alone composed of the isosceles triangle. Consequently, they are not all capable of being dissolved into one another so as to form a few large bodies composed of many small ones, or the converse; but three of them do admit of this process. For these three are all naturally compounded of one triangle, so that when the larger bodies are dissolved many small ones will form themselves from these same bodies, receiving the shapes that befit them; and conversely, when many small bodies are resolved into their triangles they will produce, when unified, one single large mass of another Kind. So let thus much be declared concerning their generation into one another.

In the next place we have to explain the form in which each Kind has come to exist and the numbers from which it is compounded. First will come that form which is primary and has the smallest components, and the element thereof is that triangle which has its hypotenuse twice as long as its lesser side. And when a pair of such triangles are joined along the line of the hypotenuse, and this is done thrice, by drawing the hypotenuses and the short sides together as to a centre, there is produced from those triangles, six in number, one equilateral

PLATO

54 ἰσόπλευρα ξυνιστάμενα τέτταρα κατὰ σύντρεις
55 ἐπιπέδους γωνίας μίαν στερεὰν γωνίαν ποιεῖ, τῆς ἀμβλυτάτης τῶν ἐπιπέδων γωνιῶν ἐφεξῆς γεγονυῖαν· τοιούτων δὲ ἀποτελεσθεισῶν τεττάρων πρῶτον εἶδος στερεόν, ὅλου περιφεροῦς διανεμητικὸν εἰς ἴσα μέρη καὶ ὅμοια, ξυνίσταται. δεύτερον δὲ ἐκ μὲν τῶν αὐτῶν τριγώνων, κατὰ δὲ ἰσόπλευρα τρίγωνα ὀκτὼ ξυστάντων, μίαν ἀπεργασαμένων στερεὰν γωνίαν· ἐκ τεττάρων ἐπιπέδων· καὶ γενομένων ἐξ τοιούτων τὸ δεύτερον αὖ σῶμα οὕτως
B ἔσχε τέλος. τὸ δὲ τρίτον ἐκ δὶς ἑξήκοντα τῶν στοιχείων ξυμπαγέντων, στερεῶν δὲ γωνιῶν δώδεκα, ὑπὸ πέντε ἐπιπέδων τριγώνων ἰσοπλεύρων περιεχομένης ἑκάστης, εἴκοσι βάσεις ἔχον ἰσοπλεύρους τριγώνους γέγονε.

Καὶ τὸ μὲν ἕτερον ἀπήλλακτο τῶν στοιχείων ταῦτα γεννῆσαν, τὸ δὲ ἰσοσκελὲς τρίγωνον ἐγέννα τὴν τοῦ τετάρτου φύσιν, κατὰ τέτταρα ξυνιστάμενον, εἰς τὸ κέντρον τὰς ὀρθὰς γωνίας ξυνάγον, ἓν ἰσόπλευρον τετράγωνον ἀπεργασάμενον· ἐξ
C δὲ τοιαῦτα ξυμπαγέντα γωνίας ὀκτὼ στερεὰς ἀπετέλεσε, κατὰ τρεῖς ἐπιπέδους ὀρθὰς ξυναρμοσθείσης ἑκάστης· τὸ δὲ σχῆμα τοῦ ξυστάντος

---

[1] As in the figure the equilateral triangle ABC is divided into 6 triangles of unequal sides by joining the vertical points A, B, C to the points of bisection of the opposite sides, viz. D, E, F. Then the hypotenuse in each such triangle is double the shortest side (e.g. AO=2FO). And <FAO=⅓ right angle; while V FOA=⅔ right angle. The "three plane angles" are thus of 60° each =180°= "the most obtuse" plane angle; so that the solid angle is one degree less, i.e. 179°.

TIMAEUS

triangle.[1] And when four equilateral triangles are combined so that three plane angles meet in a point, they form one solid angle, which comes next in order to the most obtuse of the plane angles. And when four such angles are produced, the first solid figure [2] is constructed, which divides the whole of the circumscribed sphere into equal and similar parts. And the second solid [3] is formed from the same triangles, but constructed out of eight equilateral triangles, which produce one solid angle out of four planes; and when six such solid angles have been formed, the second body in turn is completed. And the third solid [4] is composed of twice sixty of the elemental triangles conjoined, and of twelve solid angles, each contained by five plane equilateral triangles, and it has, by its production, twenty equilateral triangular bases.

Now the first of the elemental triangles ceased acting when it had generated these three solids, the substance of the fourth Kind [5] being generated by the isosceles triangle. Four of these combined, with their right angles drawn together to the centre, produced one equilateral quadrangle; and six such quadrangles, when joined together, formed eight solid angles, each composed of three plane right angles; and the shape of the body thus constructed

[2] *i.e.* the tetrahedron or pyramid (molecule of *fire*).
[3] *i.e.* the octahedron (molecule of *air*).
[4] *i.e.* the icosahedron (molecule of *water*).
[5] *i.e.* the cube, composed of 6 × 4 rectangular isosceles triangles (molecule of *earth*).

σώματος γέγονε κυβικόν, ἐξ ἐπιπέδους τετραγώνους ἰσοπλεύρους βάσεις ἔχον. ἔτι δὲ οὔσης ξυστάσεως μιᾶς πέμπτης, ἐπὶ τὸ πᾶν ὁ θεὸς αὐτῇ κατεχρήσατο ἐκεῖνο διαζωγραφῶν.

Ἃ δή τις εἰ πάντα λογιζόμενος ἐμμελῶς ἀποροῖ πότερον ἀπείρους χρὴ κόσμους εἶναι λέγειν ἢ πέρας ἔχοντας, τὸ μὲν ἀπείρους ἡγήσαιτ᾽ ἂν ὄντως D ἀπείρου τινὸς εἶναι δόγμα ὧν ἔμπειρον χρεὼν εἶναι, πότερον δὲ ἕνα ἢ πέντε αὐτοὺς ἀληθείᾳ πεφυκότας λέγειν προσήκει, μᾶλλον ἂν ταύτῃ στὰς εἰκότως διαπορήσαι. τὸ μὲν οὖν δὴ παρ᾽ ἡμῶν ἕνα αὐτὸν κατὰ τὸν εἰκότα λόγον πεφυκότα μηνύει, ἄλλος δὲ εἰς ἄλλα πῃ βλέψας ἕτερα δοξάσει. καὶ τοῦτον[1] μὲν μεθετέον, τὰ δὲ γεγονότα νῦν τῷ λόγῳ γένη διανείμωμεν εἰς πῦρ καὶ γῆν καὶ ὕδωρ καὶ ἀέρα. γῇ μὲν δὴ τὸ κυβικὸν εἶδος δῶμεν· ἀ- E κινητοτάτη γὰρ τῶν τεττάρων γενῶν γῆ καὶ τῶν σωμάτων πλαστικωτάτη, μάλιστα δὲ ἀνάγκη γεγονέναι τοιοῦτον τὸ τὰς βάσεις ἀσφαλεστάτας ἔχον· βάσις δὲ ἥ τε τῶν κατ᾽ ἀρχὰς τριγώνων ὑποτεθέντων ἀσφαλεστέρα κατὰ φύσιν, ἡ τῶν ἴσων πλευρῶν, τῆς τῶν ἀνίσων, τό τε ἐξ ἑκατέρου ξυντεθὲν ἐπίπεδον ἰσόπλευρον ἰσοπλεύρου τετράγωνον τριγώνου κατά τε μέρη καὶ καθ᾽ ὅλον στασιμωτέρως ἐξ ἀνάγκης βέβηκε. διὸ γῇ μὲν τοῦτο ἀπονέμοντες τὸν εἰκότα λόγον διασῴζομεν, ὕδατι δ᾽ αὖ

[1] τοῦτον best MS.: τούτων Zur.

[1] *i.e.* the dodecahedron. How God "used it up" is obscure: the reference may be to the 12 signs of the Zodiac.
[2] There is a play here on the two senses of ἄπειρος, " un-

TIMAEUS

was cubic, having six plane equilateral quadrangular bases. And seeing that there still remained one other compound figure, the fifth,[1] God used it up for the Universe in his decoration thereof.

Now in reasoning about all these things, a man might question whether he ought to affirm the existence of an infinite diversity of Universes or a limited number; and if he questioned aright he would conclude that the doctrine of an infinite diversity is that of a man unversed[2] in matters wherein he ought to be versed; but the question whether they ought really to be described as one Universe or five is one which might with more reason give us pause. Now our view declares the Universe to be essentially one, in accordance with the probable account; but another man, considering other facts, will hold a different opinion. Him, however, we must let pass. But as for the Kinds which have now been generated by our argument, let us assign them severally to fire and earth and water and air. To earth let us give the cubic form; for of the four Kinds earth is the most immobile and the most plastic body, and of necessity the body which has the most stable bases must be pre-eminently of this character. Now of the triangles we originally assumed, the basis formed by equal sides is of its nature more stable than that formed by unequal sides; and of the plane surfaces which are compounded of these several triangles, the equilateral quadrangle, both in its parts and as a whole, has a more stable base than the equilateral triangle. Wherefore, we are preserving the probable account when we assign this figure to earth, and of

limited " and " unskilled "; *cf. Phileb.* 17 E. The doctrine of an infinite number of worlds was held by the Atomists.

τῶν λοιπῶν τὸ δυσκινητότατον εἶδος, τὸ δ' εὐ-
κινητότατον πυρί, τὸ δὲ μέσον ἀέρι· καὶ τὸ μὲν
σμικρότατον σῶμα πυρί, τὸ δ' αὖ μέγιστον ὕδατι,
τὸ δὲ μέσον ἀέρι· καὶ τὸ μὲν ὀξύτατον αὖ πυρί, τὸ
δὲ δεύτερον ἀέρι, τὸ δὲ τρίτον ὕδατι. ταῦτ' οὖν
δὴ πάντα, τὸ μὲν ἔχον ὀλιγίστας βάσεις εὐκινη-
B τότατον ἀνάγκη πεφυκέναι τμητικώτατόν τε καὶ
ὀξύτατον ὂν πάντη πάντων, ἔτι τε ἐλαφρότατον,
ἐξ ὀλιγίστων ξυνεστὸς τῶν αὐτῶν μερῶν· τὸ δὲ
δεύτερον δευτέρως τὰ αὐτὰ ταῦτ' ἔχειν, τρίτως δὲ
τὸ τρίτον.

Ἔστω δὴ κατὰ τὸν ὀρθὸν λόγον καὶ κατὰ
τὸν εἰκότα τὸ μὲν τῆς πυραμίδος στερεὸν γεγονὸς
εἶδος πυρὸς στοιχεῖον καὶ σπέρμα· τὸ δὲ δεύτερον
κατὰ γένεσιν εἴπωμεν ἀέρος, τὸ δὲ τρίτον ὕδατος.
πάντα οὖν δὴ ταῦτα δεῖ διανοεῖσθαι σμικρὰ
C οὕτως, ὡς καθ' ἓν ἕκαστον μὲν τοῦ γένους
ἑκάστου διὰ σμικρότητα οὐδὲν ὁρώμενον ὑφ' ἡμῶν,
ξυναθροισθέντων δὲ πολλῶν τοὺς ὄγκους αὐτῶν
ὁρᾶσθαι. καὶ δὴ καὶ τὸ τῶν ἀναλογιῶν περί τε τὰ
πλήθη καὶ τὰς κινήσεις καὶ τὰς ἄλλας δυνάμεις,
πανταχῇ τὸν θεόν, ὅπηπερ ἡ τῆς ἀνάγκης ἑκοῦσα
πεισθεῖσά τε φύσις ὑπεῖκε, ταύτῃ πάντῃ δι' ἀκριβείας ἀποτελεσθεισῶν ὑπ' αὐτοῦ ξυνηρμόσθαι
ταῦτα ἀνὰ λόγον.

Ἐκ δὴ πάντων ὧν περὶ τὰ γένη προειρήκαμεν,
D ὧδ' ἂν κατὰ τὸ εἰκὸς μάλιστ' ἂν ἔχοι. γῆ μὲν
ξυντυγχάνουσα πυρὶ διαλυθεῖσά τε ὑπὸ τῆς ὀξύτη-
τος αὐτοῦ φέροιτ' ἄν, εἴτ' ἐν αὐτῷ πυρὶ λυθεῖσα
εἴτ' ἐν ἀέρος εἴτ' ἐν ὕδατος ὄγκῳ τύχοι, μέχριπερ

TIMAEUS

the remaining figures the least mobile to water, and the most mobile to fire, and the intermediate figure to air; and, further, when we assign the smallest body to fire, and the greatest to water, and the intermediate to air; and again, the first in point of sharpness to fire, the second to air, and the third to water. As regards all these forms, that which has the fewest bases must necessarily be the most mobile, since it is in all ways the sharpest and most acute of all; and it must also be the lightest, since it is composed of the fewest identical parts; and the second comes second in point of these same qualities, and the third third.

Thus, in accordance with the right account and the probable, that solid which has taken the form of a pyramid shall be the element and seed of fire; the second in order of generation we shall affirm to be air, and the third water. Now one must conceive all these to be so small that none of them, when taken singly each in its several kind, is seen by us, but when many are collected together their masses are seen. And, moreover, as regards the numerical proportions which govern their masses and motions and their other qualities, we must conceive that God realized these everywhere with exactness, in so far as the nature of Necessity submitted voluntarily or under persuasion, and thus ordered all in harmonious proportion.

From all that we have hitherto said about these Kinds, they will, in all likelihood, behave themselves as follows. Earth will keep moving when it happens to meet with fire and has been dissolved by its acuteness, whether this dissolution takes place in pure fire or in a mass of air or of water; and this

PLATO

56
ἂν αὐτῆς πῃ ξυντυχόντα τὰ μέρη πάλιν, ξυναρμοσθέντα αὐτὰ αὑτοῖς, γῆ γένοιτο· οὐ γὰρ εἰς ἄλλο γε εἶδος ἔλθοι ποτ' ἄν. ὕδωρ δὲ ὑπὸ πυρὸς μερισθέν, εἴτε καὶ ὑπ' ἀέρος, ἐγχωρεῖ γίγνεσθαι ξυστάντα ἓν μὲν πυρὸς σῶμα, δύο δὲ ἀέρος. τὰ δὲ
E ἀέρος τμήματα ἐξ ἑνὸς μέρους διαλυθέντος δύ' ἂν γενοίσθην σώματα πυρός. καὶ πάλιν, ὅταν ἀέρι πῦρ ὕδασί τε ἤ τινι γῇ περιλαμβανόμενον, ἐν πολλοῖς ὀλίγον, κινούμενον ἐν φερομένοις, μαχόμενον καὶ νικηθὲν καταθραυσθῇ, δύο πυρὸς σώματα εἰς ἓν ξυνίστασθον εἶδος ἀέρος· καὶ κρατηθέντος ἀέρος κερματισθέντος τε ἐκ δυοῖν ὅλοιν καὶ ἡμίσεος ὕδατος εἶδος ἓν ὅλον ἔσται ξυμπαγές.

57 Ὧδε γὰρ δὴ λογισώμεθα αὐτὰ πάλιν, ὡς ὅταν ἐν πυρὶ λαμβανόμενον τῶν ἄλλων ὑπ' αὐτοῦ τι γένος τῇ τῶν γωνιῶν καὶ κατὰ τὰς πλευρὰς ὀξύτητι τέμνηται, ξυστὰν μὲν εἰς τὴν ἐκείνου φύσιν πέπαυται τεμνόμενον· τὸ γὰρ ὅμοιον καὶ ταὐτὸν αὑτῷ γένος ἕκαστον οὔτε τινὰ μεταβολὴν ἐμποιῆσαι δυνατὸν οὔτε τι παθεῖν ὑπὸ τοῦ κατὰ ταὐτὰ ὁμοίως τε ἔχοντος· ἕως δ' ἂν εἰς ἄλλο τι γιγνόμενον ἧττον ὂν κρείττονι μάχηται, λυόμενον οὐ παύεται. τά
B τε αὖ σμικρότερα ὅταν ἐν τοῖς μείζοσι, πολλοῖς περιλαμβανόμενα ὀλίγα, διαθραυόμενα κατασβεννύηται, ξυνίστασθαι μὲν ἐθέλοντα εἰς τὴν τοῦ κρατοῦντος ἰδέαν πέπαυται κατασβεννύμενα γίγνεταί τε ἐκ πυρὸς ἀήρ, ἐξ ἀέρος ὕδωρ· ἐὰν δ' εἰς ταῦτα ᾖ[1] καὶ τῶν ἄλλων τι ξυνιόντα[2] γενῶν μάχηται,

[1] ταῦτα ᾖ some MSS.: αὐτὰ ἴῃ best MS., Zur.
[2] ξυνιόντα] ξυνιὸν MSS., Zur.

---

[1] The affinity of "like to like" was an axiom in early Greek thought; cf. *Lysis* 215 C ff., *Sympos.* 186 A ff.

TIMAEUS

motion will continue until the particles of earth happen to meet together somewhere and reunite one with another, when they become earth again; for assuredly earth will never change into another form. But water, when broken up by fire or even by air, is capable of becoming a compound of one corpuscle of fire with two of air; and the fractions of air which come from the dissolving of one particle will form two corpuscles of fire. And again, when a small quantity of fire is enclosed by a large quantity of air and water, or of earth, and moves within them as they rush along, and is defeated in its struggle and broken up, then two corpuscles of fire unite to make one form of air. And when air is defeated and disintegrated, from two whole forms of air and a half, one whole form of water will be compounded.

Once again let us reason out their character in this way. Whenever any of the other Kinds is caught within fire it is cut up thereby, owing to the acuteness of its angles and of the line of its sides, but when it has been re-composed into the substance of fire it ceases to be cut; for the Kind that is similar and uniform is in no case able either to cause any change in, or to suffer any affection from, a Kind which is in a uniform and similar state [1]; but so long as, in the course of its passage into another form, it is a weaker body fighting against a stronger, it is continually being dissolved. And again, whenever a few of the smaller corpuscles, being caught within a great number of larger corpuscles, are broken up and quenched, then, if they consent to be re-compounded into the shape of the victorious Kind, they cease to be quenched, and air is produced out of fire, and out of air water; but if they fight against combining with

λυόμενα οὐ παύεται, πρὶν ἢ παντάπασιν ὠθούμενα καὶ διαλυθέντα ἐκφύγῃ πρὸς τὸ ξυγγενές, ἢ νικηθέντα, ἓν ἐκ πολλῶν ὅμοιον τῷ κρατήσαντι γενόμενον, αὐτοῦ ξύνοικον μείνῃ. καὶ δὴ καὶ κατὰ ταῦτα τὰ παθήματα διαμείβεται τὰς χώρας ἅπαντα· διέστηκε μὲν γὰρ τοῦ γένους ἑκάστου τὰ πλήθη κατὰ τόπον ἴδιον διὰ τὴν τῆς δεχομένης κίνησιν, τὰ δὲ ἀνομοιούμενα ἑκάστοτε ἑαυτοῖς, ἄλλοις δὲ ὁμοιούμενα φέρεται διὰ τὸν σεισμὸν πρὸς τὸν ἐκείνων οἷς ἂν ὁμοιωθῇ τόπον.

Ὅσα μὲν οὖν ἄκρατα καὶ πρῶτα σώματα, διὰ τοιούτων αἰτιῶν γέγονε· τοῦ δ᾽ ἐν τοῖς εἴδεσιν αὐτῶν ἕτερα ἐμπεφυκέναι γένη τὴν ἑκατέρου τῶν στοιχείων αἰτιατέον ξύστασιν, μὴ μόνον ἓν ἑκατέραν μέγεθος ἔχον τὸ τρίγωνον φυτεῦσαι κατ᾽ ἀρχὰς ἀλλὰ ἐλάττω τε καὶ μείζω, τὸν ἀριθμὸν δὲ ἔχοντα τοσοῦτον, ὅσαπερ ἂν ᾖ τἀν τοῖς εἴδεσι γένη. διὸ δὴ ξυμμιγνύμενα αὐτά τε πρὸς αὑτὰ καὶ πρὸς ἄλληλα τὴν ποικιλίαν ἐστὶν ἄπειρα· ἧς δὴ δεῖ θεωροὺς γίγνεσθαι τοὺς μέλλοντας περὶ φύσεως εἰκότι λόγῳ χρήσεσθαι.

Κινήσεως οὖν στάσεώς τε πέρι, τίνα τρόπον καὶ μεθ᾽ ὧντινων γίγνεσθον, εἰ μή τις διομολογήσεται, πόλλ᾽ ἂν εἴη ἐμποδὼν τῷ κατόπισθεν λογισμῷ. τὰ μὲν οὖν ἤδη περὶ αὐτῶν εἴρηται, πρὸς δὲ

---

[1] The elements are conceived as having their proper abodes in concentric strata of space, one above another—earth in the centre, water next, then air, and fire at the circumference of the World-Sphere.

TIMAEUS

these or with any of the other Kinds, they do not cease from dissolution until either they are driven out to their own kindred, by means of this impact and dissolution, or else they are defeated and, instead of many forms, assume one form similar to the victorious Kind, and continue dwelling therewith as a united family. Moreover, it is owing to these affections that they all interchange their places; for while the bulk of each Kind keeps apart in a region of its own [1] because of the motion of the Recipient, yet those corpuscles which from time to time become dissimilar to themselves and similar to others are carried, because of the shaking, towards the region which belongs to those corpuscles whereto they have been assimilated.

Such are the causes which account for the generation of all the unmixed and primary bodies. But within these four Kinds other classes exist, whereof the cause must be sought in the construction of each of the two elemental triangles, each such construction having originally produced not merely a triangle of one definite size, but larger and smaller triangles of sizes as numerous as are the classes within the Kinds. Consequently, when these are combined amongst themselves and with one another they are infinite in their variety; and this variety must be kept in view by those who purpose to employ probable reasoning concerning Nature.

Now, unless we can arrive at some agreed conclusion concerning Motion and Rest, as to how and under what conditions they come about, our subsequent argument will be greatly hampered. The facts about them have already been stated in part;

57
ἐκείνοις ἔτι τάδε, ἐν μὲν ὁμαλότητι μηδέποτε
ἐθέλειν κίνησιν ἐνεῖναι. τὸ γὰρ κινησόμενον ἄνευ
τοῦ κινήσοντος ἢ τὸ κινῆσον ἄνευ τοῦ κινησομένου
χαλεπόν, μᾶλλον δὲ ἀδύνατον, εἶναι· κίνησις δὲ οὐκ
ἔστι τούτων ἀπόντων· ταῦτα δὲ ὁμαλὰ εἶναί ποτε
ἀδύνατον. οὕτω δὴ στάσιν μὲν ἐν ὁμαλότητι,
58 κίνησιν δὲ εἰς ἀνωμαλότητα ἀεὶ τιθῶμεν. αἰτία
δὲ ἡ ἀνισότης αὖ τῆς ἀνωμάλου φύσεως. ἀν-
ισότητος δὲ γένεσιν μὲν διεληλύθαμεν· πῶς δέ
ποτε οὐ κατὰ γένη διαχωρισθέντα ἕκαστα πέπαυται
τῆς δι' ἀλλήλων κινήσεως καὶ φορᾶς, οὐκ εἴπομεν.
ὧδε οὖν πάλιν ἐροῦμεν. ἡ τοῦ παντὸς περίοδος,
ἐπειδὴ συμπεριέλαβε τὰ γένη, κυκλοτερὴς οὖσα
καὶ πρὸς αὑτὴν πεφυκυῖα βούλεσθαι ξυνιέναι,
σφίγγει πάντα καὶ κενὴν χώραν οὐδεμίαν ἐᾷ
B λείπεσθαι. διὸ δὴ πῦρ μὲν εἰς ἅπαντα διελήλυθε
μάλιστα, ἀὴρ δὲ δεύτερον, ὡς λεπτότητι δεύτερον
ἔφυ, καὶ τἆλλα ταύτῃ· τὰ γὰρ ἐκ μεγίστων μερῶν
γεγονότα μεγίστην κενότητα ἐν τῇ ξυστάσει παρα-
λέλοιπε, τὰ δὲ σμικρότατα ἐλαχίστην. ἡ δὴ τῆς
πιλήσεως ξύνοδος τὰ σμικρὰ εἰς τὰ τῶν μεγάλων
διάκενα ξυνωθεῖ. σμικρῶν οὖν παρὰ μεγάλα τιθε-
μένων καὶ τῶν ἐλαττόνων τὰ μείζονα διακρινόντων,
τῶν δὲ μειζόνων ἐκεῖνα συγκρινόντων, πάντ' ἄνω
C κάτω μεταφέρεται πρὸς τοὺς ἑαυτῶν τόπους· μετα-
βάλλον γὰρ τὸ μέγεθος ἕκαστον καὶ τὴν τῶν τόπων
μεταβάλλει στάσιν. οὕτω δὴ διὰ ταῦτά τε ἡ τῆς
ἀνωμαλότητος διασωζομένη γένεσις ἀεὶ τὴν ἀεὶ

---

[1] *Cf.* 53 c ff.: the varying shapes and sizes of the primary triangles account for the " inequality."

[2] *i.e.* exerts a centripetal force. For this " compression " *cf.* Emped. *Frag.* 185 Τιτὰν ἠδ' αἰθὴρ σφίγγων περὶ κύκλον ἅπαντα.

TIMAEUS

but in addition thereto we must state further that motion never consents to exist within uniformity. For it is difficult, or rather impossible, for that which is to be moved to exist without that which is to move, or that which is to move without that which is to be moved; but in the absence of these there is no motion, and that these should ever be uniform is a thing impossible. Accordingly, we must always place rest in uniformity, and motion in non-uniformity; and the cause of the non-uniform nature lies in inequality. Now we have explained the origin of inequality [1]; but we have not declared how it is that these bodies are not separated according to their several Kinds, and cease not from their motion and passage one through another. Wherefore, we shall once more expound the matter as follows. The revolution of the All, since it comprehends the Kinds, compresses them all, seeing that it is circular and tends naturally to come together to itself [2]; and thus it suffers no void place to be left. Wherefore, fire most of all has permeated all things, and in a second degree air, as it is by nature second in fineness; and so with the rest; for those that have the largest constituent parts have the largest void left in their construction, and those that have the smallest the least. Thus the tightening of the compression forces together the small bodies into the void intervals of the large. Therefore, when small bodies are placed beside large, and the smaller disintegrate the larger while the larger unite the smaller, they all shift up and down towards their own proper regions; for the change in their several sizes causes their position in space also to change. And since in this way and for these reasons the production of non-uniformity is

κίνησιν τούτων οὖσαν ἐσομένην τε ἐνδελεχῶς παρέχεται.

Μετὰ δὴ ταῦτα δεῖ νοεῖν ὅτι πυρός τε γένη πολλὰ γέγονεν, οἷον φλόξ τό τε ἀπὸ τῆς φλογὸς ἀπιόν, ὃ κάει μὲν οὔ, φῶς δὲ τοῖς ὄμμασι παρέχει, τό τε φλογὸς ἀποσβεσθείσης ἐν τοῖς διαπύροις καταλειπόμενον αὐτοῦ. κατὰ ταὐτὰ δὲ ἀέρος τὸ μὲν εὐαγέστατον ἐπίκλην αἰθὴρ καλούμενος, ὁ δὲ θολερώτατος ὁμίχλη τε καὶ σκότος, ἕτερά τε ἀνώνυμα εἴδη γεγονότα διὰ τὴν τῶν τριγώνων ἀνισότητα. τὰ δὲ ὕδατος διχῇ μὲν πρῶτον, τὸ μὲν ὑγρόν, τὸ δὲ χυτὸν γένος αὐτοῦ. τὸ μὲν οὖν ὑγρὸν διὰ τὸ μετέχον εἶναι τῶν γενῶν τῶν ὑδάτων, ὅσα σμικρά, ἀνίσων ὄντων, κινητὸν αὐτό τε καθ' αὑτὸ καὶ ὑπ' ἄλλου διὰ τὴν ἀνωμαλότητα καὶ τὴν τοῦ σχήματος ἰδέαν γέγονε· τὸ δ' ἐκ μεγάλων καὶ ὁμαλῶν στασιμώτερον μὲν ἐκείνου καὶ βαρὺ πεπηγὸς ὑπὸ ὁμαλότητός ἐστιν, ὑπὸ δὲ πυρὸς εἰσιόντος καὶ διαλύοντος αὐτὸ τὴν ὁμαλότητα ⟨ἀποβάλλει, ταύτην δὲ⟩[1] ἀπολέσαν μετίσχει μᾶλλον κινήσεως, γενόμενον δὲ εὐκίνητον, ὑπὸ τοῦ πλησίον ἀέρος ὠθούμενον καὶ κατατεινόμενον ἐπὶ γῆν, τήκεσθαι μὲν τὴν τῶν ὄγκων καθαίρεσιν, ῥοὴν δὲ τὴν κατάτασιν ἐπὶ γῆν ἐπωνυμίαν ἑκατέρου τοῦ πάθους ἔλαβε. πάλιν δὲ ἐκπίπτοντος αὐτόθεν τοῦ πυρός, ἅτε οὐκ εἰς κενὸν ἐξιόντος, ὠθούμενος ὁ πλησίον ἀὴρ εὐκίνητον ὄντα ἔτι τὸν ὑγρὸν ὄγκον εἰς τὰς τοῦ πυρὸς ἕδρας ξυνωθῶν αὐτὸν αὑτῷ ξυμμίγνυσιν· ὁ δὲ ξυνωθούμενος ἀπολαμβάνων τε

[1] ἀποβάλλει, ταύτην δὲ added by ms. corr.: om. Z.

---

[1] *i.e.* metals are classed as "water," *cf.* 59 B ff.

TIMAEUS

perpetually maintained, it brings about unceasingly, both now and for the future, the perpetual motion of these bodies.

In the next place, we must observe that there are many kinds of fire : for example, there is flame ; and the kind issuing from flame, which does not burn but supplies light to the eyes ; and the kind which, when the flame is quenched, is left behind among the embers. So likewise of air, there is the most translucent kind which is called by the name of aether, and the most opaque which is mist and darkness, and other species without a name, which are produced by reason of the inequality of the triangles. The kinds of water are, primarily, two, the one being the liquid, the other the fusible[1] kind. Now the liquid kind, inasmuch as it partakes of those small particles of water which are unequal, is mobile both in itself and by external force owing to its non-uniformity and the shape of its figure. But the other kind, which is composed of large and uniform particles, is more stable than the first and is heavy, being solidified by its uniformity ; but when fire enters and dissolves it, this causes it to abandon its uniformity, and this being lost it partakes more largely in motion ; and when it has become mobile it is pushed by the adjacent air and extended upon the earth ; and for each of these modifications it has received a descriptive name—" melting " for the disintegration of its masses, and for its extension over the earth " fluidity." Again, since the fire on issuing from the water does not pass into a void but presses on the adjacent air, this in turn compresses the liquid mass which is still mobile into the abodes of the fire and combines it with itself ; and the mass, being thus

145

PLATO

59 τὴν ὁμαλότητα πάλιν, ἅτε τοῦ τῆς ἀνωμαλότητος δημιουργοῦ πυρὸς ἀπιόντος, εἰς ταὐτὸν αὑτῷ καθίσταται. καὶ τὴν μὲν τοῦ πυρὸς ἀπαλλαγὴν ψύξιν, τὴν δὲ ξύνοδον ἀπελθόντος ἐκείνου πεπηγὸς εἶναι γένος προσερρήθη.

B Τούτων δὴ πάντων, ὅσα χυτὰ προσείπομεν ὕδατα, τὸ μὲν ἐκ λεπτοτάτων καὶ ὁμαλωτάτων πυκνότατον γιγνόμενον, μονοειδὲς γένος, στίλβοντι καὶ ξανθῷ χρώματι κοινωθέν, τιμαλφέστατον κτῆμα χρυσὸς ἠθημένος διὰ πέτρας ἐπάγη. χρυσοῦ δὲ ὄζος διὰ πυκνότητα σκληρότατον ὂν καὶ μελανθέν, ἀδάμας ἐκλήθη. τὸ δ' ἐγγὺς μὲν χρυσοῦ τῶν μερῶν, εἴδη δὲ πλέονα ἑνὸς ἔχον, πυκνότητι μὲν χρυσοῦ πυκνότερον ὄν, καὶ γῆς μόριον ὀλίγον καὶ λεπτὸν μετασχόν, ὥστε σκληρότερον εἶναι, τῷ δὲ
C μεγάλα ἐντὸς αὑτοῦ διαλείμματα ἔχειν κουφότερον, τῶν λαμπρῶν πηκτῶν τε ἓν γένος ὑδάτων χαλκὸς ξυσταθεὶς γέγονε. τὸ δ' ἐκ γῆς αὐτῷ μιχθέν, ὅταν παλαιουμένῳ διαχωρίζησθον πάλιν ἀπ' ἀλλήλων, ἐκφανὲς καθ' αὑτὸ γιγνόμενον ἰὸς λέγεται.

Τἆλλα δὲ τῶν τοιούτων οὐδὲν ποικίλον ἔτι διαλογίσασθαι τὴν τῶν εἰκότων μύθων μεταδιώκοντα ἰδέαν, ἣν ὅταν τις ἀναπαύσεως ἕνεκα τοὺς περὶ τῶν ὄντων ἀεὶ καταθέμενος λόγους, τοὺς γενέσεως
D πέρι διαθεώμενος εἰκότας ἀμεταμέλητον ἡδονὴν κτᾶται, μέτριον ἂν ἐν τῷ βίῳ παιδιὰν καὶ φρόνιμον ποιοῖτο. ταύτῃ δὴ καὶ τὰ νῦν ἀφέντες τὸ μετὰ

[1] Perhaps haematite or platinum.
[2] Cf. 29 B, D, 48 C, etc.

TIMAEUS

compressed and recovering again its uniformity, because of the departure of the fire, the author of its non-uniformity, returns to its state of self-identity. And this cessation of the fire is termed " cooling," and the combination which follows on its departure " solidification."

Of all the kinds of water which we have termed " fusible," the densest is produced from the finest and most uniform particles: this is a kind of unique form, tinged with a glittering and yellow hue, even that most precious of possessions, " gold," which has been strained through stones and solidified. And the off-shoot of gold, which is very hard because of its density and black in colour, is called " adamant."[1] And the kind which closely resembles gold in its particles but has more forms than one, and in density is more dense than gold, and partakes of small and fine portions of earth so that it is harder, while it is also lighter owing to its having large interstices within it,—this particular kind of the bright and solid waters, being compounded thus, is termed " bronze." And the portion of earth that is mixed therewith becomes distinct by itself, when both grow old and separate again each from the other; and then it is named " rust."

And the rest of such phenomena it is no longer difficult to explain in full, if one aims at framing a description that is probable.[2] For as regards this, whenever for the sake of recreation a man lays aside arguments concerning eternal Realities and considers probable accounts of Becoming, gaining thereby a pleasure not to be repented of, he provides for his life a pastime that is both moderate and sensible. To this pastime let us now give free play, and

PLATO

τοῦτο τῶν αὐτῶν πέρι τὰ ἑξῆς εἰκότα δίιμεν τῇδε.

Τὸ πυρὶ μεμιγμένον ὕδωρ, ὅσον λεπτὸν ὑγρόν τε διὰ τὴν κίνησιν καὶ τὴν ὁδὸν ἣν κυλινδούμενον ἐπὶ γῆς ὑγρὸν λέγεται, μαλακόν τε αὖ τῷ τὰς βάσεις ἧττον ἑδραίους οὔσας ἢ τὰς γῆς ὑπείκειν, τοῦτο ὅταν πυρὸς ἀποχωρισθὲν ἀέρος τε μονωθῇ, γέγονε μὲν ὁμαλώτερον, ξυνέωσται δὲ ὑπὸ τῶν ἐξιόντων εἰς αὐτό, παγέν τε οὕτω τὸ μὲν ὑπὲρ γῆς μάλιστα παθὸν ταῦτα χάλαζα, τὸ δ' ἐπὶ γῆς κρύσταλλος, τὸ δὲ ἧττον ἡμιπαγές τε ὂν ἔτι, τὸ μὲν ὑπὲρ γῆς αὖ χιών, τὸ δ' ἐπὶ γῆς ξυμπαγέν, ἐκ δρόσου γενόμενον, πάχνη λέγεται.

Τὰ δὲ δὴ πλεῖστα ὑδάτων εἴδη μεμιγμένα ἀλλήλοις, ξύμπαν μὲν τὸ γένος, διὰ τῶν ἐκ γῆς φυτῶν ἠθημένα, χυμοὶ λεγόμενοι· διὰ δὲ τὰς μίξεις ἀνομοιότητα ἕκαστοι σχόντες τὰ μὲν ἄλλα πολλὰ ἀνώνυμα γένη παρέσχοντο, τέτταρα δέ, ὅσα ἔμπυρα εἴδη, διαφανῆ μάλιστα γενόμενα εἴληφεν ὀνόματα αὐτῶν, τὸ μὲν τῆς ψυχῆς μετὰ τοῦ σώματος θερμαντικὸν οἶνος, τὸ δὲ λεῖον καὶ διακριτικὸν ὄψεως διὰ ταῦτά τε ἰδεῖν λαμπρὸν καὶ στίλβον λιπαρόν τε φανταζόμενον ἐλαιηρὸν εἶδος, πίττα καὶ κίκι καὶ ἔλαιον αὐτὸ ὅσα τ' ἄλλα τῆς αὐτῆς δυνάμεως· ὅσον δὲ διαχυτικὸν μέχρι φύσεως τῶν περὶ τὸ στόμα ξυνόδων, ταύτῃ τῇ

---

[1] Alluding to a fanciful derivation of ὑγρόν from ὑπὲρ γῆν ῥέον.

TIMAEUS

proceed to expound in order the subsequent probabilities concerning these same phenomena in the following way.

The water that is mixed with fire, which is fine and fluid, is termed " fluid," owing to its motion and the way it rolls over the earth.[1] Also it is soft owing to the fact that its bases, being less stable than those of earth, give way. When this kind is separated off from fire and air and isolated it becomes more uniform, but because of their outflow it is compressed upon itself; and when it is thus solidified, the part of it above the earth which is most affected by this process is termed " hail," and the part upon the earth " ice "; and the part which is less affected and is still only half-solid is called " snow " when it is above the earth, but when it is upon the earth and solidified out of dew it is called " hoar-frost."

Now as regards most forms of water that are intermingled one with another, the kind as a whole, consisting of water that has been strained through earth-grown plants, is called " sap "; but inasmuch as the several sorts have become dissimilar owing to intermixture, most of the kinds thus produced are unnamed. Four of these kinds, however, being fiery and specially conspicuous, have received names. Of these, that which is heating to the soul as well as the body is called " wine "; that which is smooth and divisive of the vision, and therefore bright to look upon and gleaming and glistening in appearance, is the species " oil," including pitch and castor oil and olive oil itself and all the others that are of the same character; and all that kind which tends to expand the contracted parts of the mouth, so far as their nature allows, and by this property produces sweet-

PLATO

60
δυνάμει γλυκύτητα παρεχόμενον, μέλι τὸ κατὰ πάντων μάλιστα πρόσρημα ἔσχε· τὸ δὲ τῆς σαρκὸς διαλυτικὸν τῷ κάειν ἀφρῶδες γένος, ἐκ πάντων ἀφορισθὲν τῶν χυμῶν, ὀπὸς ἐπωνομάσθη.

Γῆς δὲ εἴδη, τὸ μὲν ἠθημένον δι' ὕδατος τοιῷδε τρόπῳ γίγνεται σῶμα λίθινον. τὸ ξυμμιγὲς ὕδωρ ὅταν ἐν τῇ ξυμμίξει κοπῇ, μετέβαλεν εἰς ἀέρος ἰδέαν· γενόμενος δὲ ἀὴρ εἰς τὸν ἑαυτοῦ τόπον
C ἀναθεῖ. κενὸν δ' ὑπερεῖχεν[1] αὐτῶν οὐδέν· τὸν οὖν πλησίον ἔωσεν ἀέρα. ὁ δέ, ἅτε ὢν βαρύς, ὠσθεὶς καὶ περιχυθεὶς τῷ τῆς γῆς ὄγκῳ σφόδρα ἔθλιψε ξυνέωσέ τε αὐτὸν εἰς τὰς ἕδρας ὅθεν ἀνῄειν ὁ νέος ἀήρ. ξυνωσθεῖσα δὲ ὑπ' ἀέρος ἀλύτως ὕδατι γῆ ξυνίσταται πέτρα, καλλίων μὲν ἡ τῶν ἴσων καὶ ὁμαλῶν διαφανὴς μερῶν, αἰσχίων δὲ ἡ ἐναντία. τὸ δὲ ὑπὸ πυρὸς τάχους τὸ νοτερὸν πᾶν ἐξαρ-
D πασθὲν καὶ κραυρότερον ἐκείνου ξυστάν, ᾧ γένει κέραμον ἐπωνομάκαμεν, τοῦτο γέγονεν. ἔστι δὲ ὅτε νοτίδος ὑπολειφθείσης χυτὴ γῆ γενομένη διὰ πυρός, ὅταν ψυχθῇ, γίγνεται τὸ μέλαν χρῶμα ἔχον εἶδος[2]. τὼ[3] δ' αὖ κατὰ ταὐτὰ μὲν ταῦτα ἐκ ξυμμίξεως ὕδατος ἀπομονουμένω πολλοῦ, λεπτοτέρων δὲ ἐκ γῆς μερῶν ἁλμυρώ τε ὄντε, ἡμιπαγῆ γενομένω καὶ λυτὼ πάλιν ὑφ' ὕδατος, τὸ μὲν ἐλαίου καὶ γῆς καθαρτικὸν γένος λίτρον, τὸ δ' εὐάρμοστον ἐν ταῖς κοινωνίαις ταῖς περὶ τὴν τοῦ

---

[1] ὑπερεῖχεν best mss.: ὑπῆρχεν Zur.
[2] εἶδος Hermann: λίθος mss., Zur.
[3] τὼ Schneider: τῷ mss. (the following duals also being dat. in mss.).

---

[1] Perhaps a kind of fig-juice.
[2] *i.e.* potash or saltpetre.

TIMAEUS

ness, has received as a general designation the name of "honey"; and the foamy kind, which tends to dissolve the flesh by burning, and is secreted from all the saps, is named "verjuice."[1]

Of the species of earth, that which is strained through water becomes a stony substance in the following way. When the water commingled therewith is divided in the process of mingling, it changes into the form of air; and when it has become air it rushes up to its own region; but because there was no void space above them, therefore it pressed against the adjacent air; and it, being heavy, when pressed and poured round the mass of earth, crushed it forcibly and compressed it into the spaces from which the new air was ascending. But when earth is thus compressed by the air so as to be indissoluble by water it forms "stone"; of which the fairer sort is that composed of equal and uniform parts and transparent, and the coarser sort the opposite. That kind from which all the moisture has been carried off by the rapidity of fire, and which is more brittle in its composition than the first kind, is the kind to which we have given the name of "earthenware." But sometimes, when moisture is still left in the earth and it has been fused by fire and has cooled again, it forms the species which is black in hue. On the other hand there are two kinds, which, in exactly the same manner, are isolated after the mixture from much of their water, but are composed of finer parts of earth, and are saline: when these have become semi-solid and soluble again by water, one of them is purgative of oil and earth and forms the species called "lye"[2]; and the other, which blends well with the combinations which affect the sensation of the mouth, is that

60
E στόματος αἴσθησιν ἁλῶν κατὰ λόγον νόμου θεοφιλὲς σῶμα ἐγένετο.

Τὰ δὲ κοινὰ ἐξ ἀμφοῖν ὕδατι μὲν οὐ λυτά, πυρὶ δέ, διὰ τὸ τοιόνδε οὕτω ξυμπήγνυται. γῆς ὄγκους πῦρ μὲν ἀήρ τε οὐ τήκει· τῆς γὰρ ξυστάσεως τῶν διακένων αὐτῆς σμικρομερέστερα πεφυκότα, διὰ πολλῆς εὐρυχωρίας ἰόντα, οὐ βιαζόμενα, ἄλυτον αὐτὴν ἐάσαντα ἄτηκτον παρέσχε· τὰ δὲ ὕδατος, ἐπειδὴ μείζω πέφυκε μέρη βίαιον ποιούμενα τὴν
61 διέξοδον, λύοντα αὐτὴν τήκει. γῆν μὲν γὰρ ἀξύστατον ὑπὸ βίας οὕτως ὕδωρ μόνον λύει, ξυνεστηκυῖαν δὲ πλὴν πυρὸς οὐδέν· εἴσοδος γὰρ οὐδενὶ πλὴν πυρὶ λέλειπται. τὴν δὲ ὕδατος αὖ ξύνοδον τὴν μὲν βιαιοτάτην πῦρ μόνον, τὴν δὲ ἀσθενεστέραν ἀμφότερα, πῦρ τε καὶ ἀήρ, διαχεῖτον, ὁ μὲν κατὰ τὰ διάκενα, τὸ δὲ καὶ κατὰ τὰ τρίγωνα· βίᾳ δὲ ἀέρα ξυστάντα οὐδὲν λύει πλὴν κατὰ τὸ στοιχεῖον, ἀβίαστον δὲ κατατήκει μόνον πῦρ.

Τὰ δὲ δὴ τῶν ξυμμίκτων ἐκ γῆς τε καὶ ὕδατος
B σωμάτων, μέχριπερ ἂν ὕδωρ αὐτοῦ τὰ τῆς γῆς διάκενα καὶ βίᾳ ξυμπεπιλημένα κατέχῃ, τὰ μὲν ὕδατος ἐπιόντα ἔξωθεν εἴσοδον οὐκ ἔχοντα μέρη περιρρέοντα τὸν ὅλον ὄγκον ἄτηκτον εἴασε, τὰ δὲ πυρὸς εἰς τὰ τῶν ὑδάτων διάκενα εἰσιόντα, ὅπερ ὕδωρ γῆν, τοῦθ' ὕδωρ[1] ἀπεργαζόμενα, τηχθέντι τῷ κοινῷ σώματι ῥεῖν μόνα αἴτια ξυμβέβηκε. τυγχάνει δὲ ταῦτα ὄντα, τὰ μὲν ἔλαττον ἔχοντα

[1] τοῦθ' ὕδωρ Archer-Hind conj.: τοῦτο πῦρ ἀέρα mss., Zur.

---

[1] *Cf.* Hom. *Il.* ix. 214 πάσσε δ' ἁλὸς θείοιο.

TIMAEUS

substance which is customarily termed " beloved of the gods,"[1] namely " salt."

As regards the kinds which are a blend of these two, and are dissoluble by fire and not by water, their composition is due to the following cause. Fire and air do not melt masses of earth; for, inasmuch as their particles are smaller then the interstices of its structure, they have room to pass through without forcible effort and leave the earth undissolved, with the result that it remains unmelted; whereas the particles of water, being larger, must use force to make their way out, and consequently dissolve and melt the earth. Thus earth when it is not forcibly condensed is dissolved only by water; and when it is condensed it is dissolved by fire only, since no entrance is left for anything save fire. Water, again, when most forcibly massed together is dissolved by fire only, but when massed less forcibly both by fire and air, the latter acting by way of the interstices, and the former by way of the triangles; but air when forcibly condensed is dissolved by nothing save by way of its elemental triangles, and when unforced it is melted down by fire only.

As regards the classes of bodies which are compounds of earth and water, so long as the water occupies the interspaces of earth which are forcibly contracted, the portions of water which approach from without find no entrance, but flow round the whole mass and leave it undissolved. But when portions of fire enter into the interspaces of the water they produce the same effects on water as water does on earth; consequently, they are the sole causes why the compound substance is dissolved and flows. And of these substances those which contain less water

ὕδατος ἢ γῆς τό τε περὶ τὴν ὕαλον γένος ἅπαν ὅσα τε λίθων χυτὰ εἴδη καλεῖται, τὰ δὲ πλέον ὕδατος αὖ πάντα ὅσα κηροειδῆ καὶ θυμιατικὰ σώματα ξυμπήγνυται.

Καὶ τὰ μὲν δὴ σχήμασι[1] κοινωνίαις τε καὶ μεταλλαγαῖς εἰς ἄλληλα πεποικιλμένα εἴδη σχεδὸν ἐπιδέδεικται, τὰ δὲ παθήματα αὐτῶν δι' ἃς αἰτίας γέγονε πειρατέον ἐμφανίζειν. πρῶτον μὲν οὖν ὑπάρχειν αἴσθησιν δεῖ τοῖς λεγομένοις ἀεί· σαρκὸς δὲ καὶ τῶν περὶ σάρκα γένεσιν, ψυχῆς τε ὅσον θνητόν, οὔπω διεληλύθαμεν. τυγχάνει δὲ οὔτε ταῦτα χωρὶς τῶν περὶ τὰ παθήματα ὅσα αἰσθητικά, οὔτ' ἐκεῖνα ἄνευ τούτων δυνατὰ ἱκανῶς λεχθῆναι· τὸ δὲ ἅμα σχεδὸν οὐ δυνατόν. ὑποθετέον δὴ πρότερον θάτερα, τὰ δ' ὑποτεθέντα ἐπάνιμεν αὖθις. ἵνα οὖν ἑξῆς τὰ παθήματα λέγηται τοῖς γένεσιν, ἔστω πρότερα ἡμῖν τὰ περὶ σῶμα καὶ ψυχὴν ὄντα.

Πρῶτον μὲν οὖν ᾗ πῦρ θερμὸν λέγομεν, ἴδωμεν ὧδε σκοποῦντες, τὴν διάκρισιν καὶ τομὴν αὐτοῦ περὶ τὸ σῶμα ἡμῶν γιγνομένην ἐννοηθέντες. ὅτι μὲν γὰρ ὀξύ τι τὸ πάθος, πάντες σχεδὸν αἰσθανόμεθα· τὴν δὲ λεπτότητα τῶν πλευρῶν καὶ γωνιῶν ὀξύτητα τῶν τε μορίων σμικρότητα καὶ τῆς φορᾶς τὸ τάχος, οἷς πᾶσι σφοδρὸν ὂν καὶ τομὸν ὀξέως τὸ προστυχὸν ἀεὶ τέμνει, λογιστέον ἀναμιμνησκομένοις τὴν τοῦ σχήματος αὐτοῦ γένεσιν, ὅτι μάλιστα ἐκείνη καὶ οὐκ ἄλλη φύσις διακρίνουσα ἡμῶν κατὰ σμικρά τε τὰ σώματα κερματίζουσα

[1] σχήμασι mss.: σχήματα Zur.

than earth form the whole kind known as "glass," and all the species of stone called "fusible"; while those which contain more water include all the solidified substances of the type of wax and frankincense.

And now we have explained with some fullness the Four Kinds, which are thus variegated in their shapes and combinations and permutations; but we have still to try to elucidate the Causes which account for their affective qualities. Now, first of all, the quality of sense-perceptibility must always belong to the objects under discussion; but we have not as yet described the generation of flesh and the appurtenances of flesh, nor of that portion of Soul which is mortal. But, in truth, these last cannot be adequately explained apart from the subject of the sensible affections, nor the latter without the former; while to explain both simultaneously is hardly possible. Therefore, we must assume one of the two, to begin with, and return later to discuss our assumptions. In order, then, that the affective properties may be treated next after the kinds, let us presuppose the facts about body and soul.

Firstly, then, let us consider how it is that we call fire "hot" by noticing the way it acts upon our bodies by dividing and cutting. That its property is one of sharpness we all, I suppose, perceive; but as regards the thinness of its sides and the acuteness of its angles and the smallness of its particles and the rapidity of its motion—owing to all which properties fire is intense and keen and sharply cuts whatever it encounters,—these properties we must explain by recalling the origin of its form, how that it above all others is the one substance which so divides our bodies and minces them up as to produce naturally

τοῦτο ὃ νῦν θερμὸν λέγομεν εἰκότως τὸ πάθημα καὶ τοὔνομα παρέσχε.

Τὸ δ' ἐναντίον τούτῳ κατάδηλον μέν, ὅμως δὲ μηδὲν ἐπιδεὲς ἔστω λόγου. τὰ γὰρ δὴ τῶν περὶ τὸ σῶμα ὑγρῶν μεγαλομερέστερα εἰσιόντα, τὰ σμικρότερα ἐξωθοῦντα, εἰς τὰς ἐκείνων οὐ δυνάμενα ἕδρας ἐνδῦναι, ξυνωθοῦντα ἡμῶν τὸ νοτερόν, ἐξ B ἀνωμάλου κεκινημένου τε ἀκίνητον δι' ὁμαλότητα καὶ τὴν ξύνωσιν ἀπεργαζόμενα πήγνυσι· τὸ δὲ παρὰ φύσιν ξυναγόμενον μάχεται κατὰ φύσιν αὐτὸ ἑαυτὸ εἰς τοὐναντίον ἀπωθοῦν. τῇ δὴ μάχῃ καὶ τῷ σεισμῷ τούτῳ τρόμος καὶ ῥῖγος ἐτέθη, ψυχρόν τε τὸ πάθος ἅπαν τοῦτο καὶ τὸ δρῶν αὐτὸ ἔσχεν ὄνομα.

Σκληρὸν δέ, ὅσοις ἂν ἡμῶν ἡ σὰρξ ὑπείκῃ· μαλακὸν δέ, ὅσα ἂν τῇ σαρκί· πρὸς ἄλληλά τε οὕτως. ὑπείκει δὲ ὅσον ἐπὶ σμικροῦ βαίνει· τὸ C δὲ ἐκ τετραγώνων ὂν βάσεων, ἅτε βεβηκὸς σφόδρα, ἀντιτυπώτατον εἶδος, ὅ τί τε ἂν εἰς πυκνότητα ξυνιὸν πλείστην ἀντίτονον ᾖ μάλιστα.

Βαρὺ δὲ καὶ κοῦφον μετὰ τῆς κάτω φύσεως ἄνω τε λεγομένης ἐξεταζόμενον ἂν δηλωθείη σαφέστατα. φύσει γὰρ δή τινας τόπους δύο εἶναι διειληφότας διχῇ τὸ πᾶν ἐναντίους, τὸν μὲν κάτω, πρὸς ὃν φέρεται πάνθ' ὅσα τινὰ ὄγκον σώματος ἔχει, τὸν δ' ἄνω, πρὸς ὃν ἀκουσίως

---

[1] *i.e.* θερμόν (*quasi* κερμόν) is derived from κερματίζω ("mince up" or "mint").

TIMAEUS

both that affection which we call "heat" and its very name.[1]

The opposite affection is evident, but none the less it must not lack description. When liquids with larger particles, which surround the body, enter into it they drive out the smaller particles; but as they cannot pass into their room they compress the moisture within us, so that in place of non-uniformity and motion they produce immobility and density, as a result of the uniformity and compression. But that which is being contracted contrary to nature fights, and, in accordance with its nature, thrusts itself away in the contrary direction. And to this fighting and shaking we give the names of "trembling" and "shivering"; while this affection as a whole, as well as the cause thereof, is termed "cold."

By the term "hard" we indicate all the things to which our flesh gives way; and by the term "soft" all those which give way to our flesh; and these terms are similarly used relatively to each other. Now a substance gives way when it has its base small; but when it is constructed of quadrangular bases, being very firmly based, it is a most inelastic form; and so too is everything which is of very dense composition and most rigid.

The nature of "heavy" and "light" would be shown most clearly if, along with them, we examined also the nature of "above" and "below," as they are called. That there really exist two distinct and totally opposite regions, each of which occupies one-half of the Universe,—the one termed "below," towards which move all things possessing any bodily mass, and the other "above," towards which everything goes against its will,—this is a wholly erroneous

PLATO

62
D ἔρχεται πᾶν, οὐκ ὀρθὸν οὐδαμῇ νομίζειν. τοῦ γὰρ παντὸς οὐρανοῦ σφαιροειδοῦς ὄντος, ὅσα μὲν ἀφεστῶτα ἴσον τοῦ μέσου γέγονεν ἔσχατα, ὁμοίως αὐτὰ χρὴ ἔσχατα πεφυκέναι, τὸ δὲ μέσον τὰ αὐτὰ μέτρα τῶν ἐσχάτων ἀφεστηκὸς ἐν τῷ καταντικρὺ νομίζειν δεῖ πάντων εἶναι. τοῦ δὴ κόσμου ταύτῃ πεφυκότος τί τῶν εἰρημένων ἄνω τις ἢ κάτω τιθέμενος οὐκ ἐν δίκῃ δόξει τὸ μηδὲν προσῆκον ὄνομα λέγειν; ὁ μὲν γὰρ μέσος ἐν αὐτῷ τόπος οὔτε κάτω πεφυκὼς οὔτε ἄνω λέγεσθαι δίκαιος, ἀλλ' αὐτὸ ἐν μέσῳ· ὁ δὲ πέριξ οὔτε δὴ μέσος οὔτ' ἔχων διάφορον αὑτοῦ μέρος ἕτερον θατέρου μᾶλλον·
E πρὸς τὸ μέσον ἤ τι τῶν καταντικρύ. τοῦ δὲ ὁμοίως πάντῃ πεφυκότος ποῖά τις ἐπιφέρων ὀνόματα αὐτῷ ἐναντία καὶ πῇ καλῶς ἂν ἡγοῖτο λέγειν; εἰ γάρ τι καὶ στερεὸν εἴη κατὰ μέσον
63 τοῦ παντὸς ἰσοπαλές, εἰς οὐδὲν ἄν ποτε τῶν ἐσχάτων ἐνεχθείη διὰ τὴν πάντῃ ὁμοιότητα αὐτῶν· ἀλλ' εἰ καὶ περὶ αὐτὸ πορεύοιτό τις ἐν κύκλῳ, πολλάκις ἂν στὰς ἀντίπους ταὐτὸν αὑτοῦ κάτω καὶ ἄνω προσείποι. τὸ μὲν γὰρ ὅλον, καθάπερ εἴρηται νῦν δή, σφαιροειδὲς ὂν τόπον τινὰ κάτω, τὸν δὲ ἄνω, λέγειν ἔχειν οὐκ ἔμφρονος.

Ὅθεν δὲ ὠνομάσθη ταῦτα καὶ ἐν οἷς ὄντα εἰθίσμεθα δι' ἐκεῖνα καὶ τὸν οὐρανὸν ὅλον οὕτω δι-
B αιρούμενοι λέγειν, ταῦτα διομολογητέον ὑποθεμένοις τάδ' ἡμῖν. εἴ τις ἐν τῷ τοῦ παντὸς τόπῳ καθ' ὃν ἡ τοῦ πυρὸς εἴληχε μάλιστα φύσις, οὗ καὶ πλεῖ-

---

[1] The reference here is, probably, to Democritus (Aristotle also speaks of τὸ ἄνω φύσει, *Phys.* 208 b 14).

[2] *i.e.* "above" and "below" are purely relative terms.

TIMAEUS

supposition.[1] For inasmuch as the whole Heaven is spherical, all its outermost parts, being equally distant from the centre, must really be "outermost" in a similar degree; and one must conceive of the centre, which is distant from all the outermost parts by the same measures, as being opposite to them all. Seeing, then, that the Cosmos is actually of this nature, which of the bodies mentioned can one set "above" or "below" without incurring justly the charge of applying a wholly unsuitable name? For its central region cannot rightly be termed either "above" or "below," but just "central"; while its circumference neither is central nor has it any one part more divergent than another from the centre or any of its opposite parts. But to that which is in all ways uniform, what opposite names can we suppose are rightly applicable, or in what sense? For suppose there were a solid body evenly-balanced at the centre of the Universe, it would never be carried to any of the extremities because of their uniformity in all respects; nay, even were a man to travel round it in a circle he would often call the same part of it both "above" and "below," according as he stood now at one pole, now at the opposite.[2] For seeing that the Whole is, as we said just now, spherical, the assertion that it has one region "above" and one "below" does not become a man of sense.

Now the origin of these names and their true meaning which accounts for our habit of making these verbal distinctions even about the whole Heaven, we must determine on the basis of the following principles. Suppose that a man were to take his stand in that region of the Universe in which the substance of fire has its special abode, and where also that sub-

στον ἂν ἠθροισμένον εἴη πρὸς ὃ φέρεται, ἐπαναβὰς ἐπ' ἐκεῖνο καὶ δύναμιν εἰς τοῦτο ἔχων, μέρη τοῦ πυρὸς ἀφαιρῶν ἱσταίη, τιθεὶς εἰς πλάστιγγας, αἴρων τὸν ζυγὸν καὶ τὸ πῦρ ἕλκων εἰς ἀνόμοιον ἀέρα βιαζόμενος, δῆλον ὡς τοὔλαττόν που τοῦ μείζονος C ῥᾶον βιᾶται· ῥώμῃ γὰρ μιᾷ δυοῖν ἅμα μετεωριζομένοιν τὸ μὲν ἔλαττον μᾶλλον, τὸ δὲ πλέον ἧττον ἀνάγκη που κατατεινόμενον ξυνέπεσθαι τῇ βίᾳ, καὶ τὸ μὲν πολὺ βαρὺ καὶ κάτω φερόμενον κληθῆναι, τὸ δὲ σμικρὸν ἐλαφρὸν καὶ ἄνω. ταὐτὸν δὴ τοῦτο δεῖ φωρᾶσαι δρῶντας ἡμᾶς περὶ τόνδε τὸν τόπον. ἐπὶ γὰρ γῆς βεβῶτες, γεώδη γένη διιστάμενοι καὶ γῆν ἐνίοτε αὐτήν, ἕλκομεν εἰς ἀνόμοιον ἀέρα βίᾳ καὶ παρὰ φύσιν, ἀμφότερα τοῦ ξυγγενοῦς ἀντεχό- D μενα· τὸ δὲ σμικρότερον ῥᾶον τοῦ μείζονος βιαζομένοις εἰς τὸ ἀνόμοιον πρότερον ξυνέπεται· κοῦφον οὖν αὐτὸ προσειρήκαμεν, καὶ τὸν τόπον εἰς ὃν βιαζόμεθ' ἄνω, τὸ δ' ἐναντίον τούτοις πάθος βαρὺ καὶ κάτω. ταῦτ' οὖν δὴ διαφόρως ἔχειν αὐτὰ πρὸς αὑτὰ ἀνάγκη διὰ τὸ τὰ πλήθη τῶν γενῶν τόπον ἐναντίον ἄλλα ἄλλοις κατέχειν· τὸ γὰρ ἐν ἑτέρῳ κοῦφον ὂν τόπῳ τῷ κατὰ τὸν ἐναντίον τόπον ἐλαφρῷ καὶ τῷ βαρεῖ τὸ βαρὺ τῷ τε κάτω E τὸ κάτω καὶ τῷ ἄνω τὸ ἄνω πάντ' ἐναντία καὶ πλάγια καὶ πάντως διάφορα πρὸς ἄλληλα ἀνευρεθήσεται γιγνόμενα καὶ ὄντα· τόδε γε μὴν ἕν τι διανοητέον περὶ πάντων αὐτῶν, ὡς ἡ μὲν πρὸς τὸ ξυγγενὲς ὁδὸς ἑκάστοις οὖσα βαρὺ μὲν τὸ φερόμενον ποιεῖ, τὸν δὲ τόπον εἰς ὃν τὸ τοιοῦτον

TIMAEUS

stance to which it flies is collected in largest bulk; and suppose that, having the power to do so, he were to separate portions of the fire and weigh them, putting them on scales and lifting the balance and pulling the fire by force into the dissimilar air, it is obvious that he will force the smaller mass more easily than the larger. For if two masses are lifted up simultaneously by a single effort, the smaller will necessarily yield more and the larger less, owing to its resistance, to the force exerted; and the large mass will be said to be " heavy " and moving " down," the small " light " and moving " up." Now this is just what we ought to detect ourselves doing in our region here. Standing on the earth and detaching various earthy substances, and sometimes pure earth, we pull them into the dissimilar air by force and against nature, since both these kinds cleave to their own kindred; and the smaller mass yields more easily, and follows first, as we force it into the dissimilar kind; wherefore we name it " light," and the region to which we force it " above "; and the conditions opposite thereto we name " heavy " and " below." Thus, these must necessarily differ in their mutual relations, because the main masses of the Kinds occupy regions opposite to one another; for when we compare what is light in one region with what is light in the opposite region, and the heavy with the heavy, the " below " with the below, and the " above " with the above, we shall discover that these all become and are opposite and oblique and in every way different in their mutual relations. There is, however, this one fact to be noticed about them all, that it is the passage of each kind to its kindred mass which makes the moving body heavy, and the region

φέρεται κάτω, τὰ δὲ τούτοις ἔχοντα ὡς ἑτέρως θάτερα. περὶ δὴ τούτων αὖ τῶν παθημάτων ταῦτα αἴτια εἰρήσθω.

Λείου δ' αὖ καὶ τραχέος παθήματος αἰτίαν πᾶς που κατιδὼν καὶ ἑτέρῳ δυνατὸς ἂν εἴη λέγειν· σκληρότης γὰρ ἀνωμαλότητι μιχθεῖσα, τὸ δ' ὁμαλότης πυκνότητι παρέχεται.

Μέγιστον δὲ καὶ λοιπὸν τῶν κοινῶν περὶ ὅλον τὸ σῶμα παθημάτων τὸ τῶν ἡδέων καὶ τῶν ἀλγεινῶν αἴτιον ἐν οἷς διεληλύθαμεν καὶ ὅσα διὰ τῶν τοῦ σώματος μορίων αἰσθήσεις κεκτημένα καὶ λύπας ἐν αὑτοῖς ἡδονάς θ' ἅμα ἑπομένας ἔχει. ὧδ' οὖν κατὰ παντὸς αἰσθητοῦ καὶ ἀναισθήτου παθήματος τὰς αἰτίας λαμβάνωμεν, ἀναμιμνησκόμενοι τὸ τῆς εὐκινήτου τε καὶ δυσκινήτου φύσεως ὅτι διειλόμεθα ἐν τοῖς πρόσθεν· ταύτῃ γὰρ δὴ μεταδιωκτέον πάντα, ὅσα ἐπινοοῦμεν ἑλεῖν. τὸ μὲν γὰρ κατὰ φύσιν εὐκίνητον, ὅταν καὶ βραχὺ πάθος εἰς αὐτὸ ἐμπίπτῃ, διαδίδωσι κύκλῳ μόρια ἕτερα ἑτέροις ταὐτὸν ἀπεργαζόμενα, μέχριπερ ἂν ἐπὶ τὸ φρόνιμον ἐλθόντα ἐξαγγείλῃ τοῦ ποιήσαντος τὴν δύναμιν· τὸ δ' ἐναντίον ἑδραῖον ὂν κατ' οὐδένα τε κύκλον ἰὸν πάσχει μόνον, ἄλλο δὲ οὐ κινεῖ τῶν πλησίον, ὥστε οὐ διαδιδόντων μορίων μορίοις ἄλλων ἄλλοις τὸ πρῶτον πάθος ἐν αὐτοῖς ἀκίνητον εἰς τὸ πᾶν ζῷον γενόμενον ἀναίσθητον παρέσχε τὸ παθόν. ταῦτα

---

[1] *i.e.* the attraction takes different directions, therefore " up " and " down " are relative terms.
[2] *Cf.* 54 B ff., 57 D, E.

TIMAEUS

to which such a body moves " below " ; while the opposite conditions produce the contrary results.[1] Let this, then, stand as our account of the causes of these conditions.

Of " smoothness " and " roughness " anyone might be able to discern the causes and explain them also to others. For the cause of the latter is hardness combined with irregularity, and of the former regularity combined with density.

In respect of the affections common to the whole body a very important point, which still remains, is the cause of the pleasures and pains attaching to the sense-affections we have been discussing ; and the cause also of those affections which have become perceptible by means of the bodily parts and involve in themselves concomitant pains and pleasures. Let us, then, try to grasp the causes in connexion with every perceptible and imperceptible affection in the following way, bearing in mind the distinction we previously drew [2] between mobile and immobile substances ; for it is in this way that we must track down all those facts that we intend to grasp. Whenever what is naturally mobile is impressed by even a small affection, it transmits it in a circle, the particles passing on to one another this identical impression until they reach the organ of intelligence and announce the quality of the agent. But a substance of the opposite kind, being stable and having no circular movement, is only affected in itself and does not move any other adjacent particle ; consequently, since the particles do not transmit to one another the original affection, it fails to act upon the living creature as a whole, and the result is that the affected body is non-percipient. This is the case with the

PLATO

δὲ περί τε ὀστᾶ καὶ τρίχας ἐστὶ καὶ ὅσα ἄλλα γήϊνα τὸ πλεῖστον ἔχομεν ἐν ἡμῖν μόρια· τὰ δὲ ἔμπροσθεν περὶ τὰ τῆς ὄψεως καὶ ἀκοῆς μάλιστα, διὰ τὸ πυρὸς ἀέρος τε ἐν αὐτοῖς δύναμιν ἐνεῖναι μεγίστην.

Τὸ δὴ τῆς ἡδονῆς καὶ λύπης ὧδε δεῖ διανοεῖσθαι. τὸ μὲν παρὰ φύσιν καὶ βίαιον γιγνόμενον ἀθρόον παρ' ἡμῖν πάθος ἀλγεινόν, τὸ δ' εἰς φύσιν ἀπιὸν πάλιν ἀθρόον ἡδύ, τὸ δὲ ἠρέμα καὶ κατὰ σμικρὸν ἀναίσθητον, τὸ δ' ἐναντίον, τούτοις ἐναντίως. τὸ δὲ μετ' εὐπετείας γιγνόμενον ἅπαν αἰσθητὸν μὲν ὅ τι μάλιστα, λύπης δὲ καὶ ἡδονῆς οὐ μετέχον, οἷον τὰ περὶ τὴν ὄψιν αὐτὴν παθήματα, ᾗ δὴ σῶμα ἐν τοῖς πρόσθεν ἐρρήθη καθ' ἡμέραν ξυμφυὲς ἡμῶν γίγνεσθαι. ταύτῃ γὰρ τομαὶ μὲν καὶ καύσεις καὶ ὅσα ἄλλα πάσχει λύπας οὐκ ἐμποιοῦσιν, οὐδὲ ἡδονὰς πάλιν ἐπὶ ταὐτὸν ἀπιούσης εἶδος, μέγισται δὲ αἰσθήσεις καὶ σαφέσταται καθότι τ' ἂν πάθῃ καὶ ὅσων ἂν αὐτή πῃ προσβαλοῦσα ἐφάπτηται· βία γὰρ τὸ πάμπαν οὐκ ἔνι τῇ διακρίσει τε αὐτῆς καὶ συγκρίσει. τὰ δὲ ἐκ μειζόνων μερῶν σώματα μόγις εἴκοντα τῷ δρῶντι, διαδιδόντα δὲ εἰς ὅλον τὰς κινήσεις, ἡδονὰς ἴσχει καὶ λύπας, ἀλλοτριούμενα μὲν λύπας, καθιστάμενα δὲ εἰς τὸ αὐτὸ πάλιν ἡδονάς. ὅσα δὲ κατὰ σμικρὸν τὰς ἀποχωρήσεις ἑαυτῶν καὶ κενώσεις εἴληφε, τὰς δὲ πληρώσεις ἀθρόας καὶ κατὰ μεγάλα, κενώσεως μὲν ἀναίσθητα, πληρώσεως δὲ αἰσθητικὰ γιγνόμενα,

---

[1] Cf. Rep. 583 c ff., Phileb. 31 D ff.
[2] Cf. 45 B.

TIMAEUS

bones and the hair and all our other parts that are mainly earthy; whereas the former character belongs especially to the organs of sight and of hearing, owing to the fact that they contain a very large quantity of fire and air.

Now the nature of pleasure and pain we must conceive of in this way. When an affection which is against nature and violent occurs within us with intensity it is painful, whereas the return back to the natural condition, when intense, is pleasant[1]; and an affection which is mild and gradual is imperceptible, while the converse is of a contrary character. And the affection which, in its entirety, takes place with ease is eminently perceptible, but it does not involve pain or pleasure; such, for example, are the affections of the visual stream itself, which, as we said before,[2] becomes in the daylight a body substantially one with our own. For no pains are produced therein by cuttings or burnings or any other affections, nor does its reversion to its original form produce pleasures; but it has most intense and clear perceptions concerning every object that affects it, and every object also which it strikes against or touches; for force is wholly absent both from its dilation and from its contraction. But those bodies which are composed of larger particles, since they yield with difficulty to the agent and transmit their motions to the whole, feel pleasures and pains—pains when they suffer alteration, and pleasures when they are restored to their original state. And all those bodies which undergo losses of substance and emptyings that are gradual, but replenishings that are intense and abundant, become insensitive to the emptyings but sensitive to the replenishings; conse-

PLATO

65 λύπας μὲν οὐ παρέχει τῷ θνητῷ τῆς ψυχῆς, μεγίστας δὲ ἡδονάς· ἔστι δ' ἔνδηλα περὶ τὰς εὐωδίας. ὅσα δὲ ἀπαλλοτριοῦται μὲν ἀθρόα, κατὰ σμικρὰ δὲ
B μόγις τε εἰς ταὐτὸν πάλιν ἑαυτοῖς καθίσταται, τοὐναντίον τοῖς ἔμπροσθεν πάντα ἀποδίδωσι· ταῦτα δ' αὖ περὶ τὰς καύσεις καὶ τομὰς τοῦ σώματος γιγνόμενά ἐστι κατάδηλα.

Καὶ τὰ μὲν δὴ κοινὰ τοῦ σώματος παντὸς παθήματα, τῶν τ' ἐπωνυμιῶν ὅσαι τοῖς δρῶσιν αὐτὰ γεγόνασι, σχεδὸν εἴρηται· τὰ δ' ἐν ἰδίοις μέρεσιν ἡμῶν γιγνόμενα, τά τε πάθη καὶ τὰς αἰτίας αὖ τῶν δρώντων, πειρατέον εἰπεῖν, ἄν πῃ δυνώμεθα.

C Πρῶτον οὖν ὅσα τῶν χυμῶν πέρι λέγοντες ἐν τοῖς πρόσθεν ἀπελίπομεν, ἴδι' ὄντα παθήματα περὶ τὴν γλῶτταν, ἐμφανιστέον ᾗ δυνατόν. φαίνεται δὲ καὶ ταῦτα, ὥσπερ οὖν καὶ τὰ πολλά, διὰ συγκρίσεών τέ τινων καὶ διακρίσεων γίγνεσθαι, πρὸς δὲ αὐταῖς κεχρῆσθαι μᾶλλόν τι τῶν ἄλλων τραχύτησί τε καὶ λειότησιν. ὅσα μὲν γὰρ εἰσιόντα περὶ τὰ φλέβια,
D οἷόνπερ δοκίμια τῆς γλώττης τεταμένα ἐπὶ τὴν καρδίαν, εἰς τὰ νοτερὰ τῆς σαρκὸς καὶ ἁπαλὰ ἐμπίπτοντα γήϊνα μέρη κατατηκόμενα ξυνάγει τὰ φλέβια καὶ ἀποξηραίνει, τραχύτερα μὲν ὄντα στρυφνά, ἧττον δὲ τραχύνοντα αὐστηρὰ φαίνεται. τὰ δὲ τούτων τε ῥυπτικὰ καὶ πᾶν τὸ περὶ τὴν γλῶτταν ἀποπλύνοντα πέρα μὲν τοῦ μετρίου τοῦτο δρῶντα καὶ προσεπιλαμβανόμενα ὥστε ἀποτήκειν

---

[1] Cf. 64 E.
[2] The function of the nerves is here assigned to the veins.

TIMAEUS

quently, they furnish no pains to the mortal part of the soul, but the greatest pleasures—a result which is obvious in the case of perfumes. But all those parts which undergo violent alterations, and are restored gradually and with difficulty to their original condition, produce results the opposite of those last mentioned; and it is evident that this is what occurs in the case of burnings and cuttings of the body.

And now we have given a fairly complete statement of the affections which are common to the body as a whole, and of all the names which belong to the agents which produce them. Next we must try, if haply we are able, to describe what takes place in the several parts of our bodies, both the affections themselves and the agents to which they are ascribed.

Firstly, then, we must endeavour to elucidate so far as possible those affections which we omitted in our previous account of the flavours, they being affections peculiar to the tongue. It is evident that these also, like most others, are brought about by means of certain contractions and dilations [1]; and, more than other affections, they involve also conditions of roughness and smoothness. For all the earthy particles which enter in by the small veins—which, extending as far as to the heart, serve as it were for testing-instruments [2] of the tongue,—when they strike upon the moist and soft parts of the flesh and are melted down, contract the small veins and dry them up; and these particles when more rough appear to be "astringent," when less rough "harsh." And such as act on these veins as detergents and wash out all the surface of the tongue, when they do this excessively and lay such hold on the tongue as to

PLATO

65
αὐτῆς τῆς φύσεως, οἷον ἡ τῶν λίτρων δύναμις,
E πικρὰ πάνθ' οὕτως ὠνόμασται· τὰ δὲ ὑποδεέστερα
τῆς λιτρώδους ἕξεως ἐπὶ τὸ μέτριόν τε τῇ ῥύψει
χρώμενα ἁλυκὰ ἄνευ πικρότητος τραχείας καὶ φίλα
μᾶλλον ἡμῖν φαντάζεται. τὰ δὲ τῇ τοῦ στόματος
θερμότητι κοινωνήσαντα καὶ λεαινόμενα ὑπ' αὐτοῦ,
ξυνεκπυρούμενα καὶ πάλιν αὐτὰ ἀντικάοντα τὸ
διαθερμῆναν, φερόμενά τε ὑπὸ κουφότητος ἄνω
πρὸς τὰς τῆς κεφαλῆς αἰσθήσεις, τέμνοντά τε πάνθ'
66 ὁπόσοις ἂν προσπίπτῃ, διὰ ταύτας τὰς δυνάμεις
δριμέα πάντα τοιαῦτα ἐλέχθη. τὸ δὲ αὖ τῶν[1]
προλελεπτυσμένων μὲν ὑπὸ σηπεδόνος, εἰς δὲ τὰς
στενὰς φλέβας ἐνδυομένων, καὶ τοῖς ἐνοῦσιν αὐτόθι
μέρεσι γεώδεσι καὶ ὅσα ἀέρος ξυμμετρίαν ἐχόντων,[2]
ὥστε κινήσαντα περὶ ἄλληλα ποιεῖν κυκᾶσθαι,
κυκώμενα δὲ περιπίπτειν τε καὶ εἰς ἕτερα ἐνδυόμενα
ἕτερα κοῖλα ἀπεργάζεσθαι, περιτεινόμενα τοῖς εἰσ-
ιοῦσιν—ἃ δὴ νοτίδος περὶ ἀέρα κοίλης περιταθείσης,
B τοτὲ μὲν γεώδους, τοτὲ δὲ καὶ καθαρᾶς, νοτερὰ
ἀγγεῖα ἀέρος [ὕδατα][3] κοῖλα περιφερῆ τε γενέσθαι,
καὶ τὰ μὲν τῆς καθαρᾶς διαφανεῖς περιστῆναι,
κληθείσας ὄνομα πομφόλυγας, τὰ δὲ τῆς γεώδους,
ὁμοῦ κινουμένης τε καὶ αἱρομένης, ζέσιν τε καὶ
ζύμωσιν ἐπίκλην λεχθῆναι—τὸ δὲ τούτων αἴτιον
τῶν παθημάτων ὀξὺ προσρηθῆναι.

Ξύμπασι δὲ τοῖς περὶ ταῦτα εἰρημένοις πάθος
C ἐναντίον ἀπ' ἐναντίας ἐστὶ προφάσεως, ὁπόταν ἡ
τῶν εἰσιόντων ξύστασις ἐν ὑγροῖς, οἰκεία τῇ τῆς
γλώττης ἕξει πεφυκυῖα, λεαίνῃ μὲν ἐπαλείφουσα

---
[1] τὸ δὲ αὖ τῶν Schneider: τῶν δὲ αὐτῶν MSS., Zur.
[2] ἐχόντων Lindau: ἔχοντα MSS., Zur.
[3] ὕδατα I bracket.

dissolve part of its substance—and such, for example, is the property of alkalies,—are all termed "bitter"; while those which have a property less strong than the alkaline, being detergent in a moderate degree, seem to us to be "saline," and more agreeable, as being devoid of the rough bitterness. And those which share in the heat of the mouth and are made smooth thereby, when they are fully inflamed and are themselves in turn burning the part which heated them, fly upwards because of their lightness towards the senses of the head and cut all the parts on which they impinge; and because of these properties all such are called "pungent." Again, when particles already refined by putrefaction, entering into the narrow veins, are symmetrical with the particles of earth and air contained therein, so that they cause them to circulate round one another and ferment, then, in thus fermenting they change round and pass into fresh places, and thereby create fresh hollows which envelop the entering particles. By this means, the air being veiled in a moist film, sometimes of earth, sometimes of pure moisture, moist and hollow and globular vessels of air are formed; and those formed of pure moisture are the transparent globules called by the name of "bubbles," while those of the earthy formation which moves throughout its mass and seethes are designated "boiling" and "fermenting"; and the cause of these processes is termed "acid."

An affection which is the opposite of all those last described results from an opposite condition. Whenever the composition of the particles which enter into the moist parts is naturally akin to the state of the tongue, they oil its roughened parts and smooth it,

66 τὰ τραχυνθέντα, τὰ δὲ παρὰ φύσιν ξυνεστῶτα ἢ κεχυμένα τὰ μὲν ξυνάγῃ, τὰ δὲ χαλᾷ, καὶ πάνθ' ὅ τι μάλιστα ἱδρύῃ κατὰ φύσιν, ἡδὺ καὶ προσφιλὲς παντὶ πᾶν τὸ τοιοῦτον ἴαμα τῶν βιαίων παθημάτων γιγνόμενον κέκληται γλυκύ.

D Καὶ τὰ μὲν ταύτῃ ταῦτα· περὶ δὲ δὴ τὴν τῶν μυκτήρων δύναμιν, εἴδη μὲν οὐκ ἔνι. τὸ γὰρ τῶν ὀσμῶν πᾶν ἡμιγενές, εἴδει δὲ οὐδενὶ ξυμβέβηκε ξυμμετρία πρὸς τό τινα σχεῖν ὀσμήν. ἀλλ' ἡμῶν αἱ περὶ ταῦτα φλέβες πρὸς μὲν τὰ γῆς ὕδατός τε γένη στενώτεραι ξυνέστησαν, πρὸς δὲ τὰ πυρὸς ἀέρος τε εὐρύτεραι, διὸ τούτων οὐδεὶς οὐδενὸς ὀσμῆς πώποτε ᾔσθετό τινος, ἀλλ' ἢ βρεχομένων ἢ σηπομένων ἢ τηκομένων ἢ θυμιωμένων γίγνονταί E τινων· μεταβάλλοντος γὰρ ὕδατος εἰς ἀέρα ἀέρος τε εἰς ὕδωρ ἐν τῷ μεταξὺ τούτων γεγόνασιν, εἰσὶ δὲ ὀσμαὶ ξύμπασαι καπνὸς ἢ ὁμίχλη· τούτων δὲ τὸ μὲν ἐξ ἀέρος εἰς ὕδωρ ἰὸν ὁμίχλη, τὸ δὲ ἐξ ὕδατος εἰς ἀέρα καπνός. ὅθεν λεπτότεραι μὲν ὕδατος, παχύτεραι δὲ ὀσμαὶ ξύμπασαι γεγόνασιν ἀέρος. δηλοῦνται δέ, ὁπόταν τινὸς ἀντιφραχθέντος περὶ τὴν ἀναπνοὴν ἄγῃ τις βίᾳ τὸ πνεῦμα εἰς αὑτόν· τότε γὰρ ὀσμὴ μὲν οὐδεμία ξυνδιηθεῖται, τὸ δὲ πνεῦμα τῶν ὀσμῶν ἐρημωθὲν αὐτὸ μόνον ἕπεται. δι'[1] οὖν ταῦτα ἀνώνυμα τὰ τούτων ποικίλματα 67 γέγονεν, οὐκ ἐκ πολλῶν οὐδ' ἁπλῶν εἰδῶν ὄντα, ἀλλὰ διχῇ τό θ' ἡδὺ καὶ τὸ λυπηρὸν αὐτόθι μόνω διαφανῆ λέγεσθον, τὸ μὲν τραχῦνόν τε καὶ βιαζόμενον τὸ κύτος ἅπαν, ὅσον ἡμῶν μεταξὺ κορυφῆς

[1] δι' Stobaeus: δύ' mss., Zur.

TIMAEUS

contracting the parts that are unnaturally dilated or dilating those that are contracted, and thus settling them all, so far as possible, in their natural condition ; and every such remedy of the forcible affections, being pleasant and welcome to everyone, is called " sweet."

For this subject, then, let this account suffice. Next, as regards the property of the nostrils, it does not contain fixed kinds. For the whole range of smells is a half-formed class, and no kind possesses the symmetry requisite for containing any smell ; for our veins in these organs are of too narrow a construction for the kinds of earth and of water and too wide for those of fire and air, so that no one has ever yet perceived any smell from any of these, but only from substances which are in process of being moistened or putrefied or melted or vaporized. For smells arise in the intermediate state, when water is changing into air or air into water, and they are all smoke or mist ; and of these, the passage from air to water is mist, and the passage from water to air is smoke ; whence it is that all the smells are thinner than water and thicker than air. Their nature is made clear whenever there is some block in the respiration and a man draws in his breath forcibly ; for then no accompanying smell is strained through, but the breath passes in alone by itself isolated from the smells. So for these reasons the varieties of these smells have no name, not being derived either from many or from simple forms, but are indicated by two distinctive terms only, " pleasant " and " painful " ; of which the one kind roughens and violently affects the whole of our bodily cavity which lies between the head and the

τοῦ τε ὀμφαλοῦ κεῖται, τὸ δὲ ταὐτὸν τοῦτο καταπραῦνον καὶ πάλιν ᾗ πέφυκεν ἀγαπητῶς ἀποδιδόν.

Τρίτον δὲ αἰσθητικὸν ἐν ἡμῖν μέρος ἐπισκοποῦσι τὸ περὶ τὴν ἀκοήν, δι' ἃς αἰτίας τὰ περὶ αὐτὸ ξυμβαίνει παθήματα, λεκτέον. ὅλως μὲν οὖν φωνὴν θῶμεν τὴν δι' ὤτων ὑπ' ἀέρος ἐγκεφάλου τε καὶ αἵματος μέχρι ψυχῆς πληγὴν διαδιδομένην, τὴν δὲ ὑπ' αὐτῆς κίνησιν, ἀπὸ τῆς κεφαλῆς μὲν ἀρχομένην, τελευτῶσαν δὲ περὶ τὴν τοῦ ἥπατος ἕδραν, ἀκοήν· ὅση δ' αὐτῆς ταχεῖα, ὀξεῖαν, ὅση δὲ βραδυτέρα, βαρυτέραν· τὴν δὲ ὁμοίαν ὁμαλήν τε καὶ λείαν, τὴν δ' ἐναντίαν τραχεῖαν· μεγάλην δὲ τὴν πολλήν, ὅση δ' ἐναντία, σμικράν. τὰ δὲ περὶ ξυμφωνίας αὐτῶν ἐν τοῖς ὕστερον λεχθησομένοις ἀνάγκη ῥηθῆναι.

Τέταρτον δὴ λοιπὸν ἔτι γένος ἡμῖν αἰσθητικόν, ὃ διελέσθαι δεῖ συχνὰ ἐν ἑαυτῷ ποικίλματα κεκτημένον, ἃ ξύμπαντα μὲν χρόας ἐκαλέσαμεν, φλόγα τῶν σωμάτων ἑκάστων ἀπορρέουσαν, ὄψει ξύμμετρα μόρια ἔχουσαν πρὸς αἴσθησιν· ὄψεως δὲ ἐν τοῖς πρόσθεν αὐτῶν περὶ τῶν αἰτίων τῆς γενέσεώς ἐρρήθη. τῇδ' οὖν τῶν χρωμάτων πέρι μάλιστα εἰκὸς πρέποι τ' ἂν ἐπιεικεῖ λόγῳ διεξελθεῖν· τὰ φερόμενα ἀπὸ τῶν ἄλλων μόρια ἐμπίπτοντά τε εἰς τὴν ὄψιν τὰ μὲν ἐλάττω, τὰ δὲ μείζω, τὰ δ' ἴσα τοῖς αὐτῆς τῆς ὄψεως μέρεσιν εἶναι· τὰ μὲν οὖν ἴσα ἀναίσθητα, ἃ δὴ καὶ διαφανῆ λέγομεν, τὰ δὲ μείζω

---

[1] *Cf.* 80 A.   [2] *Cf.* 45 c ff.

TIMAEUS

navel, whereas the other mollifies this same region and restores it agreeably to its natural condition.

The third organ of perception within us which we have to describe in our survey is that of hearing, and the causes whereby its affections are produced. In general, then, let us lay it down that sound is a stroke transmitted through the ears, by the action of the air upon the brain and the blood, and reaching to the soul; and that the motion caused thereby, which begins in the head and ends about the seat of the liver, is " hearing "; and that every rapid motion produces a " shrill " sound, and every slower motion a more " deep " sound; and that uniform motion produces an " even " and smooth sound and the opposite kind of motion a " harsh " sound; and that large motion produces " loud " sound, and motion of the opposite kind " soft " sound. The subject of concords of sounds must necessarily be treated in a later part of our exposition.[1]

We have still remaining a fourth kind of sensation, which we must divide up seeing that it embraces numerous varieties, which, as a whole, we call " colours." This consists of a flame which issues from the several bodies, and possesses particles so proportioned to the visual stream as to produce sensation; and as regards the visual stream, we have already stated[2] merely the causes which produced it. Concerning colours, then, the following explanation will be the most probable and worthy of a judicious account. Of the particles which fly off from the rest and strike into the visual stream some are smaller, some larger, and some equal to the particles of the stream itself; those, then, that are equal are imperceptible, and we term them " transparent ";

67 καὶ ἐλάττω, τὰ μὲν συγκρίνοντα, τὰ δὲ διακρίνοντα αὐτήν, τοῖς περὶ τὴν σάρκα θερμοῖς καὶ ψυχροῖς καὶ τοῖς περὶ τὴν γλῶτταν στρυφνοῖς καὶ ὅσα θερμαντικὰ ὄντα δριμέα ἐκαλέσαμεν ἀδελφὰ εἶναι, τά τε λευκὰ καὶ τὰ μέλανα, ἐκείνων παθήματα γεγονότα ἐν ἄλλῳ γένει ταὐτά, φανταζόμενα δὲ ἄλλα διὰ ταύτας τὰς αἰτίας. οὕτως οὖν αὐτὰ προσρητέον, τὸ μὲν διακριτικὸν τῆς ὄψεως λευκόν, τὸ δ᾽ ἐναντίον αὐτοῦ μέλαν, τὴν δ᾽ ὀξυτέραν φορὰν καὶ γένους πυρὸς ἑτέρου προσπίπτουσαν καὶ διακρίνουσαν τὴν ὄψιν μέχρι τῶν ὀμμάτων, αὐτάς τε τῶν ὀφθαλμῶν 68 τὰς διεξόδους βίᾳ διωθοῦσαν καὶ τήκουσαν, πῦρ μὲν καὶ ὕδωρ ἀθρόον, ὃ δάκρυον καλοῦμεν, ἐκεῖθεν ἐκχέουσαν, αὐτὴν δὲ οὖσαν πῦρ, ἐξ ἐναντίας ἀπαντῶσαν, καὶ τοῦ μὲν ἐκπηδῶντος πυρὸς οἷον ἀπ᾽ ἀστραπῆς, τοῦ δ᾽ εἰσιόντος καὶ περὶ τὸ νοτερὸν κατασβεννυμένου, παντοδαπῶν ἐν τῇ κυκήσει ταύτῃ γιγνομένων χρωμάτων, μαρμαρυγὰς μὲν τὸ πάθος προσείπομεν, τὸ δὲ τοῦτο ἀπεργαζόμενον λαμπρόν τε καὶ στίλβον ἐπωνομάσαμεν. τὸ δὲ τούτων αὖ μεταξὺ πυρὸς γένος, πρὸς μὲν τὸ τῶν ὀμμάτων ὑγρὸν ἀφικνούμενον καὶ κεραννύμενον αὐτῷ, στίλβον δὲ οὔ, τῇ δὲ διὰ τῆς νοτίδος αὐγῇ τοῦ πυρὸς μιγνυμένῃ χρῶμα ἔναιμον παρασχόμενον,[1] τοὔνομα ἐρυθρὸν λέγομεν. λαμπρόν τε ἐρυθρῷ λευκῷ τε μιγνύμενον ξανθὸν γέγονε· τὸ δ᾽ ὅσον μέτρον ὅσοις, οὐδ᾽ εἴ τις εἰδείη, νοῦν ἔχει τὸ λέγειν, ὧν μήτε τινὰ ἀνάγκην μήτε τὸν εἰκότα λόγον καὶ μετρίως ἄν τις

[1] παρασχόμενον Lindau: παρασχομένῃ most mss., Zur.

---

[1] Cf. 65 E.   [2] Cf. 45 c ff.
[3] *i.e.* between the kinds of fire which produce "blackness" and "brightness."

TIMAEUS

while the larger and smaller particles—of which the one kind contracts, the other dilates the visual stream—are akin to the particles of heat and cold which affect the flesh, and to the astringent particles which affect the tongue, and to all the heating particles which we call "bitter"[1]: with these "white" and "black" are really identical affections, occurring in a separate class of sensation, although they appear different for the causes stated. These, therefore, are the names we must assign to them: that which dilates the visual stream is "white"; and the opposite thereof "black"[2]; and the more rapid motion, being that of a different species of fire, which strikes upon the visual stream and dilates it as far as to the eyes, and penetrating and dissolving the very passages of the eyes causes a volume of fire and water to pour from them, which we call "tears." And this moving body, being itself fire, meets fire from the opposite direction; and as the one fire-stream is leaping out like a flash, and the other passing in and being quenched in the moisture, in the resultant mixture colours of all kinds are produced. This sensation we term "dazzling" and the object which causes it "bright" or "brilliant." Again, when the kind of fire which is midway between these[3] reaches to the liquid of the eyes and is mingled therewith, it is not brilliant but, owing to the blending of the fire's ray through the moisture, it gives off a sanguine colour, and we give it the name of "red." And "bright" colour when blended with red and white becomes "yellow." But in what proportions the colours are blended it were foolish to declare, even if one knew, seeing that in such matters one could not properly adduce any necessary ground or prob-

68
εἰπεῖν εἴη δυνατός. ἐρυθρὸν δὲ δὴ μέλανι λευκῷ τε
C κραθὲν ἁλουργόν· ὀρφνινον δέ, ὅταν τούτοις μεμιγ-
μένοις καυθεῖσί τε μᾶλλον συγκραθῇ μέλαν. πυρρὸν
δὲ ξανθοῦ τε καὶ φαιοῦ κράσει γίγνεται, φαιὸν δὲ
λευκοῦ τε καὶ μέλανος, τὸ δὲ ὠχρὸν λευκοῦ ξανθῷ
μιγνυμένου. λαμπρῷ δὲ λευκὸν ξυνελθὸν καὶ εἰς
μέλαν κατακορὲς ἐμπεσὸν κυανοῦν χρῶμα ἀποτελεῖται, κυανοῦ δὲ λευκῷ κεραννυμένου γλαυκόν,
πυρροῦ δὲ μέλανι πράσιον. τὰ δὲ ἄλλα ἀπὸ τούτων
D σχεδὸν δῆλα, αἷς ἂν ἀφομοιούμενα μίξεσι διασώζοι
τὸν εἰκότα μῦθον. εἰ δέ τις τούτων ἔργῳ σκοπού-
μενος βάσανον λαμβάνοι, τὸ τῆς ἀνθρωπίνης καὶ
θείας φύσεως ἠγνοηκὼς ἂν εἴη διάφορον, ὅτι θεὸς
μὲν τὰ πολλὰ εἰς ἓν ξυγκεραννύναι καὶ πάλιν ἐξ
ἑνὸς εἰς πολλὰ διαλύειν ἱκανῶς[1] ἐπιστάμενος ἅμα
καὶ δυνατός, ἀνθρώπων δὲ οὐδεὶς οὐδέτερα τούτων
ἱκανὸς οὔτε ἔστι νῦν οὔτ᾽ εἰσαῦθίς ποτ᾽ ἔσται.
E Ταῦτα δὴ πάντα τότε ταύτῃ πεφυκότα ἐξ ἀνάγκης
ὁ τοῦ καλλίστου τε καὶ ἀρίστου δημιουργὸς ἐν τοῖς
γιγνομένοις παρελάμβανεν, ἡνίκα τὸν αὐτάρκη τε
καὶ τὸν τελεώτατον θεὸν ἐγέννα, χρώμενος μὲν ταῖς
περὶ ταῦτα αἰτίαις ὑπηρετούσαις, τὸ δὲ εὖ τεκταινόμενος ἐν πᾶσι τοῖς γιγνομένοις αὐτός. διὸ δὴ
χρὴ δύ᾽ αἰτίας εἴδη διορίζεσθαι, τὸ μὲν ἀναγκαῖον,
τὸ δὲ θεῖον, καὶ τὸ μὲν θεῖον ἐν ἅπασι ζητεῖν
κτήσεως ἕνεκα εὐδαίμονος βίου, καθ᾽ ὅσον ἡμῶν ἡ
69 φύσις ἐνδέχεται, τὸ δὲ ἀναγκαῖον ἐκείνων χάριν,

---

[1] ἱκανῶς best MSS.: ἱκανὸς ὡς Zur.

[1] Cf. 46 D, 48 A.

TIMAEUS

able reason. Red blended with black and white makes "purple"; but when these colours are mixed and more completely burned, and black is blended therewith, the result is "violet." "Chestnut" comes from the blending of yellow and grey; and "grey" from white and black; and "ochre" from white mixed with yellow. And when white is combined with "bright" and is steeped in deep black it turns into a "dark blue" colour; and dark blue mixed with white becomes "light blue"; and chestnut with black becomes "green." As to the rest, it is fairly clear from these examples what are the mixtures with which we ought to identify them if we would preserve probability in our account. But should any inquirer make an experimental test of these facts, he would evince his ignorance of the difference between man's nature and God's—how that, whereas God is sufficiently wise and powerful to blend the many into one and to dissolve again the one into many, there exists not now, nor ever will exist hereafter, a child of man sufficient for either of these tasks.

Such, then, being the necessary nature of all these things, the Artificer of the most fair and good took them over at that time amongst things generated when He was engendering the self-sufficing and most perfect God; and their inherent properties he used as subservient causes, but Himself designed the Good in all that was being generated. Wherefore one ought to distinguish two kinds of causes,[1] the necessary and the divine, and in all things to seek after the divine for the sake of gaining a life of blessedness, so far as our nature admits thereof, and to seek the necessary for the sake of the divine, reckoning that without the

69 λογιζομένους ὡς ἄνευ τούτων οὐ δυνατὰ αὐτὰ ἐκεῖνα, ἐφ' οἷς σπουδάζομεν, μόνα κατανοεῖν οὐδ' αὖ λαβεῖν οὐδ' ἄλλως πως μετασχεῖν.

Ὅτ' οὖν δὴ τὰ νῦν οἷα τέκτοσιν ἡμῖν ὕλη παράκειται τὰ τῶν αἰτίων γένη διυλασμένα, ἐξ ὧν τὸν ἐπίλοιπον λόγον δεῖ ξυνυφανθῆναι, πάλιν ἐπ' ἀρχὴν ἐπανέλθωμεν διὰ βραχέων, ταχύ τε εἰς ταὐτὸν
B πορευθῶμεν ὅθεν δεῦρο ἀφικόμεθα, καὶ τελευτὴν ἤδη κεφαλήν τε τῷ μύθῳ πειρώμεθα ἁρμόττουσαν ἐπιθεῖναι τοῖς πρόσθεν.

Ὥσπερ οὖν καὶ κατ' ἀρχὰς ἐλέχθη, ταῦτα ἀτάκτως ἔχοντα ὁ θεὸς ἐν ἑκάστῳ τε αὐτῷ πρὸς αὑτὸ καὶ πρὸς ἄλληλα συμμετρίας ἐνεποίησεν, ὅσας τε καὶ ὅπῃ δυνατὸν ἦν ἀνάλογα καὶ σύμμετρα εἶναι. τότε γὰρ οὔτε τούτων ὅσον μὴ τύχῃ τι μετεῖχεν, οὔτε τὸ παράπαν ὀνομάσαι τῶν νῦν ὀνομαζομένων ἀξιόλογον ἦν οὐδέν, οἷον πῦρ καὶ
C ὕδωρ καὶ εἴ τι τῶν ἄλλων. ἀλλὰ πάντα ταῦτα πρῶτον διεκόσμησεν, ἔπειτα ἐκ τούτων πᾶν τόδε ξυνεστήσατο, ζῷον ἓν ζῷα ἔχον τὰ πάντα ἐν αὑτῷ θνητὰ ἀθάνατά τε. καὶ τῶν μὲν θείων αὐτὸς γίγνεται δημιουργός, τῶν δὲ θνητῶν τὴν γένεσιν τοῖς ἑαυτοῦ γεννήμασι δημιουργεῖν προσέταξεν. οἱ δὲ μιμούμενοι, παραλαβόντες ἀρχὴν ψυχῆς ἀθάνατον, τὸ μετὰ τοῦτο θνητὸν σῶμα αὐτῇ περιετόρνευσαν ὄχημά τε πᾶν τὸ σῶμα ἔδοσαν ἄλλο τε εἶδος ἐν αὐτῷ ψυχῆς προσῳκοδόμουν τὸ θνητόν,
D δεινὰ καὶ ἀναγκαῖα ἐν ἑαυτῷ παθήματα ἔχον,

---

[1] *i.e.* 47 E.   [2] *Cf.* 30 A, 42 D ff.   [3] *Cf.* 44 E.

TIMAEUS

former it is impossible to discern by themselves alone the divine objects after which we strive, or to apprehend them or in any way partake thereof.

Seeing, then, that we have now lying before us and thoroughly sifted—like wood ready for the joiner,—the various kinds of causes, out of which the rest of our account must be woven together, let us once more for a moment revert to our starting-point,[1] and thence proceed rapidly to the point from which we arrived hither. In this way we shall endeavour now to supplement our story with a conclusion and a crown in harmony with what has gone before.

As we stated at the commencement,[2] all these things were in a state of disorder, when God implanted in them proportions both severally in relation to themselves and in their relations to one another, so far as it was in any way possible for them to be in harmony and proportion. For at that time nothing partook thereof, save by accident, nor was it possible to name anything worth mentioning which bore the names we now give them, such as fire and water, or any of the other elements; but He, in the first place, set all these in order, and then out of these He constructed this present Universe, one single Living Creature containing within itself all living creatures both mortal and immortal. And He Himself acts as the Constructor of things divine, but the structure of the mortal things He commanded His own engendered sons to execute. And they, imitating Him, on receiving the immortal principle of soul, framed around it a mortal body, and gave it all the body to be its vehicle,[3] and housed therein besides another form of soul, even the mortal form, which has within it passions both fearful and unavoidable—firstly,

PLATO

69 πρῶτον μὲν ἡδονήν, μέγιστον κακοῦ δέλεαρ, ἔπειτα λύπας, ἀγαθῶν φυγάς, ἔτι δ' αὖ θάρρος καὶ φόβον, ἄφρονε ξυμβούλω, θυμὸν δὲ δυσπαραμύθητον, ἐλπίδα δ' εὐπαράγωγον· αἰσθήσει τε ἀλόγῳ καὶ ἐπιχειρητῇ παντὸς ἔρωτι ξυγκερασάμενοι ταῦτα ἀναγκαίως τὸ θνητὸν γένος ξυνέθεσαν. καὶ διὰ ταῦτα δὴ σεβόμενοι μιαίνειν τὸ θεῖον, ὅ τι μὴ πᾶσα
Ε ἦν ἀνάγκη, χωρὶς ἐκείνου κατοικίζουσιν εἰς ἄλλην τοῦ σώματος οἴκησιν τὸ θνητόν, ἰσθμὸν καὶ ὅρον διοικοδομήσαντες τῆς τε κεφαλῆς καὶ τοῦ στήθους, αὐχένα μεταξὺ τιθέντες, ἵνα εἴη χωρίς. ἐν δὴ τοῖς στήθεσι καὶ τῷ καλουμένῳ θώρακι τὸ τῆς ψυχῆς θνητὸν γένος ἐνέδουν. καὶ ἐπειδὴ τὸ μὲν ἄμεινον αὐτῆς, τὸ δὲ χεῖρον ἐπεφύκει, διοικοδομοῦσι τοῦ
70 θώρακος αὖ τὸ κύτος, διορίζοντες οἷον γυναικῶν, τὴν δὲ ἀνδρῶν χωρὶς οἴκησιν, τὰς φρένας διάφραγμα εἰς τὸ μέσον αὐτῶν τιθέντες. τὸ μετέχον οὖν τῆς ψυχῆς ἀνδρίας καὶ θυμοῦ, φιλόνικον ὄν, κατῴκισαν ἐγγυτέρω τῆς κεφαλῆς μεταξὺ τῶν φρενῶν τε καὶ αὐχένος, ἵνα τοῦ λόγου κατήκοον ὂν κοινῇ μετ' ἐκείνου βίᾳ τὸ τῶν ἐπιθυμιῶν κατέχοι γένος, ὁπότ' ἐκ τῆς ἀκροπόλεως τῷ ἐπιτάγματι καὶ λόγῳ μηδαμῇ πείθεσθαι ἑκὸν ἐθέλοι. τὴν δὲ δὴ καρδίαν
Β ἅμμα τῶν φλεβῶν καὶ πηγὴν τοῦ περιφερομένου κατὰ πάντα τὰ μέλη σφοδρῶς αἵματος εἰς τὴν δορυφορικὴν οἴκησιν κατέστησαν, ἵνα, ὅτε ζέσειε τὸ τοῦ θυμοῦ μένος, τοῦ λόγου παραγγείλαντος ὥς τις ἄδικος περὶ αὐτὰ γίγνεται πρᾶξις ἔξωθεν ἢ καί τις ἀπὸ τῶν ἔνδοθεν ἐπιθυμιῶν, ὀξέως διὰ πάντων τῶν

---
[1] Cf. 64 E.

TIMAEUS

pleasure, a most mighty lure to evil; next, pains, which put good to rout [1]; and besides these, rashness and fear, foolish counsellors both; and anger, hard to dissuade; and hope, ready to seduce. And blending these with irrational sensation and with all-daring lust, they thus compounded in necessary fashion the mortal kind of soul. Wherefore, since they scrupled to pollute the divine, unless through absolute necessity, they planted the mortal kind apart therefrom in another chamber of the body, building an isthmus and boundary for the head and chest by setting between them the neck, to the end that they might remain apart. And within the chest—or "thorax," as it is called—they fastened the mortal kind of soul. And inasmuch as one part thereof is better, and one worse, they built a division within the cavity of the thorax—as if to fence off two separate chambers, for men and for women—by placing the midriff between them as a screen. That part of the soul, then, which partakes of courage and spirit, since it is a lover of victory, they planted more near to the head, between the midriff and the neck, in order that it might hearken to the reason, and, in conjunction therewith, might forcibly subdue the tribe of the desires whensoever they should utterly refuse to yield willing obedience to the word of command from the citadel of reason. And the heart, which is the junction of the veins and the fount of the blood which circulates vigorously through all the limbs, they appointed to be the chamber of the bodyguard, to the end that when the heat of the passion boils up, as soon as reason passes the word round that some unjust action is being done which affects them, either from without or possibly even from the interior desires, every organ

στενωπῶν πᾶν ὅσον αἰσθητικὸν ἐν τῷ σώματι τῶν τε παρακελεύσεων καὶ ἀπειλῶν αἰσθανόμενον γίγνοιτο ἐπήκοον καὶ ἕποιτο πάντη καὶ τὸ βέλτιστον οὕτως ἐν αὐτοῖς πᾶσιν ἡγεμονεῖν ἐῷ. τῇ δὲ δὴ πηδήσει τῆς καρδίας ἐν τῇ τῶν δεινῶν προσδοκίᾳ καὶ τῇ τοῦ θυμοῦ ἐγέρσει, προγιγνώσκοντες ὅτι διὰ πυρὸς ἡ τοιαύτη πᾶσα ἔμελλεν οἴδησις γίγνεσθαι τῶν θυμουμένων, ἐπικουρίαν αὐτῇ μηχανώμενοι τὴν τοῦ πλεύμονος ἰδέαν ἐνεφύτευσαν, πρῶτον μὲν μαλακὴν καὶ ἄναιμον, εἶτα σήραγγας ἐντὸς ἔχουσαν οἷον σπόγγου κατατετρημένας, ἵνα τό τε πνεῦμα καὶ τὸ πῶμα δεχομένη, ψύχουσα, ἀναπνοὴν καὶ ῥᾳστώνην ἐν τῷ καύματι παρέχοι. διὸ δὴ τῆς ἀρτηρίας ὀχετοὺς ἐπὶ τὸν πλεύμονα ἔτεμον, καὶ περὶ τὴν καρδίαν αὐτὸν περιέστησαν οἷον μάλαγμα,[1] ἵν' ὁ θυμὸς ἡνίκα ἐν αὐτῇ ἀκμάζοι, πηδῶσα εἰς ὑπεῖκον καὶ ἀναψυχομένη, πονοῦσα ἧττον μᾶλλον τῷ λόγῳ μετὰ θυμοῦ δύναιτο ὑπηρετεῖν.

Τὸ δὲ δὴ σίτων τε καὶ ποτῶν ἐπιθυμητικὸν τῆς ψυχῆς καὶ ὅσων ἔνδειαν διὰ τὴν τοῦ σώματος ἴσχει φύσιν, τοῦτο εἰς τὰ μεταξὺ τῶν τε φρενῶν καὶ τοῦ πρὸς τὸν ὀμφαλὸν ὅρου κατῴκισαν, οἷον φάτνην ἐν ἅπαντι τούτῳ τῷ τόπῳ τῇ τοῦ σώματος τροφῇ τεκτηνάμενοι· καὶ κατέδησαν δὴ τὸ τοιοῦτον ἐνταῦθα ὡς θρέμμα ἄγριον, τρέφειν δὲ ξυνημμένον ἀναγκαῖον, εἴπερ τι μέλλοι τὸ θνητὸν ἔσεσθαι γένος. ἵν' οὖν ἀεὶ νεμόμενον πρὸς φάτνῃ καὶ ὅ τι πορρωτάτω τοῦ βουλευομένου κατοικοῦν, θόρυβον καὶ βοὴν ὡς ἐλαχίστην παρέχον, τὸ κράτιστον καθ'

[1] μάλαγμα Longinus: ἅλμα μαλακόν most mss., Zur.

TIMAEUS

of sense in the body might quickly perceive through all the channels both the injunctions and the threats and in all ways obey and follow them, thus allowing their best part to be the leader of them all. And as a means of relief for the leaping of the heart, in times when dangers are expected and passion is excited—since they knew that all such swelling of the passionate parts would arise from the action of fire,—they contrived and implanted the form of the lungs. This is, in the first place, soft and bloodless; and, moreover, it contains within it perforated cavities like those of a sponge, so that, when it receives the breath and the drink, it might have a cooling effect and furnish relief and comfort in the burning heat. To this end they drew the channels of the windpipe to the lungs, and placed the lungs as a kind of padding round the heart, in order that, when the passion therein should be at its height, by leaping upon a yielding substance and becoming cool, the heart might suffer less and thereby be enabled the more to be subservient to the reason in time of passion.

And all that part of the Soul which is subject to appetites for foods and drinks, and all the other wants that are due to the nature of the body, they planted in the parts midway between the midriff and the boundary at the navel, fashioning as it were a manger in all this region for the feeding of the body; and there they tied up this part of the Soul, as though it were a creature which, though savage, they must necessarily keep joined to the rest and feed, if the mortal stock were to exist at all. In order, then, that this part, feeding thus at its manger and housed as far away as possible from the counselling part, and creating the least possible turmoil and din, should

PLATO

71 ἡσυχίαν περὶ τοῦ πᾶσι κοινῇ ⟨καὶ ἰδίᾳ⟩¹ ξυμφέροντος ἐῷ βουλεύεσθαι, διὰ ταῦτ' ἐνταῦθα ἔδοσαν αὐτῷ τὴν τάξιν. εἰδότες δὲ αὐτὸ ὡς λόγου μὲν οὔτε ξυνήσειν ἔμελλεν, εἴ τέ πῃ καὶ μεταλαμβάνοι τινὸς αὐτῶν αἰσθήσεως,² οὐκ ἔμφυτον αὐτῷ τὸ μέλειν τινῶν ἔσοιτο λόγων, ὑπὸ δὲ εἰδώλων καὶ φαντασμάτων νυκτός τε καὶ μεθ' ἡμέραν μάλιστα ψυχαγωγήσοιτο, τούτῳ δὴ θεὸς ἐπιβουλεύσας αὐτῷ τὴν ἥπατος ἰδέαν ξυνέστησε καὶ ἔθηκεν εἰς τὴν ἐκεί-
B νου κατοίκησιν, πυκνὸν καὶ λεῖον καὶ λαμπρὸν καὶ γλυκὺ καὶ πικρότητα ἔχον μηχανησάμενος, ἵνα ἐν αὐτῷ τῶν διανοημάτων ἡ ἐκ τοῦ νοῦ φερομένη δύναμις, οἷον ἐν κατόπτρῳ δεχομένῳ τύπους καὶ κατιδεῖν εἴδωλα παρέχοντι, φοβοῖ μὲν αὐτό, ὁπότε μέρει τῆς πικρότητος χρωμένη ξυγγενεῖ χαλεπὴ προσενεχθεῖσα ἀπειλῇ κατὰ πᾶν ὑπομιγνῦσα ὀξέως τὸ ἧπαρ χολώδη χρώματα ἐμφαίνοι, ξυνάγουσά τε
C πᾶν ῥυσὸν καὶ τραχὺ ποιοῖ, λοβὸν δὲ καὶ δοχὰς πύλας τε τὰ μὲν ἐξ ὀρθοῦ κατακάμπτουσα καὶ ξυσπῶσα, τὰ δὲ ἐμφράττουσα συγκλείουσά τε, λύπας καὶ ἄσας παρέχοι, καὶ ὅτ' αὖ τὰ ἐναντία φαντάσματα ἀποζωγραφοῖ πραότητός τις ἐκ διανοίας ἐπίπνοια, τῆς μὲν πικρότητος ἡσυχίαν παρέχουσα τῷ μήτε κινεῖν μήτε προσάπτεσθαι τῆς ἐναντίας ἑαυτῇ φύσεως ἐθέλειν, γλυκύτητι δὲ τῇ κατ' ἐκεῖνο
D ξυμφύτῳ πρὸς αὐτὸ χρωμένη καὶ πάντα ὀρθὰ καὶ λεῖα αὐτοῦ καὶ ἐλεύθερα ἀπευθύνουσα ἵλεών τε καὶ

---
¹ καὶ ἰδίᾳ added by Burnet.
² αὐτῶν αἰσθήσεως most mss.: αὖ τῶν αἰσθήσεων Zur.

---
¹ i.e. gall.

TIMAEUS

allow the Supreme part to take counsel in peace concerning what benefits all, both individually and in the mass,—for these reasons they stationed it in that position. And inasmuch as they knew that it would not understand reason, and that, even if it did have some share in the perception of reasons, it would have no natural instinct to pay heed to any of them but would be bewitched for the most part both day and night by images and phantasms,—to guard against this God devised and constructed the form of the liver and placed it in that part's abode; and He fashioned it dense and smooth and bright and sweet, yet containing bitterness, that the power of thoughts which proceed from the mind, moving in the liver as in a mirror which receives impressions and provides visible images, should frighten this part of the soul; for when the mental power bears down upon it with stern threats, it uses a kindred portion of the liver's bitterness [1] and makes it swiftly suffuse the whole liver, so that it exhibits bilious colours, and by contraction makes it all wrinkled and rough; moreover, as regards the lobe and passages and gates [2] of the liver, the first of these it bends back from the straight and compresses, while it blocks the others and closes them up, and thus it produces pains and nausea. On the other hand, when a breath of mildness from the intellect paints on the liver appearances of the opposite kind, and calms down its bitterness by refusing to move or touch the nature opposite to itself, and using upon the liver the sweetness inherent therein rectifies all its parts so as to make them straight and smooth and free, it causes the part of the soul planted

[2] *i.e.* the right lobe, the biliary vesicle, and the *vena porta*; *cf.* Eurip. *Electra* 827 ff.

εὐήμερον ποιοῖ τὴν περὶ τὸ ἧπαρ ψυχῆς μοῖραν κατῳκισμένην, ἔν τε τῇ νυκτὶ διαγωγὴν ἔχουσαν μετρίαν, μαντείᾳ χρωμένην καθ' ὕπνον, ἐπειδὴ λόγου καὶ φρονήσεως οὐ μετεῖχε.

Μεμνημένοι γὰρ τῆς τοῦ πατρὸς ἐπιστολῆς οἱ ξυστήσαντες ἡμᾶς, ὅτε τὸ θνητὸν ἐπέστελλε γένος ὡς ἄριστον εἰς δύναμιν ποιεῖν, οὕτω δὴ κατορθοῦντες καὶ τὸ φαῦλον ἡμῶν, ἵνα ἀληθείας πῃ προσάπτοιτο, κατέστησαν ἐν τούτῳ τὸ μαντεῖον. ἱκανὸν δὲ σημεῖον ὡς μαντικὴν ἀφροσύνῃ θεὸς ἀνθρωπίνῃ δέδωκεν· οὐδεὶς γὰρ ἔννους ἐφάπτεται μαντικῆς ἐνθέου καὶ ἀληθοῦς, ἀλλ' ἢ καθ' ὕπνον τὴν τῆς φρονήσεως πεδηθεὶς δύναμιν ἢ διὰ νόσον ἢ διά τινα ἐνθουσιασμὸν παραλλάξας. ἀλλὰ ξυννοῆσαι μὲν ἔμφρονος τά τε ῥηθέντα ἀναμνησθέντα ὄναρ ἢ ὕπαρ ὑπὸ τῆς μαντικῆς τε καὶ ἐνθουσιαστικῆς φύσεως, καὶ ὅσα ἂν φαντάσματα ὀφθῇ, πάντα λογισμῷ διελέσθαι, ὅπῃ τι σημαίνει καὶ ὅτῳ μέλλοντος ἢ παρελθόντος ἢ παρόντος κακοῦ ἢ ἀγαθοῦ· τοῦ δὲ μανέντος ἔτι τε ἐν τούτῳ μένοντος οὐκ ἔργον τὰ φανέντα καὶ φωνηθέντα ὑφ' ἑαυτοῦ κρίνειν, ἀλλ' εὖ καὶ πάλαι λέγεται τὸ πράττειν καὶ γνῶναι τά τε αὑτοῦ καὶ ἑαυτὸν σώφρονι μόνῳ προσήκειν. ὅθεν δὴ καὶ τὸ τῶν προφητῶν γένος ἐπὶ ταῖς ἐνθέοις μαντείαις κριτὰς ἐπικαθιστάναι νόμος· οὓς μάντεις αὐτοὺς ὀνομάζουσί τινες, τὸ πᾶν ἠγνοηκότες ὅτι τῆς δι' αἰνιγμῶν οὗτοι φήμης καὶ

---

[1] Cf. Rep. 364 B, Laws 772 D, Phaedr. 244 A ff.

TIMAEUS

round the liver to be cheerful and serene, so that in the night it passes its time sensibly, being occupied in its slumbers with divination, seeing that in reason and intelligence it has no share.

For they who constructed us, remembering the injunction of their Father, when He enjoined upon them to make the mortal kind as good as they possibly could, rectified the vile part of us by thus establishing therein the organ of divination, that it might in some degree lay hold on truth. And that God gave unto man's foolishness the gift of divination [1] a sufficient token is this: no man achieves true and inspired divination when in his rational mind, but only when the power of his intelligence is fettered in sleep or when it is distraught by disease or by reason of some divine inspiration. But it belongs to a man when in his right mind to recollect and ponder both the things spoken in dream or waking vision by the divining and inspired nature, and all the visionary forms that were seen, and by means of reasoning to discern about them all wherein they are significant and for whom they portend evil or good in the future, the past, or the present. But it is not the task of him who has been in a state of frenzy, and still continues therein, to judge the apparitions and voices seen or uttered by himself; for it was well said of old that to do and to know one's own and oneself belongs only to him who is sound of mind. Wherefore also it is customary to set the tribe of prophets [2] to pass judgement upon these inspired divinations; and they, indeed, themselves are named "diviners" by certain who are wholly ignorant of the truth that they are not diviners but interpreters of the mysterious voice and

[1] *Cf. Laws* 871 c, Eurip. *Ion* 413 ff.

PLATO

φαντάσεως ὑποκριταί, καὶ οὔ τι μάντεις, προφῆται δὲ μαντευομένων δικαιότατα ὀνομάζοιντ' ἄν.

Ἡ μὲν οὖν φύσις ἥπατος διὰ ταῦτα τοιαύτη τε καὶ ἐν τόπῳ ᾧ λέγομεν πέφυκε, χάριν μαντικῆς. καὶ ἔτι μὲν δὴ ζῶντος ἑκάστου τὸ τοιοῦτον σημεῖα ἐναργέστερα ἔχει, στερηθὲν δὲ τοῦ ζῆν γέγονε τυφλὸν καὶ τὰ μαντεῖα ἀμυδρότερα ἔσχε τοῦ τι C σαφὲς σημαίνειν.

Ἡ δ' αὖ τοῦ γείτονος αὐτῷ ξύστασις καὶ ἕδρα σπλάγχνου γέγονεν ἐξ ἀριστερᾶς χάριν ἐκείνου, τοῦ παρέχειν αὐτὸ λαμπρὸν ἀεὶ καὶ καθαρόν, οἷον κατόπτρῳ παρεσκευασμένον καὶ ἕτοιμον ἀεὶ παρακείμενον ἐκμαγεῖον. διὸ δὴ καὶ ὅταν τινὲς ἀκαθαρσίαι γίγνωνται διὰ νόσους σώματος περὶ τὸ ἧπαρ, πάντα ἡ σπληνὸς καθαίρουσα αὐτὰ δέχεται μανότης, ἅτε κοίλου καὶ ἀναίμου ὑφανθέντος· ὅθεν πληρούμενος τῶν ἀποκαθαιρομένων μέγας καὶ D ὕπουλος αὐξάνεται, καὶ πάλιν, ὅταν καθαρθῇ τὸ σῶμα, ταπεινούμενος εἰς ταὐτὸν ξυνίζει.

Τὰ μὲν οὖν περὶ ψυχῆς, ὅσον θνητὸν ἔχει καὶ ὅσον θεῖον, καὶ ὅπῃ καὶ μεθ' ὧν καὶ δι' ἃ χωρὶς ᾠκίσθη, τὸ μὲν ἀληθές, ὡς εἴρηται, θεοῦ ξυμφήσαντος τότ' ἂν οὕτω μόνως διισχυριζοίμεθα· τό γε μὴν εἰκὸς ἡμῖν εἰρῆσθαι καὶ νῦν καὶ ἔτι μᾶλλον ἀνασκοποῦσι διακινδυνευτέον τὸ φάναι, καὶ πεφάσθω. E Τὸ δ' ἑξῆς δὴ τούτοισι κατὰ ταὐτὰ μεταδιωκτέον· ἦν δὲ τὸ τοῦ σώματος ἐπίλοιπον ᾗ

---

[1] *i.e.* in the sacrificed victim; *cf. Rep.* 364 c ff.
[2] *i.e.* the spleen, which, in relation to the liver, is concave.

TIMAEUS

apparition, for whom the most fitting name would be "prophets of things divined."

For these reasons, then, the nature of the liver is such as we have stated and situated in the region we have described, for the sake of divination. Moreover, when the individual creature is alive this organ affords signs that are fairly manifest, but when deprived of life [1] it becomes blind and the divinations it presents are too much obscured to have any clear significance.

The structure of the organ which adjoins it,[2] with its seat on the left, is for the sake of the liver, to keep it always bright and clean, as a wiper that is laid beside a mirror always prepared and ready to hand. Wherefore also, whenever any impurities due to ailments of the body occur round about the liver, the loose texture of the spleen cleanses and absorbs them all, seeing that it is woven of a stuff that is porous and bloodless: hence, when it is filled with the offscourings, the spleen grows to be large and festered; and conversely, when the body is cleansed, it is reduced and shrinks back to its primal state.

Concerning the soul, then, what part of it is mortal, what part immortal, and where and with what companions and for what reasons these have been housed apart, only if God concurred could we dare to affirm that our account is true [3]; but that our account is probable we must dare to affirm now, and to affirm still more positively as our inquiry proceeds: affirmed, therefore, let it be.

The subject which comes next to this we must investigate on the same lines; and that subject is the way in which the remainder of the body has been

[3] *Cf.* 68 D, 74 D.

PLATO

72 γέγονεν. ἐκ δὴ λογισμοῦ τοιοῦδε ξυνίστασθαι μάλιστ' ἂν αὐτὸ πάντων πρέποι. τὴν ἐσομένην ἐν ἡμῖν ποτῶν καὶ ἐδεστῶν ἀκολασίαν ᾔδεσαν οἱ ξυντιθέντες ἡμῶν τὸ γένος, καὶ ὅτι τοῦ μετρίου καὶ ἀναγκαίου διὰ μαργότητα πολλῷ χρησοίμεθα πλέονι. ἵν' οὖν μὴ φθορὰ διὰ νόσους ὀξεῖα γίγνοιτο
73 καὶ ἀτελὲς τὸ γένος εὐθὺς τὸ θνητὸν τελευτῷ, ταῦτα προορώμενοι τῇ τοῦ περιγενησομένου πώματος ἐδέσματός τε ἕξει τὴν ὀνομαζομένην κάτω κοιλίαν ὑποδοχὴν ἔθεσαν, εἵλιξάν τε πέριξ τὴν τῶν ἐντέρων γένεσιν, ὅπως μὴ ταχὺ διεκπερῶσα ἡ τροφὴ ταχὺ πάλιν τροφῆς ἑτέρας δεῖσθαι τὸ σῶμα ἀναγκάζοι, καὶ παρέχουσα ἀπληστίαν διὰ γαστριμαργίαν ἀφιλόσοφον καὶ ἄμουσον πᾶν ἀποτελοῖ τὸ γένος, ἀνυπήκοον τοῦ θειοτάτου τῶν παρ' ἡμῖν.

Τὸ δὲ ὀστῶν καὶ σαρκῶν καὶ τῆς τοιαύτης
B φύσεως πέρι πάσης ὧδε ἔσχε. τούτοις ξύμπασιν ἀρχὴ μὲν ἡ τοῦ μυελοῦ γένεσις· οἱ γὰρ τοῦ βίου δεσμοὶ τῆς ψυχῆς τῷ σώματι ξυνδουμένης ἐν τούτῳ διαδούμενοι κατερρίζουν τὸ θνητὸν γένος· αὐτὸς δὲ ὁ μυελὸς γέγονεν ἐξ ἄλλων. τῶν γὰρ τριγώνων ὅσα πρῶτα ἀστραβῆ καὶ λεῖα ὄντα πῦρ τε καὶ ὕδωρ καὶ ἀέρα καὶ γῆν δι' ἀκριβείας μάλιστα ἦν παρασχεῖν δυνατά, ταῦτα ὁ θεὸς ἀπὸ τῶν ἑαυτῶν ἕκαστα
C γενῶν χωρὶς ἀποκρίνων, μιγνὺς δ' ἀλλήλοις ξύμμετρα, πανσπερμίαν παντὶ θνητῷ γένει μηχανώμενος, τὸν μυελὸν ἐξ αὐτῶν ἀπειργάσατο, καὶ μετὰ

---

[1] *Cf.* 61 c.
[2] Literally "the lower belly," as distinct from "the upper belly" or thorax.
[3] *Cf.* 53 c ff.

TIMAEUS

generated.[1] Its construction would most fittingly be ascribed to reasoning such as this. Those who were constructing our kind were aware of the incontinence that would reside in us in respect of drinks and meats, and how that because of our greed we would consume far more than what was moderate and necessary; wherefore, lest owing to maladies swift destruction should overtake them, and the mortal kind, while still incomplete, come straightway to a complete end,—foreseeing this, the Gods set the "abdomen,"[2] as it is called, to serve as a receptacle for the holding of the superfluous meat and drink; and round about therein they coiled the structure of the entrails, to prevent the food from passing through quickly and thereby compelling the body to require more food quickly, and causing insatiate appetite, whereby the whole kind by reason of its gluttony would be rendered devoid of philosophy and of culture, and disobedient to the most divine part we possess.

As regards the bones and the flesh and all such substances the position was this. All these had their origin in the generation of the marrow. For it was in this that the bonds of life by which the Soul is bound to the body were fastened, and implanted the roots of the mortal kind; but the marrow itself was generated out of other elements. Taking all those primary triangles[3] which, being unwarped and smooth, were best able to produce with exactness fire and water and air and earth, God separated them, each apart from his own kind, and mixing them one with another in due proportion, He fashioned therefrom the marrow, devising it as a universal seed-stuff for every mortal kind. Next, He en-

PLATO

73 ταῦτα δὴ φυτεύων ἐν αὐτῷ κατέδει τὰ τῶν ψυχῶν γένη, σχημάτων τε ὅσα ἔμελλεν αὖ σχήσειν οἷά τε καθ' ἕκαστα εἴδη, τὸν μυελὸν αὐτὸν τοσαῦτα καὶ τοιαῦτα διῃρεῖτο σχήματα εὐθὺς ἐν τῇ διανομῇ τῇ κατ' ἀρχάς. καὶ τὴν μὲν τὸ θεῖον σπέρμα οἷον ἄρουραν μέλλουσαν ἕξειν ἐν αὑτῇ περιφερῆ πανταχῇ
D πλάσας ἐπωνόμασε τοῦ μυελοῦ ταύτην τὴν μοῖραν ἐγκέφαλον, ὡς ἀποτελεσθέντος ἑκάστου ζῴου τὸ περὶ τοῦτο ἀγγεῖον κεφαλὴν γενησόμενον· ὃ δ' αὖ τὸ λοιπὸν καὶ θνητὸν τῆς ψυχῆς ἔμελλε καθέξειν, ἅμα στρογγύλα καὶ προμήκη διῃρεῖτο σχήματα, μυελὸν δὲ πάντα ἐπεφήμισε, καὶ καθάπερ ἐξ ἀγκυρῶν βαλλόμενος ἐκ τούτων πάσης ψυχῆς δεσμοὺς περὶ τοῦτο ξύμπαν ἤδη τὸ σῶμα ἡμῶν ἀπειργάζετο, στέγασμα μὲν αὐτῷ πρῶτον ξυμ-
E πηγνὺς περὶ ὅλον ὀστέϊνον.

Τὸ δὲ ὀστοῦν ξυνίστησιν ὧδε. γῆν διαττήσας καθαρὰν καὶ λείαν ἐφύρασε καὶ ἔδευσε μυελῷ, καὶ μετὰ τοῦτο εἰς πῦρ αὐτὸ ἐντίθησι, μετ' ἐκεῖνο δὲ εἰς ὕδωρ βάπτει, πάλιν δὲ εἰς πῦρ, αὖθίς τε εἰς ὕδωρ· μεταφέρων δ' οὕτω πολλάκις εἰς ἑκάτερον ὑπ' ἀμφοῖν ἄτηκτον ἀπειργάσατο. καταχρώμενος δὴ τούτῳ περὶ μὲν τὸν ἐγκέφαλον αὐτοῦ σφαῖραν περιετόρνευσεν ὀστεΐνην, ταύτῃ δὲ στενὴν διέξοδον κατελίπετο· καὶ περὶ τὸν
74 διαυχένιον ἅμα καὶ νωτιαῖον μυελὸν ἐξ αὐτοῦ σφονδύλους πλάσας ὑπέτεινεν οἷον στρόφιγγας, ἀρξάμενος ἀπὸ τῆς κεφαλῆς, διὰ παντὸς τοῦ κύτους. καὶ τὸ πᾶν δὴ σπέρμα διασῴζων οὕτω

---

[1] *i.e.* the rational (νοῦς), " spirited " (θυμός), and appetitive (ἐπιθυμία) kinds or parts.   [2] *Cf.* 44 D.
[3] *i.e.* the vertebral column, cylindrical in shape.

TIMAEUS

gendered therein the various kinds of Soul[1] and bound them down; and He straightway divided the marrow itself, in His original division, into shapes corresponding in their number and their nature to the number and the nature of the shapes which should belong to the several kinds of Soul. And that portion of the marrow which was intended to receive within itself, as it were into a field, the divine seed He moulded in the shape of a perfect globe[2] and bestowed on it the name of "brain," purposing that, when each living creature should be completed, the vessel surrounding this should be called the "head." But that portion which was to contain the other and mortal part of the Soul He divided into shapes that were at once rounded and elongated,[3] and all these He designated "marrow"; and from these, as from anchors, He cast out bands of the Whole Soul, and around this He finally wrought the whole of this body of ours, when He had first built round about it for a shelter a framework all of bone.

And bone He compounded in this wise. Having sifted earth till it was pure and smooth, He kneaded it and moistened it with marrow; then He placed it in fire, and after that dipped it in water, and from this back to fire, and once again in water; and by thus transferring it many times from the one element to the other He made it so that it was soluble by neither. This, then, He used, and fashioned thereof, by turning, a bony sphere round about the brain; and therein he left a narrow opening; and around the marrow of both neck and back He moulded vertebrae of bone, and set them, like pivots, in a vertical row, throughout all the trunk, beginning from the head. And thus for preserving the whole

λιθοειδεῖ περιβόλῳ ξυνέφραξεν, ἐμποιῶν ἄρθρα, τῇ θατέρου προσχρώμενος ἐν αὑτοῖς ὡς μέσῃ ἐνισταμένῃ δυνάμει, κινήσεως καὶ κάμψεως ἕνεκα. τὴν δ' αὖ τῆς ὀστεΐνης φύσεως ἕξιν ἡγησάμενος τοῦ δέοντος κραυροτέραν εἶναι καὶ ἀκαμπτοτέραν, διάπυρόν τ' αὖ γιγνομένην καὶ πάλιν ψυχομένην σφακελίσασαν ταχὺ διαφθερεῖν τὸ σπέρμα ἐντὸς αὑτῆς, διὰ ταῦτα οὕτω τὸ τῶν νεύρων καὶ τὸ τῆς σαρκὸς γένος ἐμηχανᾶτο, ἵνα τῷ μὲν ἅπαντα τὰ μέλη ξυνδήσας ἐπιτεινομένῳ καὶ ἀνιεμένῳ περὶ τοὺς στρόφιγγας καμπτόμενον τὸ σῶμα καὶ ἐκτεινόμενον παρέχοι, τὴν δὲ σάρκα προβολὴν μὲν καυμάτων, πρόβλημα δὲ χειμώνων, ἔτι δὲ πτωμάτων οἷον τὰ πιλητὰ ἔσεσθαι ἐσθήματα,[1] σώμασι μαλακῶς καὶ πράως ὑπείκουσαν, θερμὴν δὲ νοτίδα ἐντὸς ἑαυτῆς ἔχουσαν θέρους μὲν ἀνιδίουσαν καὶ νοτιζομένην ἔξωθεν ψῦχος κατὰ πᾶν τὸ σῶμα παρέξειν οἰκεῖον, διὰ χειμῶνος δὲ πάλιν αὖ τούτῳ τῷ πυρὶ τὸν προσφερόμενον ἔξωθεν καὶ περιστάμενον πάγον ἀμυνεῖσθαι μετρίως. ταῦτα ἡμῶν διανοηθεὶς ὁ κηροπλάστης, ὕδατι μὲν καὶ πυρὶ καὶ γῇ ξυμμίξας καὶ ξυναρμόσας, ἐξ ὀξέος καὶ ἁλμυροῦ ξυνθεὶς ζύμωμα ὑπομίξας αὐτοῖς, σάρκα ἔγχυμον καὶ μαλακὴν ξυνέστησε· τὴν δὲ τῶν νεύρων φύσιν ἐξ ὀστοῦ καὶ σαρκὸς ἀζύμου κράσεως μίαν ἐξ ἀμφοῖν μέσην δυνάμει ξυνεκεράσατο, ξανθῷ χρώματι προσχρώμενος. ὅθεν συντονωτέραν μὲν καὶ γλισχροτέραν σαρκῶν, μαλακωτέραν δὲ ὀστῶν ὑγροτέραν τε ἐκτήσατο δύναμιν νεῦρα· οἷς ξυμ-

---

[1] ἐσθήματα conj. A. E. Taylor: κτήματα mss., Zur.

[1] *i.e.* the principle of plurality, *cf.* 35 B.

TIMAEUS

seed He closed it in with a ring-fence of stony substance; and therein He made joints, using as an aid the power of the Other [1] as an intermediary between them, for the sake of movement and bending. And inasmuch as He deemed that the texture of the bony substance was too hard and inflexible, and that if it were fired and cooled again it would decay and speedily destroy the seed within it, for these reasons He contrived the species known as sinew and flesh. He designed to bind all the limbs together by means of the former, which tightens and relaxes itself around the pivots, and thus cause the body to bend and stretch itself. And the flesh He designed to be a shield against the heat and a shelter against the cold; and, moreover, that in case of falls it should yield to the body softly and gently, like padded garments [2]; and, inasmuch as it contains within it warm moisture, that it should supply in summer, by its perspiration and dampness, a congenial coolness over the exterior of the whole body, and contrariwise in winter defend the body sufficiently, by means of its fire, from the frost which attacks and surrounds it from without. Wherefore, with this intent, our Modeller mixed and blended together water and fire and earth, and compounding a ferment of acid and salt mixed it in therewith, and thus moulded flesh full of sap and soft. And the substance of the sinews He compounded of a mixture of bone and unfermented flesh, forming a single substance blended of both and intermediate in quality, and he used yellow also for its colouring. Hence it is that the sinews have acquired a quality that is firmer and more rigid than flesh, but softer and more elastic than bone.

[2] *Cf.* 70 D.

74 περιλαβὼν ὁ θεὸς ὀστᾶ καὶ μυελόν, δήσας πρὸς ἄλληλα νεύροις, μετὰ ταῦτα σαρξὶ πάντα αὐτὰ κατεσκίασεν ἄνωθεν.

E Ὅσα μὲν οὖν ἐμψυχότατα τῶν ὀστῶν ἦν, ὀλιγίσταις ξυνέφραττε σαρξίν, ἃ δ' ἀψυχότατα ἐντός, πλείσταις καὶ πυκνοτάταις. καὶ δὴ καὶ κατὰ τὰς ξυμβολὰς τῶν ὀστῶν, ὅπῃ μήτινα ἀνάγκην ὁ λόγος ἀπέφαινε δεῖν αὐτὰς εἶναι, βραχεῖαν σάρκα ἔφυσεν, ἵνα μήτε ἐμποδὼν ταῖς καμπαῖσιν οὖσαι δύσφορα τὰ σώματα ἀπεργάζοιντο, ἅτε δυσκίνητα γιγνόμενα, μήτ' αὖ πολλαὶ καὶ πυκναὶ σφόδρα τε ἐν ἀλλήλαις ἐμπεπιλημέναι, διὰ στερεότητα ἀναισθησίαν ἐμποιοῦσαι, δυσμνημονευτότερα καὶ κωφότερα τὰ περὶ τὴν διάνοιαν ποιοῖεν. διὸ δὴ 75 τό τε τῶν μηρῶν καὶ κνημῶν καὶ τὸ περὶ τὴν τῶν ἰσχίων φύσιν τά τε τῶν βραχιόνων ὀστᾶ καὶ τὰ τῶν πήχεων, καὶ ὅσα ἄλλα ἡμῶν ἄναρθρα, ὅσα τε ἐντὸς ὀστᾶ δι' ὀλιγότητα ψυχῆς ἐν μυελῷ κενά ἐστι φρονήσεως, ταῦτα πάντα ξυμπεπλήρωται σαρξίν, ὅσα δ' ἔμφρονα, ἧττον, εἰ μή πού τινα αὐτὴν καθ' αὑτὴν αἰσθήσεων ἕνεκα σάρκα οὕτω ξυνέστησεν, οἷον τὸ τῆς γλώττης εἶδος· τὰ δὲ πλεῖστα ἐκείνως. ἡ γὰρ ἐξ ἀνάγκης γιγνομένη καὶ ξυντρεφομένη

B φύσις οὐδαμῇ προσδέχεται πυκνὸν ὀστοῦν καὶ σάρκα πολλὴν ἅμα τε αὐτοῖς ὀξυήκοον αἴσθησιν. μάλιστα γὰρ ἂν αὐτὰ πάντων ἔσχεν ἡ περὶ τὴν κεφαλὴν ξύστασις, εἴπερ ἅμα ξυμπίπτειν ἠθελησάτην, καὶ τὸ τῶν ἀνθρώπων γένος σαρκώδη ἔχον ἐφ' ἑαυτῷ καὶ νευρώδη κρατεράν τε κεφαλὴν βίον ἂν διπλοῦν καὶ πολλαπλοῦν καὶ ὑγιεινότερον καὶ

---

[1] *i.e.* those of the head and spine.

TIMAEUS

With these, then, God enclosed the bones and marrow, first binding them one to another with the sinews, and then shrouding them all over with flesh.

All the bones, then, that possessed most soul [1] He enclosed in least flesh, but the bones which contained least soul with most and most dense flesh; moreover, at the junctions of the bones, except where reason revealed some necessity for its existence, He made but little flesh to grow, lest by hindering the flexions it should make the bodies unwieldy, because stiff in movement, or else through its size and density, when thickly massed together, it should produce insensitiveness, owing to its rigidity, and thereby cause the intellectual parts to be more forgetful and more obtuse. Wherefore the thighs and the shins and the region of the loins and the bones of the upper and lower arm, and all our other parts which are jointless, and all those bones which are void of intelligence within, owing to the small quantity of soul in the marrow—all these are abundantly supplied with flesh; but those parts which are intelligent are supplied less abundantly—except possibly where He so fashioned the flesh that it can of itself convey sensations, as is the case with the tongue; but most of these parts He made in the way described above. For the substance which is generated by necessity and grows up with us in no wise admits of quick perception co-existing with dense bone and abundant flesh. For if these characteristics were willing to consort together, then the structure of the head would have acquired them most of all, and mankind, crowned with a head that was fleshy and sinewy and strong, would have enjoyed a life that was twice (nay, many times) as long as our present life, and

PLATO

75 ἀλυπότερον τοῦ νῦν κατεκτήσατο· νῦν δὲ τοῖς περὶ τὴν ἡμετέραν γένεσιν δημιουργοῖς, ἀναλογιζομένοις C πότερον πολυχρονιώτερον χεῖρον ἢ βραχυχρονιώτερον βέλτιον ἀπεργάσαιντο γένος, ξυνέδοξε τοῦ πλείονος βίου φαυλοτέρου δὲ τὸν ἐλάττονα ἀμείνονα ὄντα παντὶ πάντως αἱρετέον, ὅθεν δὴ μανῷ μὲν ὀστῷ, σαρξὶ δὲ καὶ νεύροις κεφαλήν, ἅτε οὐδὲ καμπὰς ἔχουσαν, οὐ ξυνεστέγασαν. κατὰ πάντ' οὖν ταῦτα εὐαισθητοτέρα μὲν καὶ φρονιμωτέρα, πολὺ δὲ ἀσθενεστέρα παντὸς ἀνδρὸς προσετέθη κεφαλὴ σώματι.

Τὰ δὲ νεῦρα διὰ ταῦτα καὶ οὕτως ὁ θεὸς ἐπ' D ἐσχάτην τὴν κεφαλὴν περιστήσας κύκλῳ περὶ τὸν τράχηλον ἐκόλλησεν ὁμοιότητι, καὶ τὰς σιαγόνας ἄκρας αὐτοῖς ξυνέδησεν ὑπὸ τὴν φύσιν τοῦ προσώπου· τὰ δ' ἄλλα εἰς ἅπαντα τὰ μέλη διέσπειρε, ξυνάπτων ἄρθρον ἄρθρῳ.

Τὴν δὲ δὴ τοῦ στόματος ἡμῶν δύναμιν ὀδοῦσι καὶ γλώττῃ καὶ χείλεσιν ἕνεκα τῶν ἀναγκαίων καὶ τῶν ἀρίστων διεκόσμησαν οἱ διακοσμοῦντες, ᾗ E νῦν διατέτακται, τὴν μὲν εἴσοδον τῶν ἀναγκαίων μηχανώμενοι χάριν, τὴν δ' ἔξοδον τῶν ἀρίστων· ἀναγκαῖον μὲν γὰρ πᾶν ὅσον εἰσέρχεται τροφὴν διδὸν τῷ σώματι, τὸ δὲ λόγων νᾶμα ἔξω ῥέον καὶ ὑπηρετοῦν φρονήσει κάλλιστον καὶ ἄριστον πάντων ναμάτων.

Τὴν δ' αὖ κεφαλὴν οὔτε μόνον ὀστεΐνην ψιλὴν δυνατὸν ἐᾶν ἦν διὰ τὴν ἐν ταῖς ὥραις ἐφ' ἑκάτερον ὑπερβολήν, οὔτ' αὖ ξυσκιασθεῖσαν. κωφὴν καὶ ἀναίσθητον διὰ τὸν τῶν σαρκῶν ὄχλον περιδεῖν γιγνο-

TIMAEUS

healthier, to boot, and more free from pain. But as it is, when the Constructors of our being were cogitating whether they should make a kind that was more long-lived and worse or more short-lived and better, they agreed that the shorter and superior life should by all means be chosen by all rather than the longer and inferior. Wherefore they covered the head closely with thin bone, but not with flesh and sinews, since it was also without flexions. For all these reasons, then, the head that was joined to the body in every man was more perceptive and more intelligent but less strong.

It was on these grounds and in this way that God set the sinews at the bottom of the head round about the neck and glued them there symmetrically; and with these He fastened the extremities of the jaws below the substance of the face; and the rest of the sinews He distributed amongst all the limbs, attaching joint to joint.

And those who fashioned the features of our mouth fashioned it with teeth and tongue and lips, even as it is fashioned now, for ends both necessary and most good, contriving it as an entrance with a view to necessary ends, and as an outlet with a view to the ends most good. For all that enters in and supplies food to the body is necessary; while the stream of speech which flows out and ministers to intelligence is of all streams the fairest and most good.

Moreover, it was not possible to leave the head to consist of bare bone only, because of the excessive variations of temperature in either direction, due to the seasons; nor yet was it possible to allow it to be shrouded up, and to become, in consequence, stupid and insensitive owing to its burdensome mass of flesh.

76 μένην. τῆς δὴ σαρκοειδοῦς φύσεως οὐ καταξηραι-
νομένης λέμμα μεῖζον περιγιγνόμενον ἐχωρίζετο,
δέρμα τὸ νῦν λεγόμενον. τοῦτο δὲ διὰ τὴν περὶ
τὸν ἐγκέφαλον νοτίδα ξυνιόν, αὐτὸ πρὸς αὑτὸ καὶ
βλαστάνον κύκλῳ περιημφίεννυε τὴν κεφαλήν. ἡ
δὲ νοτὶς ὑπὸ τὰς ῥαφὰς ἀνιοῦσα ἦρδε καὶ συν-
έκλεισεν αὐτὸ ἐπὶ τὴν κορυφήν, οἷον ἅμμα ξυναγα-
γοῦσα· τὸ δὲ τῶν ῥαφῶν παντοδαπὸν εἶδος γέγονε
διὰ τὴν τῶν περιόδων δύναμιν καὶ τῆς τροφῆς,
μᾶλλον μὲν ἀλλήλοις μαχομένων τούτων πλείους,
B ἧττον δὲ ἐλάττους. τοῦτο δὴ πᾶν τὸ δέρμα κύκλῳ
κατεκέντει πυρὶ τὸ θεῖον, τρηθέντος δὲ καὶ τῆς
ἰκμάδος ἔξω δι' αὐτοῦ φερομένης τὸ μὲν ὑγρὸν καὶ
θερμὸν ὅσον εἰλικρινὲς ἀπῄειν, τὸ δὲ μικτὸν ἐξ ὧν
καὶ τὸ δέρμα ἦν, αἰρόμενον μὲν ὑπὸ τῆς φορᾶς ἔξω
μακρὸν ἐτείνετο, λεπτότητα ἴσην ἔχον τῷ κατα-
κεντήματι, διὰ δὲ βραδυτῆτα ἀπωθούμενον ὑπὸ τοῦ
περιεστῶτος ἔξωθεν πνεύματος πάλιν ἐντὸς ὑπὸ τὸ
C δέρμα εἰλλόμενον κατερριζοῦτο· καὶ κατὰ ταῦτα δὴ
τὰ πάθη τὸ τριχῶν γένος ἐν τῷ δέρματι πέφυκε,
ξυγγενὲς μὲν ἱμαντῶδες ὂν αὐτοῦ, σκληρότερον δὲ
καὶ πυκνότερον τῇ πιλήσει τῆς ψύξεως, ἣν ἀπο-
χωριζομένη δέρματος ἑκάστη θρὶξ ψυχθεῖσα ξυνεπι-
λήθη. τούτῳ δὴ λασίαν ἡμῶν ἀπειργάσατο τὴν
κεφαλὴν ὁ ποιῶν, χρώμενος μὲν αἰτίοις τοῖς εἰρημένοις, διανοούμενος δὲ ἀντὶ σαρκὸς αὐτὸ δεῖν
εἶναι στέγασμα τῆς περὶ τὸν ἐγκέφαλον ἕνεκα
D ἀσφαλείας κοῦφον καὶ θέρους χειμῶνός τε ἱκανὸν
σκιὰν καὶ σκέπην παρέχειν, εὐαισθησίας δὲ οὐδὲν
διακώλυμα ἐμποδὼν γενησόμενον.

TIMAEUS

Accordingly, of the fleshy substance which was not being fully dried up a larger enveloping film was separated off, forming what is now called " skin." And this, having united with itself because of the moisture round the brain and spreading, formed a vesture round about the head; and this was damped by the moisture ascending under the seams and closed down over the crown, being drawn together as it were in a knot; and the seams had all kinds of shapes owing to the force of the soul's revolutions and of her food, being more in number when these are more in conflict with one another, and less when they are less in conflict. And the Deity kept puncturing all this skin round about with fire; and when the skin was pierced and the moisture flew out through it, all the liquid and heat that was pure went away, but such as was mixed with the substance whereof the skin also was composed was lifted up by the motion and extended far beyond the skin, being of a fineness to match the puncture; but since it was thrust back, because of its slowness, by the external air that surrounded it, it coiled itself round inside and rooted itself under the skin. Such, then, were the processes by which hair grew in the skin, it being a cord-like species akin to the skin but harder and denser owing to the constriction of the cold, whereby each hair as it separated off from the skin was chilled and constricted. Making use, then, of the causes mentioned our Maker fashioned the head shaggy with hair, purposing that, in place of flesh, the hair should serve as a light roofing for the part about the brain for safety's sake, providing a sufficient shade and screen alike in summer and in winter, while proving no obstacle in the way of easy perception.

Τὸ δὲ ἐν τῇ περὶ τοὺς δακτύλους καταπλοκῇ τοῦ νεύρου καὶ τοῦ δέρματος ὀστοῦ τε, ξυμμιχθὲν ἐκ τριῶν, ἀποξηρανθὲν ἓν κοινὸν ξυμπάντων σκληρὸν γέγονε δέρμα, τοῖς μὲν ξυναιτίοις τούτοις δημιουργηθέν, τῇ δ' αἰτιωτάτῃ διανοίᾳ τῶν ἔπειτα ἐσομένων ἕνεκα εἰργασμένον· ὡς γάρ ποτε ἐξ ἀνδρῶν γυναῖκες E καὶ τἆλλα θηρία γενήσοιντο, ἠπίσταντο οἱ ξυνιστάντες ἡμᾶς, καὶ δὴ καὶ τῆς τῶν ὀνύχων χρείας ὅτι πολλὰ τῶν θρεμμάτων καὶ ἐπὶ πολλὰ δεήσοιτο ᾔδεσαν, ὅθεν ἐν ἀνθρώποις εὐθὺς γιγνομένοις ὑπετυπώσαντο τὴν τῶν ὀνύχων γένεσιν. τούτῳ δὴ τῷ λόγῳ καὶ ταῖς προφάσεσι ταύταις δέρμα τρίχας ὄνυχάς τε ἐπ' ἄκροις τοῖς κώλοις ἔφυσαν.

Ἐπειδὴ δὲ πάντ' ἦν τὰ τοῦ θνητοῦ ζῴου ξυμπεφυ-
77 κότα μέρη καὶ μέλη, τὴν δὲ ζωὴν ἐν πυρὶ καὶ πνεύματι ξυνέβαινεν ἐξ ἀνάγκης ἔχειν αὐτῷ, καὶ διὰ ταῦτα ὑπὸ τούτων τηκόμενον κενούμενόν τ' ἔφθινε, βοήθειαν αὐτῷ θεοὶ μηχανῶνται· τῆς γὰρ ἀνθρωπίνης ξυγγενῆ φύσεως φύσιν ἄλλαις ἰδέαις καὶ αἰσθήσεσι κεραννύντες, ὥσθ' ἕτερον ζῷον εἶναι, φυτεύουσιν· ἃ δὴ νῦν ἥμερα δένδρα καὶ φυτὰ καὶ σπέρματα παιδευθέντα ὑπὸ γεωργίας τιθασῶς πρὸς ἡμᾶς ἔσχε, πρὶν δ' ἦν μόνα τὰ τῶν ἀγρίων γένη,
B πρεσβύτερα τῶν ἡμέρων ὄντα. πᾶν γὰρ οὖν, ὅ τι περ ἂν μετάσχῃ τοῦ ζῆν, ζῷον μὲν ἂν ἐν δίκῃ λέγοιτο ὀρθότατα· μετέχει γε μὴν τοῦτο, ὃ νῦν

---

[1] Cf. 68 E f.   [2] Cf. 90 E ff.

TIMAEUS

And at the place in the fingers where sinew and skin and bone were interlaced there was formed a material blended of these three; and this when it was dried off became a single hard skin compounded of them all; and whereas these were the auxiliary causes [1] whereby it was fashioned, it was wrought by the greatest of causes, divine Purpose, for the sake of what should come to pass hereafter. For those who were constructing us knew that out of men women should one day spring and all other animals [2]; and they understood, moreover, that many of these creatures would need for many purposes the help of nails; wherefore they impressed upon men at their very birth the rudimentary structure of finger-nails. Upon this account and with these designs they caused skin to grow into hair and nails upon the extremities of the limbs.

And when all the limbs and parts of the mortal living creature had been naturally joined together, it was so that of necessity its life consisted in fire and air; and because of this it wasted away when dissolved by these elements or left empty thereby; wherefore the Gods contrived succour for the creature. Blending it with other shapes and senses they engendered a substance akin to that of man, so as to form another living creature: such are the cultivated trees and plants and seeds which have been trained by husbandry and are now domesticated amongst us; but formerly the wild kinds only existed, these being older than the cultivated kinds. For everything, in fact, which partakes of life may justly and with perfect truth be termed a living creature. Certainly that creature which we are now describing partakes of the third kind of soul, which is seated, as

λέγομεν, τοῦ τρίτου ψυχῆς εἴδους, ὃ μεταξὺ φρενῶν ὀμφαλοῦ τε ἱδρύσθαι λόγος, ᾧ δόξης μὲν λογισμοῦ τε καὶ νοῦ μέτεστι τὸ μηδέν, αἰσθήσεως δὲ ἡδείας καὶ ἀλγεινῆς μετὰ ἐπιθυμιῶν. πάσχον γὰρ διατελεῖ πάντα, στραφέντι δ' αὐτῷ ἐν ἑαυτῷ περὶ ἑαυτό, τὴν μὲν ἔξωθεν ἀπωσαμένῳ κίνησιν, τῇ δ' οἰκείᾳ χρησαμένῳ, τῶν αὐτοῦ τι λογίσασθαι κατιδόντι φύσει[1] οὐ παραδέδωκεν ἡ γένεσις. διὸ δὴ ζῇ μὲν ἔστι τε οὐχ ἕτερον ζῴου, μόνιμον δὲ καὶ κατερριζωμένον πέπηγε διὰ τὸ τῆς ὑφ' ἑαυτοῦ κινήσεως ἐστερῆσθαι.

Ταῦτα δὴ τὰ γένη πάντα φυτεύσαντες οἱ κρείττους τοῖς ἥττοσιν ἡμῖν τροφήν, τὸ σῶμα αὐτὸ ἡμῶν διωχέτευσαν τέμνοντες οἷον ἐν κήποις ὀχετούς, ἵν' ὥσπερ ἐκ νάματος ἐπιόντος ἄρδοιτο. καὶ πρῶτον μὲν ὀχετοὺς κρυφαίους ὑπὸ τὴν ξύμφυσιν τοῦ δέρματος καὶ τῆς σαρκὸς δύο φλέβας ἔτεμον νωτιαίας, δίδυμον ὡς τὸ σῶμα ἐτύγχανε δεξιοῖς τε καὶ ἀριστεροῖς ὄν. ταύτας δὲ καθῆκαν παρὰ τὴν ῥάχιν, καὶ τὸν γόνιμον μεταξὺ λαβόντες μυελόν, ἵνα οὗτός τε ὅ τι μάλιστα θάλλοι, καὶ ἐπὶ τἆλλα εὔρους ἐντεῦθεν ἅτ' ἐπὶ κάταντες ἡ ἐπίχυσις γιγνομένη παρέχοι τὴν ὑδρείαν ὁμαλήν. μετὰ δὲ ταῦτα σχίσαντες περὶ τὴν κεφαλὴν τὰς φλέβας καὶ δι' ἀλλήλων ἐναντίας πλέξαντες διεῖσαν, τὰς μὲν ἐκ τῶν δεξιῶν ἐπὶ τἀριστερὰ τοῦ σώματος, τὰς δ' ἐκ τῶν ἀριστερῶν ἐπὶ τὰ δεξιὰ κλίναντες, ὅπως δεσμὸς ἅμα τῇ κεφαλῇ πρὸς τὸ σῶμα εἴη μετὰ τοῦ δέρματος, ἐπειδὴ νεύροις οὐκ ἦν κύκλῳ κατὰ

---

[1] φύσει MSS.: φύσιν Zur.

[1] Cf. 70 D ff., 72 E ff.

we affirm, between the midriff and the navel,[1] and which shares not at all in opinion and reasoning and mind but in sensation, pleasant and painful, together with desires. For inasmuch as it continues wholly passive and does not turn within itself around itself, repelling motion from without and using its own native motion, it is not endowed by its original constitution with a natural capacity for discerning or reflecting upon any of its own experiences. Wherefore it lives indeed and is not other than a living creature, but it remains stationary and rooted down owing to its being deprived of the power of self-movement.

And when our Superiors had generated all these kinds as nutriment for us inferior beings, they channelled out our body itself, like as if they were cutting channels in gardens, to the end that it might be irrigated as it were by an inflowing stream. And firstly, beneath the junction of the skin and flesh they cut for hidden channels two veins[2] along the back, seeing that the body was in fact double, with right side and left; and these they drew down along by the spine, keeping between them the spermatic marrow, in order that this might thrive as much as possible, and that the stream of moisture from there, being in a downward course, might flow easily to the other parts and cause the irrigation to be uniform. After this they clave the veins round the head and interlaced them, and drew them opposite ways, bending those from the right of the head to the left and those from the left to the right, in order that they, together with the skin, might serve as a bond between the head and the body, seeing that the head was not encircled

[2] *i.e.* the *aorta* and the *vena cava*. The distinction between veins and arteries was unknown in Plato's time.

PLATO

77 κορυφὴν περιειλημμένη, καὶ δὴ καὶ τὸ τῶν αἰσθήσεων πάθος ἵν᾽ ἀφ᾽ ἑκατέρων τῶν μερῶν εἰς ἅπαν τὸ σῶμα εἴη διάδηλον.[1]

78 Τὸ δ᾽ ἐντεῦθεν ἤδη τὴν ὑδραγωγίαν παρεσκεύασαν τρόπῳ τινὶ τοιῷδε, ὃν κατοψόμεθα ῥᾷον προδιομολογησάμενοι τὸ τοιόνδε, ὅτι πάντα ὅσα ἐξ ἐλαττόνων ξυνίσταται στέγει τὰ μείζω, τὰ δ᾽ ἐκ μειζόνων τὰ σμικρότερα οὐ δύναται, πῦρ δὲ πάντων γενῶν σμικρομερέστατον, ὅθεν δι᾽ ὕδατος καὶ γῆς ἀέρος τε καὶ ὅσα ἐκ τούτων ξυνίσταται διαχωρεῖ καὶ στέγειν οὐδὲν αὐτὸ δύναται. ταὐτὸν δὴ καὶ περὶ τῆς παρ᾽ ἡμῖν κοιλίας διανοητέον, ὅτι σιτία μὲν καὶ ποτὰ
B ὅταν εἰς αὐτὴν ἐμπέσῃ στέγει, πνεῦμα δὲ καὶ πῦρ σμικρομερέστερα ὄντα τῆς αὐτῆς ξυστάσεως οὐ δύναται. τούτοις οὖν κατεχρήσατο ὁ θεὸς εἰς τὴν ἐκ τῆς κοιλίας ἐπὶ τὰς φλέβας ὑδρείαν, πλέγμα ἐξ ἀέρος καὶ πυρὸς οἷον οἱ κύρτοι ξυνυφηνάμενος, διπλᾶ κατὰ τὴν εἴσοδον ἐγκύρτια ἔχον, ὧν θάτερον αὖ πάλιν διέπλεξε δίκρουν· καὶ ἀπὸ τῶν ἐγκυρτίων δὴ διετείνατο οἷον σχοίνους κύκλῳ διὰ παντὸς πρὸς τὰ ἔσχατα τοῦ πλέγματος. τὰ μὲν οὖν ἔνδον

[1] διάδηλον some mss., Galen : διαδιδόμενον Zur.

---

[1] Cf. 56 A, 58 A ff.

TIMAEUS

by sinews at the crown; and in order, also, that the sense-impressions derived from the parts on either side might be manifest to the whole body.

Thereupon they arranged the irrigation on some such plan as this—a plan which we shall perceive more easily when we have first agreed upon the following postulates. All bodies composed of smaller particles shut in the larger, but those composed of larger particles cannot shut in the smaller; and fire, because of all the elements it has the smallest particles,[1] passes through water and earth and air and all things composed thereof, and nothing can shut it in. We must conceive that the same law holds good of the action of our belly. Whenever foods and drinks flow into it it shuts them in, but air and fire, being of smaller particles than its own structure, it cannot shut in. These elements, therefore, God employed to provide irrigation from the belly to the veins, weaving out of air and fire a veil of mesh-work like unto a fish-weel, having two inner-weels at its entrance; and one of these inner-weels He wove over again so as to make it bifurcated; and from the inner-weels He stretched as it were ropes all over it in a circle up to the extremities of the veil.[2]

---

[2] A rough diagram (after Archer-Hind, based on Galen) will best serve to explain this obscure account:

$a=$ upper ἐγκύρτιον ("inner-weel").
$b=$ lower ἐγκύρτιον ("inner-weel").
$c=$ outer stratum of air.
$d=$ inner stratum of fire.
$e=$ double air-passages through nostrils.
$f=$ single food-passage through mouth.

ἐκ πυρὸς συνεστήσατο τοῦ πλοκάνου ἅπαντα, τὰ δ' ἐγκύρτια καὶ τὸ κύτος ἀεροειδῆ· καὶ λαβὼν αὐτὸ περιέστησε τῷ πλασθέντι ζώῳ τρόπον τοιόνδε. τὸ μὲν τῶν ἐγκυρτίων εἰς τὸ στόμα μεθῆκε· διπλοῦ δὲ ὄντος αὐτοῦ κατὰ μὲν τὰς ἀρτηρίας εἰς τὸν πλεύμονα καθῆκε θάτερον, τὸ δ' εἰς τὴν κοιλίαν παρὰ τὰς ἀρτηρίας· τὸ δ' ἕτερον σχίσας τὸ μέρος ἑκάτερον κατὰ τοὺς ὀχετοὺς τῆς ῥινὸς ἀφῆκε κοινόν, ὥσθ' ὅτε μὴ κατὰ στόμα ἴοι θάτερον, ἐκ τούτου πάντα καὶ τὰ ἐκείνου ῥεύματα ἀναπληροῦσθαι. τὸ δ' ἄλλο κύτος τοῦ κύρτου περὶ τὸ σῶμα ὅσον κοῖλον ἡμῶν περιέφυσε, καὶ πᾶν δὴ τοῦτο τοτὲ μὲν εἰς τὰ ἐγκύρτια ξυρρεῖν μαλακῶς, ἅτε ἀέρα ὄντα, ἐποίησε, τοτὲ δὲ ἀναρρεῖν μὲν τὰ ἐγκύρτια, τὸ δὲ πλέγμα, ὡς ὄντος τοῦ σώματος μανοῦ, δύεσθαι εἴσω δι' αὐτοῦ καὶ πάλιν ἔξω, τὰς δ' ἐντὸς τοῦ πυρὸς ἀκτῖνας διαδεδεμένας ἀκολουθεῖν ἐφ' ἑκάτερα ἰόντος τοῦ ἀέρος, καὶ τοῦτο, ἕωσπερ ἂν τὸ θνητὸν ξυνεστήκῃ ζῶον, μὴ διαπαύεσθαι γιγνόμενον. τούτῳ δὲ δὴ τῷ γένει τὸν τὰς ἐπωνυμίας θέμενον ἀναπνοὴν καὶ ἐκπνοὴν λέγομεν θέσθαι τοὔνομα. πᾶν δὲ δὴ τό τ' ἔργον καὶ τὸ πάθος τοῦθ' ἡμῶν τῷ σώματι γέγονεν ἀρδομένῳ καὶ ἀναψυχομένῳ τρέφεσθαι καὶ ζῆν· ὁπόταν γὰρ εἴσω καὶ ἔξω τῆς ἀναπνυῆς ἰούσης τὸ πῦρ ἐντὸς ξυνημμένον ἕπηται, διαιωρούμενον δὲ ἀεὶ διὰ τῆς κοιλίας εἰσελθὸν τὰ σιτία καὶ ποτὰ λάβῃ, τήκει δή,

---

[1] A mythical figure, like Adam in Gen. ii. 19-20; *cf. Cratyl.* 438-439.

TIMAEUS

Now the inward parts of the veil He constructed wholly of fire, but the inner-weels and the envelope of air; and taking this He placed it round about the living creature that was moulded in the following manner. The part consisting of the inner-weels He let down into the mouth; and since this part was twofold, He let down one inner-weel by way of the windpipe into the lungs, and the other into the belly alongside the windpipe. And cleaving the former of these weels in two He gave to both sections a common outlet by way of the channels of the nose, so that when the first conduit by way of the mouth failed to act, its streams as well should be plenished from this. The rest of the enveloping mesh-work He made to grow round all the hollow part of our body; and He caused all this at one time to flow gently into the inner-weels, seeing they were of air, and at another time the weels to flow back into it. And inasmuch as the body was porous, He caused the veil to pass in through it and out again; and the inner rays of fire that were enclosed within it He made to follow the air as it moved in either direction; whence it comes that, so long as the mortal living creature preserves its structure, this process goes on unceasingly. And to this kind of process the Giver of Titles [1] gave, as we say, the names of "inspiration" and "expiration." And the whole of this mechanism and its effects have been created in order to secure nourishment and life for our body, by means of moistening and cooling. For as the respiration goes in and out the inward fire attached thereto follows it; and whenever in its constant oscillations this fire enters in through the belly and lays hold on the meats and drinks, it dissolves them, and dividing them into small particles

καὶ κατὰ σμικρὰ διαιροῦν, διὰ τῶν ἐξόδων ᾗπερ πορεύεται διάγον, οἷον ἐκ κρήνης ἐπ' ὀχετοὺς ἐπὶ τὰς φλέβας ἀντλοῦν αὐτά, ῥεῖν ὥσπερ αὐλῶνος διὰ τοῦ σώματος τὰ τῶν φλεβῶν ποιεῖ ῥεύματα.

Πάλιν δὲ τὸ τῆς ἀναπνοῆς ἴδωμεν πάθος, αἷς χρώμενον αἰτίαις τοιοῦτον γέγονεν οἷόνπερ τὰ νῦν B ἐστίν. ὧδ' οὖν. ἐπειδὴ κενὸν οὐδέν ἐστιν, εἰς ὃ τῶν φερομένων δύναιτ' ἂν εἰσελθεῖν τι, τὸ δὲ πνεῦμα φέρεται παρ' ἡμῶν ἔξω, τὸ μετὰ τοῦτο ἤδη παντὶ δῆλον ὡς οὐκ εἰς κενόν, ἀλλὰ τὸ πλησίον ἐκ τῆς ἕδρας ὠθεῖ· τὸ δὲ ὠθούμενον ἐξελαύνει τὸ πλησίον ἀεί, καὶ κατὰ ταύτην τὴν ἀνάγκην πᾶν περιελαυννόμενον εἰς τὴν ἕδραν ὅθεν ἐξῆλθε τὸ πνεῦμα, εἰσιὸν ἐκεῖσε καὶ ἀναπληροῦν αὐτὴν ξυνέπεται τῷ πνεύματι, καὶ τοῦτο ἅμα πᾶν οἷον τροχοῦ περιαγομένου γίγνεται διὰ τὸ κενὸν μηδὲν εἶναι. C διὸ δὴ τὸ τῶν στηθῶν καὶ τὸ τοῦ πλεύμονος ἔξω μεθιὲν τὸ πνεῦμα πάλιν ὑπὸ τοῦ περὶ τὸ σῶμα ἀέρος, εἴσω διὰ μανῶν τῶν σαρκῶν δυομένου καὶ περιελαυννομένου, γίγνεται πλῆρες· αὖθις δὲ ἀποτρεπόμενος ὁ ἀὴρ καὶ διὰ τοῦ σώματος ἔξω ἰὼν εἴσω τὴν ἀναπνοὴν περιωθεῖ κατὰ τὴν τοῦ στόματος καὶ τὴν τῶν μυκτήρων δίοδον. τὴν δὲ D αἰτίαν τῆς ἀρχῆς αὐτῶν θετέον τήνδε· πᾶν ζῷον αὐτοῦ τἀντὸς περὶ τὸ αἷμα καὶ τὰς φλέβας θερμότατα ἔχει, οἷον ἐν ἑαυτῷ πηγήν τινα ἐνοῦσαν πυρός· ὃ δὴ καὶ προσῃκάζομεν τῷ τοῦ κύρτου πλέγματι, κατὰ μέσον διατεταμένον ἐκ πυρὸς πεπλέχθαι πᾶν, τὰ δὲ ἄλλα, ὅσα ἔξωθεν, ἀέρος.

---

[1] *Cf.* 58 A.

TIMAEUS

it disperses them through the outlets by which it passes and draws them off to the veins, like water drawn into channels from a spring; and thus it causes the streams of the veins to flow through the body as through a pipe.

Once again let us consider the process of respiration, and the causes in virtue of which it has come to be such as it now is. This, then, is the way of it. Inasmuch as no void exists [1] into which any of the moving bodies could enter, while the breath from us moves outwards, what follows is plain to everyone—namely, that the breath does not enter a void but pushes the adjacent body from its seat; and the body thus displaced drives out in turn the next; and by this law of necessity every such body is driven round towards the seat from which the breath went out and enters therein, filling it up and following the breath; and all this takes place as one simultaneous process, like a revolving wheel, because that no void exists. Wherefore the region of the chest and that of the lungs when they let out the breath become filled again by the air surrounding the body, which filters in through the porous flesh and circulates round. And again, when the air is repelled and passes out through the body it pushes the inspired air round and in by way of the passages of the mouth and of the nostrils. The originating cause of these processes we must assume to be this. Every living creature has its inward parts round the blood and the veins extremely hot, as it were a fount of fire residing within it; and this region we have, in fact, likened to the envelope of the fish-weel, saying that all that was extended at its middle was woven of fire, whereas all the other and outward parts were of air. Now we must agree

PLATO

79 τὸ θερμὸν δὴ κατὰ φύσιν εἰς τὴν αὑτοῦ χώραν ἔξω πρὸς τὸ ξυγγενὲς ὁμολογητέον ἰέναι· δυοῖν δὲ ταῖν διεξόδοιν οὔσαιν, τῆς μὲν κατὰ τὸ σῶμα ἔξω, Ε τῆς δὲ αὖ κατὰ τὸ στόμα καὶ τὰς ῥῖνας, ὅταν μὲν ἐπὶ θάτερα ὁρμήσῃ, θάτερα περιωθεῖ· τὸ δὲ περιωσθὲν εἰς τὸ πῦρ ἐμπῖπτον θερμαίνεται, τὸ δ' ἐξιὸν ψύχεται. μεταβαλλούσης δὲ τῆς θερμότητος καὶ τῶν κατὰ τὴν ἑτέραν ἔξοδον θερμοτέρων γιγνομένων πάλιν ἐκείνῃ ῥέπον αὖ τὸ θερμότερον μᾶλλον, πρὸς τὴν αὑτοῦ φύσιν φερόμενον, περιωθεῖ τὸ κατὰ θάτερα· τὸ δὲ τὰ αὐτὰ πάσχον καὶ τὰ αὐτὰ ἀνταποδιδὸν ἀεί, κύκλον οὕτω σαλευόμενον ἔνθα καὶ ἔνθα ἀπειργασμένον ὑπ' ἀμφοτέρων τὴν ἀναπνοὴν καὶ ἐκπνοὴν γίγνεσθαι παρέχεται.

Καὶ δὴ καὶ τὰ τῶν περὶ τὰς ἰατρικὰς σικύας **80** παθημάτων αἴτια καὶ τὰ τῆς καταπόσεως τά τε τῶν ῥιπτουμένων, ὅσα ἀφεθέντα μετέωρα καὶ ὅσα ἐπὶ γῆς φέρεται, ταύτῃ διωκτέον, καὶ ὅσοι φθόγγοι ταχεῖς τε καὶ βραδεῖς ὀξεῖς τε καὶ βαρεῖς φαίνονται, τοτὲ μὲν ἀνάρμοστοι φερόμενοι δι' ἀνομοιότητα τῆς ἐν ἡμῖν ὑπ' αὐτῶν κινήσεως, τοτὲ δὲ ξύμφωνοι δι' ὁμοιότητα. τὰς γὰρ τῶν προτέρων καὶ θαττόνων οἱ βραδύτεροι κινήσεις, ἀποπαυομένας Β ἤδη τε εἰς ὅμοιον ἐληλυθυίας αἷς ὕστερον αὐτοὶ προσφερόμενοι κινοῦσιν ἐκείνας, καταλαμβάνουσι, καταλαμβάνοντες δὲ οὐκ ἄλλην ἐπεμβάλλοντες ἀνετάραξαν κίνησιν, ἀλλ' ἀρχὴν βραδυτέρας φορᾶς

---

[1] *Cf.* 67 A ff.

TIMAEUS

that heat, by Nature's law, goes out into its own region to its kindred substance; and inasmuch as there are two outlets, the one out by way of the body, the other by way of the mouth and the nose, whenever the fire rushes in one direction it propels the air round to the other, and the air which is thus propelled round becomes heated by streaming into the fire, whereas the air which passes out becomes cooled. And as the heat changes its situation and the particles about the other outlet become hotter, the hotter body in its turn tends in that direction, and moving towards its own substance propels round the air which is at the former outlet; and thus the air, by continually undergoing and transmitting the same affections, causes inspiration and expiration to come about as a result of this double process, as it were a wheel that oscillates backwards and forwards.

Moreover, we must trace out in this way the causes of the phenomena connected with medical cupping-glasses, and the causes of deglutition, and of projectiles, whether discharged aloft or flying over the surface of the earth; and the causes also of all the sounds [1] which because of their quickness or slowness seem shrill or deep, and the movement of which is at one time discordant because of the irregularity of the motion they cause within us, and at another time concordant because of its regularity. For the slower sounds overtake the motions of the earlier and quicker sounds when the latter begin to stop and have already fallen to a speed similar to that with which the slower sounds collide with them afterwards and move them; and when the slower overtake the quicker sounds they do not perturb them by imposing upon them a different motion, but they attach to

PLATO

κατὰ τὴν τῆς θάττονος, ἀπολληγούσης δὲ ὁμοιότητα προσάψαντες μίαν ἐξ ὀξείας καὶ βαρείας ξυνεκεράσαντο πάθην, ὅθεν ἡδονὴν μὲν τοῖς ἄφροσιν, εὐφροσύνην δὲ τοῖς ἔμφροσι διὰ τὴν τῆς θείας ἁρμονίας μίμησιν ἐν θνηταῖς γενομένην φοραῖς παρέσχον.

Καὶ δὴ καὶ τὰ τῶν ὑδάτων πάντα ῥεύματα, ἔτι C δὲ τὰ τῶν κεραυνῶν πτώματα καὶ τὰ θαυμαζόμενα ἠλέκτρων περὶ τῆς ἕλξεως καὶ τῶν Ἡρακλείων λίθων, πάντων τούτων ὁλκὴ μὲν οὐκ ἔστιν οὐδενί ποτε, τὸ δὲ κενὸν εἶναι μηδὲν περιωθεῖν τε αὐτὰ ταῦτα εἰς ἄλληλα, τό τε διακρινόμενα καὶ συγκρινόμενα πρὸς τὴν αὑτῶν διαμειβόμενα ἕδραν ἕκαστα ἰέναι πάντα, τούτοις τοῖς παθήμασι πρὸς ἄλληλα συμπλεχθεῖσι τεθαυματουργημένα τῷ κατὰ τρόπον ζητοῦντι φανήσεται.

D Καὶ δὴ καὶ τὸ τῆς ἀναπνοῆς, ὅθεν ὁ λόγος ὥρμησε, κατὰ ταῦτα καὶ διὰ τούτων γέγονεν, ὥσπερ ἐν τοῖς πρόσθεν εἴρηται, τέμνοντος μὲν τὰ σιτία τοῦ πυρός, αἰωρουμένου δὲ ἐντὸς τῷ πνεύματι ξυνεπομένου, τὰς φλέβας δὲ ἐκ τῆς κοιλίας τῇ ξυναιωρήσει πληροῦντος τῷ τὰ τετμημένα αὐτόθεν ἐπαντλεῖν· καὶ διὰ ταῦτα δὴ καθ' ὅλον τὸ σῶμα πᾶσι τοῖς ζώοις τὰ τῆς τροφῆς νάματα οὕτως ἐπίρρυτα E γέγονεν. νεότμητα δὲ καὶ ἀπὸ ξυγγενῶν ὄντα, τὰ μὲν καρπῶν, τὰ δὲ χλόης, ἃ θεὸς ἐπ' αὐτὸ τοῦθ' ἡμῖν ἐφύτευσεν, εἶναι τροφήν, παντοδαπὰ μὲν χρώ-

---

[1] εὐφροσύνη (*quasi* εὐφεροσύνη), derived from φέρω, φορ ("motion"); *cf. Cratyl.* 419 D. The two kinds of sound quicker and slower, are supposed to be blended by the tim they reach the ear.

[2] *Cf.* 47 c ff.   [3] *i.e.* amber.

TIMAEUS

them the beginning of a slower motion in accord with that which was quicker but is tending to cease; and thus from shrill and deep they blend one single sensation, furnishing pleasure thereby to the unintelligent, and to the intelligent that intellectual delight [1] which is caused by the imitation of the divine harmony [2] manifested in mortal motions.

Furthermore, as regards all flowings of waters, and fallings of thunderbolts, and the marvels concerning the attraction of electron [3] and of the Heraclean stone [4]—not one of all these ever possesses any real power of attraction; but the fact that there is no void, and that these bodies propel themselves round one into another, and that according as they separate or unite they all exchange places and proceed severally each to its own region,—it is by means of these complex and reciprocal processes that such marvels are wrought, as will be evident to him who investigates them properly.

Moreover, the process of respiration—with which our account commenced—came about, as we previously stated, in this manner and by these means. The fire divides the foods, and rises through the body following after the breath; and as it rises, with the breath it fills the veins from the belly by drawing into them from thence the divided particles. And it is owing to this that in all living creatures the streams of nutriment course in this way through the whole body. And inasmuch as these nutritive particles are freshly divided and derived from kindred substances,—some from fruits, and some from cereals, which God planted for us for the express purpose of serving as food,[5]

---

[4] *i.e.* the loadstone or magnet; *cf. Ion* 533 D.

[5] *Cf.* 77 A.

80 ματα ἴσχει διὰ τὴν ξύμμιξιν, ἡ δ' ἐρυθρὰ πλείστη περὶ αὐτὰ¹ χρόα διαθεῖ, τῆς τοῦ πυρὸς τομῆς τε καὶ ἐξομόρξεως ἐν ὑγρῷ δεδημιουργημένη φύσις· ὅθεν τοῦ κατὰ τὸ σῶμα ῥέοντος τὸ χρῶμα ἔσχεν οἵαν ὄψιν διεληλύθαμεν, ὃ καλοῦμεν αἷμα, νομὴν
81 σαρκῶν καὶ ξύμπαντος τοῦ σώματος, ὅθεν ὑδρευόμενα ἕκαστα πληροῖ τὴν τοῦ κενουμένου βάσιν· ὁ δὲ τρόπος τῆς πληρώσεως ἀποχωρήσεώς τε γίγνεται, καθάπερ ἐν τῷ παντὶ παντὸς ἡ φορὰ γέγονεν, ἣν τὸ ξυγγενὲς πᾶν φέρεται πρὸς ἑαυτό. τὰ μὲν γὰρ δὴ περιεστῶτα ἐκτὸς ἡμᾶς τήκει τε ἀεὶ καὶ διανέμει πρὸς ἕκαστον εἶδος τὸ ὁμόφυλον ἀποπέμποντα,² τὰ δὲ ἔναιμα αὖ, κερματισθέντα ἐντὸς παρ' ἡμῖν καὶ περιειλημμένα ὥσπερ ὑπ' οὐρανοῦ ξυνεστῶτος ἑκάστου τοῦ ζώου, τὴν τοῦ παντὸς
B ἀναγκάζεται μιμεῖσθαι φοράν· πρὸς τὸ ξυγγενὲς οὖν φερόμενον ἕκαστον τῶν ἐντὸς μερισθέντων τὸ κενωθὲν τότε πάλιν ἀνεπλήρωσεν. ὅταν μὲν δὴ πλέον τοῦ ἐπιρρέοντος ἀπίῃ, φθίνει πᾶν, ὅταν δὲ ἔλαττον, αὐξάνεται. νέα μὲν οὖν ξύστασις τοῦ παντὸς ζώου, καινὰ τὰ τρίγωνα οἷον ἐκ δρυόχων ἔτι ἔχουσα τῶν γενῶν, ἰσχυρὰν μὲν τὴν ξύγκλεισιν αὐτῶν πρὸς ἄλληλα κέκτηται, ξυμπέπηγε δὲ ὁ πᾶς
C ὄγκος αὐτῆς ἁπαλός, ἅτε ἐκ μυελοῦ μὲν νεωστὶ γεγονυίας, τεθραμμένης δὲ ἐν γάλακτι· τὰ δὴ περιλαμβανόμενα ἐν αὐτῇ τρίγωνα ἔξωθεν ἐπεισελθόντα, ἐξ ὧν ἂν ᾖ τά τε σιτία καὶ ποτά, τῶν ἑαυτῆς τριγώνων παλαιότερα ὄντα καὶ ἀσθενέ-

---

¹ αὐτὰ Galen : αὐτὸ MSS., Zur.
² ἀποπέμποντα some MSS. : ἀποπέμπον other MSS., Zur.

¹ Cf. 68 B·C.   ² Cf. 53 D ff.

TIMAEUS

—they get all varieties of colours because of their commingling, but red is the colour that runs through them most of all, it being a natural product of the action of the fire in dividing the liquid food and imprinting itself thereon.[1] Wherefore the colour of the stream which flows through the body acquired an appearance such as we have described; and this stream we call "blood," which is the nutriment of the flesh and of the whole body, each part drawing therefrom supplies of fluid and filling up the room of the evacuated matter. And the processes of filling and evacuating take place just as the motion of everything in the Universe takes place, namely, according to the law that every kindred substance moves towards its kind. For the bodies which surround us without are always dissolving us and sending off and distributing to each species of substance what is akin thereto; while the blood-particles, again, being minced up within us and surrounded by the structure of each creature as by a Heaven, are compelled to copy the motion of the whole; hence, when each of the particles that are divided up inside moves towards its kin, it fills up again the emptied place. And when what passes out is more than the inflow every creature decays, but when less, it increases. Now when the structure of the whole creature is new, inasmuch as the triangles which form its elements [2] are still fresh, and as it were straight from the stocks, it keeps them firmly interlocked one with another, and the whole mass of it is of a soft composition, seeing that it is newly produced from marrow and nourished on milk; and as the triangles contained therein, which have invaded it from without and go to form the meats and drinks, are older and weaker than its own, it

στερα καινοῖς ἐπικρατεῖ τέμνουσα, καὶ μέγα ἀπεργάζεται τὸ ζῷον τρέφουσα ἐκ πολλῶν ὁμοίων. ὅταν δ' ἡ ῥίζα τῶν τριγώνων χαλᾷ διὰ τὸ πολλοὺς ἀγῶνας ἐν πολλῷ χρόνῳ πρὸς πολλὰ ἠγωνίσθαι, τὰ μὲν τῆς τροφῆς εἰσιόντα οὐκέτι δύναται τέμνειν εἰς ὁμοιότητα ἑαυτοῖς, αὐτὰ δὲ ὑπὸ τῶν ἔξωθεν ἐπεισιόντων εὐπετῶς διαιρεῖται. φθίνει δὴ πᾶν ζῷον ἐν τούτῳ κρατούμενον, γῆράς τε ὀνομάζεται τὸ πάθος. τέλος δέ, ἐπειδὰν τῶν περὶ τὸν μυελὸν τριγώνων οἱ ξυναρμοσθέντες μηκέτι ἀντέχωσι δεσμοὶ τῷ πόνῳ διεσταμένοι, μεθιᾶσι τοὺς τῆς ψυχῆς αὖ δεσμούς, ἡ δὲ λυθεῖσα κατὰ φύσιν μεθ' ἡδονῆς ἐξέπτατο· πᾶν γὰρ τὸ μὲν παρὰ φύσιν ἀλγεινόν, τὸ δ' ᾗ πέφυκε γιγνόμενον ἡδύ. καὶ θάνατος δὴ κατὰ ταῦτα ὁ μὲν κατὰ νόσους καὶ ὑπὸ τραυμάτων γιγνόμενος ἀλγεινὸς καὶ βίαιος, ὁ δὲ μετὰ γήρας ἰὼν ἐπὶ τέλος κατὰ φύσιν ἀπονώτατος τῶν θανάτων καὶ μᾶλλον μεθ' ἡδονῆς γιγνόμενος ἢ λύπης.

Τὸ δὲ τῶν νόσων ὅθεν ξυνίσταται, δῆλόν που καὶ παντί. τεττάρων γὰρ ὄντων γενῶν, ἐξ ὧν συμπέπηγε τὸ σῶμα, γῆς πυρὸς ὕδατός τε καὶ ἀέρος, τούτων ἡ παρὰ φύσιν πλεονεξία καὶ ἔνδεια καὶ τῆς χώρας μετάστασις ἐξ οἰκείας ἐπ' ἀλλοτρίαν γιγνομένη, πυρός τε αὖ καὶ τῶν ἑτέρων ἐπειδὴ γένη πλείονα ἑνὸς ὄντα τυγχάνει, τὸ μὴ προσῆκον ἕκαστον ἑαυτῷ προσλαμβάνειν, καὶ πάνθ' ὅσα τοιαῦτα στάσεις καὶ νόσους παρέχει· παρὰ φύσιν γὰρ ἑκάστου γιγνομένου καὶ μεθισταμένου θερ-

---

[1] *i.e.* the *radical* structure of the primary triangles; *cf.* 53 D ff.

TIMAEUS

divides and overcomes them with its own new triangles, and thus renders the creature large by feeding it on many similar substances. But when the root of the triangles [1] grows slack owing to their having fought many fights during long periods, they are no longer able to divide the entering triangles of the food and assimilate them to themselves, but are themselves easily divided by those which enter from without; and in this condition every animal is overpowered and decays; and this process is named " old age." And finally, when the bonds of the triangles in the marrow which have been fitly framed together no longer resist the strain but fall asunder, they let slip in turn the bonds of the soul, and it, when thus naturally set loose, flies out gladly; for whereas every process which is contrary to nature is painful, that which takes place naturally is pleasurable. So too, in like manner, the death which occurs in consequence of disease or by wounds is painful and violent, but that which follows on old age and constitutes a natural end is the least grievous of deaths and is accompanied by more of pleasure than of pain.

The origin of disease is plain, of course, to everybody. For seeing that there are four elements of which the body is compacted,—earth, fire, water and air,—when, contrary to nature, there occurs either an excess or a deficiency of these elements, or a transference thereof from their native region to an alien region; or again, seeing that fire and the rest have each more than one variety, every time that the body admits an inappropriate variety, then these and all similar occurrences bring about internal disorders and disease. For when any one element suffers a change of condition that is contrary to

PLATO

B μαίνεται μὲν ὅσα ἂν πρότερον ψύχηται, ξηρὰ δὲ ὄντα εἰς ὕστερον γίγνεται νοτερά, καὶ κοῦφα δὴ καὶ βαρέα, καὶ πάσας πάντη μεταβολὰς δέχεται. μόνως γὰρ δή, φαμέν, ταὐτὸν ταὐτῷ κατὰ ταὐτὸ καὶ ὡσαύτως καὶ ἀνὰ λόγον προσγιγνόμενον καὶ ἀπογιγνόμενον ἐάσει ταὐτὸν ὂν αὐτῷ σῶν καὶ ὑγιὲς μένειν· ὃ δ' ἂν πλημμελήσῃ τι τούτων ἐκτὸς ἀπιὸν ἢ προσιόν, ἀλλοιότητας παμποικίλας καὶ νόσους φθοράς τε ἀπείρους παρέξεται.

C Δευτέρων δὴ ξυστάσεων αὖ κατὰ φύσιν ξυνεστηκυιῶν δευτέρα κατανόησις νοσημάτων τῷ βουλομένῳ γίγνεται ξυννοῆσαι. μυελοῦ γὰρ ἐξ ἐκείνων ὀστοῦ τε καὶ σαρκὸς καὶ νεύρου ξυμπαγέντος, ἔτι τε αἵματος ἄλλον μὲν τρόπον, ἐκ δὲ τῶν αὐτῶν γεγονότος, τῶν μὲν ἄλλων τὰ πλεῖστα ᾗπερ τὰ πρόσθεν, τὰ δὲ μέγιστα τῶν νοσημάτων τῇδε χαλεπὰ ξυμπέπτωκεν. ὅταν ἀνάπαλιν ἡ γένεσις τούτων πορεύηται, τότε ταῦτα διαφθείρεται. κατὰ φύσιν γὰρ σάρκες μὲν καὶ νεῦρα ἐξ αἵματος γίγνεται,
D νεῦρον μὲν ἐξ ἰνῶν διὰ τὴν συγγένειαν, σάρκες δὲ ἀπὸ τοῦ παγέντος, ὃ πήγνυται χωριζόμενον ἰνῶν· τὸ δὲ ἀπὸ τῶν νεύρων καὶ σαρκῶν ἀπιὸν αὖ γλίσχρον καὶ λιπαρὸν ἅμα μὲν τὴν σάρκα κολλᾷ πρὸς τὴν τῶν ὀστῶν φύσιν αὐτό τε τὸ περὶ τὸν μυελὸν ὀστοῦν τρέφον αὔξει, τὸ δ' αὖ διὰ τὴν πυκνότητα τῶν ὀστῶν διηθούμενον καθαρώτατον γένος τῶν

---

[1] Cf. 41 D ff.   [2] Cf. 74 D.
[3] i.e. the synovial fluid.

TIMAEUS

nature, all its particles that formerly were being cooled become heated, and the dry presently become moist, and the light heavy, and they undergo every variety of change in every respect. For, as we maintain, it is only the addition or subtraction of the same substance from the same substance in the same order and in the same manner and in due proportion which will allow the latter to remain safe and sound in its sameness with itself. But whatsoever oversteps any of these conditions in its going out or its coming in will produce alterations of every variety and countless diseases and corruptions.

Again, in the structures which are naturally secondary [1] in order of construction, there is a second class of diseases to be noted by him who has a mind to take cognisance of them. For inasmuch as marrow and bone and flesh and sinew are compacted from the elements,—and blood also is formed from the same constituents, although in a different way,— most of the other maladies come about like those previously described, but the most severe of them have dangerous results for the reason following: whenever the production of these secondary substances proceeds in the reverse direction, then they are corrupted. For in the order of nature flesh and sinews arise from blood,[2] the sinew from the fibrine because of its kindred quality, and flesh from the coagulated substance which coagulates on its separation from the fibrine; and further, the substance which is derived from the sinews and flesh, being viscid and oily,[3] not only glues the flesh to the substance of the bones but also feeds and increases the bone itself which encloses the marrow, while that which is formed of the purest kind of triangles, very

τριγώνων λειότατόν τε καὶ λιπαρώτατον, λειβόμενον ἀπὸ τῶν ὀστῶν καὶ στάζον, ἄρδει τὸν μυελόν. καὶ κατὰ ταῦτα μὲν γιγνομένων ἑκάστων ὑγίεια ξυμβαίνει τὰ πολλά· νόσοι δέ, ὅταν ἐναντίως. ὅταν γὰρ τηκομένη σὰρξ ἀνάπαλιν εἰς τὰς φλέβας τὴν τηκεδόνα ἐξιῇ, τότε μετὰ πνεύματος αἷμα πολύ τε καὶ παντοδαπὸν ἐν ταῖς φλεψὶ χρώμασι καὶ πικρότησι ποικιλλόμενον, ἔτι δὲ ὀξείαις καὶ ἁλμυραῖς δυνάμεσι, χολὰς καὶ ἰχῶρας καὶ φλέγματα παντοῖα ἴσχει· παλιναίρετα γὰρ πάντα γεγονότα καὶ διεφθαρμένα τό τε αἷμα αὐτὸ πρῶτον διόλλυσι, καὶ αὐτὰ οὐδεμίαν τροφὴν ἔτι τῷ σώματι παρέχοντα φέρεται πάντῃ διὰ τῶν φλεβῶν, τάξιν τῶν κατὰ φύσιν οὐκέτ᾿ ἴσχοντα περιόδων, ἐχθρὰ μὲν αὐτὰ αὑτοῖς διὰ τὸ μηδεμίαν ἀπόλαυσιν ἑαυτῶν ἔχειν, τῷ ξυνεστῶτι δὲ τοῦ σώματος καὶ μένοντι κατὰ χώραν πολέμια, διολλύντα καὶ τήκοντα. ὅσον μὲν οὖν ἂν παλαιότατον ὂν τῆς σαρκὸς τακῇ, δύσπεπτον γιγνόμενον μελαίνει μὲν ὑπὸ παλαιᾶς ξυγκαύσεως, διὰ δὲ τὸ πάντῃ διαβεβρῶσθαι πικρὸν ὂν παντὶ χαλεπὸν προσπίπτει τοῦ σώματος, ὅσον ἂν μήπω διεφθαρμένον ᾖ. καὶ τοτὲ μὲν ἀντὶ τῆς πικρότητος ὀξύτητα ἔσχε τὸ μέλαν χρῶμα, ἀπολεπτυνθέντος μᾶλλον τοῦ πικροῦ· τοτὲ δὲ ἡ πικρότης αὖ βαφεῖσα αἵματι χρῶμα ἔσχεν ἐρυθρώτερον, τοῦ δὲ μέλανος τούτῳ ξυγκεραννυμένου χλοῶδες[1] ἔτι δὲ ξυμμίγνυται ξανθὸν χρῶμα μετὰ τῆς πικρό-

---

[1] χλοῶδες Galen: χολῶδες mss., Zur.

TIMAEUS

smooth and very oily, filters through the density of the bones, and, as it oozes and drips from the bones, moistens the marrow. Now when each of these substances is produced in this order, health as a rule results; but if in the reverse order, disease. For whenever the flesh is decomposed and sends its decomposed matter back again into the veins, then, uniting with the air, the blood in the veins, which is large in volume and of every variety, is diversified by colours and bitter flavours, as well as by sharp and saline properties, and contains bile and serum and phlegm of every sort. For when all the substances become reversed and corrupted, they begin by destroying the blood itself, and then they themselves cease to supply any nourishment to the body; for they move through the veins in all directions and no longer preserve the order of their natural revolutions, being at enmity with themselves because they have no enjoyment of themselves, and being at war also with the established and regular constitution of the body, which they corrupt and dissolve. Therefore all the oldest part of the flesh that is decomposed becomes tough and is blackened by the continued combustion; and because it is eaten away on every side it is bitter, and therefore dangerous in its attack on any part of the body that is not as yet corrupted. And at one time the black matter acquires a sharpness in place of its bitterness, when the bitter substance becomes more diluted; and at another time the bitter substance acquires a redder colour through being dipped in blood, while if the black matter is blended with this it turns greenish; and again, whenever new flesh also is decomposed by the fire

PLATO

τητος, ὅταν νέα ξυντακῇ σὰρξ ὑπὸ τοῦ περὶ τὴν φλόγα πυρός.

Καὶ τὸ μὲν κοινὸν ὄνομα πᾶσι τούτοις ἤ τινες ἰατρῶν που χολὴν ἐπωνόμασαν ἢ καί τις ὢν δυνατὸς εἰς πολλὰ μὲν καὶ ἀνόμοια βλέπειν, ὁρᾶν δ' ἐν αὑτοῖς ἓν γένος ἐνὸν ἄξιον ἐπωνυμίας πᾶσι· τὰ δ' ἄλλα ὅσα χολῆς εἴδη λέγεται, κατὰ τὴν χρόαν ἔσχε λόγον αὐτῶν ἕκαστον ἴδιον.

Ἰχὼρ δέ, ὁ μὲν αἵματος ὀρὸς πρᾷος, ὁ δὲ μελαίνης χολῆς ὀξείας τε ἄγριος, ὅταν ξυμμιγνύηται διὰ θερμότητα ἁλμυρᾷ δυνάμει· καλεῖται δὲ ὀξὺ φλέγμα τὸ τοιοῦτον. τὸ δ' αὖ μετὰ ἀέρος τηκόμενον ἐκ νέας καὶ ἁπαλῆς σαρκός, τούτου δὲ ἀνεμωθέντος καὶ ξυμπεριληφθέντος ὑπὸ ὑγρότητος, καὶ πομφολύγων ξυστασῶν ἐκ τοῦ πάθους τούτου καθ' ἑκάστην μὲν ἀοράτων διὰ σμικρότητα, ξυναπασῶν δὲ τὸν ὄγκον παρεχομένων ὁρατόν, χρῶμα ἐχουσῶν διὰ τὴν τοῦ ἀφροῦ γένεσιν ἰδεῖν λευκόν, ταύτην πᾶσαν τηκεδόνα ἁπαλῆς σαρκὸς μετὰ πνεύματος ξυμπλακεῖσαν λευκὸν εἶναι φλέγμα φαμέν.

Φλέγματος δ' αὖ νέου ξυνισταμένου ὀρὸς ἱδρὼς καὶ δάκρυον, ὅσα τε ἄλλα τοιαῦτα σώματος καθ' ἡμέραν χεῖται καθαιρομένου.[1] καὶ ταῦτα μὲν δὴ πάντα νόσων ὄργανα γέγονεν, ὅταν αἷμα μὴ ἐκ τῶν σιτίων καὶ ποτῶν πληθύσῃ κατὰ φύσιν, ἀλλ' ἐξ ἐναντίων τὸν ὄγκον παρὰ τοὺς τῆς φύσεως λαμβάνῃ νόμους.

Διακρινομένης μὲν οὖν ὑπὸ νόσων τῆς σαρκὸς ἑκάστης, μενόντων δὲ τῶν πυθμένων αὐταῖς

---

[1] σώματος ... καθαιρομένου] σῶμα τὸ ... καθαιρόμενον Zur.

TIMAEUS

of the inflammation, a yellow matter is commingled with the bitter substance.

To all these humours the general designation "bile" has been given,[1] either by certain physicians or by someone who was capable of surveying a number of dissimilar cases and discerning amongst them one single type [2] worthy to give its name to them all. All the rest that are counted as species of bile have gained their special descriptions in each case from their colours.

Serum is of two kinds: one is the mild whey of the blood; the other, being derived from black and acid bile, is malignant whenever it is imbued with a saline quality through the action of heat; and this kind is termed "acid phlegm." Another kind involves air and is produced by dissolution from new and tender flesh. And when this is inflated and enclosed by a fluid, and when as a result of this process bubbles [3] are formed which individually are invisible because of their small size but in the aggregate form a mass which is visible, and which possess a colour which appears white owing to the foam created,—then we describe all this decomposition of tender flesh intermixed with air as "white phlegm."

And the whey of phlegm that is newly formed is "sweat" and "tears," and all other such humours as pour forth in the daily purgings of the body. And all these are factors in disease, whenever the blood is not replenished naturally from meats and drinks but receives its mass from opposite substances contrary to Nature's laws.

Now, when the flesh in any part is being decomposed by disease, but the bases thereof still remain

[1] *Cf.* 71 B.   [2] *Cf.* 68 D.   [3] *Cf.* 66 B.

ἡμίσεια τῆς ξυμφορᾶς ἡ δύναμις· ἀνάληψιν γὰρ ἔτι μετ' εὐπετείας ἴσχει· τὸ δὲ δὴ σάρκας ὀστοῖς ξυνδοῦν ὁπότ' ἂν νοσήσῃ, καὶ μηκέτι αὐτὸ ἐκείνων ἅμα[1] καὶ νεύρων ἀποχωριζόμενον ὀστῷ μὲν τροφή, σαρκὶ δὲ πρὸς ὀστοῦν γίγνηται δεσμός, ἀλλ' ἐκ λιπαροῦ καὶ λείου καὶ γλίσχρου τραχὺ καὶ ἁλμυρὸν αὐχμῆσαν ὑπὸ κακῆς διαίτης γένηται, τότε ταῦτα πάσχον πᾶν τὸ τοιοῦτον καταψήχεται μὲν αὐτὸ πάλιν ὑπὸ τὰς σάρκας καὶ τὰ νεῦρα, ἀφιστάμενον ἀπὸ τῶν ὀστῶν, αἱ δ' ἐκ τῶν ῥιζῶν ξυνεκπίπτουσαι τά τε νεῦρα γυμνὰ καταλείπουσι καὶ μεστὰ ἅλμης, αὐταὶ δὲ πάλιν εἰς τὴν αἵματος φορὰν ἐμπεσοῦσαι τὰ πρόσθεν ῥηθέντα νοσήματα πλείω ποιοῦσι.

Χαλεπῶν δὲ τούτων περὶ τὰ σώματα παθημάτων γιγνομένων μείζω ἔτι γίγνεται τὰ πρὸ τούτων, ὅταν ὀστοῦν διὰ πυκνότητα σαρκὸς ἀναπνοὴν μὴ λαμβάνον ἱκανήν, ὑπ' εὐρῶτος θερμαινόμενον, σφακελίσαν μήτε τὴν τροφὴν καταδέχηται πάλιν τε αὐτὸ εἰς ἐκείνην ἐναντίως ἴῃ ψηχόμενον, ἡ δ' εἰς σάρκας, σὰρξ δὲ εἰς αἷμα ἐμπίπτουσα τραχύτερα πάντα τῶν πρόσθεν τὰ νοσήματα ἀπεργάζηται. τὸ δ' ἔσχατον πάντων, ὅταν ἡ τοῦ μυελοῦ φύσις ἀπ' ἐνδείας ἤ τινος ὑπερβολῆς νοσήσῃ, τὰ μέγιστα καὶ κυριώτατα πρὸς θάνατον τῶν νοσημάτων ἀποτελεῖ, πάσης ἀνάπαλιν τῆς τοῦ σώματος φύσεως ἐξ ἀνάγκης ῥυείσης.

Τρίτον δ' αὖ νοσημάτων εἶδος τριχῇ δεῖ δια-

---

[1] αὐτὸ ἐκείνων ἅμα] αὖ τὸ ἐξ ἰνῶν αἷμα Zur. (ἅμα conj. Stallbaum).

[1] *Cf.* 82 D.

TIMAEUS

firm, the force of the attack is reduced by half, for it still admits of easy recovery; but whenever the substance which binds the flesh to the bones [1] becomes diseased and no longer separates itself at once from them and from the sinews, so as to provide food for the bone and to serve as a bond between flesh and bone, but becomes rough and saline instead of being oily and smooth and viscid, owing to its being starved by a bad regimen,—then, every such substance, as it undergoes these affections, moulders away beneath the flesh and the sinews and withdraws from the bones; while the flesh falls away with it from the roots and leaves the sinews bare and full of saline matter, and by falling back itself into the stream of the blood it augments the maladies previously described.

But although these bodily ailments are severe, still more grave are those which precede them, whenever the bone by reason of the density of the flesh fails to receive sufficient inspiration, and becoming heated because of its mouldiness decays and does not admit its nutriment, but, on the contrary, falls back itself, as it crumbles, into its nutriment which then passes into flesh, and this flesh falling into the blood causes all such maladies to be more violent than those previously described. And the most extreme case of all occurs whenever the substance of the marrow becomes diseased either from deficiency or from excess; for this results in the gravest of diseases and the most potent in causing death, inasmuch as the whole substance of the body, by the force of necessity, streams in the reverse direction.

A third class of diseases takes place, as we must

PLATO

84
D νοεῖσθαι γιγνόμενον, τὸ μὲν ὑπὸ πνεύματος, τὸ δὲ φλέγματος, τὸ δὲ χολῆς. ὅταν μὲν γὰρ ὁ τῶν πνευμάτων τῷ σώματι ταμίας πλεύμων μὴ καθαρὰς παρέχῃ τὰς διεξόδους ὑπὸ ῥευμάτων φραχθείς, ἔνθα μὲν οὐκ ἰόν, ἔνθα δὲ πλεῖον ἢ τὸ προσῆκον πνεῦμα εἰσιὸν τὰ μὲν οὐ τυγχάνοντα ἀναψυχῆς σήπει, τὰ δὲ τῶν φλεβῶν διαβιαζόμενον καὶ ξυνεπιστρέφον αὐτὰ τῆκόν τε τὸ σῶμα εἰς τὸ μέσον αὐτοῦ διάφραγμά τ' ἴσχον ἐναπολαμβάνεται,
E καὶ μυρία δὴ νοσήματα ἐκ τούτων ἀλγεινὰ μετὰ πλήθους ἱδρῶτος ἀπείργασται. πολλάκις δ' ἐν τῷ σώματι διακριθείσης σαρκὸς πνεῦμα ἐγγενόμενον καὶ ἀδυνατοῦν ἔξω πορευθῆναι τὰς αὐτὰς τοῖς ἐπεισεληλυθόσιν ὠδῖνας παρέσχε, μεγίστας δέ, ὅταν περὶ τὰ νεῦρα καὶ τὰ ταύτῃ φλέβια περιστὰν καὶ ἀνοιδῆσαν τούς τε ἐπιτόνους καὶ τὰ ξυνεχῆ νεῦρα οὕτως εἰς τὸ ἐξόπισθεν κατατείνῃ τούτοις· ἃ δὴ καὶ ἀπ' αὐτοῦ τῆς ξυντονίας τοῦ παθήματος τὰ νοσήματα τέτανοί τε καὶ ὀπισθότονοι προσερρήθησαν. ὧν καὶ τὸ φάρμακον χαλεπόν· πυρετοὶ γὰρ οὖν δὴ τὰ τοιαῦτα ἐπιγιγνόμενοι[1] μάλιστα λύουσι.

85 Τὸ δὲ λευκὸν φλέγμα διὰ τὸ τῶν πομφολύγων πνεῦμα χαλεπὸν ἀποληφθέν, ἔξω δὲ τοῦ σώματος ἀναπνοὰς ἴσχον ἠπιώτερον μέν, καταποικίλλει δὲ τὸ σῶμα λεύκας ἀλφούς τε καὶ τὰ τούτων ξυγγενῆ νοσήματα ἀποτίκτον. μετὰ χολῆς δὲ μελαίνης κερασθὲν ἐπὶ τὰς περιόδους τε τὰς ἐν τῇ κεφαλῇ θειοτάτας οὔσας ἐπισκεδαννύμενον καὶ
B ξυνταράττον αὐτάς, καθ' ὕπνον μὲν ἰὸν πραότερον, ἐγρηγορόσι δὲ ἐπιτιθέμενον δυσαπαλλακτότερον·

---

[1] ἐπιγιγνόμενοι one ms.: ἐγγιγνόμενοι other mss., Zur.

TIMAEUS

conceive, in three ways, being due partly to air, partly to phlegm, and partly to bile. Whenever the lungs, which are the dispensers of air to the body, fail to keep their outlets clean through being blocked up with rheums, then the air, being unable to pass one way while entering by another way in more than its proper volume, causes the parts deprived of respiration to rot, but forces and distorts the vessels of the veins, and as it thus dissolves the body it is itself shut off within the centre thereof which contains the midriff; and as a result of this countless diseases of a painful kind are produced, accompanied by much sweating. And often, when the flesh is disintegrated, air which is enclosed in the body and is unable to pass out brings about the same pangs as those caused by the air that enters from without; and these pangs are most severe when the air surrounds the sinews and the adjacent veins and by its swelling up strains backwards the tendons and the sinews attached to them; hence it is actually from this process of intense strain that these maladies have derived their names of "tetanus" and "opisthotonus." Of these maladies the cure also is severe; for what does most to relieve them is, in fact, an attack of fever.

White phlegm, also, is dangerous when it is blocked inside because of the air in its bubbles; but when it has air-vents outside the body it is milder, although it marks the body with spots by breeding white scabs and tetters and the maladies akin thereto. And when this phlegm is blended with black bile and spreads over the revolutions of the head, which are the most divine, and perturbs them, its action is more gentle during sleep, but when it attacks persons who are awake it is harder to shake off; and because it

νόσημα δὲ ἱερᾶς ὂν φύσεως ἐνδικώτατα ἱερὸν λέγεται. φλέγμα δ' ὀξὺ καὶ ἁλμυρὸν πηγὴ πάντων νοσημάτων, ὅσα γίγνεται καταρροϊκά· διὰ δὲ τοὺς τόπους εἰς οὓς ῥεῖ παντοδαποὺς ὄντας παντοῖα ὀνόματα εἴληφεν.

Ὅσα δὲ φλεγμαίνειν λέγεται τοῦ σώματος, ἀπὸ τοῦ κάεσθαί τε καὶ φλέγεσθαι, διὰ χολὴν γέγονε πάντα. λαμβάνουσα μὲν οὖν ἀναπνοὴν ἔξω παντοῖ' ἀναπέμπει φύματα ζέουσα, καθειργνυμένη δ' ἐντὸς πυρίκαυτα νοσήματα πολλὰ ἐμποιεῖ, μέγιστον δέ, ὅταν αἵματι καθαρῷ ξυγκερασθεῖσα τὸ τῶν ἰνῶν γένος ἐκ τῆς ἑαυτῶν διαφορῇ τάξεως, αἳ διεσπάρησαν μὲν εἰς αἷμα, ἵνα συμμέτρως λεπτότητος ἴσχοι καὶ πάχους καὶ μήτε διὰ θερμότητα ὡς ὑγρὸν ἐκ μανοῦ τοῦ σώματος ἐκρέοι, μήτ' αὖ πυκνότερον δυσκίνητον ὂν μόγις ἀναστρέφοιτο ἐν ταῖς φλεψί. καιρὸν δὴ τούτων ἶνες τῇ τῆς φύσεως γενέσει φυλάττουσιν· ἃς ὅταν τις καὶ τεθνεῶτος αἵματος ἐν ψύξει τε ὄντος πρὸς ἀλλήλας ξυναγάγῃ, διαχεῖται πᾶν τὸ λοιπὸν αἷμα, ἐαθεῖσαι δὲ ταχὺ μετὰ τοῦ περιεστῶτος αὐτὸ ψύχους ξυμπηγνύασι. ταύτην δὴ τὴν δύναμιν ἐχουσῶν ἰνῶν ἐν αἵματι χολὴ φύσει παλαιὸν αἷμα γεγονυῖα καὶ πάλιν ἐκ τῶν σαρκῶν εἰς τοῦτο τετηκυῖα, θερμὴ καὶ ὑγρὰ κατ' ὀλίγον τὸ πρῶτον ἐμπίπτουσα πήγνυται διὰ τὴν τῶν ἰνῶν δύναμιν, πηγνυμένη δὲ καὶ βίᾳ κατασβεννυμένη χειμῶνα καὶ τρόμον ἐντὸς παρέχει. πλείων δ' ἐπιρρέουσα, τῇ παρ' αὑτῆς θερμότητι

---

[1] *i.e.* epilepsy; *cf. Laws* 916 A.
[2] *Cf.* 82 D.

TIMAEUS

is a disease of the sacred substance it is most justly termed "the sacred disease."[1] Phlegm that is sharp and saline is the fount of all the maladies which are of the nature of catarrhs; and these have received all kinds of names because the regions into which they flow are of all varieties.

All those diseases which are called inflammations, owing to the burning and inflaming of the body which they involve, are caused by bile. This, when it gains an external outlet, boils and sends up all kinds of eruptions; but when it is confined inside it produces many burning diseases; and of these the gravest occurs when the bile, being mixed with pure blood, displaces the matter of the fibrine from its proper position. For this fibrine is dispersed through the blood in order that the blood may have a due proportion of both rarity and density, and may neither flow out from the porous body through being liquefied by heat, nor yet prove immobile through its density and circulate with difficulty in the veins. Of these qualities the fibrine preserves the due amount owing to the nature of its formation.[2] Even when anyone collects together the fibrine of blood that is dead and in process of cooling, all the rest of the blood turns liquid; but if the fibrine is left alone as it is, it acts in combination with the surrounding cold and rapidly congeals the blood. As the fibrine, then, has this property, bile, which is naturally formed of old blood and dissolved again into blood from flesh, penetrates the blood gradually at first, while it is hot and moist, and is congealed by this property of the fibrine; and as it becomes congealed and forcibly chilled it causes internal cold and shivering. But when the bile flows in with more volume, it overpowers the fibrine by the

κρατήσασα τὰς ἶνας εἰς ἀταξίαν ζέσασα διέσεισε, καὶ ἐὰν μὲν ἱκανὴ διὰ τέλους κρατῆσαι γένηται, πρὸς τὸ τοῦ μυελοῦ διαπεράσασα γένος κάουσα ἔλυσε τὰ τῆς ψυχῆς αὐτόθεν οἷον νεὼς πείσματα μεθῆκέ τε ἐλευθέραν, ὅταν δ' ἐλάττων ᾖ τό τε σῶμα ἀντίσχῃ τηκόμενον, αὐτὴ κρατηθεῖσα ἢ κατὰ πᾶν τὸ σῶμα ἐξέπεσεν, ἢ διὰ τῶν φλεβῶν εἰς τὴν κάτω ξυνωσθεῖσα ἢ τὴν ἄνω κοιλίαν, οἷον φυγὰς ἐκ πόλεως στασιασάσης ἐκ τοῦ σώματος ἐκπίπτουσα, διαρροίας καὶ δυσεντερίας καὶ τὰ τοιαῦτα νοσήματα πάντα παρέσχετο.

Τὸ μὲν οὖν ἐκ πυρὸς ὑπερβολῆς μάλιστα νοσῆσαν σῶμα ξυνεχῆ καύματα καὶ πυρετοὺς ἀπεργάζεται, τὸ δ' ἐξ ἀέρος ἀμφημερινούς, τριταίους δ' ὕδατος διὰ τὸ νωθέστερον ἀέρος καὶ πυρὸς αὐτὸ εἶναι· τὸ δὲ γῆς, τετάρτως ὂν νωθέστατον τούτων, ἐν τετραπλασίαις περιόδοις χρόνου καθαιρόμενον, τεταρταίους πυρετοὺς ποιῆσαν ἀπαλλάττεται μόγις.

Καὶ τὰ μὲν περὶ τὸ σῶμα νοσήματα ταύτῃ ξυμβαίνει γιγνόμενα, τὰ δὲ περὶ ψυχὴν διὰ σώματος ἕξιν τῇδε. νόσον μὲν δὴ ψυχῆς ἄνοιαν ξυγχωρητέον, δύο δ' ἀνοίας γένη, τὸ μὲν μανίαν, τὸ δὲ ἀμαθίαν. πᾶν οὖν ὅ τι πάσχων τις πάθος ὁπότερον αὐτῶν ἴσχει, νόσον προσρητέον· ἡδονὰς δὲ καὶ λύπας ὑπερβαλλούσας τῶν νόσων μεγίστας θετέον τῇ ψυχῇ· περιχαρὴς γὰρ ἄνθρωπος ὢν ἢ καὶ τἀναντία ὑπὸ λύπης πάσχων, σπεύδων τὸ μὲν ἑλεῖν ἀκαίρως, τὸ δὲ φυγεῖν, οὔτε ὁρᾶν οὔτε ἀκούειν

---

[1] *Cf.* 73 D, 81 D.
[2] *i.e.* the fever recurs after an interval of two days.
[3] *Cf.* Rep. 571 D, Soph. 228 A, Laws 689 A ff.

TIMAEUS

heat it contains, and shakes it into disorder by its boiling up; and should it be capable of thus overpowering the fibrine continuously, it penetrates to the substance of the marrow and loosens from thence, by burning, the mooring-ropes of the soul,[1] as it were of a ship, and sets it free. But when the bile is in smaller quantity and the body resists dissolution, then the bile itself is overpowered, and either it is ejected over the whole surface of the body, or else it is forced through the veins into the lower or the upper belly, being ejected from the body like fugitives from a city in revolt; and it produces diarrhoea and dysentery and all suchlike maladies.

When a body has become diseased mainly from an excess of fire, it produces constant inflammations and fevers; when from air, quotidian fevers; when from water, tertian fevers, because that element is more sluggish than air or fire; and when from earth, which is the fourth and most sluggish of the elements and is purged in four-fold periods of time,[2] it causes quartan fevers and is cured with difficulty.

Such is the manner in which diseases of the body come about; and those of the soul which are due to the condition of the body arise in the following way. We must agree that folly is a disease of the soul[3]; and of folly there are two kinds, the one of which is madness, the other ignorance. Whatever affection a man suffers from, if it involves either of these conditions it must be termed "disease"; and we must maintain that pleasures and pains in excess are the greatest of the soul's diseases. For when a man is overjoyed or contrariwise suffering excessively from pain, being in haste to seize on the one and avoid the other beyond measure, he is unable either to see or

PLATO

86 ὀρθὸν οὐδὲν δύναται, λυττᾷ δὲ καὶ λογισμοῦ μετασχεῖν ἥκιστα τότε δὴ δυνατός ἐστι. τὸ δὲ σπέρμα ὅτῳ πολὺ καὶ ῥυῶδες περὶ τὸν μυελὸν γίγνεται, καὶ καθαπερεὶ δένδρον πολυκαρπότερον τοῦ ξυμμέτρου πεφυκὸς ᾖ, πολλὰς μὲν καθ' ἕκαστον ὠδῖνας, πολλὰς δ' ἡδονὰς κτώμενος ἐν ταῖς ἐπιθυμίαις καὶ τοῖς περὶ τὰ τοιαῦτα τόκοις, ἐμμανὴς τὸ
D πλεῖστον γιγνόμενος τοῦ βίου διὰ τὰς μεγίστας ἡδονὰς καὶ λύπας, νοσοῦσαν καὶ ἄφρονα ἴσχων ὑπὸ τοῦ σώματος τὴν ψυχήν, οὐχ ὡς νοσῶν ἀλλ' ὡς ἑκὼν κακὸς [κακῶς]¹ δοξάζεται· τὸ δὲ ἀληθές, ἡ περὶ τὰ ἀφροδίσια ἀκολασία κατὰ τὸ πολὺ μέρος διὰ τὴν ἑνὸς γένους ἕξιν ὑπὸ μανότητος ὀστῶν ἐν σώματι ῥυώδη καὶ ὑγραίνουσαν νόσος ψυχῆς γέγονε. καὶ σχεδὸν δὴ πάντα ὁπόσα ἡδονῶν ἀκράτεια κατ'² ὄνειδος ὡς ἑκόντων λέγεται τῶν κακῶν, οὐκ ὀρθῶς ὀνειδίζεται· κακὸς μὲν γὰρ ἑκὼν οὐδείς,
E διὰ δὲ πονηρὰν ἕξιν τινὰ τοῦ σώματος καὶ ἀπαίδευτον τροφὴν ὁ κακὸς γίγνεται κακός, παντὶ δὲ ταῦτα ἐχθρὰ καὶ ἄκοντι³ προσγίγνεται. καὶ πάλιν δὴ τὸ περὶ τὰς λύπας ἡ ψυχὴ κατὰ ταὐτὰ διὰ σῶμα πολλὴν ἴσχει κακίαν.

Ὅπου γὰρ ἂν οἱ τῶν ὀξέων καὶ τῶν ἁλυκῶν φλεγμάτων καὶ ὅσοι πικροὶ καὶ χολώδεις χυμοὶ κατὰ τὸ σῶμα πλανηθέντες ἔξω μὲν μὴ λάβωσιν
87 ἀναπνοήν, ἐντὸς δὲ εἱλόμενοι τὴν ἀφ' αὑτῶν ἀτμίδα τῇ τῆς ψυχῆς φορᾷ ξυμμίξαντες ἀνακερασθῶσι, παντοδαπὰ νοσήματα ψυχῆς ἐμποιοῦσι, μᾶλλον καὶ ἧττον, καὶ ἐλάττω καὶ πλείω· πρός

---
¹ κακῶς omitted by best mss.
² κατ' H. Richards: καὶ mss., Zur.
³ ἄκοντι Galen: κακόν τι most mss. and Zur.

TIMAEUS

to hear anything correctly, and he is at such a time distraught and wholly incapable of exercising reason. And whenever a man's seed grows to abundant volume in his marrow,[1] as it were a tree that is overladen beyond measure with fruit, he brings on himself time after time many pangs and many pleasures owing to his desires and the issue thereof, and comes to be in a state of madness for the most part of his life because of those greatest of pleasures and pains, and keeps his soul diseased and senseless by reason of the action of his body. Yet such a man is reputed to be voluntarily wicked and not diseased; although, in truth, this sexual incontinence, which is due for the most part to the abundance and fluidity of one substance because of the porosity of the bones, constitutes a disease of the soul. And indeed almost all those affections which are called by way of reproach "incontinence in pleasure," as though the wicked acted voluntarily, are wrongly so reproached; for no one is voluntarily wicked,[2] but the wicked man becomes wicked by reason of some evil condition of body and unskilled nurture, and these are experiences which are hateful to everyone and involuntary. And again, in respect of pains likewise the soul acquires much evil because of the body.

For whenever the humours which arise from acid and saline phlegms, and all humours that are bitter and bilious wander through the body and find no external vent but are confined within, and mingle their vapour with the movement of the soul and are blended therewith, they implant diseases of the soul of all kinds, varying in intensity and in extent; and

[1] *Cf.* 73 c, 91 c.
[2] For this Socratic dictum *cf. Protag.* 345 D ff., *Laws* 731 c ff.

τε τοὺς τρεῖς τόπους ἐνεχθέντα τῆς ψυχῆς, πρὸς ὃν ἂν ἕκαστ' αὐτῶν προσπίπτῃ, ποικίλλει μὲν εἴδη δυσκολίας καὶ δυσθυμίας παντοδαπά, ποικίλλει δὲ θρασύτητός τε καὶ δειλίας, ἔτι δὲ λήθης ἅμα καὶ δυσμαθίας. πρὸς δὲ τούτοις, ὅταν οὕτω κακῶς παγέντων πολιτεῖαι κακαὶ καὶ λόγοι κατὰ πόλεις ἰδίᾳ τε καὶ δημοσίᾳ λεχθῶσιν, ἔτι δὲ μαθήματα μηδαμῇ τούτων ἰατικὰ ἐκ νέων μανθάνηται, ταύτῃ κακοὶ πάντες οἱ κακοὶ διὰ δύο ἀκουσιώτατα γιγνόμεθα. ὧν αἰτιατέον μὲν τοὺς φυτεύοντας ἀεὶ τῶν φυτευομένων μᾶλλον καὶ τοὺς τρέφοντας τῶν τρεφομένων, προθυμητέον μήν, ὅπῃ τις δύναται, καὶ διὰ τροφῆς καὶ δι' ἐπιτηδευμάτων μαθημάτων τε φυγεῖν μὲν κακίαν, τοὐναντίον δὲ ἑλεῖν. ταῦτα μὲν οὖν δὴ τρόπος ἄλλος λόγων.

Τὸ δὲ τούτων ἀντίστροφον αὖ, τὸ περὶ τὰς τῶν σωμάτων καὶ διανοήσεων θεραπείας, αἷς αἰτίαις σώζεται, πάλιν εἰκὸς καὶ πρέπον ἀνταποδοῦναι· δικαιότερον γὰρ τῶν ἀγαθῶν πέρι μᾶλλον ἢ τῶν κακῶν ἴσχειν λόγον. πᾶν δὴ τὸ ἀγαθὸν καλόν, τὸ δὲ καλὸν οὐκ ἄμετρον· καὶ ζῷον οὖν τὸ τοιοῦτον ἐσόμενον ξύμμετρον θετέον. ξυμμετριῶν δὲ τὰ μὲν σμικρὰ διαισθανόμενοι ξυλλογιζόμεθα, τὰ δὲ κυριώτατα καὶ μέγιστα ἀλογίστως ἔχομεν. πρὸς γὰρ ὑγιείας καὶ νόσους ἀρετάς τε καὶ κακίας οὐδεμία ξυμμετρία καὶ ἀμετρία μείζων ἢ ψυχῆς αὐτῆς πρὸς σῶμα αὐτό· ὧν οὐδὲν σκοποῦμεν, οὐδ' ἐννοοῦμεν

---

[1] Cf. 73 D ff.

TIMAEUS

as these humours penetrate to the three regions [1] of the Soul, according to the region which they severally attack, they give rise to all varieties of bad temper and bad spirits, and they give rise to all manner of rashness and cowardice, and of forgetfulness also, as well as of stupidity. Furthermore, when, with men in such an evil condition, the political administration also is evil, and the speech in the cities, both public and private, is evil; and when, moreover, no lessons that would cure these evils are anywhere learnt from childhood,—thus it comes to pass that all of us who are wicked become wicked owing to two quite involuntary causes. And for these we must always blame the begetters more than the begotten, and the nurses more than the nurslings; yet each man must endeavour, as best he can, by means of nurture and by his pursuits and studies to flee the evil and to pursue the good. This, however, forms a separate subject of discussion.

Again, it is reasonable and proper to set forth in turn the subject complementary to the foregoing, namely the remedial treatment of body and mind, and the causes which conserve this. For what is good merits description more than what is evil. All that is good is fair, and the fair is not void of due measure; wherefore also the living creature that is to be fair must be symmetrical. Of symmetries we distinguish and reason about such as are small, but of the most important and the greatest we have no rational comprehension. For with respect to health and disease, virtue and vice, there is no symmetry or want of symmetry greater than that which exists between the soul itself and the body itself. But as regards these, we wholly fail to perceive or reflect

PLATO

87 ὅτι ψυχὴν ἰσχυρὰν καὶ πάντη μεγάλην ἀσθενέστερον καὶ ἔλαττον εἶδος ὅταν ὀχῇ, καὶ ὅταν αὖ τοὐναντίον ξυμπαγῆτον τούτω, οὐ καλὸν ὅλον τὸ ζῷον—ἀξύμμετρον γὰρ ταῖς μεγίσταις ξυμμετρίαις—, τὸ δὲ ἐναντίως ἔχον πάντων θεαμάτων τῷ δυναμένῳ
E καθορᾶν κάλλιστον καὶ ἐρασμιώτατον. οἷον οὖν ὑπερσκελὲς ἢ καί τινα ἑτέραν ὑπέρεξιν ἄμετρον ἑαυτῷ τι σῶμα ὂν ἅμα μὲν αἰσχρόν, ἅμα δ' ἐν τῇ κοινωνίᾳ τῶν πόνων πολλοὺς μὲν κόπους, πολλὰ δὲ σπάσματα καὶ διὰ τὴν παραφορότητα πτώματα παρέχον μυρίων κακῶν αἴτιον ἑαυτῷ, ταὐτὸν δὴ διανοητέον καὶ περὶ τοῦ ξυναμφοτέρου, ζῷον ὃ καλοῦμεν, ὡς ὅταν τε ἐν αὐτῷ ψυχὴ κρείττων οὖσα
88 σώματος περιθύμως ἴσχῃ, διασείουσα πᾶν αὐτὸ ἔνδοθεν νόσων ἐμπίπλησι, καὶ ὅταν εἴς τινας μαθήσεις καὶ ζητήσεις ξυντόνως ἴῃ, κατατήκει, διδαχάς τ' αὖ καὶ μάχας ἐν λόγοις ποιουμένη δημοσίᾳ καὶ ἰδίᾳ [δι']¹ ἐρίδων καὶ φιλονεικίας γιγνομένων διάπυρον αὐτὸ ποιοῦσα σαλεύει,² καὶ ῥεύματα ἐπάγουσα, τῶν λεγομένων ἰατρῶν ἀπατῶσα τοὺς πλείστους, τἀναίτια³ αἰτιᾶσθαι ποιεῖ.

Σῶμά τε ὅταν αὖ μέγα καὶ ὑπέρψυχον σμικρᾷ
B ξυμφυὲς ἀσθενεῖ τε διανοίᾳ γένηται, διττῶν ἐπιθυμιῶν οὐσῶν φύσει κατ' ἀνθρώπους, διὰ σῶμα μὲν τροφῆς, διὰ δὲ τὸ θειότατον τῶν ἐν ἡμῖν φρονήσεως, αἱ τοῦ κρείττονος κινήσεις κρατοῦσαι καὶ τὸ μὲν σφέτερον αὔξουσαι, τὸ δὲ τῆς ψυχῆς κωφὸν καὶ

---

¹ δι' bracketed by Madvig.
² σαλεύει some MSS.: λύει best MSS., Zur.
³ τἀναίτια some MSS.: τἀναντία best MSS., Zur.

---

¹ Cf. 44 E, 69 C.

TIMAEUS

that, whenever a weaker and inferior type of body is the vehicle [1] of a soul that is strong and in all ways great,—or conversely, when each of these two is of the opposite kind,—then the creature as a whole is not fair, seeing that it is unsymmetrical in respect of the greatest of symmetries; whereas a creature in the opposite condition is of all sights, for him who has eyes to see, the fairest and most admirable. A body, for example, which is too long in the legs, or otherwise disproportioned owing to some excess, is not only ugly, but, when joint effort is required, it is also the source of much fatigue and many sprains and falls by reason of its clumsy motion, whereby it causes itself countless evils. So likewise we must conceive of that compound of soul and body which we call the "living creature." Whenever the soul within it is stronger than the body and is in a very passionate state, it shakes up the whole body from within and fills it with maladies; and whenever the soul ardently pursues some study or investigation, it wastes the body; and again, when the soul engages, in public or in private, in teachings and battles of words carried on with controversy and contention, it makes the body inflamed and shakes it to pieces, and induces catarrhs; and thereby it deceives the majority of so-called physicians and makes them ascribe the malady to the wrong cause.

And, on the other hand, when a large and overbearing body is united to a small and weak intellect, inasmuch as two desires naturally exist amongst men, —the desire of food for the body's sake, and the desire of wisdom for the sake of the most divine part we have,—the motions of the stronger part prevail and augment their own power, but they make that of

PLATO

δυσμαθὲς ἀμνῆμόν τε ποιοῦσαι τὴν μεγίστην νόσον ἀμαθίαν ἐναπεργάζονται.

Μία δὴ σωτηρία πρὸς ἄμφω, μήτε τὴν ψυχὴν ἄνευ σώματος κινεῖν μήτε σῶμα ἄνευ ψυχῆς, ἵνα ἀμυνομένω γίγνησθον ἰσορρόπω καὶ ὑγιῆ. τὸν δὴ μαθηματικὸν ἤ τινα ἄλλην σφόδρα μελέτην διανοίᾳ κατεργαζόμενον καὶ τὴν τοῦ σώματος ἀποδοτέον κίνησιν, γυμναστικῇ προσομιλοῦντα, τόν τε αὖ σῶμα ἐπιμελῶς πλάττοντα τὰς τῆς ψυχῆς ἀνταποδοτέον κινήσεις, μουσικῇ καὶ πάσῃ φιλοσοφίᾳ προσχρώμενον, εἰ μέλλει δικαίως τις ἅμα μὲν καλός, ἅμα δὲ ἀγαθὸς ὀρθῶς κεκλήσεσθαι.

Κατὰ δὲ ταὐτὰ ταῦτα καὶ τὰ μέρη θεραπευτέον, τὸ τοῦ παντὸς ἀπομιμούμενον εἶδος. τοῦ γὰρ σώματος ὑπὸ τῶν εἰσιόντων καομένου τε ἐντὸς καὶ ψυχομένου, καὶ πάλιν ὑπὸ τῶν ἔξωθεν ξηραινομένου καὶ ὑγραινομένου καὶ τὰ τούτοις ἀκόλουθα πάσχοντος ὑπ' ἀμφοτέρων τῶν κινήσεων, ὅταν μέν τις ἡσυχίαν ἄγον τὸ σῶμα παραδιδῷ ταῖς κινήσεσι, κρατηθὲν διώλετο, ἐὰν δὲ ἥν τε τροφὸν καὶ τιθήνην τοῦ παντὸς προσείπομεν μιμῆταί τις, καὶ τὸ σῶμα μάλιστα μὲν μηδέποτε ἡσυχίαν ἄγειν ἐᾷ, κινῇ δὲ καὶ σεισμοὺς ἀεί τινας ἐμποιῶν αὐτῷ διὰ παντὸς τὰς ἐντὸς καὶ ἐκτὸς ἀμύνηται κατὰ φύσιν κινήσεις, καὶ μετρίως σείων τά τε περὶ τὸ σῶμα πλανώμενα παθήματα καὶ μέρη κατὰ ξυγγενείας εἰς τάξιν κατακοσμῇ πρὸς ἄλληλα, κατὰ τὸν πρόσθεν λόγον ὃν περὶ τοῦ παντὸς ἐλέγομεν, οὐκ ἐχθρὸν παρ' ἐχθρὸν τιθέμενον ἐάσει πολέμους ἐντίκτειν τῷ

---

[1] Cf. *Laws* 728 E.
[2] *i.e.* "music" in the wide sense of "mental culture."
[3] Cf. 49 A, 52 D.
[4] Cf. 33 A.

TIMAEUS

the soul obtuse and dull of wit and forgetful, and thereby they produce within it that greatest of diseases, ignorance.

From both these evils the one means of salvation is this—neither to exercise the soul without the body nor the body without the soul,[1] so that they may be evenly matched and sound of health. Thus the student of mathematics, or of any other subject, who works very hard with his intellect must also provide his body with exercise by practising gymnastics; while he who is diligent in moulding his body must, in turn, provide his soul with motion by cultivating music [2] and philosophy in general, if either is to deserve to be called truly both fair and good.

The various parts, likewise, must be treated in the same manner, in imitation of the form of the Universe. For as the body is inflamed or chilled within by the particles that enter it, and again is dried or moistened by those without, and suffers the affections consequent on both these motions, whenever a man delivers his body, in a state of rest, to these motions, it is overpowered and utterly perishes; whereas if a man imitates that which we have called the nurturer and nurse of the Universe,[3] and never, if possible, allows the body to be at rest but keeps it moving, and by continually producing internal vibrations defends it in nature's way against the inward and outward motions, and by means of moderate vibrations arranges the affections and particles which stray about in the body in their due reciprocal order,[4] according to their affinities,—as described in the previous account which we have given of the Universe—then he will not suffer foe set beside foe

PLATO

88 σώματι καὶ νόσους, ἀλλὰ φίλον παρὰ φίλον τεθὲν ὑγίειαν ἀπεργαζόμενον παρέξει.

89 Τῶν δ' αὖ κινήσεων ἡ ἐν ἑαυτῷ ὑφ' αὑτοῦ ἀρίστη κίνησις—μάλιστα γὰρ τῇ διανοητικῇ καὶ τῇ τοῦ παντὸς κινήσει ξυγγενής—, ἡ δὲ ὑπ' ἄλλου χείρων· χειρίστη δὲ ἡ κειμένου τοῦ σώματος καὶ ἄγοντος ἡσυχίαν δι' ἑτέρων αὐτὸ κατὰ μέρη κινοῦσα. διὸ δὴ τῶν καθάρσεων καὶ ξυστάσεων τοῦ σώματος ἡ μὲν διὰ τῶν γυμνασίων ἀρίστη, δευτέρα δὲ ἡ διὰ τῶν αἰωρήσεων κατά τε τοὺς πλοῦς καὶ ὅπῃ περ ἂν ὀχήσεις ἄκοποι γίγνωνται· τρίτον δὲ εἶδος κινήσεως σφόδρα ποτὲ ἀναγκαζο-
B μένῳ χρήσιμον, ἄλλως δὲ οὐδαμῶς τῷ νοῦν ἔχοντι προσδεκτέον, τὸ τῆς φαρμακευτικῆς καθάρσεως γιγνόμενον ἰατρικόν. τὰ γὰρ νοσήματα, ὅσα μὴ μεγάλους ἔχει κινδύνους, οὐκ ἐρεθιστέον φαρμακείαις. πᾶσα γὰρ ξύστασις νόσων τρόπον τινὰ τῇ τῶν ζῴων φύσει προσέοικε. καὶ γὰρ ἡ τούτων ξύνοδος ἔχουσα τεταγμένους τοῦ βίου γίγνεται χρόνους τοῦ τε γένους ξύμπαντος, καὶ καθ' αὑτὸ τὸ ζῷον εἱμαρμένον ἕκαστον ἔχον τὸν βίον φύεται,
C χωρὶς τῶν ἐξ ἀνάγκης παθημάτων· τὰ γὰρ τρίγωνα εὐθὺς κατ' ἀρχὰς ἑκάστου δύναμιν ἔχοντα ξυνίσταται μέχρι τινὸς χρόνου [δυνατὰ][1] ἐξαρκεῖν, οὗ βίον οὐκ ἄν ποτέ τις εἰς τὸ πέραν ἔτι βιῴη. τρόπος οὖν ὁ αὐτὸς καὶ τῆς περὶ τὰ νοσήματα ξυστάσεως· ἣν ὅταν τις παρὰ τὴν εἱμαρμένην τοῦ χρόνου φθείρῃ φαρμακείαις, ἅμα ἐκ σμικρῶν μεγάλα καὶ

---

[1] δυνατὰ bracketed by Lindau.

[1] Cf. *Laws* 789 c.

TIMAEUS

to breed war in the body and disease, but he will cause friend to be set beside friend so as to produce sound health.

Further, as concerns the motions, the best motion of a body is that caused by itself in itself; for this is most nearly akin to the motion of intelligence and the motion of the Universe. Motion due to the agency of another is less good; and the least good motion is that which is imparted to a body lying in a state of rest and which moves it piecemeal and by means of others. Wherefore the motion that is best for purgings and renovations of the body consists in gymnastic exercises; and second-best is the motion provided by swaying vehicles,[1] such as boats or any conveyances that produce no fatigue; while the third kind of motion, although useful for one who is absolutely driven to it, is by no means acceptable, under any other conditions, to a man of sense, it being the medical kind of purging by means of drugs. For no diseases which do not involve great danger ought to be irritated by drugging. For in its structure every disease resembles in some sort the nature of the living creature. For, in truth, the constitution of these creatures has prescribed periods of life for the species as a whole, and each individual creature likewise has a naturally predestined term of life, apart from the accidents due to necessity. For from the very beginning the triangles of each creature are constructed with a capacity for lasting until a certain time, beyond which no one could ever continue to live. With respect to the structure of diseases also the same rule holds good: whenever anyone does violence thereto by drugging, in despite of the predestined period of time, diseases many and grave, in place of

PLATO

πολλὰ ἐξ ὀλίγων νοσήματα φιλεῖ γίγνεσθαι. διὸ παιδαγωγεῖν δεῖ διαίταις πάντα τὰ τοιαῦτα, καθ' ὅσον ἂν ᾖ τῳ σχολή, ἀλλ' οὐ φαρμακεύοντα κακὸν δύσκολον ἐρεθιστέον.

Καὶ περὶ μὲν τοῦ κοινοῦ ζῴου καὶ τοῦ κατὰ τὸ σῶμα αὐτοῦ μέρους, ᾗ τις ἂν καὶ διαπαιδαγωγῶν καὶ διαπαιδαγωγούμενος ὑφ' αὑτοῦ μάλιστ' ἂν κατὰ λόγον ζῴη, ταύτῃ λελέχθω. τὸ δὲ δὴ παιδαγωγῆσον αὐτὸ μᾶλλόν που καὶ πρότερον παρασκευαστέον εἰς δύναμιν ὅ τι κάλλιστον καὶ ἄριστον εἰς τὴν παιδαγωγίαν εἶναι. δι' ἀκριβείας μὲν οὖν περὶ τούτων διελθεῖν ἱκανὸν ἂν γένοιτο αὐτὸ καθ' αὑτὸ μόνον ἔργον· τὸ δ' ἐν παρέργῳ κατὰ τὰ πρόσθεν ἑπόμενος ἄν τις οὐκ ἀπὸ τρόπου τῇδε σκοπῶν ὧδε τῷ λόγῳ διαπεράναιτ' ἄν. καθάπερ εἴπομεν πολλάκις, ὅτι τρία τριχῇ ψυχῆς ἐν ἡμῖν εἴδη κατῴκισται, τυγχάνει δὲ ἕκαστον κινήσεις ἔχον, οὕτω κατὰ ταὐτὰ καὶ νῦν ὡς διὰ βραχυτάτων ῥητέον, ὅτι τὸ μὲν αὐτῶν ἐν ἀργίᾳ διάγον καὶ τῶν ἑαυτοῦ κινήσεων ἡσυχίαν ἄγον ἀσθενέστατον ἀνάγκη γίγνεσθαι, τὸ δ' ἐν γυμνασίοις ἐρρωμενέστατον· διὸ φυλακτέον, ὅπως ἂν ἔχωσι τὰς κινήσεις πρὸς ἄλληλα συμμέτρους. τὸ δὲ περὶ τοῦ κυριωτάτου παρ' ἡμῖν ψυχῆς εἴδους διανοεῖσθαι δεῖ τῇδε, ὡς ἄρα αὐτὸ δαίμονα θεὸς ἑκάστῳ δέδωκε, τοῦτο ὃ δή φαμεν οἰκεῖν μὲν ἡμῶν ἐπ' ἄκρῳ τῷ σώματι, πρὸς δὲ τὴν ἐν οὐρανῷ ξυγγένειαν ἀπὸ γῆς ἡμᾶς αἴρειν ὡς ὄντας φυτὸν οὐκ ἔγγειον ἀλλ'

---

[1] Education is the theme of *Rep.* vii. and *Laws* vii. and xii. *ad fin.*   [2] *Cf.* 69 D, 79 D, 87 A.

TIMAEUS

few and slight, are wont to occur. Wherefore one ought to control all such diseases, so far as one has the time to spare, by means of dieting rather than irritate a fractious evil by drugging.

Concerning both the composite living creature and the bodily part of it, how a man should both guide and be guided by himself so as to live a most rational life, let our statement stand thus. But first and with special care we must make ready the part which is to be the guide to the best of our power, so that it may be as fair and good as possible for the work of guidance. Now to expound this subject alone in accurate detail would in itself be a sufficient task.[1] But treating it merely as a side-issue, if we follow on the lines of our previous exposition, we may consider the matter and state our conclusions not inaptly in the following terms. We have frequently asserted [2] that there are housed within us in three regions three kinds of soul, and that each of these has its own motions; so now likewise we must repeat, as briefly as possible, that the kind which remains in idleness and stays with its own motions in repose necessarily becomes weakest, whereas the kind which exercises itself becomes strongest; wherefore care must be taken that they have their motions relatively to one another in due proportion. And as regards the most lordly kind of our soul, we must conceive of it in this wise: we declare that God has given to each of us, as his daemon,[3] that kind of soul which is housed in the top of our body and which raises us—seeing that we are not an earthly but a heavenly plant—up from earth towards our kindred in the heaven. And herein

---

[3] *i.e.* " genius " or " guardian-angel "; *cf. Laws* 732 c, 877 A.

PLATO

90 οὐράνιον, ὀρθότατα λέγοντες· ἐκεῖθεν γὰρ ὅθεν ἡ πρώτη τῆς ψυχῆς γένεσις ἔφυ τὸ θεῖον τὴν κεφαλὴν
B καὶ ῥίζαν ἡμῶν ἀνακρεμαννὺν ὀρθοῖ πᾶν τὸ σῶμα.

Τῷ μὲν οὖν περὶ τὰς ἐπιθυμίας ἢ φιλονεικίας τετευτακότι καὶ ταῦτα διαπονοῦντι σφόδρα πάντα τὰ δόγματα ἀνάγκη θνητὰ ἐγγεγονέναι, καὶ παντάπασι καθ' ὅσον μάλιστα δυνατὸν θνητῷ γίγνεσθαι, τούτου μηδὲ σμικρὸν ἐλλείπειν, ἅτε τὸ τοιοῦτον ηὐξηκότι· τῷ δὲ περὶ φιλομάθειαν καὶ περὶ τὰς ἀληθεῖς φρονήσεις ἐσπουδακότι καὶ ταῦτα μάλιστα
C τῶν αὑτοῦ γεγυμνασμένῳ, φρονεῖν μὲν ἀθάνατα καὶ θεῖα, ἄνπερ ἀληθείας ἐφάπτηται, πᾶσα ἀνάγκη που, καθ' ὅσον δ' αὖ μετασχεῖν ἀνθρωπίνη φύσις ἀθανασίας ἐνδέχεται, τούτου μηδὲν μέρος ἀπολείπειν, ἅτε δὲ ἀεὶ θεραπεύοντα τὸ θεῖον ἔχοντά τε αὐτὸν εὖ κεκοσμημένον τὸν δαίμονα ξύνοικον ἐν αὑτῷ διαφερόντως εὐδαίμονα εἶναι. θεραπεία δὲ δὴ παντὶ παντὸς[1] μία, τὰς οἰκείας ἑκάστῳ τροφὰς καὶ κινήσεις ἀποδιδόναι. τῷ δ' ἐν ἡμῖν θείῳ ξυγγενεῖς
D εἰσὶ κινήσεις αἱ τοῦ παντὸς διανοήσεις καὶ περιφοραί. ταύταις δὴ ξυνεπόμενον ἕκαστον δεῖ, τὰς περὶ τὴν γένεσιν ἐν τῇ κεφαλῇ διεφθαρμένας ἡμῶν περιόδους ἐξορθοῦντα διὰ τὸ καταμανθάνειν τὰς τοῦ παντὸς ἁρμονίας τε καὶ περιφοράς, τῷ κατανοουμένῳ τὸ κατανοοῦν ἐξομοιῶσαι κατὰ τὴν ἀρχαίαν φύσιν, ὁμοιώσαντα δὲ τέλος ἔχειν τοῦ προτεθέντος ἀνθρώποις ὑπὸ θεῶν ἀρίστου βίου πρός τε τὸν παρόντα καὶ τὸν ἔπειτα χρόνον.

[1] παντὸς most mss.: πάντως Zur.

---

[1] *Cf.* Sympos. 212 A.
[2] Literally, "with a good daemon" (a play on δαίμων and εὐδαίμων).

246

TIMAEUS

we speak most truly; for it is by suspending our head and root from that region whence the substance of our soul first came that the Divine Power keeps upright our whole body.

Whoso, then, indulges in lusts or in contentions and devotes himself overmuch thereto must of necessity be filled with opinions that are wholly mortal, and altogether, so far as it is possible to become mortal, fall not short of this in even a small degree, inasmuch as he has made great his mortal part. But he who has seriously devoted himself to learning and to true thoughts, and has exercised these qualities above all his others, must necessarily and inevitably think thoughts that are immortal and divine, if so be that he lays hold on truth, and in so far as it is possible for human nature to partake of immortality,[1] he must fall short thereof in no degree; and inasmuch as he is for ever tending his divine part and duly magnifying that daemon who dwells along with him, he must be supremely blessed.[2] And the way of tendance of every part by every man is one—namely, to supply each with its own congenial food and motion; and for the divine part within us the congenial motions are the intellections and revolutions of the Universe.[3] These each one of us should follow, rectifying the revolutions within our head, which were distorted at our birth, by learning the harmonies and revolutions of the Universe, and thereby making the part that thinks like unto the object of its thought, in accordance with its original nature, and having achieved this likeness attain finally to that goal of life which is set before men by the gods as the most good both for the present and for the time to come.

[3] *Cf.* 37 A ff.

Καὶ δὴ καὶ τὰ νῦν ἡμῖν ἐξ ἀρχῆς παραγγελθέντα διεξελθεῖν περὶ τοῦ παντὸς μέχρι γενέσεως ἀνθρωπίνης σχεδὸν ἔοικε τέλος ἔχειν. τὰ γὰρ ἄλλα ζῷα ᾗ γέγονεν αὖ, διὰ βραχέων ἐπιμνηστέον, ὃ μή τις ἀνάγκη μηκύνειν· οὕτω γὰρ ἐμμετρώτερός τις ἂν αὑτῷ δόξειε περὶ τοὺς τούτων λόγους εἶναι. τῇδ' οὖν τὸ τοιοῦτον ἔστω λεγόμενον.

Τῶν γενομένων ἀνδρῶν ὅσοι δειλοὶ καὶ τὸν βίον ἀδίκως διῆλθον, κατὰ λόγον τὸν εἰκότα γυναῖκες μετεφύοντο ἐν τῇ δευτέρᾳ γενέσει. καὶ κατ' ἐκεῖνον δὴ τὸν χρόνον διὰ ταῦτα θεοὶ τὸν τῆς ξυνουσίας ἔρωτα ἐτεκτήναντο, ζῷον τὸ μὲν ἐν ἡμῖν, τὸ δ' ἐν ταῖς γυναιξὶ συστήσαντες ἔμψυχον, τοιῷδε τρόπῳ ποιήσαντες ἑκάτερον. τὴν τοῦ ποτοῦ διέξοδον, ᾗ διὰ τοῦ πλεύμονος τὸ πῶμα ὑπὸ τοὺς νεφροὺς εἰς τὴν κύστιν ἐλθὸν καὶ τῷ πνεύματι θλιφθὲν ξυνεκπέμπει δεχομένη, ξυνέτρησαν εἰς τὸν ἐκ τῆς κεφαλῆς κατὰ τὸν αὐχένα καὶ διὰ τῆς ῥάχεως μυελὸν ξυμπεπηγότα, ὃν δὴ σπέρμα ἐν τοῖς πρόσθεν λόγοις εἴπομεν· ὁ δέ, ἅτ' ἔμψυχος ὢν καὶ λαβὼν ἀναπνοήν, τοῦθ' ᾗπερ ἀνέπνευσε, τῆς ἐκροῆς ζωτικὴν ἐπιθυμίαν ἐμποιήσας αὐτῷ, τοῦ γεννᾶν ἐρῶν[1] ἀπετέλεσε. διὸ δὴ τῶν μὲν ἀνδρῶν τὸ περὶ τὴν τῶν αἰδοίων φύσιν ἀπειθές τε καὶ αὐτοκρατὲς γεγονός, οἷον ζῷον ἀνυπήκοον τοῦ λόγου, πάντων δι' ἐπιθυμίας οἰστρώδεις ἐπιχειρεῖ κρατεῖν. αἱ δ' ἐν ταῖς γυναιξὶν αὖ

[1] ἐρῶν] ἔρωτα MSS., Zur.

[1] *Cf.* 41 D ff.   [2] *Cf.* 73 c, 86 c.

TIMAEUS

And now the task prescribed for us at the beginning to give a description of the Universe up to the production of mankind, would appear to be wellnigh completed. For as regards the mode in which the rest of living creatures have been produced we must make but a brief statement, seeing that there is no need to speak at length; for by such brevity we will feel ourselves to be preserving a right proportion in our handling of these subjects. Wherefore let this matter be treated as follows.

According to the probable account, all those creatures generated as men who proved themselves cowardly and spent their lives in wrong-doing were transformed, at their second incarnation,[1] into women. And it was for this reason that the gods at that time contrived the love of sexual intercourse by constructing an animate creature of one kind in us men, and of another kind in women; and they made these severally in the following fashion. From the passage of egress for the drink, where it receives and joins in discharging the fluid which has come through the lungs beneath the kidneys into the bladder and has been compressed by the air, they bored a hole into the condensed marrow which comes from the head down by the neck and along the spine—which marrow, in our previous account,[2] we termed "seed." And the marrow, inasmuch as it is animate and has been granted an outlet, has endowed the part where its outlet lies with a love for generating by implanting therein a lively desire for emission. Wherefore in men the nature of the genital organs is disobedient and self-willed, like a creature that is deaf to reason, and it attempts to dominate all because of its frenzied lusts. And in women again,

μήτραί τε καὶ ὑστέραι λεγόμεναι διὰ ταὐτὰ ταῦτα, ζῶον ἐπιθυμητικὸν ἐνὸν τῆς παιδοποιίας, ὅταν ἄκαρπον παρὰ τὴν ὥραν χρόνον πολὺν γίγνηται, χαλεπῶς ἀγανακτοῦν φέρει, καὶ πλανώμενον πάντῃ κατὰ τὸ σῶμα, τὰς τοῦ πνεύματος διεξόδους ἀποφράττον, ἀναπνεῖν οὐκ ἐῶν εἰς ἀπορίας τὰς ἐσχάτας ἐμβάλλει καὶ νόσους παντοδαπὰς ἄλλας παρέχει, μέχριπερ ἂν ἑκατέρων ἡ ἐπιθυμία καὶ ὁ ἔρως συναγαγόντες,[1] οἷον ἀπὸ δένδρων καρπὸν καταδρέψαντες,[2] ὡς εἰς ἄρουραν τὴν μήτραν ἀόρατα ὑπὸ σμικρότητος καὶ ἀδιάπλαστα ζῷα κατασπείραντες, καὶ πάλιν διακρίναντες, μεγάλα ἐντὸς ἐκθρέψωνται καὶ μετὰ τοῦτο εἰς φῶς ἀγαγόντες ζῴων ἀποτελέσωσι γένεσιν.

Γυναῖκες μὲν οὖν καὶ τὸ θῆλυ πᾶν οὕτω γέγονε.

Τὸ δὲ τῶν ὀρνέων φῦλον μετερρυθμίζετο, ἀντὶ τριχῶν πτερὰ φύον, ἐκ τῶν ἀκάκων ἀνδρῶν, κούφων δέ, καὶ μετεωρολογικῶν μέν, ἡγουμένων δὲ δι' ὄψεως τὰς περὶ τούτων ἀποδείξεις βεβαιοτάτας εἶναι δι' εὐήθειαν. τὸ δ' αὖ πεζὸν καὶ θηριῶδες γέγονεν ἐκ τῶν μηδὲν προσχρωμένων φιλοσοφίᾳ μηδὲ ἀθρούντων τῆς περὶ τὸν οὐρανὸν φύσεως πέρι μηδέν, διὰ τὸ μηκέτι ταῖς ἐν τῇ κεφαλῇ χρῆσθαι περιόδοις, ἀλλὰ τοῖς περὶ τὰ στήθη τῆς ψυχῆς ἡγεμόσιν ἕπεσθαι μέρεσιν. ἐκ τούτων οὖν τῶν ἐπιτηδευμάτων τά τε ἐμπρόσθια κῶλα καὶ τὰς κεφαλὰς εἰς γῆν ἑλκόμενα ὑπὸ ξυγγενείας ἤρεισαν, προμήκεις τε καὶ παντοίας ἔσχον

---

[1] συναγαγόντες best mss.: ἐξαγαγόντες Zur.
[2] καταδρέψαντες best mss.: κἆτα δρέψαντες Zur.

---

[1] Cf. Rep. 529 A ff.   [2] Cf. 69 E ff.

TIMAEUS

owing to the same causes, whenever the matrix or womb, as it is called,—which is an indwelling creature desirous of child-bearing,—remains without fruit long beyond the due season, it is vexed and takes it ill; and by straying all ways through the body and blocking up the passages of the breath and preventing respiration it casts the body into the uttermost distress, and causes, moreover, all kinds of maladies; until the desire and love of the two sexes unite them. Then, culling as it were the fruit from trees, they sow upon the womb, as upon ploughed soil, animalcules that are invisible for smallness and unshapen; and these, again, they mould into shape and nourish to a great size within the body; after which they bring them forth into the light and thus complete the generation of the living creature.

In this fashion, then, women and the whole female sex have come into existence.

And the tribe of birds are derived by transformation, growing feathers in place of hair, from men who are harmless but light-minded[1]—men, too, who, being students of the worlds above, suppose in their simplicity that the most solid proofs about such matters are obtained by the sense of sight. And the wild species of animal that goes on foot is derived from those men who have paid no attention at all to philosophy nor studied at all the nature of the heavens, because they ceased to make use of the revolutions within the head and followed the lead of those parts of the soul which are in the breast.[2] Owing to these practices they have dragged their front limbs and their head down to the earth, and there planted them, because of their kinship therewith; and they have acquired elongated heads of

τὰς κορυφάς, ὅπῃ συνεθλίφθησαν ὑπὸ ἀργίας ἑκάστων αἱ περιφοραί. τετράπουν τε τὸ γένος αὐτῶν ἐκ ταύτης ἐφύετο καὶ πολύπουν τῆς προφάσεως, θεοῦ βάσεις ὑποτιθέντος πλείους τοῖς μᾶλλον ἄφροσιν, ὡς μᾶλλον ἐπὶ γῆν ἕλκοιντο. τοῖς δ᾽ ἀφρονεστάτοις αὐτῶν τούτων καὶ παντάπασι πρὸς γῆν πᾶν τὸ σῶμα κατατεινομένοις ὡς οὐδὲν ἔτι ποδῶν χρείας οὔσης, ἄποδα αὐτὰ καὶ ἰλυσπώμενα ἐπὶ γῆς ἐγέννησαν. τὸ δὲ τέταρτον γένος ἔνυδρον γέγονεν ἐκ τῶν μάλιστα ἀνοητοτάτων καὶ ἀμαθεστάτων, οὓς οὐδ᾽ ἀναπνοῆς καθαρᾶς ἔτι ἠξίωσαν οἱ μεταπλάττοντες, ὡς τὴν ψυχὴν ὑπὸ πλημμελείας πάσης ἀκαθάρτως ἐχόντων, ἀλλ᾽ ἀντὶ λεπτῆς καὶ καθαρᾶς ἀναπνοῆς ἀέρος εἰς ὕδατος θολερὰν καὶ βαθεῖαν ἔωσαν ἀνάπνευσιν· ὅθεν ἰχθύων ἔθνος καὶ τὸ τῶν ὀστρέων ξυναπάντων τε ὅσα ἔνυδρα γέγονε, δίκην ἀμαθίας ἐσχάτης ἐσχάτας οἰκήσεις εἰληχότων. καὶ κατὰ ταῦτα δὴ πάντα τότε καὶ νῦν διαμείβεται τὰ ζῶα εἰς ἄλληλα, νοῦ καὶ ἀνοίας ἀποβολῇ καὶ κτήσει μεταβαλλόμενα.

Καὶ δὴ καὶ τέλος περὶ τοῦ παντὸς νῦν ἤδη τὸν λόγον ἡμῖν φῶμεν ἔχειν· θνητὰ γὰρ καὶ ἀθάνατα ζῶα λαβὼν καὶ ξυμπληρωθεὶς ὅδε ὁ κόσμος οὕτω, ζῷον ὁρατὸν τὰ ὁρατὰ περιέχον, εἰκὼν τοῦ νοητοῦ θεὸς αἰσθητός, μέγιστος καὶ ἄριστος κάλλιστός τε καὶ τελεώτατος γέγονεν εἷς οὐρανὸς ὅδε μονογενὴς ὤν.

---

[1] *Cf.* 30 c ff., 31 B

TIMAEUS

every shape, according as their several revolutions have been distorted by disuse. On this account also their race was made four-footed and many-footed, since God set more supports under the more foolish ones, so that they might be dragged down still more to the earth. And inasmuch as there was no longer any need of feet for the most foolish of these same creatures, which stretched with their whole body along the earth, the gods generated these footless and wriggling upon the earth. And the fourth kind, which lives in the water, came from the most utterly thoughtless and stupid of men, whom those that remoulded them deemed no longer worthy even of pure respiration, seeing that they were unclean of soul through utter wickedness; wherefore in place of air, for refined and pure respiring, they thrust them into water, there to respire its turbid depths. Thence have come into being the tribe of fishes and of shell-fish and all creatures of the waters, which have for their portion the extremest of all abodes in requital for the extremity of their witlessness. Thus, both then and now, living creatures keep passing into one another in all these ways, as they undergo transformation by the loss or by the gain of reason and unreason.

And now at length we may say that our discourse concerning the Universe has reached its termination. For this our Cosmos has received the living creatures both mortal and immortal and been thereby fulfilled; it being itself a visible Living Creature embracing the visible creatures, a perceptible God made in the image of the Intelligible, most great and good and fair and perfect in its generation—even this one Heaven sole of its kind.[1]

CRITIAS

INTRODUCTION TO THE *CRITIAS*

THE *Critias* was planned as a sequel to the *Republic* and *Timaeus*. According to the programme laid down in the prefatory chapters of the *Timaeus* (see esp. 27 A ff.) the discourse of Timaeus on the creation of the world and its inhabitants was to be followed by an account of the ideal citizens of the primeval State of Athens, and how they proved their excellence in war and in peace. Accordingly the *Critias*, after a preface which serves to link it on to the *Timaeus*, commences with a detailed description of that ancient Athens, followed by a parallel account of its great rival, the State of Atlantis; but before it reaches its main theme—the war itself and how Athens served the civilized world by its civic virtue—the dialogue comes, unhappily, to an abrupt end. Why we are left with but a fragment must remain a puzzle for the literary historian; but it is a noteworthy fact that in the third book of the *Laws* Plato has given us an account of the real Athens of the historic age and its struggle against Persia, which (it is plausible to suppose) he may have substituted for the imaginative history of his original design. In any case, the moral lesson of that book is the lesson which Plato would have drawn for us in the completed *Critias*: that it is " righteousness which exalteth a nation," while " pride goeth before destruction and a haughty

INTRODUCTION TO THE *CRITIAS*

spirit before a fall." The princes of Atlantis, with their grandiose empire, were the prototypes of the Great King in their overweening pride and presumption; and their overthrow at the hands of the men who had been trained in moderation, self-control and respect for the laws of Heaven was a prophecy of the later victory of the "men of Marathon" over the invading hosts of Mardonius. Thus we may console ourselves for the loss of the greater portion of the *Critias* by the reflexion that its main theme—the defeat of barbarism—is expounded elsewhere by Plato with convincing eloquence.

ΚΡΙΤΙΑΣ

ΤΑ ΤΟΥ ΔΙΑΛΟΓΟΥ ΠΡΟΣΩΠΑ

ΤΙΜΑΙΟΣ, ΚΡΙΤΙΑΣ, ΣΩΚΡΑΤΗΣ, ΕΡΜΟΚΡΑΤΗΣ

106 ΤΙ. Ὡς ἄσμενος, ὦ Σώκρατες, οἷον ἐκ μακρᾶς ἀναπεπαυμένος ὁδοῦ, νῦν οὕτως ἐκ τῆς τοῦ λόγου διαπορείας ἀγαπητῶς ἀπήλλαγμαι. τῷ δὲ πρὶν μὲν πάλαι ποτ' ἔργῳ, νῦν δὲ λόγοις ἄρτι θεῷ γεγονότι προσεύχομαι, τῶν ῥηθέντων ὅσα μὲν ἐρρήθη μετρίως, σωτηρίαν ἡμῖν αὐτὸν αὐτῶν
B διδόναι, παρὰ μέλος δὲ εἴ τι περὶ αὐτῶν ἄκοντες εἴπομεν, δίκην τὴν πρέπουσαν ἐπιτιθέναι. δίκη δὲ ὀρθὴ τὸν πλημμελοῦντα ἐμμελῆ ποιεῖν· ἵν' οὖν τὸ λοιπὸν τοὺς περὶ θεῶν γενέσεως ὀρθῶς λέγωμεν λόγους, φάρμακον ἡμῖν αὐτὸν τελεώτατον καὶ ἄριστον φαρμάκων ἐπιστήμην εὐχόμεθα διδόναι. προσευξάμενοι δὲ παραδίδομεν κατὰ τὰς ὁμολογίας Κριτίᾳ τὸν ἑξῆς λόγον.

ΚΡ. Ἀλλ', ὦ Τίμαιε, δέχομαι μέν, ᾧ δὲ καὶ σὺ
C κατ' ἀρχὰς ἐχρήσω συγγνώμην αἰτούμενος ὡς περὶ μεγάλων μέλλων λέγειν, ταὐτὸν καὶ νῦν ἐγὼ

---

[1] *i.e.* the Universe, *cf. Tim.* 92 c, 27 c.
[2] See *Tim.* 27 A, B.

CRITIAS

CHARACTERS

TIMAEUS, CRITIAS, SOCRATES, HERMOCRATES

TIM. How gladly do I now welcome my release, Socrates, from my protracted discourse, even as a traveller who takes his rest after a long journey! And I make my prayer to that God who has recently been created by our speech [1] (although in reality created of old), that he will grant to us the conservation of all our sayings that have been rightly said, and, if unwittingly we have spoken aught discordantly, that he will impose the fitting penalty. And the correct penalty is to bring into tune him that is out of tune. In order, then, that for the future we may declare the story of the birth of the gods aright, we pray that he will grant to us that medicine which of all medicines is the most perfect and most good, even knowledge; and having made our prayer, we deliver over to Critias, in accordance with our compact,[2] the task of speaking next in order.

CRIT. And I accept the task, Timaeus; but the request which you yourself made at the beginning, when you asked for indulgence on the ground of the magnitude of the theme you were about to expound, that same request I also make now on my own behalf,

τοῦτο παραιτοῦμαι, μειζόνως δὲ αὐτοῦ τυχεῖν ἔτι
μᾶλλον ἀξιῶ περὶ τῶν μελλόντων ῥηθήσεσθαι.
καίτοι σχεδὸν μὲν οἶδα παραίτησιν εὖ μάλα φιλό-
τιμον καὶ τοῦ δέοντος ἀγροικοτέραν μέλλων παρ-
αιτεῖσθαι, ῥητέον δὲ ὅμως. ὡς μὲν γὰρ οὐκ εὖ τὰ
παρὰ σοῦ λεχθέντα εἴρηται, τίς ἂν ἐπιχειρήσειεν
ἔμφρων λέγειν; ὅτι δὲ τὰ ῥηθησόμενα πλείονος
συγγνώμης δεῖται χαλεπώτερα ὄντα, τοῦτο πειρατέον πῃ διδάξαι. περὶ θεῶν γάρ, ὦ Τίμαιε,
λέγοντά τι πρὸς ἀνθρώπους δοκεῖν ἱκανῶς λέγειν
B ῥᾷον ἢ περὶ θνητῶν πρὸς ἡμᾶς. ἡ γὰρ ἀπειρία
καὶ σφόδρα ἄγνοια τῶν ἀκουόντων περὶ ὧν ἂν
οὕτως ἔχωσι πολλὴν εὐπορίαν παρέχεσθον τῷ μέλλοντι λέγειν τι περὶ αὐτῶν· περὶ δὲ δὴ θεῶν ἴσμεν
ὡς ἔχομεν. ἵνα δὲ σαφέστερον ὃ λέγω δηλώσω,
τῇδέ μοι συνεπισπέσθε. μίμησιν μὲν γὰρ δὴ καὶ
ἀπεικασίαν τὰ παρὰ πάντων ἡμῶν ῥηθέντα χρεών
που γενέσθαι· τὴν δὲ τῶν γραφέων εἰδωλοποιίαν
περὶ τὰ θεῖά τε καὶ τὰ ἀνθρώπινα σώματα γιγνο-
C μένην ἴδωμεν ῥᾳστώνης τε πέρι καὶ χαλεπότητος
πρὸς τὸ τοῖς ὁρῶσι δοκεῖν ἀποχρώντως μεμι-
μῆσθαι, καὶ κατοψόμεθα, ὅτι γῆν μὲν καὶ ὄρη καὶ
ποταμοὺς καὶ ὕλην οὐρανόν τε ξύμπαντα καὶ τὰ
περὶ αὐτὸν ὄντα καὶ ἰόντα πρῶτον μὲν ἀγαπῶμεν
ἄν τίς τι καὶ βραχὺ πρὸς ὁμοιότητα αὐτῶν ἀπο-
μιμεῖσθαι δυνατὸς ᾖ, πρὸς δὲ τούτοις, ἅτε οὐδὲν
εἰδότες ἀκριβὲς περὶ τῶν τοιούτων, οὔτε ἐξετάζομεν οὔτε ἐλέγχομεν τὰ γεγραμμένα, σκιαγραφίᾳ δὲ
D ἀσαφεῖ καὶ ἀπατηλῷ χρώμεθα περὶ αὐτά· τὰ δὲ

---

[1] Critias speaks as a sceptic.

CRITIAS

and I claim indeed to be granted a still larger measure of indulgence in respect of the discourse I am about to deliver. I am sufficiently aware that the request I am about to make is decidedly presumptuous and less civil than is proper, but none the less it must be uttered. For as regards the exposition you gave, what man in his senses would attempt to deny its excellence? But what I must somehow endeavour to show is that the discourse now to be delivered calls for greater indulgence because of its greater difficulty. For it is easier, Timaeus, to appear to speak satisfactorily to men about the gods, than to us about mortals. For when the listeners are in a state of inexperience and complete ignorance about a matter, such a state of mind affords great opportunities to the person who is going to discourse on that matter; and we know what our state is concerning knowledge of the gods.[1] But in order that I may explain my meaning more clearly, pray follow me further. The accounts given by us all must be, of course, of the nature of imitations and representations; and if we look at the portraiture of divine and of human bodies as executed by painters, in respect of the ease or difficulty with which they succeed in imitating their subjects in the opinion of onlookers, we shall notice in the first place that as regards the earth and mountains and rivers and woods and the whole of heaven, with the things that exist and move therein, we are content if a man is able to represent them with even a small degree of likeness; and further, that, inasmuch as we have no exact knowledge about such objects, we do not examine closely or criticize the paintings, but tolerate, in such cases, an inexact and deceptive sketch. On the other hand, whenever a

107
ἡμέτερα ὁπόταν τις ἐπιχειρῇ σώματα ἀπεικάζειν, ὀξέως αἰσθανόμενοι τὸ παραλειπόμενον διὰ τὴν ἀεὶ ξύνοικον κατανόησιν χαλεποὶ κριταὶ γιγνόμεθα τῷ μὴ πάσας πάντως τὰς ὁμοιότητας ἀποδιδόντι. ταὐτὸν δὴ καὶ κατὰ τοὺς λόγους ἰδεῖν δεῖ γιγνόμενον, ὅτι τὰ μὲν οὐράνια καὶ θεῖα ἀγαπῶμεν καὶ σμικρῶς εἰκότα λεγόμενα, τὰ δὲ θνητὰ
Ε καὶ ἀνθρώπινα ἀκριβῶς ἐξετάζομεν. ἐκ δὴ τοῦ παραχρῆμα νῦν λεγόμενα, τὸ πρέπον ἂν μὴ δυνώμεθα πάντως ἀποδιδόναι, συγγιγνώσκειν χρεών· οὐ γὰρ ὡς ῥᾴδια τὰ θνητὰ ἀλλ' ὡς χαλεπὰ πρὸς δόξαν ὄντα ἀπεικάζειν δεῖ διανοεῖσθαι. ταῦτα δὴ
108 βουλόμενος ὑμᾶς ὑπομνῆσαι, καὶ τὸ τῆς συγγνώμης οὐκ ἔλαττον ἀλλὰ μεῖζον αἰτῶν περὶ τῶν μελλόντων ῥηθήσεσθαι, πάντα ταῦτα εἴρηκα, ὦ Σώκρατες. εἰ δὴ δικαίως αἰτεῖν φαίνομαι τὴν δωρεάν, ἑκόντες δίδοτε.

ΣΩ. Τί δ' οὐ μέλλομεν, ὦ Κριτία, διδόναι; καὶ πρός γ' ἔτι τρίτῳ δεδόσθω ταὐτὸν τοῦτο Ἑρμοκράτει παρ' ἡμῶν. δῆλον γάρ, ὡς ὀλίγον ὕστερον, ὅταν αὐτὸν δέῃ λέγειν, παραιτήσεται καθάπερ
Β ὑμεῖς· ἵν' οὖν ἑτέραν ἀρχὴν ἐκπορίζηται καὶ μὴ τὴν αὐτὴν ἀναγκασθῇ λέγειν, ὡς ὑπαρχούσης αὐτῷ συγγνώμης εἰς τότε οὕτω λεγέτω. προλέγω γε μήν, ὦ φίλε Κριτία, σοὶ τὴν τοῦ θεάτρου διάνοιαν, ὅτι θαυμαστῶς ὁ πρότερος εὐδοκίμηκεν ἐν αὐτῷ ποιητής, ὥστε τῆς συγγνώμης δεήσει τινός σοι παμπόλλης, εἰ μέλλεις αὐτὰ δυνατὸς γενέσθαι παραλαβεῖν.

CRITIAS

painter tries to render a likeness of our own bodies, we quickly perceive what is defective because of our constant familiar acquaintance with them, and become severe critics of him who fails to bring out to the full all the points of similarity. And precisely the same thing happens, as we should notice, in the case of discourses: in respect of what is celestial and divine we are satisfied if the account possesses even a small degree of likelihood, but we examine with precision what is mortal and human. To an account given now on the spur of the moment indulgence must be granted, should we fail to make it a wholly fitting representation; for one must conceive of mortal objects as being difficult, and not easy, to represent satisfactorily. It is because I wish to remind you of these facts, and crave a greater rather than a less measure of indulgence for what I am about to say, that I have made all these observations, Socrates. If, therefore, I seem justified in craving this boon, pray grant it willingly.

soc. And why should we hesitate to grant it, Critias? Nay, what is more, the same boon shall be granted by us to a third, Hermocrates. For it is plain that later on, before long, when it is his duty to speak, he will make the same request as you. So, in order that he may provide a different prelude and not be compelled to repeat the same one, let him assume, when he comes to speak, that he already has our indulgence. I forewarn you, however, my dear Critias, of the mind of your audience,—how that the former poet won marvellous applause from it, so that you will require an extraordinary measure of indulgence if you are to prove capable of following in his steps.

ερ. Ταὐτὸν μήν, ὦ Σώκρατες, κἀμοὶ παραγγέλλεις ὅπερ τῷδε. ἀλλὰ γὰρ ἀθυμοῦντες ἄνδρες οὔπω τρόπαιον ἔστησαν, ὦ Κριτία· προϊέναι τε οὖν ἐπὶ τὸν λόγον ἀνδρείως χρή, καὶ τὸν Παίωνά τε καὶ τὰς Μούσας ἐπικαλούμενον τοὺς παλαιοὺς πολίτας ἀγαθοὺς ὄντας ἀναφαίνειν τε καὶ ὑμνεῖν.

κρ. Ὦ φίλε Ἑρμόκρατες, τῆς ὑστέρας τεταγμένος, ἐπίπροσθεν ἔχων ἄλλον, ἔτι θαρρεῖς. τοῦτο μὲν οὖν οἷόν ἐστιν, αὐτό σοι τάχα δηλώσει· παραμυθουμένῳ δ' οὖν καὶ παραθαρρύνοντί σοι πειστέον, καὶ πρὸς οἷς θεοῖς εἶπες, τούς τε ἄλλους κλητέον καὶ δὴ καὶ τὰ μάλιστα Μνημοσύνην. σχεδὸν γὰρ τὰ μέγιστα ἡμῖν τῶν λόγων ἐν ταύτῃ τῇ θεῷ πάντ' ἐστί· μνησθέντες γὰρ ἱκανῶς καὶ ἀπαγγείλαντες τά ποτε ῥηθέντα ὑπὸ τῶν ἱερέων καὶ δεῦρο ὑπὸ Σόλωνος κομισθέντα, σχεδὸν οἶδ' ὅτι τῷδε τῷ θεάτρῳ δόξομεν τὰ προσήκοντα μετρίως ἀποτετελεκέναι. τοῦτ' οὖν αὐτὸ ἤδη δραστέον, καὶ μελλητέον οὐδὲν ἔτι.

Πάντων δὴ πρῶτον μνησθῶμεν ὅτι τὸ κεφάλαιον ἦν ἐνάκις χίλια ἔτη, ἀφ' οὗ γεγονὼς ἐμηνύθη πόλεμος τοῖς θ' ὑπὲρ Ἡρακλείας στήλας ἔξω κατοικοῦσι καὶ τοῖς ἐντὸς πᾶσιν· ὃν δεῖ νῦν διαπεραίνειν. τῶν μὲν οὖν ἥδε ἡ πόλις ἄρξασα καὶ πάντα τὸν πόλεμον διαπολεμήσασα ἐλέγετο, τῶν δ' οἱ τῆς Ἀτλαντίδος νήσου βασιλεῖς, ἣν δὴ Λιβύης καὶ Ἀσίας μείζω νῆσον οὖσαν ἔφαμεν εἶναί ποτε, νῦν δὲ ὑπὸ σεισμῶν δῦσαν ἄπορον πηλὸν τοῖς ἐνθένδε ἐκπλέουσιν ἐπὶ τὸ πέραν[1] πέλαγος, ὥστε

---

[1] πέραν v. Heusde: πᾶν mss., Zur.

[1] i.e. Apollo, as god of victory.
[2] The goddess of Memory.

CRITIAS

HERM. And in truth, Socrates, you are giving me the same warning as Critias. But men of faint heart never yet set up a trophy, Critias; wherefore you must go forward to your discoursing manfully, and, invoking the aid of Paion[1] and the Muses, exhibit and celebrate the excellence of your ancient citizens.

CRIT. You, my dear Hermocrates, are posted in the last rank, with another man before you, so you are still courageous. But experience of our task will of itself speedily enlighten you as to its character. However, I must trust to your consolation and encouragement, and in addition to the gods you mentioned I must call upon all the rest and especially upon Mnemosynê.[2] For practically all the most important part of our speech depends upon this goddess; for if I can sufficiently remember and report the tale once told by the priests and brought hither by Solon, I am wellnigh convinced that I shall appear to the present audience to have fulfilled my task adequately. This, then, I must at once proceed to do, and procrastinate no longer.

Now first of all we must recall the fact that 9000 is the sum of years[3] since the war occurred, as is recorded, between the dwellers beyond the pillars of Heracles and all that dwelt within them[4]; which war we have now to relate in detail. It was stated that this city of ours was in command of the one side and fought through the whole of the war, and in command of the other side were the kings of the island of Atlantis, which we said was an island larger than Libya and Asia once upon a time, but now lies sunk by earthquakes and has created a barrier of impassable mud which prevents those who are sailing out from here

[3] *Cf. Tim.* 23 E.   [4] *Cf. Tim.* 24 E.

PLATO

μηκέτι πορεύεσθαι, κωλυτὴν παρασχεῖν. τὰ μὲν
δὴ πολλὰ ἔθνη βάρβαρα, καὶ ὅσα Ἑλλήνων ἦν
γένη τότε, καθ᾽ ἕκαστα ἡ τοῦ λόγου διέξοδος οἷον
ἀνειλλομένη τὸ προστυχὸν ἑκασταχοῦ δηλώσει· τὸ
δὲ Ἀθηναίων τε τῶν τότε καὶ τῶν ἐναντίων,
οἷς διεπολέμησαν, ἀνάγκη κατ᾽ ἀρχὰς διελθεῖν
πρῶτα, τήν τε δύναμιν ἑκατέρων καὶ τὰς πολιτείας.
αὐτῶν δὲ τούτων τὰ τῇδε ἔμπροσθεν προτιμητέον
εἰπεῖν.

B Θεοὶ γὰρ ἅπασαν γῆν ποτὲ κατὰ τοὺς τόπους δι-
ελάγχανον, οὐ κατ᾽ ἔριν· οὐ γὰρ ἂν ὀρθὸν ἔχοι
λόγον θεοὺς ἀγνοεῖν τὰ πρέποντα ἑκάστοις αὑτῶν,
οὐδ᾽ αὖ γιγνώσκοντας τὸ μᾶλλον ἄλλοις προσῆκον
τοῦτο ἑτέρους αὑτοῖς δι᾽ ἐρίδων ἐπιχειρεῖν κτᾶσθαι.
δίκης δὴ κλήροις τὸ φίλον λαγχάνοντες κατῴκιζον
τὰς χώρας, καὶ κατοικίσαντες, οἷον νομῆς ποίμνια,
κτήματα καὶ θρέμματα ἑαυτῶν ἡμᾶς ἔτρεφον,
πλὴν οὐ σώμασι σώματα βιαζόμενοι, καθάπερ
C ποιμένες κτήνη πληγῇ νέμοντες, ἀλλ᾽ ᾗ μάλιστα
εὔστροφον ζῷον, ἐκ πρύμνης ἀπευθύνοντες οἷον
οἴακι πειθοῖ ψυχῆς ἐφαπτόμενοι κατὰ τὴν αὐτῶν
διάνοιαν, οὕτως ἄγοντες τὸ θνητὸν πᾶν ἐκυβέρνων.
ἄλλοι μὲν οὖν κατ᾽ ἄλλους τόπους κληρουχήσαντες
θεῶν ἐκεῖνα ἐκόσμουν, Ἥφαιστος δὲ κοινὴν καὶ
Ἀθηνᾶ φύσιν ἔχοντες, ἅμα μὲν ἀδελφὴν ἐκ ταὐτοῦ
πατρός, ἅμα δὲ φιλοσοφίᾳ φιλοτεχνίᾳ τε ἐπὶ τὰ
αὐτὰ ἐλθόντες, οὕτω μίαν ἄμφω λῆξιν τήνδε τὴν

---

[1] Cf. *Tim.* 25 D.
[2] This contradicts the myth in *Menex.* 237 C, D, relating the strife between Poseidon and Athena.

CRITIAS

to the ocean beyond from proceeding further.¹ Now as regards the numerous barbaric tribes and all the Hellenic nations that then existed, the sequel of our story, when it is, as it were, unrolled, will disclose what happened in each locality; but the facts about the Athenians of that age and the enemies with whom they fought we must necessarily describe first, at the outset,—the military power, that is to say, of each and their forms of government. And of these two we must give the priority in our account to the state of Athens.

Once upon a time the gods were taking over by lot the whole earth according to its regions,—not according to the results of strife ² : for it would not be reasonable to suppose that the gods were ignorant of their own several rights, nor yet that they attempted to obtain for themselves by means of strife a possession to which others, as they knew, had a better claim. So by just allotments they received each one his own, and they settled their countries; and when they had thus settled them, they reared us up, even as herdsmen rear their flocks, to be their cattle and nurslings; only it was not our bodies that they constrained by bodily force, like shepherds guiding their flocks with stroke of staff, but they directed from the stern where the living creature is easiest to turn about, laying hold on the soul by persuasion, as by a rudder, according to their own disposition; and thus they drove and steered all the mortal kind. Now in other regions others of the gods had their allotments and ordered the affairs, but inasmuch as Hephaestus and Athena were of a like nature, being born of the same father, and agreeing, moreover, in their love of wisdom and of craftsmanship, they both took for their

109
χώραν εἰλήχατον ὡς οἰκείαν καὶ πρόσφορον ἀρετῇ
D καὶ φρονήσει πεφυκυῖαν, ἄνδρας δὲ ἀγαθοὺς ἐμ-
ποιήσαντες αὐτόχθονας ἐπὶ νοῦν ἔθεσαν τὴν τῆς
πολιτείας τάξιν· ὧν τὰ μὲν ὀνόματα σέσωται, τὰ
δὲ ἔργα διὰ τὰς τῶν παραλαμβανόντων φθορὰς
καὶ τὰ μήκη τῶν χρόνων ἠφανίσθη· τὸ γὰρ περι-
λειπόμενον ἀεὶ γένος, ὥσπερ καὶ πρόσθεν ἐρρήθη,
κατελείπετο ὄρειον καὶ ἀγράμματον, τῶν ἐν τῇ
χώρᾳ δυναστῶν τὰ ὀνόματα ἀκηκοὸς μόνον καὶ
βραχέα πρὸς αὐτοῖς τῶν ἔργων. τὰ μὲν οὖν
E ὀνόματα τοῖς ἐκγόνοις ἐτίθεντο ἀγαπῶντες, τὰς
δὲ ἀρετὰς καὶ τοὺς νόμους τῶν ἔμπροσθεν οὐκ
εἰδότες, εἰ μὴ σκοτεινὰς περὶ ἑκάστων τινὰς ἀκοάς,
ἐν ἀπορίᾳ δὲ τῶν ἀναγκαίων ἐπὶ πολλὰς γενεὰς
ὄντες καὶ αὐτοὶ καὶ παῖδες, πρὸς οἷς ἠπόρουν τὸν
110 νοῦν ἔχοντες, τούτων πέρι καὶ τοὺς λόγους ποιούμενοι, τῶν ἐν τοῖς πρόσθεν καὶ πάλαι ποτὲ
γεγονότων ἠμέλουν. μυθολογία γὰρ ἀναζήτησίς
τε τῶν παλαιῶν μετὰ σχολῆς ἅμ᾽ ἐπὶ τὰς πόλεις
ἔρχεσθον, ὅταν ἴδητόν τισιν ἤδη τοῦ βίου τἀναγκαῖα
κατεσκευασμένα, πρὶν δὲ οὔ.

Ταύτῃ δὴ τὰ τῶν παλαιῶν ὀνόματα ἄνευ τῶν
ἔργων διασέσωται. λέγω δὲ αὐτὰ τεκμαιρόμενος,
ὅτι Κέκροπός τε καὶ Ἐρεχθέως καὶ Ἐριχθονίου
καὶ Ἐρυσίχθονος τῶν τε ἄλλων τὰ πλεῖστα,
B ὅσαπερ καὶ Θησέως τῶν ἄνω περὶ τῶν ὀνομάτων
ἑκάστων ἀπομνημονεύεται, τούτων ἐκείνους τὰ
πολλὰ ἐπονομάζοντας τοὺς ἱερέας Σόλων ἔφη τὸν
τότε διηγεῖσθαι πόλεμον, καὶ τὰ τῶν γυναικῶν κατὰ

---

[1] *Cf. Tim.* 23 A.

CRITIAS

joint portion this land of ours as being naturally congenial and adapted for virtue and for wisdom, and therein they planted as native to the soil men of virtue and ordained to their mind the mode of government. And of these citizens the names are preserved, but their works have vanished owing to the repeated destruction of their successors and the length of the intervening periods. For, as was said before,[1] the stock that survived on each occasion was a remnant of unlettered mountaineers which had heard the names only of the rulers, and but little besides of their works. So though they gladly passed on these names to their descendants, concerning the mighty deeds and the laws of their predecessors they had no knowledge, save for some invariably obscure reports; and since, moreover, they and their children for many generations were themselves in want of the necessaries of life, their attention was given to their own needs and all their talk was about them; and in consequence they paid no regard to the happenings of bygone ages. For legendary lore and the investigation of antiquity are visitants that come to cities in company with leisure, when they see that men are already furnished with the necessaries of life, and not before.

In this way, then, the names of the ancients, without their works, have been preserved. And for evidence of what I say I point to the statement of Solon, that the Egyptian priests, in describing the war of that period, mentioned most of those names—such as those of Cecrops and Erechtheus and Erichthonius and Erysichthon and most of the other names which are recorded of the various heroes before Theseus—and in like manner also the names of the

269

110 ταὐτά. καὶ δὴ καὶ τὸ τῆς θεοῦ σχῆμα καὶ ἄγαλμα
[ὡς κοινὰ τότ' ἦν τὰ ἐπιτηδεύματα ταῖς τε γυναιξὶ
καὶ τοῖς ἀνδράσι τὰ περὶ τὸν πόλεμον, οὕτω κατ'
ἐκεῖνον τὸν νόμον ὡπλισμένην τὴν θεὸν ἀνάθημα
εἶναι τοῖς τότε,][1] ἔνδειγμα ὅτι πάνθ' ὅσα ξύννομα
C ζῷα θήλεα καὶ ὅσα ἄρρενα, τὴν προσήκουσαν
ἀρετὴν ἑκάστῳ γένει πᾶν κοινῇ δυνατὸν ἐπιτηδεύειν
πέφυκεν.

"Ὤικει δὲ δὴ τότ' ἐν τῇδε τῇ χώρᾳ τὰ μὲν ἄλλα
ἔθνη τῶν πολιτῶν περὶ τὰς δημιουργίας ὄντα καὶ
τὴν ἐκ τῆς γῆς τροφήν, τὸ δὲ μάχιμον ὑπ' ἀνδρῶν
θείων κατ' ἀρχὰς ἀφορισθὲν ᾤκει χωρίς, πάντα
εἰς τροφὴν καὶ παίδευσιν τὰ προσήκοντα ἔχον,
ἴδιον μὲν αὐτῶν οὐδεὶς οὐδὲν κεκτημένος, ἅπαντα
D δὲ πάντων κοινὰ νομίζοντες αὐτῶν, πέρα δὲ ἱκανῆς
τροφῆς οὐδὲν ἀξιοῦντες παρὰ τῶν ἄλλων δέχεσθαι
πολιτῶν, καὶ πάντα δὴ τὰ χθὲς λεχθέντα ἐπιτηδεύματα ἐπιτηδεύοντες, ὅσα περὶ τῶν ὑποτεθέντων
ἐρρήθη φυλάκων. καὶ δὴ καὶ τὸ περὶ τῆς χώρας
ἡμῶν πιθανὸν καὶ ἀληθὲς ἐλέγετο, πρῶτον μὲν
τοὺς ὅρους αὐτὴν ἐν τῷ τότ' ἔχειν ἀφωρισμένους
πρὸς τὸν Ἰσθμὸν καὶ τὸ κατὰ τὴν ἄλλην ἤπειρον
μέχρι τοῦ Κιθαιρῶνος καὶ Πάρνηθος τῶν ἄκρων,
E καταβαίνειν δὲ τοὺς ὅρους ἐν δεξιᾷ τὴν Ὠρωπίαν
ἔχοντας, ἐν ἀριστερᾷ δὲ πρὸς θαλάττης ἀφορίζοντας
τὸν Ἀσωπόν, ἀρετῇ δὲ πᾶσαν γῆν ὑπὸ τῆς ἐνθάδε
ὑπερβάλλεσθαι, διὸ καὶ δυνατὴν εἶναι τότε τρέφειν
τὴν χώραν στρατόπεδον πολὺ τῶν περὶ γῆν ἀργὸν
ἔργων.[2] μέγα δὲ τεκμήριον ἀρετῆς· τὸ γὰρ νῦν
αὐτῆς λείψανον ἐνάμιλλόν ἐστι πρὸς ἡντινοῦν τῷ

---

[1] ὡς ... τότε bracketed by W.-Möllendorff.
[2] περὶ γῆν ἀργὸν ἔργων best ms.: περιοίκων Zur.

CRITIAS

women. Moreover, the habit and figure of the goddess indicate that in the case of all animals, male and female, that herd together, every species is naturally capable of practising as a whole and in common its own proper excellence.

Now at that time there dwelt in this country not only the other classes of the citizens who were occupied in the handicrafts and in the raising of food from the soil, but also the military class, which had been separated off at the commencement by divine heroes and dwelt apart.[1] It was supplied with all that was required for its sustenance and training, and none of its members possessed any private property, but they regarded all they had as the common property of all; and from the rest of the citizens they claimed to receive nothing beyond a sufficiency of sustenance; and they practised all those pursuits which were mentioned yesterday,[2] in the description of our proposed " Guardians." Moreover, what was related [3] about our country was plausible and true, namely, that, in the first place, it had its boundaries at that time marked off by the Isthmus, and on the inland side reaching to the heights of Cithaeron and Parnes; and that the boundaries ran down with Oropia on the right, and on the seaward side they shut off the Asopus on the left; and that all other lands were surpassed by ours in goodness of soil, so that it was actually able at that period to support a large host which was exempt from the labours of husbandry. And of its goodness a strong proof is this: what is now left of our soil rivals any other in being all-

---

[1] *Cf. Tim.* 24 B.
[2] *Cf. Rep.* 376 c ff.; *Tim.* 17 D ff.
[3] *i.e.* by the Egyptians.

PLATO

πάμφορον εὔκαρπόν τε εἶναι καὶ τοῖς ζῴοις πᾶσιν εὔβοτον. τότε δὲ πρὸς τῷ κάλλει καὶ παμπλήθη ταῦτα ἔφερε. πῶς οὖν δὴ τοῦτο πιστόν, καὶ κατὰ τί λείψανον τῆς τότε γῆς ὀρθῶς ἂν λέγοιτο; πᾶσα ἀπὸ τῆς ἄλλης ἠπείρου μακρὰ προτείνουσα εἰς τὸ πέλαγος οἷον ἄκρα κεῖται· τὸ δὴ τῆς θαλάττης ἀγγεῖον περὶ αὐτὴν τυγχάνει πᾶν ἀγχιβαθὲς ὄν. πολλῶν οὖν γεγονότων καὶ μεγάλων κατακλυσμῶν ἐν τοῖς ἐνακισχιλίοις ἔτεσι, τοσαῦτα γὰρ πρὸς τὸν νῦν ἀπ᾽ ἐκείνου τοῦ χρόνου γέγονεν ἔτη, τὸ τῆς γῆς ἐν τούτοις τοῖς χρόνοις καὶ πάθεσιν ἐκ τῶν ὑψηλῶν ἀπορρέον οὔτε χῶμα, ὡς ἐν ἄλλοις τόποις, προχοῖ λόγου ἄξιον ἀεί τε κύκλῳ περιρρέον εἰς βάθος ἀφανίζεται· λέλειπται δή, καθάπερ ἐν ταῖς μικραῖς νήσοις, πρὸς τὰ τότε τὰ νῦν οἷον νοσήσαντος σώματος ὀστᾶ, περιερρυηκυίας τῆς γῆς ὅση πίειρα καὶ μαλακή, τοῦ λεπτοῦ σώματος τῆς χώρας μόνον λειφθέντος. τότε δὲ ἀκέραιος οὖσα τά τε ὄρη γηλόφους ὑψηλοὺς εἶχε, καὶ τὰ φελλέως νῦν ὀνομασθέντα πεδία πλήρη γῆς πιείρας ἐκέκτητο, καὶ πολλὴν ἐν τοῖς ὄρεσιν ὕλην εἶχεν, ἧς καὶ νῦν ἔτι φανερὰ τεκμήρια· τῶν γὰρ ὀρῶν ἔστιν ἃ νῦν μὲν ἔχει μελίτταις μόναις τροφήν, χρόνος δ᾽ οὐ πάμπολυς ὅτε δένδρα, ὧν[1] αὐτόθεν εἰς οἰκοδομήσεις τὰς μεγίστας ἐρεψίμων τμηθέντων στεγάσματ᾽ ἐστὶν ἔτι σᾶ. πολλὰ δ᾽ ἦν ἄλλ᾽ ἥμερα ὑψηλὰ δένδρα, νομὴν δὲ βοσκήμασιν ἀμήχανον ἔφερε. καὶ δὴ καὶ τὸ κατ᾽ ἐνιαυτὸν ὕδωρ ἐκαρποῦτο ἐκ

[1] δένδρα, ὧν] δένδρων mss., Zur.

---

[1] *Cf.* 108 E.
[2] φελλεύς means a porous stone, like lava, or a field of stony soil.

CRITIAS

productive and abundant in crops and rich in pasturage for all kinds of cattle; and at that period, in addition to their fine quality it produced these things in vast quantity. How, then, is this statement plausible, and what residue of the land then existing serves to confirm its truth? The whole of the land lies like a promontory jutting out from the rest of the continent far into the sea; and all the cup of the sea round about it is, as it happens, of a great depth. Consequently, since many great convulsions took place during the 9000 years—for such was the number of years from that time to this [1]—the soil which has kept breaking away from the high lands during these ages and these disasters, forms no pile of sediment worth mentioning, as in other regions, but keeps sliding away ceaselessly and disappearing in the deep. And, just as happens in small islands, what now remains compared with what then existed is like the skeleton of a sick man, all the fat and soft earth having wasted away, and only the bare framework of the land being left. But at that epoch the country was unimpaired, and for its mountains it had high arable hills, and in place of the "moorlands," [2] as they are now called, it contained plains full of rich soil; and it had much forest-land in its mountains, of which there are visible signs even to this day; for there are some mountains which now have nothing but food for bees, but they had trees no very long time ago, and the rafters from those felled there to roof the largest buildings are still sound. And besides, there were many lofty trees of cultivated species; and it produced boundless pasturage for flocks. Moreover, it was enriched by the yearly

PLATO

111

D Διός, οὐχ ὡς νῦν ἀπολλῦσα ῥέον ἀπὸ ψιλῆς τῆς
γῆς εἰς θάλατταν, ἀλλὰ πολλὴν ἔχουσα καὶ εἰς
αὑτὴν καταδεχομένη, τῇ κεραμίδι στεγούσῃ γῇ
διαταμιευομένη, τὸ καταποθὲν ἐκ τῶν ὑψηλῶν
ὕδωρ εἰς τὰ κοῖλα ἀφιεῖσα, κατὰ πάντας τοὺς
τόπους παρείχετο ἄφθονα κρηνῶν καὶ ποταμῶν
νάματα, ὧν καὶ νῦν ἔτι ἐπὶ ταῖς πηγαῖς ταῖς πρότερον οὔσαις ἱερὰ λελειμμένα ἐστὶ σημεῖα ὅτι περὶ
αὐτῆς ἀληθῆ λέγεται τὰ νῦν.

E Τὰ μὲν οὖν τῆς ἄλλης χώρας φύσει τε οὕτως εἶχε,
καὶ διεκεκόσμητο ὡς εἰκὸς ὑπὸ γεωργῶν μὲν
ἀληθινῶν καὶ πραττόντων αὐτὸ τοῦτο, φιλοκάλων δὲ
καὶ εὐφυῶν, γῆν δὲ ἀρίστην καὶ ὕδωρ ἀφθονώτατον
ἐχόντων καὶ ὑπὲρ τῆς γῆς ὥρας μετριώτατα κεκρα-
μένας· τὸ δ' ἄστυ κατῳκισμένον ὧδ' ἦν ἐν τῷ τότε
χρόνῳ. πρῶτον μὲν τὸ τῆς ἀκροπόλεως εἶχε τότε

112 οὐχ ὡς τὰ νῦν ἔχει. νῦν μὲν γὰρ μία γενομένη νὺξ
ὑγρὰ διαφερόντως γῆς αὐτὴν ψιλὴν περιτήξασα
πεποίηκε, σεισμῶν ἅμα καὶ πρὸ τῆς ἐπὶ Δευκαλίωνος
φθορᾶς τρίτου πρότερον ὕδατος ἐξαισίου γενομένου·
τὸ δὲ πρὶν ἐν ἀνωτέρω[1] χρόνῳ μέγεθος μὲν ἦν πρὸς
τὸν Ἠριδανὸν καὶ τὸν Ἰλισὸν ἀποβεβηκυῖα καὶ
περιειληφυῖα ἐντὸς τὴν Πύκνα καὶ τὸν Λυκαβητ-
τὸν ὅρον ἐκ τοῦ καταντικρὺ τῆς Πυκνὸς ἔχουσα,
γεώδης δ' ἦν πᾶσα καὶ πλὴν ὀλίγων ἐπίπεδος
B ἄνωθεν. ᾠκεῖτο δὲ τὰ μὲν ἔξωθεν, ὑπ' αὐτὰ τὰ
πλάγια αὐτῆς, ὑπὸ τῶν δημιουργῶν καὶ τῶν

---

[1] ἀνωτέρω] ἑτέρῳ MSS., Zur.

---

[1] *Cf. Tim.* 22 A, 23 A, B.
[2] The Eridanus ran on the N., the Ilissus on the S. side of Athens. The Pnyx was a hill W. of the Acropolis; the Lycabettus a larger hill to the N.E. of the city.

CRITIAS

rains from Zeus, which were not lost to it, as now, by flowing from the bare land into the sea; but the soil it had was deep, and therein it received the water, storing it up in the retentive loamy soil; and by drawing off into the hollows from the heights the water that was there absorbed, it provided all the various districts with abundant supplies of spring-waters and streams, whereof the shrines which still remain even now, at the spots where the fountains formerly existed, are signs which testify that our present description of the land is true.

Such, then, was the natural condition of the rest of the country, and it was ornamented as you would expect from genuine husbandmen who made husbandry their sole task, and who were also men of taste and of native talent, and possessed of most excellent land and a great abundance of water, and also, above the land, a climate of most happily tempered seasons. And as to the city, this is the way in which it was laid out at that time. In the first place, the acropolis, as it existed then, was different from what it is now. For as it is now, the action of a single night of extraordinary rain has crumbled it away and made it bare of soil, when earthquakes occurred simultaneously with the third of the disastrous floods which preceded the destructive deluge in the time of Deucalion.[1] But in its former extent, at an earlier period, it went down towards the Eridanus and the Ilissus, and embraced within it the Pnyx, and had the Lycabettus as its boundary over against the Pnyx[2]; and it was all rich in soil and, save for a small space, level on the top. And its outer parts, under its slopes, were inhabited by the craftsmen and by such of the

PLATO

γεωργῶν ὅσοι πλησίον ἐγεώργουν· τὰ δ' ἐπάνω τὸ μάχιμον αὐτὸ καθ' αὑτὸ μόνον γένος περὶ τὸ τῆς Ἀθηνᾶς Ἡφαίστου τε ἱερὸν κατῳκήκειν, οἷον μιᾶς οἰκίας κῆπον ἑνὶ περιβόλῳ προσπεριβεβλημένοι. τὰ γὰρ πρὸς βορρᾶ αὐτῆς ᾤκουν οἰκίας κοινὰς καὶ ξυσσίτια χειμερινὰ κατασκευασάμενοι καὶ πάντα ὅσα πρέποντα ἦν τῇ κοινῇ πολιτείᾳ δι' οἰκοδομήσεων ὑπάρχειν αὐτῶν καὶ τῶν ἱερέων,[1] ἄνευ χρυσοῦ καὶ ἀργύρου· τούτοις γὰρ οὐδὲν οὐδαμόσε προσεχρῶντο, ἀλλὰ τὸ μέσον ὑπερηφανίας καὶ ἀνελευθερίας μεταδιώκοντες κοσμίας ᾠκοδομοῦντο οἰκήσεις, ἐν αἷς αὐτοί τε καὶ ἐκγόνων ἔκγονοι καταγηρῶντες ἄλλοις ὁμοίοις τὰς αὐτὰς ἀεὶ παρεδίδοσαν· τὰ δὲ πρὸς νότου, κήπους καὶ γυμνάσια συσσίτιά τε ἀνέντες οἷα θέρους κατεχρῶντο ἐπὶ ταῦτα αὑτοῖς. κρήνη δ' ἦν μία κατὰ τὸν τῆς νῦν ἀκροπόλεως τόπον, ἧς ἀποσβεσθείσης ὑπὸ τῶν σεισμῶν τὰ νῦν νάματα σμικρὰ κύκλῳ καταλέλειπται, τοῖς δὲ τότε πᾶσι παρεῖχεν ἄφθονον ῥεῦμα, εὐκρὰς οὖσα πρὸς χειμῶνά τε καὶ θέρος. τούτῳ δὴ κατῴκουν τῷ σχήματι, τῶν μὲν αὑτῶν πολιτῶν φύλακες, τῶν δὲ ἄλλων Ἑλλήνων ἡγεμόνες ἑκόντων, πλῆθος δὲ διαφυλάττοντες ὅ τι μάλιστα ταὐτὸν ἑαυτῶν εἶναι πρὸς τὸν ἀεὶ χρόνον ἀνδρῶν καὶ γυναικῶν, τὸ δυνατὸν πολεμεῖν ἤδη καὶ τὸ ἔτι, περὶ δύο μάλιστα ὄντες μυριάδας.

Οὗτοι μὲν οὖν δὴ τοιοῦτοί τε ὄντες αὐτοὶ καὶ τινα τοιοῦτον ἀεὶ τρόπον τήν τε ἑαυτῶν καὶ τὴν Ἑλλάδα δίκῃ διοικοῦντες, ἐπὶ πᾶσαν Εὐρώπην

[1] ἱερέων Hermann: ἱερῶν mss., Zur.

[1] Cf. Rep. 416 D ff.; Laws, 801 B.

CRITIAS

husbandmen as had their farms close by; but on the topmost part only the military class by itself had its dwellings round about the temple of Athene and Hephaestus, surrounding themselves with a single ring-fence, which formed, as it were, the enclosure of a single dwelling. On the northward side of it they had established their public dwellings and winter mess-rooms, and all the arrangements in the way of buildings which were required for the community life of themselves and the priests; but all was devoid of gold or silver, of which they made no use anywhere[1]; on the contrary, they aimed at the mean between luxurious display and meanness, and built themselves tasteful houses, wherein they and their children's children grew old and handed them on in succession unaltered to others like themselves. As for the southward parts, when they vacated their gardens and gymnasia and mess-rooms as was natural in summer, they used them for these purposes. And near the place of the present Acropolis there was one spring—which was choked up by the earthquakes so that but small tricklings of it are now left round about; but to the men of that time it afforded a plentiful stream for them all, being well tempered both for winter and summer. In this fashion, then, they dwelt, acting as guardians of their own citizens and as leaders, by their own consent, of the rest of the Greeks; and they watched carefully that their own numbers, of both men and women, who were neither too young nor too old to fight, should remain for all time as nearly as possible the same, namely, about 20,000.

So it was that these men, being themselves of the character described and always justly administering in some such fashion both their own land and Hellas,

καὶ Ἀσίαν κατά τε σωμάτων κάλλη καὶ κατὰ τὴν
τῶν ψυχῶν παντοίαν ἀρετὴν ἐλλόγιμοί τε ἦσαν
καὶ ὀνομαστότατοι πάντων τῶν τότε· τὰ δὲ δὴ
τῶν ἀντιπολεμησάντων αὐτοῖς οἷα ἦν ὥς τε ἀπ᾽
ἀρχῆς ἐγένετο, μνήμης ἂν μὴ στερηθῶμεν ὧν ἔτι
παῖδες ὄντες ἠκούσαμεν, εἰς τὸ μέσον αὐτὰ νῦν
ἀποδώσομεν ὑμῖν τοῖς φίλοις εἶναι κοινά.

Τὸ δ᾽ ἔτι βραχὺ πρὸ τοῦ λόγου δεῖ δηλῶσαι, μὴ
πολλάκις ἀκούοντες Ἑλληνικὰ βαρβάρων ἀνδρῶν
ὀνόματα θαυμάζητε· τὸ γὰρ αἴτιον αὐτῶν πεύσεσθε.
Σόλων ἅτ᾽ ἐπινοῶν εἰς τὴν αὑτοῦ ποίησιν κατα-
χρήσασθαι τῷ λόγῳ, διαπυνθανόμενος τὴν τῶν
ὀνομάτων δύναμιν, εὗρε τούς τε Αἰγυπτίους τοὺς
πρώτους ἐκείνους αὐτὰ γραψαμένους εἰς τὴν αὑτῶν
φωνὴν μετενηνοχότας, αὐτός τε αὖ πάλιν ἑκάστου
τὴν διάνοιαν ὀνόματος ἀναλαμβάνων εἰς τὴν
ἡμετέραν ἄγων φωνὴν ἀπεγράφετο. καὶ ταῦτά γε
δὴ τὰ γράμματα παρὰ τῷ πάππῳ τ᾽ ἦν καὶ ἔτ᾽
ἐστὶ παρ᾽ ἐμοὶ νῦν διαμεμελέτηταί τε ὑπ᾽ ἐμοῦ
παιδὸς ὄντος. ἂν οὖν ἀκούητε τοιαῦτα οἷα καὶ
τῇδε ὀνόματα, μηδὲν ὑμῖν ἔστω θαῦμα· τὸ γὰρ
αἴτιον αὐτῶν ἔχετε. μακροῦ δὲ δὴ λόγου τοιάδε
τις ἦν ἀρχὴ τότε.

Καθάπερ ἐν τοῖς πρόσθεν ἐλέχθη περὶ τῆς τῶν
θεῶν λήξεως, ὅτι κατενείμαντο γῆν πᾶσαν ἔνθα
μὲν μείζους λήξεις, ἔνθα δὲ καὶ ἐλάττους, ἱερὰ
θυσίας τε αὐτοῖς κατασκευάζοντες, οὕτω δὴ καὶ
τὴν νῆσον Ποσειδῶν τὴν Ἀτλαντίδα λαχὼν ἐκγόνους ἑαυτοῦ κατῴκισεν ἐκ θνητῆς γυναικὸς

---

[1] *Cf.* Tim. 21 A ff.   [2] *Cf.* 109 B.

CRITIAS

were famous throughout all Europe and Asia both for their bodily beauty and for the perfection of their moral excellence, and were of all men then living the most renowned. And now, if we have not lost recollection of what we heard when we were still children,[1] we will frankly impart to you all, as friends, our story of the men who warred against our Athenians, what their state was and how it originally came about.

But before I begin my account, there is still a small point which I ought to explain, lest you should be surprised at frequently hearing Greek names given to barbarians. The reason of this you shall now learn. Since Solon was planning to make use of the story for his own poetry, he had found, on investigating the meaning of the names, that those Egyptians who had first written them down had translated them into their own tongue. So he himself in turn recovered the original sense of each name and, rendering it into our tongue, wrote it down so. And these very writings were in the possession of my grandfather and are actually now in mine, and when I was a child I learnt them all by heart. Therefore if the names you hear are just like our local names, do not be at all astonished; for now you know the reason for them. The story then told was a long one, and it began something like this.

Like as we previously stated [2] concerning the allotments of the Gods, that they portioned out the whole earth, here into larger allotments and there into smaller, and provided for themselves shrines and sacrifices, even so Poseidon took for his allotment the island of Atlantis and settled therein the children whom he had begotten of a mortal woman in a

γεννήσας ἔν τινι τόπῳ τοιῷδε τῆς νήσου. πρὸς θαλάττης μέν, κατὰ δὲ μέσον πάσης πεδίον ἦν, ὃ δὴ πάντων πεδίων κάλλιστον ἀρετῇ τε ἱκανὸν γενέσθαι λέγεται. πρὸς τῷ πεδίῳ δ' αὖ κατὰ μέσον σταδίους ὡς πεντήκοντα ἀφεστὸς ἦν ὄρος βραχὺ πάντῃ. τούτῳ δ' ἦν ἔνοικος τῶν ἐκεῖ κατὰ ἀρχὰς ἐκ γῆς ἀνδρῶν γεγονότων Εὐήνωρ μὲν ὄνομα, γυναικὶ δὲ συνοικῶν Λευκίππῃ· Κλειτὼ δὲ μονογενῆ θυγατέρα ἐγεννησάσθην. ἤδη δ' εἰς ἀνδρὸς ὥραν ἡκούσης τῆς κόρης ἥ τε μήτηρ τελευτᾷ καὶ ὁ πατήρ, αὐτῆς δὲ εἰς ἐπιθυμίαν Ποσειδῶν ἐλθὼν ξυμμίγνυται, καὶ τὸν γήλοφον, ἐν ᾧ κατῴκιστο, ποιῶν εὐερκῆ περιρρήγνυσι κύκλῳ, θαλάττης γῆς τε ἐναλλὰξ ἐλάττους μείζους τε περὶ ἀλλήλους ποιῶν τροχούς, δύο μὲν γῆς, θαλάττης δὲ τρεῖς οἷον τορνεύων ἐκ μέσης τῆς νήσου, πάντῃ ἴσον ἀφεστῶτας, ὥστε ἄβατον ἀνθρώποις εἶναι· πλοῖα γὰρ καὶ τὸ πλεῖν οὔπω τότ' ἦν. αὐτὸς δὲ τήν τε ἐν μέσῳ νῆσον οἷα δὴ θεὸς εὐμαρῶς διεκόσμησεν, ὕδατα μὲν διττὰ ὑπὸ γῆς ἄνω πηγαῖα κομίσας, τὸ μὲν θερμόν, ψυχρὸν δὲ ἐκ κρήνης ἀπορρέον ἕτερον, τροφὴν δὲ παντοίαν καὶ ἱκανὴν ἐκ τῆς γῆς ἀναδιδούς. παίδων δὲ ἀρρένων πέντε γενέσεις διδύμους γεννησάμενος ἐθρέψατο, καὶ τὴν νῆσον τὴν Ἀτλαντίδα πᾶσαν δέκα μέρη κατανείμας τῶν μὲν πρεσβυτάτων τῷ προτέρῳ γενομένῳ τήν τε μητρῴαν οἴκησιν καὶ τὴν κύκλῳ λῆξιν, πλείστην καὶ ἀρίστην οὖσαν, ἀπένειμε, βασιλέα τε τῶν ἄλλων κατέστησε, τοὺς δὲ ἄλλους ἄρχοντας, ἑκάστῳ δὲ ἀρχὴν πολλῶν ἀνθρώπων καὶ

---

[1] *i.e.* "autochthons," *cf.* 109 c, *Menex.* 237 b.

CRITIAS

region of the island of the following description. Bordering on the sea and extending through the centre of the whole island there was a plain, which is said to have been the fairest of all plains and highly fertile; and, moreover, near the plain, over against its centre, at a distance of about 50 stades, there stood a mountain that was low on all sides. Thereon dwelt one of the natives originally sprung from the earth,[1] Evenor by name, with his wife Leucippe; and they had for offspring an only-begotten daughter, Cleito. And when this damsel was now come to marriageable age, her mother died and also her father; and Poseidon, being smitten with desire for her, wedded her; and to make the hill whereon she dwelt impregnable he broke it off all round about; and he made circular belts of sea and land enclosing one another alternately, some greater, some smaller, two being of land and three of sea, which he carved as it were out of the midst of the island; and these belts were at even distances on all sides, so as to be impassable for man; for at that time neither ships nor sailing were as yet in existence. And Poseidon himself set in order with ease, as a god would, the central island, bringing up from beneath the earth two springs of waters, the one flowing warm from its source, the other cold, and producing out of the earth all kinds of food in plenty. And he begat five pairs of twin sons and reared them up; and when he had divided all the island of Atlantis into ten portions, he assigned to the first-born of the eldest sons his mother's dwelling and the allotment surrounding it, which was the largest and best; and him he appointed to be king over the rest, and the others to be rulers, granting to each the rule over many men and a large

τόπον πολλῆς χώρας ἔδωκεν. ὀνόματα δὲ πᾶσιν ἔθετο, τῷ μὲν πρεσβυτάτῳ καὶ βασιλεῖ τοῦτο οὗ δὴ καὶ πᾶσα ἡ νῆσος τό τε πέλαγος ἔσχεν ἐπωνυμίαν, Ἀτλαντικὸν λεχθέν, ὅτι τοὔνομ᾽ ἦν τῷ πρώτῳ βασιλεύσαντι τότε Ἄτλας· τῷ δὲ διδύμῳ μετ᾽ ἐκεῖνόν τε γενομένῳ, λῆξιν δὲ ἄκρας τῆς νήσου πρὸς Ἡρακλείων στηλῶν εἰληχότι ἐπὶ τὸ τῆς Γαδειρικῆς νῦν χώρας κατ᾽ ἐκεῖνον τὸν τόπον ὀνομαζομένης, Ἑλληνιστὶ μὲν Εὔμηλον, τὸ δ᾽ ἐπιχώριον Γάδειρον, ὅπερ ἂν τὴν ἐπίκλην ταύτῃ[1] [ὄνομα] παράσχοι. τοῖν δὲ δευτέροιν γενομένοιν τὸν μὲν Ἀμφήρη, τὸν δὲ Εὐαίμονα ἐκάλεσε· τρίτοις δέ, Μνησέα μὲν τῷ προτέρῳ γενομένῳ, τῷ δὲ μετὰ τοῦτον Αὐτόχθονα· τῶν δὲ τετάρτων Ἐλάσιππον μὲν τὸν πρότερον, Μήστορα δὲ τὸν ὕστερον· ἐπὶ δὲ τοῖς πέμπτοις τῷ μὲν ἔμπροσθεν Ἀζάης ὄνομα ἐτέθη, τῷ δ᾽ ὑστέρῳ Διαπρεπής. οὗτοι δὴ πάντες αὐτοί τε καὶ ἔκγονοι τούτων ἐπὶ γενεὰς πολλὰς ᾤκουν ἄρχοντες μὲν πολλῶν ἄλλων κατὰ τὸ πέλαγος νήσων, ἔτι δέ, ὥσπερ καὶ πρότερον ἐρρήθη, μέχρι τε Αἰγύπτου καὶ Τυρρηνίας τῶν ἐντὸς δεῦρο ἐπάρχοντες.

Ἄτλαντος δὴ πολὺ μὲν ἄλλο καὶ τίμιον γίγνεται γένος, βασιλεὺς δὲ ὁ πρεσβύτατος ἀεὶ τῷ πρεσβυτάτῳ τῶν ἐκγόνων παραδιδοὺς ἐπὶ γενεὰς πολλὰς τὴν βασιλείαν διέσωζον, πλοῦτον μὲν κεκτημένοι πλήθει τοσοῦτον, ὅσος οὔτε πω πρόσθεν ἐν δυναστείαις τισὶ βασιλέων γέγονεν οὔτε ποτὲ ὕστερον γενέσθαι ῥᾴδιος, κατεσκευασμένα δὲ πάντα ἦν αὐτοῖς ὅσα ἐν πόλει καὶ ὅσα κατὰ τὴν ἄλλην χώραν ἦν ἔργον κατα-

[1] ἐπίκλην ταύτῃ best ms.: ἐπίκλησιν ταύτην Zur.: [ὄνομα] bracketed by W.-Möllendorff.

CRITIAS

tract of country. And to all of them he gave names, giving to him that was eldest and king the name after which the whole island was called and the sea spoken of as the Atlantic, because the first king who then reigned had the name of Atlas. And the name of his younger twin-brother, who had for his portion the extremity of the island near the pillars of Heracles up to the part of the country now called Gadeira after the name of that region, was Eumelus in Greek, but in the native tongue Gadeirus,—which fact may have given its title to the country. And of the pair that were born next he called the one Ampheres and the other Evaemon; and of the third pair the elder was named Mneseus and the younger Autochthon; and of the fourth pair, he called the first Elasippus and the second Mestor; and of the fifth pair, Azaes was the name given to the elder, and Diaprepês to the second. So all these, themselves and their descendants, dwelt for many generations bearing rule over many other islands throughout the sea, and holding sway besides, as was previously stated,[1] over the Mediterranean peoples as far as Egypt and Tuscany.

Now a large family of distinguished sons sprang from Atlas; but it was the eldest, who, as king, always passed on the sceptre to the eldest of his sons, and thus they preserved the sovereignty for many generations; and the wealth they possessed was so immense that the like had never been seen before in any royal house nor will ever easily be seen again; and they were provided with everything of which provision was needed either in the city or throughout

[1] *Cf. Tim.* 25 A, B.

PLATO

σκευάσασθαι. πολλὰ μὲν γὰρ διὰ τὴν ἀρχὴν αὐτοῖς προσῄειν ἔξωθεν, πλεῖστα δὲ ἡ νῆσος αὐτὴ παρείχετο εἰς τὰς τοῦ βίου κατασκευάς, πρῶτον μὲν ὅσα ὑπὸ μεταλλείας ὀρυττόμενα στερεὰ καὶ ὅσα τηκτὰ γέγονε, καὶ τὸ νῦν ὀνομαζόμενον μόνον, τότε δὲ πλέον ὀνόματος ἦν, τὸ γένος ἐκ γῆς ὀρυττόμενον ὀρειχάλκου κατὰ τόπους πολλοὺς τῆς νήσου, πλὴν χρυσοῦ τιμιώτατον ἐν τοῖς τότε ὄν· καὶ ὅσα ὕλη πρὸς τὰ τεκτόνων διαπονήματα παρέχεται, πάντα φέρουσα ἄφθονα, τά τε αὖ περὶ τὰ ζῷα ἱκανῶς ἥμερα καὶ ἄγρια τρέφουσα. καὶ δὴ καὶ ἐλεφάντων ἦν ἐν αὐτῇ γένος πλεῖστον· νομὴ γὰρ τοῖς τε ἄλλοις ζῴοις, ὅσα καθ' ἕλη καὶ λίμνας καὶ ποταμούς, ὅσα τ' αὖ κατ' ὄρη καὶ ὅσα ἐν τοῖς πεδίοις νέμεται, ξύμπασι παρῆν ἄδην, καὶ τούτῳ κατὰ ταὐτὰ τῷ ζῴῳ, μεγίστῳ πεφυκότι καὶ πολυβορωτάτῳ. πρὸς δὴ τούτοις, ὅσα εὐώδη τρέφει πού γῆ τὰ νῦν, ῥιζῶν ἢ χλόης ἢ ξύλων ἢ χυλῶν στακτῶν εἴτε ἀνθῶν ἢ καρπῶν, ἔφερέ τε ταῦτα καὶ ἔφερβεν εὖ. ἔτι δὲ τὸν ἥμερον καρπόν, τόν τε ξηρόν, ὃς ἡμῖν τροφῆς ἕνεκά ἐστι, καὶ ὅσοις χάριν τοῦ σίτου προσχρώμεθα—καλοῦμεν δὲ αὐτοῦ τὰ μέρη ὄσπρια—καὶ τὸν ὅσος ξύλινος, πώματα καὶ βρώματα καὶ ἀλείμματα φέρων, παιδιᾶς τε ὃς ἕνεκα ἡδονῆς τε γέγονε δυσθησαύριστος ἀκροδρύων καρπός, ὅσα τε παραμύθια πλησμονῆς μεταδόρπια ἀγαπητὰ κάμνοντι τίθεμεν, ἅπαντα ταῦτα ἡ τότε

---

[1] *i.e.* "mountain-copper"; a "sparkling" metal (116 c) hard to identify (*cf.* Hesiod, *Sc.* 122).
[2] *i.e.* the vine (*cf.* Hom. *Od.* v. 69).
[3] *i.e.* corn.
[4] Perhaps the olive, or coco-palm.
[5] Perhaps the pomegranate, or apple (*cf. Laws* 819 A, B).

CRITIAS

the rest of the country. For because of their headship they had a large supply of imports from abroad, and the island itself furnished most of the requirements of daily life,—metals, to begin with, both the hard kind and the fusible kind, which are extracted by mining, and also that kind which is now known only by name but was more than a name then, there being mines of it in many places of the island,—I mean "orichalcum,"[1] which was the most precious of the metals then known, except gold. It brought forth also in abundance all the timbers that a forest provides for the labours of carpenters; and of animals it produced a sufficiency, both of tame and wild. Moreover, it contained a very large stock of elephants; for there was an ample food-supply not only for all the other animals which haunt the marshes and lakes and rivers, or the mountains or the plains, but likewise also for this animal, which of its nature is the largest and most voracious. And in addition to all this, it produced and brought to perfection all those sweet-scented stuffs which the earth produces now, whether made of roots or herbs or trees, or of liquid gums derived from flowers or fruits. The cultivated fruit[2] also, and the dry,[3] which serves us for nutriment, and all the other kinds that we use for our meals—the various species of which are comprehended under the name "vegetables,"—and all the produce of trees which affords liquid and solid food and unguents,[4] and the fruit of the orchard-trees, so hard to store, which is grown for the sake of amusement and pleasure,[5] and all the after-dinner fruits[6] that we serve up as welcome remedies for the sufferer from repletion,—all these that hallowed island, as it

[6] Perhaps the citron.

115

ποτὲ οὖσα ὑφ' ἡλίῳ νῆσος ἱερὰ καλά τε καὶ θαυμαστὰ καὶ πλήθεσιν ἄπειρα ἔφερε. ταῦτα οὖν λαμβάνοντες πάντα παρὰ τῆς γῆς κατεσκευάζοντο C τά τε ἱερὰ καὶ τὰς βασιλικὰς οἰκήσεις καὶ τοὺς λιμένας καὶ τὰ νεώρια καὶ ξύμπασαν τὴν ἄλλην χώραν, τοιᾷδε ἐν τάξει διακοσμοῦντες.

Τοὺς τῆς θαλάττης τροχούς, οἳ περὶ τὴν ἀρχαίαν ἦσαν μητρόπολιν, πρῶτον μὲν ἐγεφύρωσαν, ὁδοὺ ἔξω καὶ ἐπὶ τὰ βασίλεια ποιούμενοι. τὰ δὲ βασίλεια ἐν ταύτῃ τῇ τοῦ θεοῦ καὶ τῶν προγόνων κατοικήσει κατ' ἀρχὰς ἐποιήσαντο εὐθύς, ἕτερος δὲ παρ' ἑτέρου δεχόμενος, κεκοσμημένα κοσμῶν, D ὑπερεβάλλετο εἰς δύναμιν ἀεὶ τὸν ἔμπροσθεν, ἕως εἰς ἔκπληξιν μεγέθεσι κάλλεσί τε ἔργων ἰδεῖν τὴν οἴκησιν ἀπειργάσαντο. διώρυχα μὲν γὰρ ἐκ τῆς θαλάττης ἀρχόμενοι τρίπλεθρον τὸ πλάτος, ἑκατὸν δὲ ποδῶν βάθος, μῆκος δὲ πεντήκοντα σταδίων, ἐπὶ τὸν ἐξωτάτω τροχὸν συνέτρησαν, καὶ τὸν ἀνάπλουν ἐκ τῆς θαλάττης ταύτῃ πρὸς ἐκεῖνον ὡς εἰς λιμένα ἐποιήσαντο, διελόντες στόμα ναυσὶ ταῖς μεγίσταις ἱκανὸν εἰσπλεῖν. καὶ δὴ καὶ τοὺς τῆς E γῆς τροχούς, οἳ τοὺς τῆς θαλάττης διεῖργον, κατὰ τὰς γεφύρας διεῖλον ὅσον μιᾷ τριήρει διέκπλουν εἰς ἀλλήλους, καὶ κατεστέγασαν ἄνωθεν ὥστε τὸν ὑπόπλουν κάτωθεν εἶναι· τὰ γὰρ τῶν τῆς γῆς τροχῶν χείλη βάθος εἶχεν ἱκανὸν ὑπερέχον τῆς θαλάττης. ἦν δὲ ὁ μὲν μέγιστος τῶν τροχῶν, εἰς ὃν ἡ θάλαττα συνετέτρητο, τριστάδιος τὸ πλάτος, ὁ δ' ἑξῆς τῆς γῆς ἴσος ἐκείνῳ· τοῖν δὲ δευτέροιν ὁ

---

[1] *i.e.* Poseidon.

CRITIAS

lay then beneath the sun, produced in marvellous beauty and endless abundance. And thus, receiving from the earth all these products, they furnished forth their temples and royal dwellings, their harbours and their docks, and all the rest of their country, ordering all in the fashion following.[1]

First of all they bridged over the circles of sea which surrounded the ancient metropolis, making thereby a road towards and from the royal palace. And they had built the palace at the very beginning where the settlement was first made by their God[1] and their ancestors; and as each king received it from his predecessor, he added to its adornment and did all he could to surpass the king before him, until finally they made of it an abode amazing to behold for the magnitude and beauty of its workmanship. For, beginning at the sea, they bored a channel right through to the outermost circle, which was three plethra in breadth, one hundred feet in depth, and fifty stades[2] in length; and thus they made the entrance to it from the sea like that to a harbour by opening out a mouth large enough for the greatest ships to sail through. Moreover, through the circles of land, which divided those of sea, over against the bridges they opened out a channel leading from circle to circle, large enough to give passage to a single trireme; and this they roofed over above so that the sea-way was subterranean; for the lips of the land-circles were raised a sufficient height above the level of the sea. The greatest of the circles into which a boring was made for the sea was three stades in breadth, and the circle of land next to it was of equal

[2] The *plethron* was about 100 ft.; the *stade* (= 6 plethra) about 600 ft.

μὲν ὑγρὸς δυοῖν σταδίοιν πλάτος, ὁ δὲ ξηρὸς ἴσος αὖ πάλιν τῷ πρόσθεν ὑγρῷ· σταδίου δὲ ὁ περὶ αὐτὴν τὴν ἐν μέσῳ νῆσον περιθέων. ἡ δὲ νῆσος, ἐν ᾗ τὰ βασίλεια ἦν, πέντε σταδίων τὴν διάμετρον εἶχε. ταύτην δὴ κύκλῳ καὶ τοὺς τροχοὺς καὶ τὴν γέφυραν πλεθριαίαν τὸ πλάτος οὖσαν ἔνθεν καὶ ἔνθεν λιθίνῳ περιεβάλλοντο τείχει, πύργους καὶ πύλας ἐπὶ τῶν γεφυρῶν κατὰ τὰς τῆς θαλάττης διαβάσεις ἑκασταχόσε ἐπιστήσαντες. τὸν δὲ λίθον ἔτεμνον ὑπὸ τῆς νήσου κύκλῳ τῆς ἐν μέσῳ καὶ ὑπὸ τῶν τροχῶν ἔξωθεν καὶ ἐντός, τὸν μὲν λευκόν, τὸν δὲ μέλανα, τὸν δὲ ἐρυθρὸν ὄντα· τέμνοντες δὲ ἅμα ἀπειργάζοντο νεωσοίκους κοίλους διπλοῦς ἐντός, κατηρεφεῖς αὐτῇ τῇ πέτρᾳ. καὶ τῶν οἰκοδομημάτων τὰ μὲν ἁπλᾶ, τὰ δὲ μιγνύντες τοὺς λίθους ποικίλα ὕφαινον παιδιᾶς χάριν, ἡδονὴν αὐτοῖς ξύμφυτον ἀπονέμοντες· καὶ τοῦ μὲν περὶ τὸν ἐξωτάτω τροχὸν τείχους χαλκῷ περιελάμβανον πάντα τὸν περίδρομον, οἷον ἀλοιφῇ προσχρώμενοι, τοῦ δ᾽ ἐντὸς καττιτέρῳ περιέτηκον, τὸν δὲ περὶ αὐτὴν τὴν ἀκρόπολιν ὀρειχάλκῳ μαρμαρυγὰς ἔχοντι πυρώδεις.

Τὰ δὲ δὴ τῆς ἀκροπόλεως ἐντὸς βασίλεια κατεσκευασμένα ὧδ᾽ ἦν. ἐν μέσῳ μὲν ἱερὸν ἅγιον αὐτόθι τῆς τε Κλειτοῦς καὶ τοῦ Ποσειδῶνος ἄβατον ἀφεῖτο, περιβόλῳ χρυσῷ περιβεβλημένον, τοῦτ᾽ ἐν ᾧ κατ᾽ ἀρχὰς ἐφίτυσαν καὶ ἐγέννησαν τὸ τῶν δέκα βασιλειδῶν γένος· ἔνθα καὶ κατ᾽ ἐνιαυτὸν ἐκ πασῶν τῶν δέκα λήξεων ὡραῖα αὐτόσε ἀπ-

breadth; and of the second pair of circles that of water was two stades in breadth and that of dry land equal again to the preceding one of water; and the circle which ran round the central island itself was of a stade's breadth. And this island, wherein stood the royal palace, was of five stades in diameter. Now the island and the circles and the bridge, which was a plethrum in breadth, they encompassed round about, on this side and on that, with a wall of stone; and upon the bridges on each side, over against the passages for the sea, they erected towers and gates. And the stone they quarried beneath the central island all round, and from beneath the outer and inner circles, some of it being white, some black and some red; and while quarrying it they constructed two inner docks, hollowed out and roofed over by the native rock. And of the buildings some they framed of one simple colour, in others they wove a pattern of many colours by blending the stones for the sake of ornament so as to confer upon the buildings a natural charm. And they covered with brass, as though with a plaster, all the circumference of the wall which surrounded the outermost circle; and that of the inner one they coated with tin; and that which encompassed the acropolis itself with orichalcum which sparkled like fire.

The royal palace within the acropolis was arranged in this manner. In the centre there stood a temple sacred to Cleito and Poseidon, which was reserved as holy ground, and encircled with a wall of gold; this being the very spot where at the beginning they had generated and brought to birth the family of the ten royal lines. Thither also they brought year by year from all the ten allotments their seasonable offerings

ἐτέλουν ἱερὰ ἐκείνων ἑκάστῳ. τοῦ δὲ Ποσειδῶνος αὐτοῦ νεὼς ἦν, σταδίου μὲν μῆκος, εὖρος δὲ τρίπλεθρος, ὕψος δ' ἐπὶ τούτοις σύμμετρον ἰδεῖν, εἶδος δέ τι βαρβαρικὸν ἔχων.[1] πάντα δὲ ἔξωθεν περιήλειψαν τὸν νεὼν ἀργύρῳ, πλὴν τῶν ἀκρωτηρίων, τὰ δὲ ἀκρωτήρια χρυσῷ. τὰ δὲ ἐντός, τὴν μὲν ὀροφὴν ἐλεφαντίνην ἰδεῖν πᾶσαν χρυσῷ καὶ ἀργύρῳ καὶ ὀρειχάλκῳ πεποικιλμένην, τὰ δὲ ἄλλα πάντα τῶν τοίχων τε καὶ κιόνων καὶ ἐδάφους ὀρειχάλκῳ περιέλαβον. χρυσᾶ δὲ ἀγάλματα ἐνέστησαν, τὸν μὲν θεὸν ἐφ' ἅρματος ἑστῶτα ἓξ ὑποπτέρων ἵππων ἡνίοχον, αὐτόν τε ὑπὸ μεγέθους τῇ κορυφῇ τῆς ὀροφῆς ἐφαπτόμενον, Νηρῇδας δὲ ἐπὶ δελφίνων ἑκατὸν κύκλῳ· τοσαύτας γὰρ ἐνόμιζον αὐτὰς οἱ τότε εἶναι· πολλὰ δ' ἄλλα ἀγάλματα ἰδιωτῶν ἀναθήματα ἐνῆν. περὶ δὲ τὸν νεὼν ἔξωθεν εἰκόνες ἁπάντων ἕστασαν ἐκ χρυσοῦ, τῶν γυναικῶν καὶ αὐτῶν, ὅσοι τῶν δέκα ἐγεγόνεσαν βασιλέων, καὶ πολλὰ ἕτερα ἀναθήματα μεγάλα τῶν τε βασιλέων καὶ ἰδιωτῶν ἔξ αὐτῆς τε τῆς πόλεως καὶ τῶν ἔξωθεν ὅσων ἐπῆρχον. βωμός τε δὴ ξυνεπόμενος ἦν τὸ μέγεθος καὶ τὸ τῆς ἐργασίας ταύτῃ τῇ κατασκευῇ, καὶ τὰ βασίλεια κατὰ τὰ αὐτὰ πρέποντα μὲν τῷ τῆς ἀρχῆς μεγέθει, πρέποντα δὲ τῷ περὶ τὰ ἱερὰ κόσμῳ.

Ταῖς δὲ δὴ κρήναις, τῇ τοῦ ψυχροῦ καὶ τῇ τοῦ θερμοῦ νάματος, πλῆθος μὲν ἄφθονον ἐχούσαις, ἡδονῇ δὲ καὶ ἀρετῇ τῶν ὑδάτων ἑκατέρου πρὸς[2]

---

[1] ἔχων Stephens: ἔχοντος mss., Zur.
[2] ἑκατέρου πρὸς Ast: πρὸς ἑκατέρου mss., Zur.

CRITIAS

to do sacrifice to each of those princes. And the temple of Poseidon himself was a stade in length, three plethra in breadth, and of a height which appeared symmetrical therewith; and there was something of the barbaric in its appearance. All the exterior of the temple they coated with silver, save only the pinnacles, and these they coated with gold. As to the interior, they made the roof all of ivory in appearance, variegated with gold and silver and orichalcum, and all the rest of the walls and pillars and floors they covered with orichalcum. And they placed therein golden statues, one being that of the God standing on a chariot and driving six winged steeds, his own figure so tall as to touch the ridge of the roof, and round about him a hundred Nereids on dolphins (for that was the number of them as men then believed)[1]; and it contained also many other images, the votive offerings of private men. And outside, round about the temple, there stood images in gold of all the princes, both themselves and their wives, as many as were descended from the ten kings, together with many other votive offerings both of the kings and of private persons not only from the State itself but also from all the foreign peoples over whom they ruled. And the altar, in respect of its size and its workmanship, harmonized with its surroundings; and the royal palace likewise was such as befitted the greatness of the kingdom, and equally befitted the splendour of the temples.

The springs they made use of, one kind being of cold, another of warm water,[2] were of abundant volume, and each kind was wonderfully well adapted

---

[1] The usual tradition made them 50 in number; *cf.* Hes. *Theog.* 240 ff.; Pind. *Isthm.* v. 6.   [2] *Cf.* 113 E.

PLATO

τὴν χρῆσιν θαυμαστοῦ πεφυκότος, ἐχρῶντο περιστήσαντες οἰκοδομήσεις καὶ δένδρων φυτεύσεις πρεπούσας ὕδασι, δεξαμενάς τε αὖ τὰς μὲν ὑπαιθρίους, τὰς δὲ χειμερινὰς τοῖς θερμοῖς λουτροῖς ὑποστέγους περιτιθέντες, χωρὶς μὲν βασιλικάς, χωρὶς δὲ ἰδιωτικάς, ἔτι δὲ γυναιξὶν ἄλλας καὶ ἑτέρας ἵπποις καὶ τοῖς ἄλλοις ὑποζυγίοις, τὸ πρόσφορον τῆς κοσμήσεως ἑκάστοις ἀπονέμοντες. τὸ δὲ ἀπορρέον ἦγον ἐπὶ τὸ τοῦ Ποσειδῶνος ἄλσος, δένδρα παντοδαπὰ κάλλος ὕψος τε δαιμόνιον ὑπὸ ἀρετῆς τῆς γῆς ἔχον,[1] καὶ ἐπὶ τοὺς ἔξω κύκλους δι' ὀχετῶν κατὰ τὰς γεφύρας ἐπωχέτευον. οὗ δὴ πολλὰ μὲν ἱερὰ καὶ πολλῶν θεῶν, πολλοὶ δὲ κῆποι καὶ γυμνάσια ἐκεχειρούργητο, τὰ μὲν ἀνδρῶν, τὰ δὲ ἵππων χωρὶς ἐν ἑκατέρᾳ τῇ τῶν τροχῶν νήσῳ τά τε ἄλλα καὶ κατὰ μέσην τὴν μείζω τῶν νήσων ἐξῃρημένος ἱππόδρομος ἦν αὐτοῖς, σταδίου τὸ πλάτος ἔχων, τὸ δὲ μῆκος περὶ τὸν κύκλον ὅλον ἀφεῖτο εἰς ἅμιλλαν τοῖς ἵπποις. δορυφορικαὶ δὲ περὶ αὐτὸν ἔνθεν τε καὶ ἔνθεν οἰκήσεις ἦσαν τῷ πλήθει τῶν δορυφόρων· τοῖς δὲ πιστοτέροις ἐν τῷ σμικροτέρῳ τροχῷ καὶ πρὸς τῆς ἀκροπόλεως μᾶλλον ὄντι διετέτακτο ἡ φρουρά, τοῖς δὲ πάντων διαφέρουσι πρὸς πίστιν ἐντὸς τῆς ἀκροπόλεως περὶ τοὺς βασιλέας αὐτοὺς ἦσαν οἰκήσεις δεδομέναι.

Τὰ δὲ νεώρια τριήρων μεστὰ ἦν καὶ σκευῶν

---

[1] ἔχον] ἔχοντα some MSS., Zur.

[1] Cf. *Laws* 761 A ff. for the importance attached to water-supplies.

CRITIAS

for use because of the natural taste and excellence of its waters; and these they surrounded with buildings and with plantations of trees such as suited the waters; and, moreover, they set reservoirs round about, some under the open sky, and others under cover to supply hot baths in the winter; they put separate baths for the kings and for the private citizens, besides others for women, and others again for horses and all other beasts of burden, fitting out each in an appropriate manner.[1] And the outflowing water they conducted to the sacred grove of Poseidon, which contained trees of all kinds that were of marvellous beauty and height because of the richness of the soil; and by means of channels they led the water to the outer circles over against the bridges. And there they had constructed many temples for gods, and many gardens and many exercising grounds, some for men and some set apart for horses, in each of the circular belts of island; and besides the rest they had in the centre of the large island [2] a racecourse laid out for horses, which was a stade in width, while as to length, a strip which ran round the whole circumference was reserved for equestrian contests. And round about it, on this side and on that, were barracks for the greater part of the spearmen [3]; but the guard-house of the more trusty of them was posted in the smaller circle, which was nearer the acropolis; while those who were the most trustworthy of all had dwellings granted to them within the acropolis round about the persons of the kings.

And the shipyards were full of triremes and all the

---

[2] *i.e.* the larger of the circular belts of land (*cf.* 113 D).
[3] The technical term for the body-guard of a tyrant (*cf. Rep.* 567 D, 575 B).

293

117 ὅσα τριήρεσι προσήκει, πάντα δὲ ἐξηρτυμένα ἱκανῶς.

Καὶ τὰ μὲν δὴ περὶ τὴν τῶν βασιλέων οἴκησιν οὕτω κατεσκεύαστο· διαβάντι δὲ τοὺς λιμένας ἔξω E τρεῖς ὄντας ἀρξάμενον ἀπὸ τῆς θαλάττης ᾔειν ἐν κύκλῳ τεῖχος, πεντήκοντα σταδίους τοῦ μεγίστου τροχοῦ τε καὶ λιμένος ἀπέχον πανταχῇ, καὶ συνέκλειεν εἰς ταὐτὸν πρὸς τὸ τῆς διώρυχος στόμα τὸ πρὸς θαλάττης. τοῦτο δὴ πᾶν συνῳκεῖτο μὲν ὑπὸ πολλῶν καὶ πυκνῶν οἰκήσεων, ὁ δὲ ἀνάπλους καὶ ὁ μέγιστος λιμὴν ἔγεμε πλοίων καὶ ἐμπόρων ἀφικνουμένων πάντοθεν, φωνὴν καὶ θόρυβον παντοδαπὸν κτύπον τε μεθ᾽ ἡμέραν καὶ διὰ νυκτὸς ὑπὸ πλήθους παρεχομένων.

Τὸ μὲν οὖν ἄστυ καὶ τὸ περὶ τὴν ἀρχαίαν οἴκησιν σχεδὸν ὡς τότ᾽ ἐλέχθη νῦν διεμνημόνευται· τῆς δ᾽ 118 ἄλλης χώρας ὡς ἡ φύσις εἶχε καὶ τὸ τῆς διακοσμήσεως εἶδος ἀπομνημονεῦσαι πειρατέον. πρῶτον μὲν οὖν ὁ τόπος ἅπας ἐλέγετο σφόδρα τε ὑψηλὸς καὶ ἀπότομος ἐκ θαλάττης, τὸ δὲ περὶ τὴν πόλιν πᾶν πεδίον, ἐκείνην μὲν περιέχον, αὐτὸ δὲ κύκλῳ περιεχόμενον ὄρεσι μέχρι πρὸς τὴν θάλατταν καθειμένοις, λεῖον καὶ ὁμαλές, πρόμηκες δὲ πᾶν, ἐπὶ μὲν θάτερα τρισχιλίων σταδίων, κατὰ δὲ μέσον ἀπὸ θαλάττης ἄνω δισχιλίων. ὁ δὲ τόπος οὗτος B ὅλης τῆς νήσου πρὸς νότον ἐτέτραπτο, ἀπὸ τῶν ἄρκτων κατάβορρος. τὰ δὲ περὶ αὐτὸν ὄρη τότε ὑμνεῖτο πλῆθος καὶ μέγεθος καὶ κάλλος παρὰ πάντα τὰ νῦν ὄντα γεγονέναι, πολλὰς μὲν κώμας καὶ

tackling that belongs to triremes, and they were all amply equipped.

Such then was the state of things round about the abode of the kings. And after crossing the three outer harbours, one found a wall which began at the sea and ran round in a circle, at a uniform distance of fifty stades from the largest circle and harbour, and its ends converged at the seaward mouth of the channel. The whole of this wall had numerous houses built on to it, set close together; while the sea-way and the largest harbour were filled with ships and merchants coming from all quarters, which by reason of their multitude caused clamour and tumult of every description and an unceasing din night and day.

Now as regards the city and the environs of the ancient dwelling we have now wellnigh completed the description as it was originally given. We must endeavour next to repeat the account of the rest of the country, what its natural character was, and in what fashion it was ordered. In the first place, then, according to the account, the whole region rose sheer out of the sea to a great height, but the part about the city was all a smooth plain, enclosing it round about, and being itself encircled by mountains which stretched as far as to the sea; and this plain had a level surface and was as a whole rectangular in shape, being 3000 stades long on either side and 2000 stades wide at its centre, reckoning upwards from the sea. And this region, all along the island, faced towards the South and was sheltered from the Northern blasts. And the mountains which surrounded it were at that time celebrated as surpassing all that now exist in number, magnitude and beauty; for they had upon

πλουσίας περιοίκων ἐν ἑαυτοῖς ἔχοντα, ποταμοὺς δὲ καὶ λίμνας καὶ λειμῶνας τροφὴν τοῖς πᾶσιν ἡμέροις καὶ ἀγρίοις ἱκανὴν θρέμμασιν, ὕλην δὲ καὶ πλήθει καὶ γένεσι ποικίλην ξύμπασί τε τοῖς ἔργοις καὶ πρὸς ἕκαστα ἄφθονον.

Ὧδε οὖν τὸ πεδίον φύσει καὶ ὑπὸ βασιλέων πολλῶν ἐν πολλῷ χρόνῳ διεπεπόνητο. τετράγωνον μὲν αὖθ' ὑπῆρχε τὰ πλεῖστ' ὀρθὸν καὶ πρόμηκες· ὅ τι δ' ἐνέλειπε, κατεύθυντο τάφρου κύκλῳ περιορυχθείσης. τὸ δὲ βάθος καὶ πλάτος τό τε μῆκος αὐτῆς ἄπιστον μὲν [τὸ]¹ λεχθέν, ὡς χειροποίητον ἔργον, πρὸς τοῖς ἄλλοις διαπονήμασι τοσοῦτον εἶναι, ῥητέον δὲ ὅ γε ἠκούσαμεν· πλέθρου μὲν γὰρ βάθος ὀρώρυκτο, τὸ δὲ πλάτος ἁπάντῃ σταδίου, περὶ δὲ πᾶν τὸ πεδίον ὀρυχθεῖσα συνέβαινεν εἶναι τὸ μῆκος σταδίων μυρίων. τὰ δ' ἐκ τῶν ὀρῶν καταβαίνοντα ὑποδεχομένη ῥεύματα καὶ περὶ τὸ πεδίον κυκλωθεῖσα, πρὸς τὴν πόλιν ἔνθεν τε καὶ ἔνθεν ἀφικομένη, ταύτῃ πρὸς θάλατταν μεθεῖτο ἐκρεῖν. ἄνωθεν δὲ ἀπ' αὐτῆς τὸ πλάτος μάλιστα ἑκατὸν ποδῶν διώρυχες εὐθεῖαι τετμημέναι κατὰ τὸ πεδίον πάλιν εἰς τὴν τάφρον τὴν πρὸς θαλάττης ἀφεῖντο, ἑτέρα δὲ ἀφ' ἑτέρας αὐτῶν σταδίους ἑκατὸν ἀπεῖχεν· ᾗ δὴ τήν τ' ἐκ τῶν ὀρῶν ὕλην κατῆγον εἰς τὸ ἄστυ καὶ τἆλλα δὲ ὡραῖα πλοίοις κατεκομίζοντο, διάπλους ἐκ τῶν διωρύχων εἰς ἀλλήλας τε πλαγίας καὶ πρὸς τὴν πόλιν τεμόντας. καὶ δὶς δὴ τοῦ ἐνιαυτοῦ τὴν γῆν ἐκαρποῦντο, χειμῶνος μὲν τοῖς ἐκ Διὸς ὕδασι χρώμενοι, θέρους δὲ

---

¹ τὸ omitted by some MSS.

¹ The sides of the plain being 2000 and 3000 stades (118 A above).

CRITIAS

them many rich villages of country folk, and streams and lakes and meadows which furnished ample nutriment to all the animals both tame and wild, and timber of various sizes and descriptions, abundantly sufficient for the needs of all and every craft.

Now as a result of natural forces, together with the labours of many kings which extended over many ages, the condition of the plain was this. It was originally a quadrangle, rectilinear for the most part, and elongated; and what it lacked of this shape they made right by means of a trench dug round about it. Now, as regards the depth of this trench and its breadth and length, it seems incredible that it should be so large as the account states, considering that it was made by hand, and in addition to all the other operations, but none the less we must report what we heard: it was dug out to the depth of a plethrum and to a uniform breadth of a stade, and since it was dug round the whole plain its consequent length was 10,000 stades.[1] It received the streams which came down from the mountains and after circling round the plain, and coming towards the city on this side and on that, it discharged them thereabouts into the sea. And on the inland side of the city channels were cut in straight lines, of about 100 feet in width, across the plain, and these discharged themselves into the trench on the seaward side, the distance between each being 100 stades. It was in this way that they conveyed to the city the timber from the mountains and transported also on boats the seasons' products, by cutting transverse passages from one channel to the next and also to the city. And they cropped the land twice a year, making use of the rains from Heaven in the winter, and the waters that issue from

118 ὅσα γῆ φέρει, τὰ ἐκ τῶν διωρύχων ἐπάγοντες
νάματα.

Πλῆθος δέ, τῶν μὲν ἐν τῷ πεδίῳ χρησίμων
πρὸς πόλεμον ἀνδρῶν ἐτέτακτο τὸν κλῆρον ἕκαστον
119 παρέχειν ἄνδρα ἡγεμόνα, τὸ δὲ τοῦ κλήρου μέγεθος
εἰς δέκα δεκάκις ἦν στάδια, μυριάδες δὲ ξυμπάντων τῶν κλήρων ἦσαν ἕξ· τῶν δὲ ἐκ τῶν
ὀρῶν καὶ τῆς ἄλλης χώρας ἀπέραντος μὲν ἀριθμὸς
ἀνθρώπων ἐλέγετο, κατὰ δὲ τόπους καὶ κώμας
εἰς τούτους τοὺς κλήρους πρὸς τοὺς ἡγεμόνας
ἅπαντες διενενέμηντο. τὸν οὖν ἡγεμόνα ἦν τεταγ-
μένον εἰς τὸν πόλεμον παρέχειν ἕκτον μὲν ἅρματος
πολεμιστηρίου μόριον εἰς μύρια ἅρματα, ἵππους
B δὲ δύο καὶ ἀναβάτας, ἔτι δὲ ξυνωρίδα χωρὶς
δίφρου καταβάτην τε σμικρασπίδα καὶ τὸν ἀμφοῖν
μετεπιβάτην¹ τοῖν ἵπποιν ἡνίοχον ἔχουσαν, ὁπλίτας
δὲ δύο καὶ τοξότας σφενδονήτας τε ἑκατέρους δύο,
γυμνῆτας δὲ λιθοβόλους καὶ ἀκοντιστὰς τρεῖς
ἑκατέρους, ναύτας δὲ τέτταρας εἰς πλήρωμα διακοσίων καὶ χιλίων νεῶν. τὰ μὲν οὖν πολεμιστήρια
οὕτω διετέτακτο τῆς βασιλικῆς πόλεως, τῶν δὲ
ἐννέα ἄλλα ἄλλως, ἃ μακρὸς ἂν χρόνος εἴη λέγειν.

C Τὰ δὲ τῶν ἀρχῶν καὶ τιμῶν ὧδ᾽ εἶχεν ἐξ ἀρχῆς
διακοσμηθέντα. τῶν δέκα βασιλέων εἷς ἕκαστος
ἐν μὲν τῷ καθ᾽ αὑτὸν μέρει κατὰ τὴν αὑτοῦ πόλιν
τῶν ἀνδρῶν καὶ τῶν πλείστων νόμων ἦρχε, κολάζων
καὶ ἀποκτιννὺς ὅντιν᾽ ἐθελήσειεν· ἡ δὲ ἐν ἀλλήλοις
ἀρχὴ καὶ κοινωνία κατὰ ἐπιστολὰς ἦν τὰς τοῦ

¹ μετεπιβάτην] μετ᾽ ἐπιβάτην MSS., Zur.

CRITIAS

the earth in summer, by conducting the streams from the trenches.

As regards their man-power, it was ordained that each allotment should furnish one man as leader of all the men in the plain who were fit to bear arms; and the size of the allotment was about ten times ten stades, and the total number of all the allotments was 60,000; and the number of the men in the mountains and in the rest of the country was countless, according to the report, and according to their districts and villages they were all assigned to these allotments under their leaders. So it was ordained that each such leader should provide for war the sixth part of a war-chariot's equipment, so as to make up 10,000 chariots in all, together with two horses and mounted men; also a pair of horses without a car, and attached thereto a combatant[1] with a small shield and for charioteer the rider who springs from horse to horse; and two hoplites; and archers and slingers, two of each; and light-armed slingers and javelin-men, three of each; and four sailors towards the manning of twelve hundred ships. Such then were the military dispositions of the royal City; and those of the other nine varied in various ways, which it would take a long time to tell.

Of the magistracies and posts of honour the disposition, ever since the beginning, was this. Each of the ten kings ruled over the men and most of the laws in his own particular portion and throughout his own city, punishing and putting to death whomsoever he willed. But their authority over one another and their mutual relations were governed by

[1] This "combatant" (*desultor*) jumped off the chariot to fight on foot.

PLATO

119 Ποσειδῶνος, ὡς ὁ νόμος αὐτοῖς παρέδωκε καὶ γράμματα ὑπὸ τῶν πρώτων ἐν στήλῃ γεγραμμένα ὀρειχαλκίνῃ, ἣ κατὰ μέσην τὴν νῆσον ἔκειτο ἐν D ἱερῷ Ποσειδῶνος, οἳ δὴ δι' ἐνιαυτοῦ πέμπτου, τοτὲ δὲ ἐναλλὰξ ἕκτου, συνελέγοντο, τῷ τε ἀρτίῳ καὶ τῷ περιττῷ μέρος ἴσον ἀπονέμοντες, ξυλλεγόμενοι δὲ περί τε τῶν κοινῶν ἐβουλεύοντο καὶ ἐξήταζον εἴ τίς τι παραβαίνοι, καὶ ἐδίκαζον. ὅτε δὲ δικάζειν μέλλοιεν, πίστεις ἀλλήλοις τοιάσδε ἐδίδοσαν πρότερον. ἀφέτων ὄντων ταύρων ἐν τῷ τοῦ Ποσειδῶνος ἱερῷ μόνοι γιγνόμενοι δέκα ὄντες, ἐπευξάμενοι τῷ θεῷ τὸ κεχαρισμένον αὐτῷ θῦμα E ἑλεῖν, ἄνευ σιδήρου ξύλοις καὶ βρόχοις ἐθήρευον, ὃν δὲ ἕλοιεν τῶν ταύρων πρὸς τὴν στήλην προσαγαγόντες κατὰ κορυφὴν αὐτῆς ἔσφαττον κατὰ τῶν γραμμάτων. ἐν δὲ τῇ στήλῃ πρὸς τοῖς νόμοις ὅρκος ἦν μεγάλας ἀρὰς ἐπευχόμενος τοῖς ἀπειθοῦσιν. ὅτ' οὖν κατὰ τοὺς αὑτῶν νόμους θύσαντες καθ-
120 αγίζοιεν πάντα τοῦ ταύρου τὰ μέλη, κρατῆρα κεράσαντες ὑπὲρ ἑκάστου θρόμβον ἐνέβαλλον αἵματος, τὰ δ' ἄλλ' εἰς τὸ πῦρ ἔφερον, περικαθήραντες τὴν στήλην· μετὰ δὲ τοῦτο χρυσαῖς φιάλαις ἐκ τοῦ κρατῆρος ἀρυτόμενοι, κατὰ τοῦ πυρὸς σπένδοντες ἐπώμνυσαν δικάσειν τε κατὰ τοὺς ἐν τῇ στήλῃ νόμους καὶ κολάσειν εἴ τίς τι πρότερον παραβεβηκὼς εἴη, τό τε αὖ μετὰ τοῦτο μηδὲν τῶν γραμμάτων ἑκόντες παραβήσεσθαι, μηδὲ ἄρξειν

---

[1] For the sacrifice of bulls to Poseidon *cf.* Hom. *Odyss.* iii. 6. Ταύρεος was also a local epithet of the Sea-god (Hes. *Sc.* 104).

the precepts of Poseidon, as handed down to them by the law and by the records inscribed by the first princes on a pillar of orichalcum, which was placed within the temple of Poseidon in the centre of the island; and thither they assembled every fifth year, and then alternately every sixth year—giving equal honour to both the even and the odd—and when thus assembled they took counsel about public affairs and inquired if any had in any way transgressed and gave judgement. And when they were about to give judgement they first gave pledges one to another of the following description. In the sacred precincts of Poseidon there were bulls at large [1]; and the ten princes, being alone by themselves, after praying to the God that they might capture a victim well-pleasing unto him, hunted after the bulls with staves and nooses but with no weapon of iron; and whatsoever bull they captured they led up to the pillar and cut its throat over the top of the pillar, raining down blood on the inscription. And inscribed upon the pillar, besides the laws, was an oath which invoked mighty curses upon them that disobeyed. When, then, they had done sacrifice according to their laws and were consecrating all the limbs of the bull, they mixed a bowl of wine and poured in on behalf of each one a gout of blood, and the rest they carried to the fire, when they had first purged the pillars round about. And after this they drew out from the bowl with golden ladles, and making libation over the fire swore to give judgement according to the laws upon the pillar and to punish whosoever had committed any previous transgression; and, moreover, that henceforth they would not transgress any of the writings willingly, nor govern nor submit to any

B μηδὲ ἄρχοντι πείσεσθαι πλὴν κατὰ τοὺς τοῦ πατρὸς ἐπιτάττοντι νόμους. ταῦτα ἐπευξάμενος ἕκαστος αὐτῶν αὑτῷ καὶ τῷ ἀφ' αὑτοῦ γένει, πιὼν καὶ ἀναθεὶς τὴν φιάλην εἰς τὸ ἱερὸν τοῦ θεοῦ, περὶ τὸ δεῖπνον καὶ τἀναγκαῖα διατρίψας, ἐπειδὴ γίγνοιτο σκότος καὶ τὸ πῦρ ἐψυγμένον τὸ περὶ τὰ θύματα εἴη, πάντες οὕτως ἐνδύντες ὅ τι καλλίστην κυανῆν στολήν, ἐπὶ τὰ τῶν ὁρκωμοσίων καύματα χαμαὶ καθίζοντες, νύκτωρ, πᾶν τὸ περὶ τὸ ἱερὸν
C ἀποσβεννύντες πῦρ, ἐδικάζοντό τε καὶ ἐδίκαζον, εἴ τίς τι παραβαίνειν αὐτῶν αἰτιῷτό τινα· δικάσαντες δὲ τὰ δικασθέντα, ἐπειδὴ φῶς γένοιτο, ἐν χρυσῷ πίνακι γράψαντες μετὰ τῶν στολῶν μνημεῖα ἀνετίθεσαν. νόμοι δὲ πολλοὶ μὲν ἄλλοι περὶ τὰ γέρα τῶν βασιλέων ἑκάστων ἦσαν ἴδιοι, τὰ δὲ μέγιστα μήτε ποτὲ ὅπλα ἐπ' ἀλλήλους οἴσειν βοηθήσειν τε πάντας, ἄν πού τις αὐτῶν ἔν τινι πόλει τὸ βασιλικὸν καταλύειν ἐπιχειρῇ γένος, κοινῇ δέ, καθάπερ οἱ
D πρόσθεν, βουλευόμενοι τὰ δόξαντα περὶ πολέμου καὶ τῶν ἄλλων πράξεων, ἡγεμονίαν ἀποδιδόντες τῷ Ἀτλαντικῷ γένει. θανάτου δὲ τὸν βασιλέα τῶν συγγενῶν μηδενὸς εἶναι κύριον, ἂν μὴ τῶν δέκα τοῖς ὑπὲρ ἥμισυ δοκῇ.

Ταύτην δὴ τοσαύτην καὶ τοιαύτην δύναμιν ἐν ἐκείνοις τότε οὖσαν τοῖς τόποις ὁ θεὸς ἐπὶ τούσδε αὖ τοὺς τόπους ξυντάξας ἐκόμισεν ἔκ τινος τοιᾶσδε, ὡς λόγος, προφάσεως. ἐπὶ πολλὰς μὲν γενεάς,
E μέχριπερ ἡ τοῦ θεοῦ φύσις αὐτοῖς ἐξήρκει, κατ-

CRITIAS

governor's edict save in accordance with their father's laws. And when each of them had made this invocation both for himself and for his seed after him, he drank of the cup and offered it up as a gift in the temple of the God; and after spending the interval in supping and necessary business, when darkness came on and the sacrificial fire had died down, all the princes robed themselves in most beautiful sable vestments, and sate on the ground beside the cinders of the sacramental victims throughout the night, extinguishing all the fire that was round about the sanctuary; and there they gave and received judgement, if any of them accused any of committing any transgression. And when they had given judgement, they wrote the judgements, when it was light, upon a golden tablet, and dedicated them together with their robes as memorials. And there were many other special laws concerning the peculiar rights of the several princes, whereof the most important were these: that they should never take up arms against one another, and that, should anyone attempt to overthrow in any city their royal house, they should all lend aid, taking counsel in common, like their forerunners, concerning their policy in war and other matters, while conceding the leadership to the royal branch of Atlas; and that the king had no authority to put to death any of his brother-princes save with the consent of more than half of the ten.

Such was the magnitude and character of the power which existed in those regions at that time; and this power the God set in array and brought against these regions of ours on some such pretext as the following, according to the story. For many generations, so long as the inherited nature of the God remained

120 ἤκοοί τε ἦσαν τῶν νόμων καὶ πρὸς τὸ ξυγγενὲς θεῖον
φιλοφρόνως εἶχον· τὰ γὰρ φρονήματα ἀληθινὰ καὶ
πάντη μεγάλα ἐκέκτηντο, πραότητι μετὰ φρονήσεως πρός τε τὰς ἀεὶ ξυμβαινούσας τύχας καὶ
πρὸς ἀλλήλους χρώμενοι, διὸ πλὴν ἀρετῆς πάντα
ὑπερορῶντες σμικρὰ ἡγοῦντο τὰ παρόντα καὶ
121 ῥᾳδίως ἔφερον οἷον ἄχθος τὸν τοῦ χρυσοῦ τε καὶ
τῶν ἄλλων κτημάτων ὄγκον, ἀλλ' οὐ μεθύοντες
ὑπὸ τρυφῆς διὰ πλοῦτον ἀκράτορες αὑτῶν ὄντες
ἐσφάλλοντο, νήφοντες δὲ ὀξὺ καθεώρων ὅτι καὶ
ταῦτα πάντα ἐκ φιλίας τῆς κοινῆς μετὰ ἀρετῆς
αὐξάνεται, τῇ δὲ τούτων σπουδῇ καὶ τιμῇ φθίνει
ταῦτά τε αὐτὰ κἀκείνη ξυναπόλλυται τούτοις. ἐκ
δὴ λογισμοῦ τε τοιούτου καὶ φύσεως θείας παρα-
μενούσης πάντ' αὐτοῖς ηὐξήθη, ἃ πρὶν διήλθομεν.
ἐπεὶ δ' ἡ τοῦ θεοῦ μὲν μοῖρα ἐξίτηλος ἐγίγνετο ἐν
αὐτοῖς πολλῷ τῷ θνητῷ καὶ πολλάκις ἀνακεραν-
B νυμένη, τὸ δὲ ἀνθρώπινον ἦθος ἐπεκράτει, τότε
ἤδη τὰ παρόντα φέρειν ἀδυνατοῦντες ἠσχημόνουν,
καὶ τῷ δυναμένῳ μὲν ὁρᾶν αἰσχροὶ κατεφαίνοντο,
τὰ κάλλιστα ἀπὸ τῶν τιμιωτάτων ἀπολλύντες, τοῖς
δὲ ἀδυνατοῦσιν ἀληθινὸν πρὸς εὐδαιμονίαν βίον
ὁρᾶν τότε δὴ μάλιστα πάγκαλοί τε μακάριοί τε
ἐδοξάζοντο εἶναι, πλεονεξίας ἀδίκου καὶ δυνάμεως
ἐμπιπλάμενοι. θεὸς δὲ ὁ θεῶν Ζεὺς ἐν νόμοις

strong in them, they were submissive to the laws and kindly disposed to their divine kindred. For the intents of their hearts were true and in all ways noble, and they showed gentleness joined with wisdom in dealing with the changes and chances of life and in their dealings one with another. Consequently they thought scorn of everything save virtue and lightly esteemed their rich possessions, bearing with ease the burden, as it were, of the vast volume of their gold and other goods; and thus their wealth did not make them drunk with pride so that they lost control of themselves and went to ruin; rather, in their soberness of mind they clearly saw that all these good things are increased by general amity combined with virtue, whereas the eager pursuit and worship of these goods not only causes the goods themselves to diminish but makes virtue also to perish with them. As a result, then, of such reasoning and of the continuance of their divine nature all their wealth had grown to such a greatness as we previously described. But when the portion of divinity within them was now becoming faint and weak through being ofttimes blended with a large measure of mortality, whereas the human temper was becoming dominant, then at length they lost their comeliness, through being unable to bear the burden of their possessions, and became ugly to look upon, in the eyes of him who has the gift of sight; for they had lost the fairest of their goods from the most precious of their parts; but in the eyes of those who have no gift of perceiving what is the truly happy life, it was then above all that they appeared to be superlatively fair and blessed, filled as they were with lawless ambition and power. And Zeus, the God of gods, who reigns by Law,

121 βασιλεύων, ἅτε δυνάμενος καθορᾶν τὰ τοιαῦτα, ἐννοήσας γένος ἐπιεικὲς ἀθλίως διατιθέμενον, δίκην αὐτοῖς ἐπιθεῖναι βουληθείς, ἵνα γένοιντο ἐμμελέ-
C στεροι σωφρονισθέντες, ξυνήγειρε θεοὺς πάντας εἰς τὴν τιμιωτάτην αὐτῶν οἴκησιν, ἣ δὴ κατὰ μέσον παντὸς τοῦ κόσμου βεβηκυῖα καθορᾷ πάντα ὅσα γενέσεως μετείληφε, καὶ ξυναγείρας εἶπεν. . . .

CRITIAS

inasmuch as he has the gift of perceiving such things, marked how this righteous race was in evil plight, and desired to inflict punishment upon them, to the end that when chastised they might strike a truer note. Wherefore he assembled together all the gods into that abode which they honour most, standing as it does at the centre of all the Universe, and beholding all things that partake of generation; and when he had assembled them, he spake thus : . . .

CLEITOPHON

INTRODUCTION TO THE *CLEITOPHON*

In Bekker's edition of Plato the *Cleitophon* is classed among the works *incerti auctoris*; and it is certainly doubtful whether it comes from the pen of Plato. If it does, it may perhaps be regarded as a kind of fragmentary preface to the *Republic*, inasmuch as both treat of the same subject, the nature of justice; and in the *Republic* also (340 A-B) Cleitophon appears as an adherent of the Sophist Thrasymachus. But from various peculiarities of style and vocabulary it seems more probable that it is a later composition written by someone who had the *Republic* and the *Meno* (with its discussion of the teachableness of virtue) in mind; and it would certainly be strange to find Plato leaving us with the impression, uncorrected, that Socrates was really ignorant of the true nature of justice, which is the kernel of Cleitophon's concluding criticism.

ΚΛΕΙΤΟΦΩΝ

ΤΑ ΤΟΥ ΔΙΑΛΟΓΟΥ ΠΡΟΣΩΠΑ

ΣΩΚΡΑΤΗΣ, ΚΛΕΙΤΟΦΩΝ

406   Κλειτοφῶντα τὸν Ἀριστωνύμου τις ἡμῖν διηγεῖτο ἔναγχος, ὅτι Λυσίᾳ διαλεγόμενος τὰς μὲν μετὰ Σωκράτους διατριβὰς ψέγοι, τὴν Θρασυμάχου δὲ ξυνουσίαν ὑπερεπαινοῖ.

ΚΛ. Ὅστις, ὦ Σώκρατες, οὐκ ὀρθῶς ἀπεμνημόνευέ σοι τοὺς ἐμοὶ περὶ σοῦ γενομένους λόγους πρὸς Λυσίαν· τὰ μὲν γὰρ ἔγωγε οὐκ ἐπῄνουν σε, τὰ δὲ καὶ ἐπῄνουν. ἐπεὶ δὲ δῆλος εἶ μεμφόμενος μέν μοι, προσποιούμενος δὲ μηδὲν φροντίζειν, ἥδιστ᾽ ἄν σοι διεξέλθοιμι αὐτοὺς αὐτός, ἐπειδὴ καὶ μόνω τυγχάνομεν ὄντε, ἵνα ἧττόν με ἡγῇ πρὸς σὲ φαύλως ἔχειν. νῦν γὰρ ἴσως οὐκ ὀρθῶς ἀκήκοας, ὥστε φαίνει πρὸς ἐμὲ ἔχειν τραχυτέρως τοῦ δέοντος. εἰ δέ μοι δίδως παρρησίαν, ἥδιστ᾽ ἂν δεξαίμην καὶ ἐθέλω λέγειν.

ΣΩ. Ἀλλ᾽ αἰσχρὸν μὴν σοῦ γε ὠφελεῖν με προ-
407 θυμουμένου μὴ ὑπομένειν· δῆλον γὰρ ὡς γνοὺς ὅπη

CLEITOPHON

CHARACTERS

SOCRATES, CLEITOPHON

soc. It was told us recently by someone about Cleitophon, the son of Aristonymus, that in a conversation he had with Lysias he was finding fault with the instructions of Socrates and praising to the skies the lectures of Thrasymachus.

CLEIT. That was a man, Socrates, who gave you a false report of the talk I had about you with Lysias. For I was really praising you for some things, though not for others. But since it is plain that you are reproaching me, though you pretend to be quite indifferent, I should be delighted to repeat to you myself what I said, now that we happen to be alone, so that you may be less inclined to suspect me of holding a poor opinion of you. For at present it seems that you have heard what is not true, with the result that you appear to be more vexed with me than I deserve. So if you give me leave to speak I shall avail myself of it most gladly, as I want to explain.

soc. Well, now, it would be indeed unhandsome of me not to put up with it when you are so anxious to do me a benefit. For obviously, when I have been

PLATO

χείρων εἰμὶ καὶ βελτίων, τὰ μὲν ἀσκήσω καὶ διώξομαι, τὰ δὲ φεύξομαι κατὰ κράτος.

κλ. Ἀκούοις ἄν. ἐγὼ γάρ, ὦ Σώκρατες, σοὶ συγγιγνόμενος πολλάκις ἐξεπληττόμην ἀκούων, καί μοι ἐδόκεις παρὰ τοὺς ἄλλους ἀνθρώπους κάλλιστα λέγειν, ὁπότε ἐπιτιμῶν τοῖς ἀνθρώποις, ὥσπερ ἐπὶ μηχανῆς τραγικῆς θεός, ὕμνεις λέγων Ποῖ φέρεσθε, ὤνθρωποι, καὶ ἀγνοεῖτε οὐδὲν τῶν δεόντων πράττοντες, οἵτινες χρημάτων μὲν πέρι τὴν πᾶσαν σπουδὴν ἔχετε, ὅπως ὑμῖν ἔσται, τῶν δ' υἱέων οἷς ταῦτα παραδώσετε, ὅπως ἐπιστήσονται χρῆσθαι δικαίως τούτοις, ἀμελεῖτε, καὶ οὔτε διδασκάλους αὐτοῖς εὑρίσκετε τῆς δικαιοσύνης, εἴπερ μαθητόν· εἰ δὲ μελετητόν τε καὶ ἀσκητόν, οἵτινες ἐξασκήσουσι καὶ ἐκμελετήσουσιν ἱκανῶς· οὐδέ γ' ἔτι πρότερον ὑμᾶς αὐτοὺς οὕτως ἐθεραπεύσατε. ἀλλ' ὁρῶντες γράμματα καὶ μουσικὴν καὶ γυμναστικὴν ὑμᾶς τε αὐτοὺς καὶ τοὺς παῖδας ὑμῶν ἱκανῶς μεμαθηκότας, ἃ δὴ παιδείαν ἀρετῆς εἶναι τελέαν ἥγησθε, κἄπειτα οὐδὲν ἧττον κακοὺς γιγνομένους περὶ τὰ χρήματα, πῶς οὐ καταφρονεῖτε τῆς νῦν παιδεύσεως οὐδὲ ζητεῖτε οἵτινες ὑμᾶς παύσουσι ταύτης τῆς ἀμουσίας; καί τοι διά γε ταύτην τὴν πλημμέλειαν καὶ ῥᾳθυμίαν, ἀλλ' οὐ διὰ τὴν ἐν τῷ ποδὶ πρὸς τὴν λύραν ἀμετρίαν, καὶ ἀδελφὸς ἀδελφῷ καὶ πόλεις πόλεσιν ἀμέτρως καὶ ἀναρμόστως προσφερόμεναι στασιάζουσι καὶ πολεμοῦντες τὰ ἔσχατα δρῶσι καὶ πάσχουσιν. ὑμεῖς δέ φατε οὐ δι' ἀπαιδευσίαν οὐδὲ δι' ἄγνοιαν, ἀλλ' ἑκόντας τοὺς ἀδίκους ἀδίκους εἶναι· πάλιν δ' αὖ

---

[1] *Cf. Meno* 70 A, *Euthyd.* 282 B ff.

CLEITOPHON

taught my good points and my bad, I shall practise and pursue the one and eschew the other with all my might.

CLEIT. Listen, then. When I was attending your lectures, Socrates, I was oftentimes amazed at what I heard, and you seemed to me to surpass all other men in the nobleness of your discourse, when you rebuked mankind and chanted these words like a God on the tragic stage: "Whither haste ye, O men? Yea, verily ye know not that ye are doing none of the things ye ought, seeing that ye spend your whole energy on wealth and the acquiring of it; while as to your sons to whom ye will bequeath it, ye neglect to ensure that they shall understand how to use it justly, and ye find for them no teachers of justice, if so be that it is teachable [1]—or if it be a matter of training and practice, instructors who can efficiently practise and train them—nor have ye even begun by reforming yourselves in this respect. Yet when ye perceive that ye yourselves and your children, though adequately instructed in letters and music and gymnastic —which ye, forsooth, regard as a complete education in virtue—are in consequence none the less vicious in respect of wealth, how is it that ye do not contemn this present mode of education nor search for teachers who will put an end to this your lack of culture? Yet truly it is because of this dissonance and sloth, and not because of failure to keep in step with the lyre that brother with brother and city with city clash together without measure or harmony and are at strife, and in their warring perpetrate and suffer the uttermost horrors. But ye assert that the unjust are unjust not because of their lack of education and lack of knowledge but voluntarily, while on the other hand

PLATO

407
τολμᾶτε λέγειν ὡς αἰσχρὸν καὶ θεομισὲς ἡ ἀδικία·
πῶς οὖν δή τις τό γε τοιοῦτον κακὸν ἑκὼν αἱροῖτ᾽
ἄν; ἥττων ὃς ἂν ᾖ, φατέ, τῶν ἡδονῶν. οὐκοῦν καὶ
τοῦτ᾽ ἀκούσιον, εἴπερ τὸ νικᾶν ἑκούσιον; ὥστε
ἐκ παντὸς τρόπου τό γε ἀδικεῖν ἀκούσιον ὁ λόγος
αἱρεῖ, καὶ δεῖν ἐπιμέλειαν τῆς νῦν πλείω ποιεῖσθαι
E πάντ᾽ ἄνδρα ἰδίᾳ θ᾽ ἅμα καὶ δημοσίᾳ ξυμπάσας
τὰς πόλεις.

Ταῦτ᾽ οὖν, ὦ Σώκρατες, ἐγὼ ὅταν ἀκούω
σοῦ θαμὰ λέγοντος, καὶ μάλα ἄγαμαι καὶ
θαυμαστῶς ὡς ἐπαινῶ. καὶ ὁπόταν αὖ φῇς τὸ
ἐφεξῆς τούτῳ, τοὺς ἀσκοῦντας μὲν τὰ σώματα,
τῆς δὲ ψυχῆς ἠμεληκότας ἕτερόν τι πράττειν
τοιοῦτον, τοῦ μὲν ἄρξοντος ἀμελεῖν, περὶ δὲ τὸ
ἀρξόμενον ἐσπουδακέναι. καὶ ὅταν λέγῃς ὡς ὅτῳ
τις μὴ ἐπίσταται χρῆσθαι, κρεῖττον ἐᾶν τὴν τούτου
χρῆσιν· εἰ δή τις μὴ ἐπίσταται ὀφθαλμοῖς χρῆσθαι
μηδὲ ὠσὶ μηδὲ ξύμπαντι τῷ σώματι, τούτῳ μήτε
ἀκούειν μήθ᾽ ὁρᾶν μήτ᾽ ἄλλην χρείαν μηδεμίαν
χρῆσθαι τῷ σώματι κρεῖττον ἢ ὁπῃοῦν χρῆσθαι.
408 καὶ δὴ καὶ περὶ τέχνην ὡσαύτως· ὅστις γὰρ δὴ μὴ
ἐπίσταται τῇ ἑαυτοῦ λύρᾳ χρῆσθαι, δῆλον ὡς οὐδὲ
τῇ τοῦ γείτονος, οὐδὲ ὅστις μὴ τῇ τῶν ἄλλων, οὐδὲ
τῇ ἑαυτοῦ, οὐδ᾽ ἄλλῳ τῶν ὀργάνων οὐδὲ κτημάτων
οὐδενί. καὶ τελευτᾷ δὴ καλῶς ὁ λόγος οὗτός σοι,
ὡς ὅστις ψυχῇ μὴ ἐπίσταται χρῆσθαι, τούτῳ τὸ
ἄγειν ἡσυχίαν τῇ ψυχῇ καὶ μὴ ζῆν κρεῖττον ἢ ζῆν

CLEITOPHON

ye have the face to affirm that injustice is a foul thing, and hateful to Heaven. Then how, pray, could any man voluntarily choose an evil of such a kind? Any man, you reply, who is mastered by his pleasures. But is not this condition also involuntary, if the act of mastering be voluntary? Thus in every way the argument proves that unjust action is involuntary, and that every man privately and all the cities publicly ought to pay more attention than they do now to this matter."

So then, Socrates, when I hear you constantly making these speeches I admire you immensely and praise you to the skies. So too when you state the next point in your argument, that those who train their bodies but neglect their souls are guilty of another action of the same sort—neglecting the part that should rule, and attending to that which should be ruled. Also when you declare that whatsoever object a man knows not how to make use of, it is better for him to refrain from making use thereof; thus, suppose a man knows not how to use his eyes or his ears or the whole of his body, it is better for such a man not to hear nor to see nor to employ his body for any other use rather than to use it in any way whatsoever. So too, likewise, with respect to art: it is surely plain that a man who does not know how to use his own lyre does not know either how to use his neighbour's, and that one who does not know how to use the lyre of others does not know how to use his own either,—nor yet any other instrument or chattel. Moreover, the conclusion of this argument of yours is a fine one,—how that for every man who knows not how to make use of his soul it is better to have his soul at rest and not to live, than to live

πράττοντι καθ' αὑτόν· εἰ δέ τις ἀνάγκη ζῆν εἴη, δούλῳ ἄμεινον ἢ ἐλευθέρῳ διάγειν τῷ τοιούτῳ τὸν βίον ἐστὶν ἄρα, καθάπερ πλοίου παραδόντι τὰ πηδάλια τῆς διανοίας ἄλλῳ, τῷ μαθόντι τὴν τῶν ἀνθρώπων κυβερνητικήν, ἣν δὴ σὺ πολιτικήν, ὦ Σώκρατες, ἐπονομάζεις πολλάκις, τὴν αὐτὴν δὴ ταύτην δικαστικήν τε καὶ δικαιοσύνην ὡς ἔστι λέγων.

Τούτοις δὴ τοῖς λόγοις καὶ ἑτέροις τοιούτοις παμπόλλοις καὶ παγκάλως λεγομένοις, ὡς διδακτὸν ἀρετὴ καὶ πάντων ἑαυτοῦ δεῖ μάλιστα ἐπιμελεῖσθαι, σχεδὸν οὔτ' ἀντεῖπον πώποτε οὔτ' οἶμαι μήποθ' ὕστερον ἀντείπω· προτρεπτικωτάτους τε γὰρ ἡγοῦμαι καὶ ὠφελιμωτάτους, καὶ ἀτεχνῶς ὥσπερ καθεύδοντας ἐπεγείρειν ἡμᾶς. προσεῖχον δὴ τὸν νοῦν τὸ μετὰ ταῦτα ὡς ἀκουσόμενος, ἐπανερωτῶν οὔ τι σὲ τὸ πρῶτον, ὦ Σώκρατες, ἀλλὰ τῶν ἡλικιωτῶν τε καὶ συνεπιθυμητῶν ἢ ἑταίρων σῶν, ἢ ὅπως δεῖ πρὸς σὲ περὶ αὐτῶν τὸ τοιοῦτον ὀνομάζειν. τούτων γὰρ τούς τι μάλιστα εἶναι δοξαζομένους ὑπὸ σοῦ πρώτους ἐπανηρώτων, πυνθανόμενος τίς ὁ μετὰ ταῦτ' εἴη λόγος, καὶ κατὰ σὲ τρόπον τινὰ ὑποτείνων αὐτοῖς Ὦ βέλτιστοι, ἔφην, ὑμεῖς, πῶς ποτε νῦν ἀποδεχόμεθα τὴν Σωκράτους προτροπὴν ἡμῶν ἐπ' ἀρετήν; ὡς ὄντος μόνου τούτου, ἐπεξελθεῖν δὲ οὐκ ἔνι τῷ πράγματι καὶ λαβεῖν αὐτὸ τελέως, ἀλλ' ἡμῖν παρὰ πάντα δὴ τὸν βίον ἔργον τοῦτ' ἔσται, τοὺς μήπω προτετραμμένους προτρέπειν, καὶ ἐκείνους αὖ ἑτέρους; ἢ δεῖ τὸν Σωκράτη καὶ ἀλλήλους ἡμᾶς

CLEITOPHON

acting according to his own caprice; but if it is necessary for him to live, it is better after all for such an one to spend his life as a slave rather than a free man, handing over the rudder of his will, as it were of a ship, to another man who has learnt the art of steering men—which is the name that you, Socrates, frequently give to politics, when you declare that this very same art is that of judging and justice.

Against these arguments and others of a like kind, exceedingly numerous and couched in exceedingly noble language, showing that virtue can be taught and that a man should care above all else for himself, I have hardly uttered a word up till now, nor do I suppose that I ever shall utter a word against them in the future, for I regard them as most valuable admonitions and most useful, literally capable of waking us up, as it were, out of our slumber. So I gave my attention with a view to hear what was to follow next, although I did not at first question you yourself, Socrates, but some of your contemporaries and fellow-students or companions—or whatever name one ought to give to the relation in which they stand towards you. Of these I questioned first those who are specially held in regard by yourself, asking them what was your next argument, and propounding the matter to them somewhat after your own fashion: "I ask you, my very good Sirs, in what sense do we now accept the exhortation to virtue which Socrates has given us. Are we to regard it as all there is, and suppose it to be impossible to pursue the object further and grasp it fully; and is this to be our life-long task, just to exhort those who have not as yet been exhorted, and that they in turn should exhort others? Or, when we have agreed that this is exactly

τὸ μετὰ τοῦτ' ἐπανερωτᾶν, ὁμολογήσαντας τοῦτ' E αὐτὸ ἀνθρώπῳ πρακτέον εἶναι, τί τοὐντεῦθεν; πῶς ἄρχεσθαι δεῖν φαμεν δικαιοσύνης πέρι μαθήσεως; ὥσπερ ἂν εἴ τις ἡμᾶς προὔτρεπε τοῦ σώματος ἐπιμέλειαν ποιεῖσθαι, μηδὲν προνοοῦντας ὁρῶν καθάπερ παῖδας ὡς ἔστι τις γυμναστικὴ καὶ ἰατρική, κἄπειτα ὠνείδιζε, λέγων ὡς αἰσχρὸν πυρῶν μὲν καὶ κριθῶν καὶ ἀμπέλων ἐπιμέλειαν πᾶσαν ποιεῖσθαι, καὶ ὅσα τοῦ σώματος ἕνεκα διαπονούμεθά τε καὶ κτώμεθα, τούτου δ' αὐτοῦ μηδεμίαν τέχνην μηδὲ μηχανήν, ὅπως ὡς βέλτιστον ἔσται τὸ σῶμα, ἐξευρίσκειν, καὶ ταῦτα οὖσαν. εἰ δ' ἐπανηρόμεθα τὸν ταῦθ' ἡμᾶς προτρέποντα 409 Λέγεις δὲ εἶναι τίνας ταύτας τὰς τέχνας; εἶπεν ἂν ἴσως ὅτι γυμναστικὴ καὶ ἰατρική. καὶ νῦν δὴ τίνα φαμὲν εἶναι τὴν ἐπὶ τῇ τῆς ψυχῆς ἀρετῇ τέχνην; λεγέσθω. Ὁ δὴ δοκῶν αὐτῶν ἐρρωμενέστατος εἶναι πρὸς ταῦτα ἀποκρινόμενος εἰπέ μοι ταύτην τὴν τέχνην εἶναι ἥνπερ ἀκούεις σὺ λέγοντος, ἔφη, Σωκράτους, οὐκ ἄλλην ἢ δικαιοσύνην. Εἰπόντος δ' ἐμοῦ Μή μοι τὸ ὄνομα μόνον εἴπῃς, ἀλλ' ὧδε. ἰατρική πού τις λέγεται τέχνη· ταύτης δ' ἐστὶ διττὰ τὰ ἀποτελούμενα, τὸ μὲν B ἰατροὺς ἀεὶ πρὸς τοῖς οὖσιν ἑτέρους ἐξεργάζεσθαι, τὸ δὲ ὑγίειαν· ἔστι δὲ τούτων θάτερον οὐκέτι τέχνη, τῆς τέχνης δὲ τῆς διδασκούσης τε καὶ διδασκομένης ἔργον, ὃ δὴ λέγομεν ὑγίειαν. καὶ τεκτονικῆς δὲ κατὰ ταὐτὰ οἰκία τε καὶ τεκτονικὴ τὸ μὲν ἔργον,

CLEITOPHON

what a man should do, ought we to ask Socrates, and one another, the further question—"What is the next step?" What do we say is the way in which we ought to begin the study of justice? Just as if a man were exhorting us to devote care to our bodies, observing that we like children had as yet no notion of the existence of the arts of gymnastics and medicine; and were then to reproach us and say that it is disgraceful to spend all one's care on wheat and barley and vines and all the goods which we labour to acquire for the sake of the body, and yet make no effort to discover some art or device for securing that the body itself shall be in the best possible condition—and that in spite of the fact that such an art exists. Suppose then that we had put to the man who was thus exhorting us this further question —"What arts do you say these are?" His answer, no doubt, would be—"Gymnastics and medicine." So now, in the case before us, what do we say is the art which deals with the virtue of the soul? Let it be stated." Then he who was reputed to be their most powerful exponent of these matters answered me and said that this art is precisely that which, said he, you hear Socrates describing,—nothing else than justice. I then replied—"Do not explain to me merely its name, but like this :—There is an art called medicine; and of this the effects are two-fold, the one being to produce constantly new doctors in addition to those already existing, and the other to produce health. And of these the latter result is no longer in itself an art but an effect of that art which both teaches and is taught, which effect we term 'health.' So likewise the operations of the joiner's art are a house and joinery, of which the one

τὸ δὲ δίδαγμα. τῆς δὴ δικαιοσύνης ὡσαύτως τὸ μὲν δικαίους ἔστω ποιεῖν, καθάπερ ἐκεῖ τοὺς τεχνίτας ἑκάστους· τὸ δ' ἕτερον, ὃ δύναται ποιεῖν ἡμῖν ἔργον ὁ δίκαιος, τί τοῦτό φαμεν; εἰπέ. Οὗτος μὲν, ὡς οἶμαι, τὸ συμφέρον ἀπεκρίνατο, ἄλλος δὲ τὸ δέον, ἕτερος δὲ τὸ ὠφέλιμον, ὁ δὲ τὸ λυσιτελοῦν. ἐπανήειν δὴ ἐγὼ λέγων ὅτι Κἀκεῖ τά γε ὀνόματα ταῦτ' ἐστὶν ἐν ἑκάστῃ τῶν τεχνῶν, ὀρθῶς πράττειν, λυσιτελοῦντα, ὠφέλιμα καὶ τἆλλα τὰ τοιαῦτα· ἀλλὰ πρὸς ὅ τι ταῦτα πάντα τείνει, ἐρεῖ τὸ ἴδιον ἑκάστη τέχνη, οἷον ἡ τεκτονικὴ τὸ εὖ, τὸ καλῶς, τὸ δεόντως, ὥστε τὰ ξύλινα, φήσει, σκεύη γίγνεσθαι, ἃ δὴ οὐκ ἔστι τέχνη. λεγέσθω δὴ καὶ τὸ τῆς δικαιοσύνης ὡσαύτως. Τελευτῶν ἀπεκρίνατό τις ὦ Σώκρατές μοι τῶν σῶν ἑταίρων, ὃς δὴ κομψότατα ἔδοξεν εἰπεῖν, ὅτι τοῦτ' εἴη τὸ τῆς δικαιοσύνης ἴδιον ἔργον, ὃ τῶν ἄλλων οὐδεμιᾶς, φιλίαν ἐν ταῖς πόλεσι ποιεῖν. οὗτος δ' αὖ ἐρωτώμενος τὴν φιλίαν ἀγαθόν τ' ἔφη εἶναι καὶ οὐδέποτε κακόν· τὰς δὲ τῶν παίδων φιλίας καὶ τὰς τῶν θηρίων, ἃς ἡμεῖς τοῦτο τοὔνομα ἐπονομάζομεν, οὐκ ἀπεδέχετο εἶναι φιλίας ἐπανερωτώμενος· συνέβαινε γὰρ αὐτῷ τὰ πλείω τὰς τοιαύτας βλαβερὰς ἢ ἀγαθὰς εἶναι. φεύγων δὴ τὸ τοιοῦτον οὐδὲ φιλίας ἔφη τὰς τοιαύτας εἶναι, ψευδῶς δὲ ὀνομάζειν αὐτὰς τοὺς οὕτως ὀνομάζοντας· τὴν δὲ

---

[1] *Cf. Rep.* 351 D.

CLEITOPHON

is an effect, the other a doctrine. In like manner let it be granted that the one effect of justice is to produce just men, as of the other arts their several artists; but as to the other, the operation which the just man is capable of performing for us, what do we say that is? Tell us." The reply of your exponent was, I think, "The beneficial"; while another said "The right"; a third "The useful"; and yet another "The profitable." So I resumed my inquiry and said: "In the former case also we find these names in each one of the arts—doing 'the right,' 'the profitable,' 'the useful,' and the rest of such terms; but as regards the object at which all these operations aim, each art will declare that which is peculiar to itself; for example, the art of joinery will assert that the result of good, beautiful, and right action is the production of wooden vessels, which in themselves are not an art. So let the operation of justice be stated in the same way." Finally, Socrates, one of your companions, who was reputed to be a most accomplished speaker, made answer that the peculiar effect of justice, which was effected by no other art, was to produce friendship in States.[1] And he, in turn, when questioned declared that friendship is a good thing and never an evil; while as to the friendships of children and those of wild beasts, which we call by this name, he refused to admit—when questioned upon the point—that they were friendships; since, as a result of the argument, he was forced to say that such relations were for the most part harmful rather than good. So to avoid such an admission he denied that such relations were friendships at all, and said that those who give them this name name them falsely; and real and true

PLATO

409
ὄντως καὶ ἀληθῶς φιλίαν εἶναι σαφέστατα ὁμόνοιαν. τὴν δὲ ὁμόνοιαν ἐρωτώμενος εἰ ὁμοδοξίαν εἶναι λέγοι ἢ ἐπιστήμην, τὴν μὲν ὁμοδοξίαν ἠτίμαζεν· ἠναγκάζοντο γὰρ πολλαὶ καὶ βλαβεραὶ γίγνεσθαι ὁμοδοξίαι ἀνθρώπων, τὴν δὲ φιλίαν ἀγαθὸν ὡμολογήκει πάντως εἶναι καὶ δικαιοσύνης ἔργον, ὥστε ταὐτὸν ἔφησεν εἶναι ὁμόνοιαν ἐπιστήμην οὖσαν, ἀλλ' οὐ δόξαν.

Ὅτε δὴ ἐνταῦθα ἦμεν τοῦ λόγου ἀποροῦντες,
410 οἱ παρόντες ἱκανοὶ ἦσαν ἐπιπλήττειν τε αὐτῷ καὶ λέγειν ὅτι περιδεδράμηκεν εἰς ταὐτὸν ὁ λόγος τοῖς πρώτοις, καὶ ἔλεγον ὅτι Καὶ ἡ ἰατρικὴ ὁμόνοιά τίς ἐστι καὶ ἅπασαι αἱ τέχναι, καὶ περὶ ὅτου εἰσὶν ἔχουσι λέγειν· τὴν δὲ ὑπὸ σοῦ λεγομένην δικαιοσύνην ἢ ὁμόνοιαν, ὅποι τείνουσά ἐστι, διαπέφευγε, καὶ ἄδηλον αὐτῆς ὅ τί ποτ' ἔστι τὸ ἔργον.

Ταῦτα, ὦ Σώκρατες, ἐγὼ τελευτῶν καὶ σὲ αὐτὸν ἠρώτων, καὶ εἶπές μοι δικαιοσύνης εἶναι
B τοὺς μὲν ἐχθροὺς βλάπτειν, τοὺς δὲ φίλους εὖ ποιεῖν. ὕστερον δὲ ἐφάνη βλάπτειν γε οὐδέποτε ὁ δίκαιος οὐδένα· πάντα γὰρ ἐπ' ὠφελείᾳ παντὸς[1] δρᾶν. ταῦτα δὲ οὐχ ἅπαξ οὐδὲ δὶς ἀλλὰ πολὺν δὴ ὑπομείνας χρόνον [καὶ][2] λιπαρῶν ἀπείρηκα, νομίσας σε τὸ μὲν προτρέπειν εἰς ἀρετῆς ἐπιμέλειαν κάλλιστ' ἀνθρώπων δρᾶν, δυοῖν δὲ θάτερον, ἢ τοσοῦτον μόνον δύνασθαι, μακρότερον δὲ οὐδέν, ὃ γένοιτ' ἂν καὶ περὶ ἄλλην ἡντιναοῦν τέχνην, οἷον μὴ ὄντα κυβερνήτην καταμελετῆσαι

---

[1] παντὸς] πάντας MSS., Zur.
[2] καὶ bracketed by Baumann.

---

[1] Cf. Rep. 433 c.

CLEITOPHON

friendship, he said, is most exactly described as "unanimity." And when asked about "unanimity," whether he declared it to be "unity of opinion"[1] or "knowledge," he rejected the expression "unity of opinion," for of necessity many cases of "unity of opinion" occurred amongst men that were harmful, whereas he had agreed that friendship was wholly a good thing and an effect of justice; consequently he affirmed that unanimity was the same, and was not opinion, but knowledge.

Now when we were at this point in the argument and at our wits' end, the bystanders were ready to fall upon the man and to cry that the argument had circled round to the same point as at first; and they declared that: "Medicine also is a kind of 'unanimity,' as are all the arts, and they are able to explain what it is they deal with; but as for the 'justice' or 'unanimity' which you talk of, it has no comprehension of what its own aim is, and what the effect of it is remains quite obscure."

Finally, Socrates, I put these questions to you yourself also, and you told me that it belonged to justice to injure one's enemies and to do well to one's friends. But later on it appeared that the just man never injures anyone, for in all his acts he aims at benefiting all. So after repeated questionings—not once only or twice but spending quite a long time at it—I gave it up, concluding that though you were better than any man at the task of exhorting men to devote themselves to virtue, yet of these two alternatives one must be true: either you are capable of effecting thus much only and nothing more,—a thing which might happen also in respect of any other art whatsoever, as for example a man who was no

325

τὸν ἔπαινον περὶ αὐτῆς, ὡς πολλοῦ τοῖς ἀνθρώποις ἀξία, καὶ περὶ τῶν ἄλλων τεχνῶν ὡσαύτως· ταὐτὸν δὴ καὶ σοί τις ἐπενέγκοι τάχ᾽ ἂν περὶ δικαιοσύνης, ὡς οὐ μᾶλλον ὄντι δικαιοσύνης ἐπιστήμονι, διότι καλῶς αὐτὴν ἐγκωμιάζεις. οὐ μὴν τό γε ἐμὸν οὕτως ἔχει· δυοῖν δὲ θάτερον, ἢ οὐκ εἰδέναι σε ἢ οὐκ ἐθέλειν αὐτῆς ἐμοὶ κοινωνεῖν. διὰ ταῦτα δὴ καὶ πρὸς Θρασύμαχον, οἶμαι, πορεύσομαι καὶ ἄλλοσε ὅποι δύναμαι, ἀπορῶν· ἐπεὶ εἴ γ᾽ ἐθέλεις σὺ τούτων μὲν ἤδη παύσασθαι πρὸς ἐμὲ τῶν λόγων τῶν προτρεπτικῶν, οἷον δέ, εἰ περὶ γυμναστικῆς προτετραμμένος ἢ τοῦ σώματος δεῖν μὴ ἀμελεῖν, τὸ ἐφεξῆς ἂν τῷ προτρεπτικῷ λόγῳ ἔλεγες οἷον τὸ σῶμά μου φύσει ὂν οἵας θεραπείας δεῖται· καὶ νῦν δὴ ταὐτὸν γιγνέσθω. θὲς τὸν Κλειτοφῶντα ὁμολογοῦντα ὡς ἔστι καταγέλαστον τῶν μὲν ἄλλων ἐπιμέλειαν ποιεῖσθαι, ψυχῆς δέ, ἧς ἕνεκα τἆλλα διαπονούμεθα, ταύτης ἠμεληκέναι· καὶ τἆλλα πάντα οἴου με νῦν οὕτως εἰρηκέναι τὰ τούτοις ἑξῆς, ἃ καὶ νῦν δὴ διῆλθον. καί σου δεόμενος λέγω μηδαμῶς ἄλλως ποιεῖν, ἵνα μή, καθάπερ νῦν, τὰ μὲν ἐπαινῶ σε πρὸς Λυσίαν καὶ πρὸς τοὺς ἄλλους, τὰ δέ τι καὶ ψέγω. μὴ μὲν γὰρ προτετραμμένῳ σε ἀνθρώπῳ, ὦ Σώκρατες, ἄξιον εἶναι τοῦ παντὸς φήσω, προτετραμμένῳ δὲ σχεδὸν καὶ ἐμπόδιον τοῦ πρὸς τέλος ἀρετῆς ἐλθόντα εὐδαίμονα γενέσθαι.

CLEITOPHON

steersman might practise composing an eulogy of that art as one of high value to mankind, and so too with all the other arts; so against you too one might perhaps bring the same charge in regard to justice, that you are none the more an expert about justice because you eulogize it finely. Not that this is the complaint I make myself; but it must be one or other of these two alternatives,—either you do not possess the knowledge or else you refuse to let me share it. Consequently, methinks I will betake myself, in my perplexity, to Thrasymachus and to everyone else I can. However, if you are really willing to refrain at last from addressing to me these hortatory discourses, and just as you would have followed up the hortatory discourse, suppose you had been exhorting me in respect of gymnastics that I should not neglect my body, by explaining the nature of the body and the nature of the treatment it requires—so let the same course be followed in the present case. Assume that Cleitophon agrees that it is ridiculous to expend care on everything else and to neglect the soul, for the sake of which all the other labour is incurred; and suppose also that I have made all the other subsequent statements which I rehearsed just now. And I entreat you, as I speak, by no means to act otherwise, lest I should do, as I do now, praise you in part to Lysias and to the others, and also in part blame you. For I shall maintain, Socrates, that while you are of untold value to a man who has not been exhorted, to him who has been exhorted you are almost an actual hindrance in the way of his attaining the goal of virtue and becoming a happy man.

MENEXENUS

INTRODUCTION TO THE *MENEXENUS*

The *Menexenus* is an interesting little work, not so much for the matter it contains as for the literary problems which it raises. Sandwiched between two short pieces of dialogue it gives us what purports to be a funeral oration composed by Aspasia and reported by Socrates,—an oration which challenges comparison with the famous Periclean oration recorded by Thucydides, since Aspasia (according to Socrates) was the real author of them both. The difficulty of understanding Plato's motive and purpose in the *Menexenus* lies in the apparent contrast between the bantering and satirical tone of the opening dialogue, in which Socrates disparages the orators and makes light of their art, and the patriotic and moral sentiments which are expressed with every appearance of good faith in the main body of the oration. But in spite of much in the way of patriotism that may have been sincerely felt by Plato—recurring as it does in the *Timaeus*, *Critias* and *Laws*—it is obvious that the oration itself is largely intended as an illustration of the most glaring defect of current oratory, its indifference to *truth*. It is the same defect which is criticized sharply by Socrates in the *Symposium*. This being so, we need not wonder at the historical misstatements with which the oration abounds, nor at its exaggerated encomium of Athens and her heroic sons ; nor

INTRODUCTION TO THE *MENEXENUS*

should it even amaze us, in such a connexion, that Aspasia and Socrates are supposed to be cognizant of Greek history down to the peace of Antalcidas (387 B.C.), a dozen years after Socrates died ! But none the less, regarded as an Epideictic display, and a study in formal rhetoric, Plato would have us believe, no doubt, that this exhibition of the versatility of Socrates is quite worthy to stand beside the best efforts of the disciples of Gorgias. And we may well be content to leave it there, putting down the *Menexenus* as one of Plato's literary experiments, a *parergon* of no great moment.

Menexenus himself, we may observe, was a young and well-born Athenian, a friend of Lysis and Ctesippus (*Lysis* 206 D), as well as an admirer of Socrates (*Phaedo* 59 B).

As in the rest of this volume, the text is based on that of the Zurich edition.

ΜΕΝΕΞΕΝΟΣ

ΤΑ ΤΟΥ ΔΙΑΛΟΓΟΥ ΠΡΟΣΩΠΑ

ΣΩΚΡΑΤΗΣ, ΜΕΝΕΞΕΝΟΣ

234   1. Ἐξ ἀγορᾶς ἢ πόθεν Μενέξενος;

ΜΕ. Ἐξ ἀγορᾶς, ὦ Σώκρατες, καὶ ἀπὸ τοῦ βουλευτηρίου.

ΣΩ. Τί μάλιστα σὺ πρὸς βουλευτήριον; ἢ δῆλα δὴ ὅτι παιδεύσεως καὶ φιλοσοφίας ἐπὶ τέλει ἡγεῖ εἶναι, καὶ ὡς ἱκανῶς ἤδη ἔχων ἐπὶ τὰ μείζω ἐπινοεῖς τρέπεσθαι, καὶ ἄρχειν ἡμῶν, ὦ θαυμάσιε, ἐπιχειρεῖς τῶν πρεσβυτέρων τηλικοῦτος ὤν, ἵνα
B μὴ ἐκλίπῃ ὑμῶν ἡ οἰκία ἀεί τινα ἡμῶν ἐπιμελητὴν παρεχομένη;

ΜΕ. Ἐὰν σύ γε, ὦ Σώκρατες, ἐᾷς καὶ συμβουλεύῃς ἄρχειν, προθυμήσομαι· εἰ δὲ μή, οὔ. νῦν μέντοι ἀφικόμην πρὸς τὸ βουλευτήριον πυθόμενος ὅτι ἡ βουλὴ μέλλει αἱρεῖσθαι ὅστις ἐρεῖ ἐπὶ τοῖς ἀποθανοῦσι· ταφὰς γὰρ οἶσθα ὅτι μέλλουσι ποιεῖν.

ΣΩ. Πάνυ γε. ἀλλὰ τίνα εἵλοντο;

ΜΕ. Οὐδένα, ἀλλὰ ἀνεβάλοντο εἰς τὴν αὔριον. οἶμαι μέντοι Ἀρχῖνον ἢ Δίωνα αἱρεθήσεσθαι.

MENEXENUS

CHARACTERS
SOCRATES, MENEXENUS

soc. From the agora, Menexenus, or where from?

MEN. From the agora, Socrates, and the Council Chamber.

soc. And what was it took you specially to the Council Chamber? But of course it was because you deem yourself to be at the end of your education and philosophic studies, and being sufficiently versed, as you think, in these, you are minded to turn to graver matters; and you at your age, my marvellous youth, are attempting to govern us older men, lest your house should ever fail in providing us with a succession of managers.

MEN. Certainly if you, Socrates, allow and counsel me to govern, I shall do so gladly; but otherwise not. This time, however, I went to the Council Chamber because I had learnt that the Council was going to select someone to make an oration over the dead; for you know that they propose to arrange for funeral rites.

soc. Yes, I do. And whom did they select?

MEN. Nobody: they postponed it till to-morrow. I fancy, however, that Archinus will be selected, or Dion.

PLATO

2. ΣΩ. Καὶ μήν, ὦ Μενέξενε, πολλαχῇ κινδυνεύει καλὸν εἶναι τὸ ἐν πολέμῳ ἀποθνήσκειν. καὶ γὰρ ταφῆς καλῆς τε καὶ μεγαλοπρεποῦς τυγχάνει, καὶ ἐὰν πένης τις ὢν τελευτήσῃ, καὶ ἐπαίνου αὖ ἔτυχε, καὶ ἐὰν φαῦλος ᾖ, ὑπ᾽ ἀνδρῶν σοφῶν τε καὶ οὐκ εἰκῇ ἐπαινούντων, ἀλλὰ ἐκ πολλοῦ χρόνου λόγους παρεσκευασμένων, οἳ οὕτω καλῶς ἐπαινοῦσιν ὥστε καὶ τὰ προσόντα καὶ τὰ μὴ περὶ ἑκάστου λέγοντες, κάλλιστά πως τοῖς ὀνόμασι ποικίλλοντες, γοητεύουσιν ἡμῶν τὰς ψυχάς, καὶ τὴν πόλιν ἐγκωμιάζοντες κατὰ πάντας τρόπους καὶ τοὺς τετελευτηκότας ἐν τῷ πολέμῳ καὶ τοὺς προγόνους ἡμῶν ἅπαντας τοὺς ἔμπροσθεν καὶ αὐτοὺς ἡμᾶς τοὺς ἔτι ζῶντας ἐπαινοῦντες, ὥστ᾽ ἔγωγε, ὦ Μενέξενε, γενναίως πάνυ διατίθεμαι ἐπαινούμενος ὑπ᾽ αὐτῶν, καὶ ἑκάστοτε ἕστηκα ἀκροώμενος καὶ κηλούμενος, ἡγούμενος ἐν τῷ παραχρῆμα μείζων καὶ γενναιότερος καὶ καλλίων γεγονέναι. καὶ οἷα δὴ τὰ πολλὰ ἀεὶ μετ᾽ ἐμοῦ ξένοι τινὲς ἕπονται καὶ ξυνακροῶνται, πρὸς οὓς ἐγὼ σεμνότερος ἐν τῷ παραχρῆμα γίγνομαι· καὶ γὰρ ἐκεῖνοι ταὐτὰ ταῦτα δοκοῦσί μοι πάσχειν καὶ πρὸς ἐμὲ καὶ πρὸς τὴν ἄλλην πόλιν, θαυμασιωτέραν αὐτὴν ἡγεῖσθαι εἶναι ἢ πρότερον, ὑπὸ τοῦ λέγοντος ἀναπειθόμενοι. καί μοι αὕτη ἡ σεμνότης παραμένει ἡμέρας πλείω ἢ τρεῖς· οὕτως ἔναυλος ὁ λόγος τε καὶ ὁ φθόγγος παρὰ τοῦ λέγοντος ἐνδύεται εἰς τὰ ὦτα, ὥστε μόγις τετάρτῃ ἢ πέμπτῃ ἡμέρᾳ ἀναμιμνήσκομαι ἐμαυτοῦ καὶ αἰσθάνομαι οὗ γῆς εἰμί, τέως δὲ οἶμαι μόνον οὐκ ἐν μακάρων νήσοις οἰκεῖν· οὕτως ἡμῖν οἱ ῥήτορες δεξιοί εἰσιν.

3. ΜΕ. Ἀεὶ σὺ προσπαίζεις, ὦ Σώκρατες, τοὺς

MENEXENUS

soc. In truth, Menexenus, to fall in battle seems to be a splendid thing in many ways. For a man obtains a splendid and magnificent funeral even though at his death he be but a poor man; and though he be but a worthless fellow, he wins praise, and that by the mouth of accomplished men who do not praise at random, but in speeches prepared long beforehand. And they praise in such splendid fashion, that, what with their ascribing to each one both what he has and what he has not, and the variety and splendour of their diction, they bewitch our souls; and they eulogize the State in every possible fashion, and they praise those who died in the war and all our ancestors of former times and ourselves who are living still; so that I myself, Menexenus, when thus praised by them feel mightily ennobled, and every time I listen fascinated I am exalted and imagine myself to have become all at once taller and nobler and more handsome. And as I am generally accompanied by some strangers, who listen along with me, I become in their eyes also all at once more majestic; for they also manifestly share in my feelings with regard both to me and to the rest of our City, believing it to be more marvellous than before, owing to the persuasive eloquence of the speaker. And this majestic feeling remains with me for over three days: so persistently does the speech and voice of the orator ring in my ears that it is scarcely on the fourth or fifth day that I recover myself and remember that I really am here on earth, whereas till then I almost imagined myself to be living in the Islands of the Blessed,—so expert are our orators.

MEN. You are always deriding the orators, Socrates.

235

ῥήτορας. νῦν μέντοι οἶμαι ἐγὼ τὸν αἱρεθέντα οὐ πάνυ εὐπορήσειν· ἐξ ὑπογυίου γὰρ παντάπασιν ἡ αἵρεσις γέγονεν, ὥστε ἴσως ἀναγκασθήσεται ὁ λέγων ὥσπερ αὐτοσχεδιάζειν.

D ΣΩ. Πόθεν, ὦ 'γαθέ; εἰσὶν ἑκάστοις τούτων λόγοι παρεσκευασμένοι, καὶ ἅμα οὐδὲ αὐτοσχεδιάζειν τά γε τοιαῦτα χαλεπόν. εἰ μὲν γὰρ δέοι Ἀθηναίους ἐν Πελοποννησίοις εὖ λέγειν ἢ Πελοποννησίους ἐν Ἀθηναίοις, ἀγαθοῦ ἂν ῥήτορος δέοι τοῦ πείσοντος καὶ εὐδοκιμήσοντος· ὅταν δέ τις ἐν τούτοις ἀγωνίζηται, οὕσπερ καὶ ἐπαινεῖ, οὐδὲν μέγα δοκεῖν εὖ λέγειν.

ΜΕ. Οὐκ οἴει, ὦ Σώκρατες;

ΣΩ. Οὐ μέντοι μὰ Δία.

E ΜΕ. Ἦ οἴει οἷός τ' ἂν εἶναι αὐτὸς εἰπεῖν, εἰ δέοι καὶ ἕλοιτό σε ἡ βουλή;

ΣΩ. Καὶ ἐμοὶ μέν γε, ὦ Μενέξενε, οὐδὲν θαυμαστὸν οἵῳ τ' εἶναι εἰπεῖν, ᾧ τυγχάνει διδάσκαλος οὖσα οὐ πάνυ φαύλη περὶ ῥητορικῆς, ἀλλ' ἥπερ καὶ ἄλλους πολλοὺς καὶ ἀγαθοὺς πεποίηκε ῥήτορας, ἕνα δὲ καὶ διαφέροντα τῶν Ἑλλήνων, Περικλέα τὸν Ξανθίππου.

ΜΕ. Τίς αὕτη; ἢ δῆλον ὅτι Ἀσπασίαν λέγεις;

ΣΩ. Λέγω γάρ, καὶ Κόννον γε τὸν Μητροβίου·
236 οὗτοι γάρ μοι δύο εἰσὶ διδάσκαλοι, ὁ μὲν μουσικῆς, ἡ δὲ ῥητορικῆς. οὕτω μὲν οὖν τρεφόμενον ἄνδρα οὐδὲν θαυμαστὸν δεινὸν εἶναι λέγειν. ἀλλὰ καὶ ὅστις ἐμοῦ κάκιον ἐπαιδεύθη, μουσικὴν μὲν ὑπὸ Λάμπρου παιδευθείς, ῥητορικὴν δὲ ὑπ' Ἀντι-

---

[1] Aspasia, of Miletus, famous as the mistress of the Athenian statesman Pericles (*circa* 430 B.C.).

MENEXENUS

And truly I think that this time the selected speaker will not be too well prepared; for the selection is being made without warning, so that the speaker will probably be driven to improvise his speech.

soc. Why so, my good sir? Each one of these men has speeches ready made; and what is more, it is in no wise difficult to improvise such things. For if it were a question of eulogizing Athenians before an audience of Peloponnesians, or Peloponnesians before Athenians, there would indeed be need of a good orator to win credence and credit; but when a man makes his effort in the presence of the very men whom he is praising, it is no difficult matter to win credit as a fine speaker.

men. You think not, Socrates?

soc. Yes, by Zeus, I certainly do.

men. And do you think that you yourself would be able to make the speech, if required and if the Council were to select you?

soc. That I should be able to make the speech would be nothing wonderful, Menexenus; for she who is my instructor is by no means weak in the art of rhetoric; on the contrary, she has turned out many fine orators, and amongst them one who surpassed all other Greeks, Pericles, the son of Xanthippus.

men. Who is she? But you mean Aspasia,[1] no doubt.

soc. I do; and also Connus the son of Metrobius; for these are my two instructors, the one in music, the other in rhetoric. So it is not surprising that a man who is trained like me should be clever at speaking. But even a man less well taught than I, who had learnt his music from Lamprus and his rhetoric

φῶντος τοῦ Ῥαμνουσίου, ὅμως κἂν οὗτος οἷός τ' εἴη Ἀθηναίους γε ἐν Ἀθηναίοις ἐπαινῶν εὐδοκιμεῖν.

4. ΜΕ. Καὶ τί ἂν ἔχοις εἰπεῖν, εἰ δέοι σε λέγειν;

ΣΩ. Αὐτὸς μὲν παρ' ἐμαυτοῦ ἴσως οὐδέν, Ἀσπασίας δὲ καὶ χθὲς ἠκροώμην περαινούσης ἐπιτάφιον λόγον περὶ αὐτῶν τούτων. ἤκουσε γὰρ ἅπερ σὺ λέγεις, ὅτι μέλλοιεν Ἀθηναῖοι αἱρεῖσθαι τὸν ἐροῦντα· ἔπειτα τὰ μὲν ἐκ τοῦ παραχρῆμά μοι διῄει, οἷα δέοι λέγειν, τὰ δὲ πρότερον ἐσκεμμένη, ὅτε μοι δοκεῖ συνετίθει τὸν ἐπιτάφιον λόγον, ὃν Περικλῆς εἶπε, περιλείμματ' ἄττα ἐξ ἐκείνου συγκολλῶσα.

ΜΕ. Ἦ καὶ μνημονεύσαις ἂν ἃ ἔλεγεν Ἀσπασία;

ΣΩ. Εἰ μὴ ἀδικῶ γε· ἐμάνθανόν γέ τοι παρ' αὐτῆς, καὶ ὀλίγου πληγὰς ἔλαβον ὅτι ἐπελανθανόμην.

ΜΕ. Τί οὖν οὐ διῆλθες;

ΣΩ. Ἀλλ' ὅπως μή μοι χαλεπανεῖ ἡ διδάσκαλος, ἂν ἐξενέγκω αὐτῆς τὸν λόγον.

ΜΕ. Μηδαμῶς, ὦ Σώκρατες, ἀλλ' εἰπέ, καὶ πάνυ μοι χαριεῖ, εἴτε Ἀσπασίας βούλει λέγειν εἴτε ὁτουοῦν· ἀλλὰ μόνον εἰπέ.

ΣΩ. Ἀλλ' ἴσως μου καταγελάσει, ἄν σοι δόξω πρεσβύτης ὢν ἔτι παίζειν.

ΜΕ. Οὐδαμῶς, ὦ Σώκρατες, ἀλλ' εἰπὲ παντὶ τρόπῳ.

5. ΣΩ. Ἀλλὰ μέντοι σοί γε δεῖ χαρίζεσθαι, ᾧ

MENEXENUS

from Antiphon the Rhamnusian,[1]—even such a one, I say, could none the less win credit by praising Athenians before an Athenian audience.

MEN. What, then, would you have to say, if you were required to speak?

SOC. Nothing, perhaps, myself of my own invention; but I was listening only yesterday to Aspasia going through a funeral speech for these very people. For she had heard the report you mention, that the Athenians are going to select the speaker; and thereupon she rehearsed to me the speech in the form it should take, extemporizing in part, while other parts of it she had previously prepared, as I imagine, at the time when she was composing the funeral oration which Pericles delivered; and from this she patched together sundry fragments.

MEN. Could you repeat from memory that speech of Aspasia?

SOC. Yes, if I am not mistaken; for I learnt it, to be sure, from her as she went along, and I nearly got a flogging whenever I forgot.

MEN. Why don't you repeat it then?

SOC. But possibly my teacher will be vexed with me if I publish abroad her speech.

MEN. Never fear, Socrates; only tell it and you will gratify me exceedingly, whether it is Aspasia's that you wish to deliver or anyone else's; only say on.

SOC. But you will probably laugh me to scorn if I, at my age, seem to you to be playing like a child.

MEN. Not at all, Socrates; but by all means say on.

SOC. Nay, then, I must surely gratify you; for

---

[1] Antiphon, born in 480 B.C., was the first of the ten great Attic Orators.

D γε¹ κἂν ὀλίγου, εἴ με κελεύοις ἀποδύντα ὀρχήσασθαι, χαρισαίμην ἄν, ἐπειδή γε μόνω ἐσμέν. ἀλλ' ἄκουε. ἔλεγε γάρ, ὡς ἐγῷμαι, ἀρξαμένη λέγειν ἀπ' αὐτῶν τῶν τεθνεώτων οὑτωσί.

Ἔργῳ μὲν ἡμῖν οἵδε ἔχουσι τὰ προσήκοντα σφίσιν αὐτοῖς, ὧν τυχόντες πορεύονται τὴν εἱμαρμένην πορείαν, προπεμφθέντες κοινῇ μὲν ὑπὸ τῆς πόλεως, ἰδίᾳ δὲ ὑπὸ τῶν οἰκείων· λόγῳ δὲ δὴ τὸν λειπόμενον κόσμον ὅ τε νόμος προστάττει ἀπο-
E δοῦναι τοῖς ἀνδράσι καὶ χρή· ἔργων γὰρ εὖ πραχθέντων λόγῳ καλῶς ῥηθέντι μνήμη καὶ κόσμος τοῖς πράξασι γίγνεται παρὰ τῶν ἀκουσάντων. δεῖ δὴ τοιούτου τινὸς λόγου, ὅστις τοὺς μὲν τετελευτηκότας ἱκανῶς ἐπαινέσεται, τοῖς δὲ ζῶσιν εὐμενῶς παραινέσεται, ἐκγόνοις μὲν καὶ ἀδελφοῖς μιμεῖσθαι τὴν τῶνδε ἀρετὴν παρακελευόμενος, πατέρας δὲ καὶ μητέρας καὶ εἴ τινες τῶν ἄνωθεν ἔτι προγόνων λείπονται, τούτους δὲ παραμυθού-
237 μενος. τίς οὖν ἂν ἡμῖν τοιοῦτος λόγος φανείη; ἢ πόθεν ἂν ὀρθῶς ἀρξαίμεθα ἄνδρας ἀγαθοὺς ἐπαινοῦντες, οἳ ζῶντές τε τοὺς ἑαυτῶν εὔφραινον δι' ἀρετήν, καὶ τὴν τελευτὴν ἀντὶ τῆς τῶν ζώντων σωτηρίας ἠλλάξαντο; δοκεῖ μοι χρῆναι, κατὰ φύσιν ὥσπερ ἀγαθοὶ ἐγένοντο, οὕτω καὶ ἐπαινεῖν αὐτούς. ἀγαθοὶ δ' ἐγένοντο διὰ τὸ φῦναι ἐξ ἀγαθῶν. τὴν εὐγένειαν οὖν πρῶτον αὐτῶν ἐγκωμιάζωμεν, δεύτερον δὲ τροφήν τε καὶ παιδείαν·
B ἐπὶ δὲ τούτοις τὴν τῶν ἔργων πρᾶξιν ἐπιδείξωμεν, ὡς καλὴν καὶ ἀξίαν τούτων ἀπεφήναντο.

6. Τῆς δ' εὐγενείας πρῶτον ὑπῆρξε τοῖσδε ἡ τῶν

¹ ᾧ γε conj. Stallbaum : ὥστε MSS., Zur.

MENEXENUS

indeed I would almost gratify you if you were to bid me strip and dance, now that we two are alone. Listen then. In her speech, I believe, she began by making mention of the dead men themselves in this wise:

In respect of deeds, these men have received at our hands what is due unto them, endowed wherewith they travel their predestined road; for they have been escorted forth in solemn procession publicly by the City and privately by their kinsfolk. But in respect of words, the honour that remains still due to these heroes the law enjoins us, and it is right, to pay in full. For it is by means of speech finely spoken that deeds nobly done gain for their doers from the hearers the meed of memory and renown. And the speech required is one which will adequately eulogize the dead and give kindly exhortation to the living, appealing to their children and their brethren to copy the virtues of these heroes, and to their fathers and mothers and any still surviving ancestors offering consolation. Where then could we discover a speech like that? Or how could we rightly commence our laudation of these valiant men, who in their lifetime delighted their friends by their virtue, and purchased the safety of the living by their deaths? We ought, in my judgement, to adopt the natural order in our praise, even as the men themselves were natural in their virtue. And virtuous they were because they were sprung from men of virtue. Firstly, then, let us eulogize their nobility of birth, and secondly their nurture and training: thereafter we shall exhibit the character of their exploits, how nobly and worthily they wrought them.

Now as regards nobility of birth, their first claim

PLATO

προγόνων γένεσις οὐκ ἔπηλυς οὖσα, οὐδὲ τοὺς ἐκγόνους τούτους ἀποφηναμένη μετοικοῦντας ἐν τῇ χώρᾳ ἄλλοθεν σφῶν ἡκόντων, ἀλλ' αὐτόχθονας καὶ τῷ ὄντι ἐν πατρίδι οἰκοῦντας καὶ ζῶντας, καὶ τρεφομένους οὐχ ὑπὸ μητρυιᾶς ὡς ἄλλοι, ἀλλ' ὑπὸ μητρὸς τῆς χώρας ἐν ᾗ ᾤκουν, καὶ νῦν κεῖσθαι τελευτήσαντας ἐν οἰκείοις τόποις τῆς τεκούσης καὶ θρεψάσης καὶ ὑποδεξαμένης. δικαιότατον δὴ κοσμῆσαι πρῶτον τὴν μητέρα αὐτήν· οὕτω γὰρ συμβαίνει ἅμα καὶ ἡ τῶνδε εὐγένεια κοσμουμένη.

7. Ἔστι δὲ ἀξία ἡ χώρα καὶ ὑπὸ πάντων ἀνθρώπων ἐπαινεῖσθαι, οὐ μόνον ὑφ' ἡμῶν, πολλαχῇ μὲν καὶ ἄλλῃ, πρῶτον δὲ καὶ μέγιστον ὅτι τυγχάνει οὖσα θεοφιλής. μαρτυρεῖ δὲ ἡμῶν τῷ λόγῳ ἡ τῶν ἀμφισβητησάντων περὶ αὐτὴν θεῶν ἔρις τε καὶ κρίσις.[1] ἣν δὴ θεοὶ ἐπῄνεσαν, πῶς οὐχ ὑπ' ἀνθρώπων γε ξυμπάντων δικαία ἐπαινεῖσθαι; δεύτερος δὲ ἔπαινος δικαίως ἂν αὐτῆς εἴη, ὅτι ἐν ἐκείνῳ τῷ χρόνῳ, ἐν ᾧ ἡ πᾶσα γῆ ἀνεδίδου καὶ ἔφυε ζῶα παντοδαπά, θηρία τε καὶ βοτά, ἐν τούτῳ ἡ ἡμετέρα θηρίων μὲν ἀγρίων ἄγονος καὶ καθαρὰ ἐφάνη, ἐξελέξατο δὲ τῶν ζῴων καὶ ἐγέννησεν ἄνθρωπον, ὃ συνέσει τε ὑπερέχει τῶν ἄλλων καὶ δίκην καὶ θεοὺς μόνον νομίζει. μέγα δὲ τεκμήριον τούτῳ τῷ λόγῳ, ὅτι ἥδε ἔτεκεν ἡ γῆ τοὺς τῶνδέ τε καὶ ἡμετέρους προγόνους· πᾶν γὰρ τὸ τεκὸν τροφὴν ἔχει ἐπιτηδείαν ᾧ ἂν τέκῃ· ᾧ καὶ γυνὴ δήλη τεκοῦσά τε ἀληθῶς καὶ μή, [ἀλλ' ὑποβαλλομένη],[1] ἐὰν μὴ ἔχῃ πηγὰς τροφῆς τῷ γεννωμένῳ. ὃ δὴ

---

[1] ἀλλ' ὑποβαλλομένη bracketed by Hartman.

[1] Athena and Poseidon, see Ovid, *Metam.* vi. 70 ff.

MENEXENUS

thereto is this—that the forefathers of these men were not of immigrant stock, nor were these their sons declared by their origin to be strangers in the land sprung from immigrants, but natives sprung from the soil living and dwelling in their own true fatherland; and nurtured also by no stepmother, like other folk, but by that mother-country wherein they dwelt, which bare them and reared them and now at their death receives them again to rest in their own abodes. Most meet it is that first we should celebrate that Mother herself; for by so doing we shall also celebrate therewith the noble birth of these heroes.

Our country is deserving of praise, not only from us but from all men, on many grounds, but first and foremost because she is god-beloved. The strife of the gods [1] who contended over her and their judgement testify to the truth of our statement. And how should not she whom the gods praised deserve to be praised by all mankind? And a second just ground of praise would be this,—that during that period in which the whole earth was putting forth and producing animals of every kind, wild and tame, our country showed herself barren and void of wild animals, but chose for herself and gave birth to man, who surpasses all other animals in intelligence and alone of animals regards justice and the gods. And we have a signal proof of this statement in that this land of ours has given birth to the forefathers both of these men and of ourselves. For every creature that brings forth possesses a suitable supply of nourishment for its offspring; and by this test it is manifest also whether a woman be truly a mother or no, if she possesses no founts of nourishment for her child.

237

καὶ ἡ ἡμετέρα γῆ τε καὶ μήτηρ ἱκανὸν τεκμήριον παρέχεται ὡς ἀνθρώπους γεννησαμένη· μόνη γὰρ ἐν τῷ τότε καὶ πρώτη τροφὴν ἀνθρωπείαν ἤνεγκε

238 τὸν τῶν πυρῶν καὶ κριθῶν καρπόν, ᾧ κάλλιστα καὶ ἄριστα τρέφεται τὸ ἀνθρώπειον γένος, ὡς τῷ ὄντι τοῦτο τὸ ζῷον αὐτὴ γεννησαμένη. μᾶλλον δὲ ὑπὲρ γῆς ἢ γυναικὸς προσήκει δέχεσθαι τοιαῦτα τεκμήρια· οὐ γὰρ γῆ γυναῖκα μεμίμηται κυήσει καὶ γεννήσει, ἀλλὰ γυνὴ γῆν. τούτου δὲ τοῦ καρποῦ οὐκ ἐφθόνησεν, ἀλλ' ἔνειμε καὶ τοῖς ἄλλοις. μετὰ δὲ τοῦτο ἐλαίου γένεσιν, πόνων ἀρωγήν, ἀνῆκε τοῖς ἐκγόνοις. θρεψαμένη δὲ καὶ αὐξήσασα

B πρὸς ἥβην ἄρχοντας καὶ διδασκάλους αὐτῶν θεοὺς ἐπηγάγετο· ὧν τὰ μὲν ὀνόματα πρέπει ἐν τῷ τοιῷδε ἐᾶν· ἴσμεν γάρ· οἳ τὸν βίον ἡμῶν κατεσκεύασαν πρός τε τὴν καθ' ἡμέραν δίαιταν τέχνας πρώτους παιδευσάμενοι καὶ πρὸς τὴν ὑπὲρ τῆς χώρας φυλακὴν ὅπλων κτῆσίν τε καὶ χρῆσιν διδαξάμενοι.

8. Γεννηθέντες δὲ καὶ παιδευθέντες οὕτως οἱ τῶνδε πρόγονοι ᾤκουν πολιτείαν κατασκευασά-

C μενοι ἧς ὀρθῶς ἔχει διὰ βραχέων ἐπιμνησθῆναι. πολιτεία γὰρ τροφὴ ἀνθρώπων ἐστί, καλὴ μὲν ἀγαθῶν, ἡ δὲ ἐναντία κακῶν. ὡς οὖν ἐν καλῇ πολιτείᾳ ἐτράφησαν οἱ πρόσθεν ἡμῶν ἀναγκαῖον δηλῶσαι, δι' ἣν δὴ κἀκεῖνοι ἀγαθοὶ καὶ οἱ νῦν εἰσίν, ὧν οἵδε τυγχάνουσιν ὄντες οἱ τετελευτηκότες. ἡ

MENEXENUS

Now our land, which is also our mother, furnishes to the full this proof of her having brought forth men; for, of all the lands that then existed, she was the first and the only one to produce human nourishment, namely the grain of wheat and barley, whereby the race of mankind is most richly and well nourished, inasmuch as she herself was the true mother of this creature. And proofs such as this one ought to accept more readily on behalf of a country than on behalf of a woman; for it is not the country that imitates the woman in the matter of conception and birth, but the woman the country. But this her produce of grain she did not begrudge to the rest of men, but dispensed it to them also. And after it she brought to birth for her children the olive, sore labour's balm. And when she had nurtured and reared them up to man's estate, she introduced gods to be their governors and tutors; the names of whom it behoves us to pass over in this discourse, since we know them; and they set in order our mode of life, not only in respect of daily business, by instructing us before all others in the arts, but also in respect of the guardianship of our country, by teaching us how to acquire and handle arms.

Such being the manner of their birth and of their education, the ancestors of these men framed for themselves and lived under a civic polity which it is right for us briefly to describe. For a polity is a thing which nurtures men, good men when it is noble, bad men when it is base. It is necessary, then, to demonstrate that the polity wherein our forefathers were nurtured was a noble one, such as caused goodness not only in them but also in their descendants of the present age, amongst whom we number these

PLATO

γὰρ αὐτὴ πολιτεία καὶ τότε ἦν καὶ νῦν, [ἀριστοκρατία],[1] ἐν ᾗ νῦν τε πολιτευόμεθα καὶ τὸν ἀεὶ χρόνον ἐξ ἐκείνου ὡς τὰ πολλά. καλεῖ δὲ ὁ μὲν αὐτὴν δημοκρατίαν, ὁ δὲ ἄλλο, ᾧ ἂν χαίρῃ· ἔστι D δὲ τῇ ἀληθείᾳ μετ' εὐδοξίας πλήθους ἀριστοκρατία. βασιλεῖς μὲν γὰρ ἀεὶ ἡμῖν εἰσίν· οὗτοι δὲ τοτὲ μὲν ἐκ γένους, τοτὲ δὲ αἱρετοί· ἐγκρατὲς δὲ τῆς πόλεως τὰ πολλὰ τὸ πλῆθος, τὰς δὲ ἀρχὰς δίδωσι καὶ κράτος τοῖς ἀεὶ δόξασιν ἀρίστοις εἶναι, καὶ οὔτε ἀσθενείᾳ οὔτε πενίᾳ οὔτ' ἀγνωσίᾳ πατέρων ἀπελήλαται οὐδεὶς οὐδὲ τοῖς ἐναντίοις τετίμηται, ὥσπερ ἐν ἄλλαις πόλεσιν, ἀλλὰ εἷς ὅρος, ὁ δόξας σοφὸς ἢ ἀγαθὸς εἶναι κρατεῖ καὶ ἄρχει. αἰτία δὲ ἡμῖν E τῆς πολιτείας ταύτης ἡ ἐξ ἴσου γένεσις. αἱ μὲν γὰρ ἄλλαι πόλεις ἐκ παντοδαπῶν κατεσκευασμέναι ἀνθρώπων εἰσὶ καὶ ἀνωμάλων, ὥστε αὐτῶν ἀνώμαλοι καὶ αἱ πολιτεῖαι, τυραννίδες τε καὶ ὀλιγαρχίαι· οἰκοῦσιν οὖν ἔνιοι μὲν δούλους, οἱ δὲ δεσπότας ἀλλήλους νομίζοντες· ἡμεῖς δὲ καὶ οἱ 239 ἡμέτεροι, μιᾶς μητρὸς πάντες ἀδελφοὶ φύντες, οὐκ ἀξιοῦμεν δοῦλοι οὐδὲ δεσπόται ἀλλήλων εἶναι, ἀλλ' ἡ ἰσογονία ἡμᾶς ἡ κατὰ φύσιν ἰσονομίαν ἀναγκάζει ζητεῖν κατὰ νόμον, καὶ μηδενὶ ἄλλῳ ὑπείκειν ἀλλήλοις ἢ ἀρετῆς δόξῃ καὶ φρονήσεως.

9. Ὅθεν δὴ ἐν πάσῃ ἐλευθερίᾳ τεθραμμένοι οἱ τῶνδέ τε πατέρες καὶ ⟨οἱ⟩[2] ἡμέτεροι καὶ αὐτοὶ

---

[1] ἀριστοκρατία bracketed by H. Richards.
[2] οἱ added by one MS.

---

[1] *i.e.* "rule of the best."

MENEXENUS

men who are fallen. For it is the same polity which existed then and exists now, under which polity we are living now and have been living ever since that age with hardly a break. One man calls it " democracy," another man, according to his fancy, gives it some other name; but it is, in very truth, an " aristocracy "[1] backed by popular approbation. Kings we always have[2]; but these are at one time hereditary, at another selected by vote. And while the most part of civic affairs are in the control of the populace, they hand over the posts of government and the power to those who from time to time are deemed to be the best men; and no man is debarred by his weakness or poverty or by the obscurity of his parentage, or promoted because of the opposite qualities, as is the case in other States. On the contrary, the one principle of selection is this: the man that is deemed to be wise or good rules and governs. And the cause of this our polity lies in our equality of birth. For whereas all other States are composed of a heterogeneous collection of all sorts of people, so that their polities also are heterogeneous, tyrannies as well as oligarchies, some of them regarding one another as slaves, others as masters; we and our people, on the contrary, being all born of one mother, claim to be neither the slaves of one another nor the masters; rather does our natural birth-equality drive us to seek lawfully legal equality, and to yield to one another in no respect save in reputation for virtue and understanding.

Wherefore the forefathers of these men and of us, and these men themselves, having been reared up thus

[2] A reference to the " Archons," one of whom was called *Basileus* (" King ").

PLATO

οὗτοι, καὶ καλῶς φύντες, πολλὰ δὴ καὶ καλὰ ἔργα ἀπεφήναντο εἰς πάντας ἀνθρώπους καὶ ἰδίᾳ καὶ δημοσίᾳ, οἰόμενοι δεῖν ὑπὲρ τῆς ἐλευθερίας καὶ Ἕλλησιν ὑπὲρ Ἑλλήνων μάχεσθαι καὶ βαρβάροις ὑπὲρ ἁπάντων τῶν Ἑλλήνων. Εὐμόλπου μὲν οὖν καὶ Ἀμαζόνων ἐπιστρατευσάντων ἐπὶ τὴν χώραν καὶ τῶν ἔτι προτέρων ὡς ἠμύναντο, καὶ ὡς ἤμυναν Ἀργείοις πρὸς Καδμείους καὶ Ἡρακλείδαις πρὸς Ἀργείους, ὅ τε χρόνος βραχὺς ἀξίως διηγήσασθαι, ποιηταί τε αὐτῶν ἤδη ἱκανῶς τὴν ἀρετὴν ἐν μουσικῇ ὑμνήσαντες εἰς πάντας μεμηνύκασιν· ἐὰν οὖν ἡμεῖς ἐπιχειρῶμεν τὰ αὐτὰ λόγῳ ψιλῷ κοσμεῖν, τάχ' ἂν δεύτεροι φαινοίμεθα. ταῦτα μὲν οὖν διὰ ταῦτα δοκεῖ μοι ἐᾶν, ἐπειδὴ καὶ ἔχει τὴν ἀξίαν· ὧν δὲ οὔτε ποιητής πω δόξαν ἀξίαν ἐπ' ἀξίοις λαβὼν ἔχει ἔτι τ' ἐστὶν ἐν ἀμνηστίᾳ, τούτων πέρι μοι δοκεῖ χρῆναι ἐπιμνησθῆναι ἐπαινοῦντά τε καὶ προμνώμενον ἄλλοις ἐς ᾠδάς τε καὶ τὴν ἄλλην ποίησιν αὐτὰ θεῖναι πρεπόντως τῶν πραξάντων. ἔστι δὲ τούτων ὧν λέγω πρῶτα.

Πέρσας ἡγουμένους τῆς Ἀσίας καὶ δουλουμένους τὴν Εὐρώπην ἔσχον οἱ τῆσδε τῆς χώρας ἔκγονοι, γονεῖς δὲ ἡμέτεροι, ὧν καὶ δίκαιον καὶ χρὴ πρῶτον μεμνημένους ἐπαινέσαι αὐτῶν τὴν ἀρετήν. δεῖ δὴ αὐτὴν ἰδεῖν, εἰ μέλλει τις καλῶς ἐπαινεῖν, ἐν

---

[1] Eumolpus, a Thracian bard and chieftain, son of Poseidon, said to have aided the Eleusinians in invading Attica.

[2] The Amazons, a race of female warriors in Pontus, said to have attacked Athens and been driven back to Asia by the hero Theseus.

[3] *i.e.* in the war of " the Seven against Thebes " (of which city Cadmus was the founder).

MENEXENUS

in complete freedom, and being nobly born, achieved before all men many noble deeds both individual and national, deeming it their duty to fight in the cause of freedom alike with Greeks on behalf of Greeks and with barbarians on behalf of the whole of Greece. The story of how they repulsed Eumolpus [1] and the Amazons,[2] and still earlier invaders, when they marched upon our country, and how they defended the Argives against the Cadmeians [3] and the Heracleidae against the Argives,[4] is a story which our time is too short to relate as it deserves, and already their valour has been adequately celebrated in song by poets who have made it known throughout the world; consequently, if we should attempt to magnify the same achievements in plain prose, we should probably find ourselves outmatched. These exploits, therefore, for these reasons I judge that we should pass over, seeing also that they have their due meed of praise; but those exploits for which as yet no poet has received worthy renown for worthy cause, and which lie still buried in oblivion, I ought, as I think, to celebrate, not only praising them myself but providing material also for others to build up into odes and other forms of poetry in a manner worthy of the doers of those deeds. And of the deeds whereof I speak the first were these:

The Persians were in command of Asia, and were enslaving Europe, when they came in contact with the children of this land, our own parents, of whom it is right and proper that we should make mention first and celebrate their valour. But if we are to celebrate it fitly, in order to visualize it we must place

[4] The Athenians aided " the sons of Heracles " against Eurystheus, King of Tiryns in Argolis.

PLATO

239 ἐκείνῳ τῷ χρόνῳ γενόμενον λόγῳ, ὅτε πᾶσα μὲν ἡ Ἀσία ἐδούλευε τρίτῳ ἤδη βασιλεῖ, ὧν ὁ μὲν πρῶτος Κῦρος ἐλευθερώσας Πέρσας τοὺς αὑτοῦ πολίτας τῷ αὑτοῦ φρονήματι ἅμα καὶ τοὺς δε-
E σπότας Μήδους ἐδουλώσατο καὶ τῆς ἄλλης Ἀσίας μέχρι Αἰγύπτου ἦρξεν, ὁ δὲ υἱὸς αὐτοῦ Αἰγύπτου τε καὶ Λιβύης ὅσον οἷόν τε ἦν ἐπιβαίνειν, τρίτος δὲ Δαρεῖος πεζῇ μὲν μέχρι Σκυθῶν τὴν ἀρχὴν ὡρίσατο, ναυσὶ δὲ τῆς τε θαλάττης ἐκράτει καὶ
240 τῶν νήσων, ὥστε μηδὲ ἀξιοῦν ἀντίπαλον αὑτῷ μηδένα εἶναι· αἱ δὲ γνῶμαι δεδουλωμέναι ἁπάντων ἀνθρώπων ἦσαν· οὕτω πολλὰ καὶ μεγάλα καὶ μάχιμα γένη καταδεδουλωμένη ἦν ἡ Περσῶν ἀρχή.

10. Αἰτιασάμενος δὲ Δαρεῖος ἡμᾶς τε καὶ Ἐρετριέας Σάρδεσιν ἐπιβουλεῦσαι [προφασιζόμενος],[1] πέμψας μυριάδας μὲν πεντήκοντα ἔν τε πλοίοις καὶ ναυσί, ναῦς δὲ τριακοσίας, Δᾶτιν δὲ
B ἄρχοντα, εἶπεν ἥκειν ἄγοντα Ἐρετριέας καὶ Ἀθηναίους, εἰ βούλοιτο τὴν ἑαυτοῦ κεφαλὴν ἔχειν· ὁ δὲ πλεύσας εἰς Ἐρέτριαν ἐπ᾽ ἄνδρας οἳ τῶν τότε Ἑλλήνων ἐν τοῖς εὐδοκιμώτατοι[2] ἦσαν τὰ πρὸς τὸν πόλεμον καὶ οὐκ ὀλίγοι, τούτους ἐχειρώσατο μὲν ἐν τρισὶν ἡμέραις, διηρευνήσατο δὲ αὐτῶν πᾶσαν τὴν χώραν, ἵνα μηδεὶς ἀποφύγοι, τοιούτῳ τρόπῳ· ἐπὶ τὰ ὅρια ἐλθόντες τῆς Ἐρετρικῆς οἱ στρατιῶται αὐτοῦ, ἐκ θαλάττης εἰς θάλατταν
C διαστάντες, συνάψαντες τὰς χεῖρας διῆλθον ἅπασαν

---

[1] προφασιζόμενος bracketed by Cobet.
[2] εὐδοκιμώτατοι Hirschig: εὐδοκιμωτάτοις mss., Zur.

MENEXENUS

ourselves, in thought, at that epoch when the whole of Asia was already in bondage to the third of the Persian kings. Cyrus,[1] the first of these kings, had by his own spirited action set free his fellow-countrymen, the Persians, and not only enslaved the Medes, their masters, but also gained command of the rest of Asia, as far as to Egypt. His son[2] ruled over Egypt and as much of Libya as he could traverse; while the third king, Darius, extended his empire by land as far as to the Scythians, and by his navy controlled the sea and the islands, so that none so much as thought of disputing his sway. Thus the minds of all men were enslaved; so many were the mighty and warlike nations which had fallen under the yoke of the Persian Empire.

Then Darius, accusing us and the Eretrians of having plotted against Sardis, dispatched fifty myriads of men in transports and warships, together with three hundred ships of war, and Datis as their commander; and him the king ordered to bring back the Eretrians and Athenians in captivity, if he wished to keep his own head. He then sailed to Eretria against men who were amongst the most famous warriors in Greece at that time, and by no means few in number; them he overpowered within three days, and lest any should escape he made a thorough search of the whole of their country; and his method was this. His soldiers marched to the limits of Eretria and posted themselves at intervals from sea to sea; then they joined hands and passed through the whole of the country,

---

[1] Cyrus overthrew the Medes in 559, and reigned till 529 B.C.

[2] Cambyses, son of Cyrus, 529–522 B.C.

PLATO

τὴν χώραν, ἵν᾽ ἔχοιεν τῷ βασιλεῖ εἰπεῖν, ὅτι οὐδεὶς σφᾶς ἀποπεφευγὼς εἴη. τῇ δ᾽ αὐτῇ διανοίᾳ κατηγάγοντο ἐξ Ἐρετρίας εἰς Μαραθῶνα, ὡς ἕτοιμόν σφισιν ὂν καὶ Ἀθηναίους ἐν τῇ αὐτῇ ταύτῃ ἀνάγκῃ ζεύξαντας Ἐρετριεῦσιν ἄγειν.

Τούτων δὲ τῶν μὲν πραχθέντων, τῶν δ᾽ ἐπιχειρουμένων οὔτ᾽ Ἐρετριεῦσιν ἐβοήθησεν Ἑλλήνων οὐδεὶς οὔτε Ἀθηναίοις πλὴν Λακεδαιμονίων· οὗτοι δὲ τῇ ὑστεραίᾳ τῆς μάχης ἀφίκοντο· οἱ δ᾽ ἄλλοι πάντες ἐκπεπληγμένοι, ἀγαπῶντες τὴν ἐν τῷ παρόντι σωτηρίαν, ἡσυχίαν ἦγον. ἐν τούτῳ δὴ ἄν τις γενόμενος γνοίη οἷοι ἄρα ἐτύγχανον ὄντες τὴν ἀρετὴν οἱ Μαραθῶνι δεξάμενοι τὴν τῶν βαρβάρων δύναμιν καὶ κολασάμενοι τὴν ὑπερηφανίαν ὅλης τῆς Ἀσίας καὶ πρῶτοι στήσαντες τρόπαια τῶν βαρβάρων, ἡγεμόνες καὶ διδάσκαλοι τοῖς ἄλλοις γενόμενοι ὅτι οὐκ ἄμαχος εἴη ἡ Περσῶν δύναμις, ἀλλὰ πᾶν πλῆθος καὶ πᾶς πλοῦτος ἀρετῇ ὑπείκει. ἐγὼ μὲν οὖν ἐκείνους τοὺς ἄνδρας φημὶ οὐ μόνον τῶν σωμάτων τῶν ἡμετέρων πατέρας εἶναι, ἀλλὰ καὶ τῆς ἐλευθερίας τῆς τε ἡμετέρας καὶ ξυμπάντων τῶν ἐν τῇδε τῇ ἠπείρῳ· εἰς ἐκεῖνο γὰρ τὸ ἔργον ἀποβλέψαντες καὶ τὰς ὑστέρας μάχας ἐτόλμησαν διακινδυνεύειν οἱ Ἕλληνες ὑπὲρ τῆς σωτηρίας, μαθηταὶ τῶν Μαραθῶνι γενόμενοι.

11. Τὰ μὲν οὖν ἀριστεῖα τῷ λόγῳ ἐκείνοις ἀναθετέον, τὰ δὲ δευτερεῖα τοῖς περὶ Σαλαμῖνα καὶ ἐπ᾽ Ἀρτεμισίῳ ναυμαχήσασι καὶ νικήσασι. καὶ γὰρ τούτων τῶν ἀνδρῶν πολλὰ μὲν ἄν τις ἔχοι

---

[1] *Cf.* Hdt. v. 99 ff.; *Laws* iii. 698 c ff. The expedition of Datis took place in 490 B.C.

MENEXENUS

in order that they might be able to report to the king that not a man had escaped out of their hands.[1] With the same design they sailed off from Eretria to Marathon, supposing that they would have an easy task in leading the Athenians captive under the same yoke of bondage as the Eretrians.

And while these actions were being accomplished in part, and in part attempted, not one of the Greeks lent aid to the Eretrians nor yet to the Athenians, save only the Lacedaemonians (and they arrived on the day after the battle); all the rest were terror-stricken, and, hugging their present security, made no move. It is by realizing this position of affairs that we can appreciate what manner of men those were, in point of valour, who awaited the onset of the barbarians' power, chastised all Asia's insolent pride, and were the first to rear trophies of victory over the barbarians; whereby they pointed the way to the others and taught them to know that the Persian power was not invincible, since there is no multitude of men or money but courage conquers it. I, therefore, affirm that those men were the begetters not merely of our bodies but of our freedom also, and the freedom of all the dwellers in this continent; for it was the example of that exploit of theirs which fired the Greeks with courage to risk the later battles in the cause of salvation, learning their lesson from the men of Marathon.

To them, therefore, we award in this our speech the first prize for valour, and the second to those who fought and won the sea-fights off Salamis and at Artemisium.[2] And truly concerning these men also

[2] These battles took place during Xerxes' invasion of Greece in 480 B.C.

διελθεῖν, καὶ οἷα ἐπιόντα ὑπέμειναν κατά τε γῆν καὶ κατὰ θάλατταν, καὶ ὡς ἠμύναντο ταῦτα· ὃ δέ μοι δοκεῖ καὶ ἐκείνων κάλλιστον εἶναι, τούτου μνησθήσομαι, ὅτι τὸ ἑξῆς ἔργον τοῖς Μαραθῶνι διεπράξαντο. οἱ μὲν γὰρ Μαραθῶνι τοσοῦτον μόνον ἐπέδειξαν τοῖς Ἕλλησιν, ὅτι κατὰ γῆν οἷόν B τε ἀμύνεσθαι τοὺς βαρβάρους ὀλίγοις πολλούς, ναυσὶ δὲ ἔτι ἦν ἄδηλον καὶ δόξαν εἶχον Πέρσαι ἄμαχοι εἶναι κατὰ θάλατταν καὶ πλήθει καὶ πλούτῳ καὶ τέχνῃ καὶ ῥώμῃ. τοῦτο δὴ ἄξιον ἐπαινεῖν τῶν ἀνδρῶν τῶν τότε ναυμαχησάντων, ὅτι τὸν ἐχόμενον φόβον διέλυσαν τῶν Ἑλλήνων καὶ ἔπαυσαν φοβουμένους πλῆθος νεῶν τε καὶ ἀνδρῶν. ὑπ᾽ ἀμφοτέρων δὴ ξυμβαίνει, τῶν τε Μαραθῶνι μαχεσαμένων καὶ τῶν ἐν Σαλαμῖνι C ναυμαχησάντων, παιδευθῆναι τοὺς ἄλλους Ἕλληνας, ὑπὸ μὲν τῶν κατὰ γῆν, ὑπὸ δὲ τῶν κατὰ θάλατταν μαθόντας καὶ ἐθισθέντας μὴ φοβεῖσθαι τοὺς βαρβάρους.

12. Τρίτον δὲ λέγω τὸ ἐν Πλαταιαῖς ἔργον καὶ ἀριθμῷ καὶ ἀρετῇ γενέσθαι τῆς Ἑλληνικῆς σωτηρίας, κοινὸν ἤδη τοῦτο Λακεδαιμονίων τε καὶ Ἀθηναίων.

Τὸ μὲν οὖν μέγιστον καὶ χαλεπώτατον οὗτοι πάντες ἤμυναν, καὶ διὰ ταύτην τὴν ἀρετὴν D νῦν τε ὑφ᾽ ἡμῶν ἐγκωμιάζονται καὶ εἰς τὸν ἔπειτα χρόνον ὑπὸ τῶν ὕστερον. μετὰ δὲ τοῦτο πολλαὶ μὲν πόλεις τῶν Ἑλλήνων ἔτι ἦσαν μετὰ τοῦ βαρβάρου, αὐτὸς δὲ ἠγγέλλετο βασιλεὺς διανοεῖσθαι ὡς ἐπιχειρήσων πάλιν ἐπὶ τοὺς Ἕλ-

---

[1] At Plataea the Persians under Mardonius were defeated in 479 B.C.

MENEXENUS

one might have much to relate, regarding the manner of onsets they endured both by land and sea, and how they repelled them; but the achievement I shall mention is that which was, in my judgement, the noblest that they performed, in that it followed up the achievement of the men of Marathon. For whereas the men of Marathon had only proved to the Greeks thus much,—that it was possible to repel the barbarians by land though few against many, yet the prospect in a sea-fight remained still doubtful, and the Persians still retained the reputation of being invincible by sea, in virtue of their numbers and their wealth, their naval skill and strength. For this, then, the men who fought those sea-fights merit our praise, that they delivered the Greeks from the second of their fears, and put an end to the terrors inspired by multitudes of ships and men. So it came about, by the action of both—the soldiers who fought at Marathon and the sailors who fought at Salamis—, that the rest of the Greeks were trained and accustomed to have no fear of the barbarians, neither by land, as our soldiers taught them, nor yet, as our sailors taught them, by sea.

The exploit at Plataea[1] I put third both in order and in merit of those which secured the salvation of Greece; and in this exploit, at last, the Lacedaemonians co-operated with the Athenians.

By the action of all these men the greatest and most formidable danger was warded off, and because of this their valour we pronounce their eulogy now, as our successors will in the time to come. But, in the period that followed, many cities of the Greeks were still in league with the barbarian, and of the king himself it was reported that he was purposing

λήνας. δίκαιον δὴ καὶ τούτων ἡμᾶς ἐπιμνησθῆναι, οἳ τοῖς τῶν προτέρων ἔργοις τέλος τῆς σωτηρίας ἐπέθεσαν ἀνακαθηράμενοι καὶ ἐξελάσαντες πᾶν τὸ βάρβαρον ἐκ τῆς θαλάττης. ἦσαν δὲ οὗτοι οἵ τε ἐπ' Εὐρυμέδοντι ναυμαχήσαντες καὶ οἱ εἰς Κύπρον στρατεύσαντες καὶ οἱ εἰς Αἴγυπτον πλεύσαντες καὶ ἄλλοσε πολλαχόσε, ὧν χρὴ μεμνῆσθαι καὶ χάριν αὐτοῖς εἰδέναι, ὅτι βασιλέα ἐποίησαν δείσαντα τῇ ἑαυτοῦ σωτηρίᾳ τὸν νοῦν προσέχειν, ἀλλὰ μὴ τῇ τῶν Ἑλλήνων ἐπιβουλεύειν φθορᾷ.

13. Καὶ οὗτος μὲν δὴ πάσῃ τῇ πόλει διηντλήθη ὁ πόλεμος ὑπὲρ ἑαυτῶν τε καὶ τῶν ἄλλων ὁμοφώνων πρὸς τοὺς βαρβάρους· εἰρήνης δὲ γενομένης καὶ τῆς πόλεως τιμωμένης ἦλθεν ἐπ' αὐτήν, ὃ δὴ φιλεῖ ἐκ τῶν ἀνθρώπων τοῖς εὖ πράττουσι προσπίπτειν, πρῶτον μὲν ζῆλος, ἀπὸ ζήλου δὲ φθόνος. ὃ καὶ τήνδε τὴν πόλιν ἄκουσαν ἐν πολέμῳ τοῖς Ἕλλησι κατέστησε. μετὰ δὲ τοῦτο γενομένου πολέμου, συνέβαλον μὲν ἐν Τανάγρᾳ ὑπὲρ τῆς Βοιωτῶν ἐλευθερίας Λακεδαιμονίοις μαχόμενοι, ἀμφισβητησίμου δὲ τῆς μάχης γενομένης, διέκρινε τὸ ὕστερον ἔργον· οἱ μὲν γὰρ ᾤχοντο ἀπιόντες, καταλείποντες [Βοιωτοὺς]¹ οἷς ἐβοήθουν, οἱ δ' ἡμέτεροι τρίτῃ ἡμέρᾳ ἐν Οἰνοφύτοις νικήσαντες τοὺς ἀδίκως φεύγοντας δικαίως κατήγαγον. οὗτοι δὴ πρῶτοι μετὰ τὸν Περσικὸν πόλεμον, Ἕλλησιν

¹ Βοιωτοὺς bracketed by Bekker.

---

[1] The Athenians, under Cimon, defeated the Persian forces, both by land and sea, at the river Eurymedon, in Pamphylia, in 468 (*cf.* Thucyd. i. 100).

[2] These naval operations (against Persia) took place about 461–458 B.C.   [3] B.C. 457.

MENEXENUS

to renew his attempt against the Greeks. Wherefore it is right that we should make mention also of those men who put the finishing touch to the work of salvation executed by their predecessors by sweeping away the whole of the barbarian power and driving it clean off the seas. These were the men who fought the sea-fight at the Eurymedon,[1] the men who served in the expedition against Cyprus, the men who voyaged to Egypt and to many another quarter,[2] —men whom we ought to hold in memory and render them thanks, seeing that they put the king in fear and caused him to give his whole mind to his own safety in place of plotting the destruction of Greece.

Now this war was endured to the end by all our citizens who warred against the barbarians in defence of all the other Greek-speaking peoples as well as themselves. But when peace was secured and our city was held in honour, there followed the usual consequence which the successful suffer at the hands of men; for it was assailed by jealousy first, and after jealousy by envy; and thereby our city was plunged against its will into war with the Greeks. Thereupon, when war had broken out, they encountered the Lacedaemonians at Tanagra[3] while fighting in defence of the liberties of the Boeotians; and though the battle itself was indecisive, it was decided by the subsequent result. For whereas the enemy retired and made off, deserting those whom they had come to assist, our men won a victory after a two days' battle at Oenophyta,[4] and rightfully restored those who were wrongfully exiled. These were the first of our men who, after the Persian

[4] Oenophyta was fought two months after Tanagra, Thucyd. i. 108.

ἤδη ὑπὲρ τῆς ἐλευθερίας βοηθοῦντες πρὸς Ἕλληνας, ἄνδρες ἀγαθοὶ γενόμενοι καὶ ἐλευθερώσαντες οἷς ἐβοήθουν, ἐν τῷδε τῷ μνήματι τιμηθέντες ὑπὸ τῆς πόλεως πρῶτοι ἐτέθησαν.

Μετὰ δὲ ταῦτα πολλοῦ πολέμου γενομένου, καὶ πάντων τῶν Ἑλλήνων ἐπιστρατευσάντων καὶ τεμόντων τὴν χώραν καὶ ἀναξίαν χάριν ἐκτινόντων τῇ πόλει, νικήσαντες αὐτοὺς ναυμαχίᾳ οἱ ἡμέτεροι καὶ λαβόντες αὐτῶν τοὺς ἡγεμόνας Λακεδαιμονίους ἐν τῇ Σφαγίᾳ, ἐξὸν αὐτοῖς διαφθεῖραι ἐφείσαντο καὶ ἀπέδοσαν καὶ εἰρήνην ἐποιήσαντο, ἡγούμενοι πρὸς μὲν τὸ ὁμόφυλον μέχρι νίκης δεῖν πολεμεῖν, καὶ μὴ δι' ὀργὴν ἰδίαν πόλεως τὸ κοινὸν τῶν Ἑλλήνων διολλύναι, πρὸς δὲ τοὺς βαρβάρους μέχρι διαφθορᾶς. τούτους δὴ ἄξιον ἐπαινέσαι τοὺς ἄνδρας, οἳ τοῦτον τὸν πόλεμον πολεμήσαντες ἐνθάδε κεῖνται, ὅτι ἐπέδειξαν, εἴ τις ἄρα ἠμφεσβήτει ὡς ἐν τῷ προτέρῳ πολέμῳ τῷ πρὸς τοὺς βαρβάρους ἄλλοι τινὲς εἶεν ἀμείνους Ἀθηναίων, ὅτι οὐκ ἀληθῆ ἀμφισβητοῖεν. οὗτοι γὰρ ἐνταῦθα ἔδειξαν, στασιασάσης τῆς Ἑλλάδος περιγενόμενοι τῷ πολέμῳ, τοὺς προεστῶτας τῶν ἄλλων Ἑλλήνων χειρωσάμενοι, μεθ' ὧν τότε τοὺς βαρβάρους ἐνίκων κοινῇ, τούτους νικῶντες ἰδίᾳ.

14. Τρίτος δὲ πόλεμος μετὰ ταύτην τὴν εἰρήνην ἀνέλπιστός τε καὶ δεινὸς ἐγένετο, ἐν ᾧ πολλοὶ καὶ ἀγαθοὶ τελευτήσαντες ἐνθάδε κεῖνται, πολλοὶ μὲν ἀμφὶ Σικελίαν πλεῖστα τρόπαια στήσαντες ὑπὲρ τῆς Λεοντίνων ἐλευθερίας, οἷς βοηθοῦντες διὰ τοὺς

---

[1] *i.e.* Sphacteria. These events took place in 425 B.C., the seventh year of the Peloponnesian War.

MENEXENUS

war and now helping Greeks against Greeks in the cause of freedom, proved themselves men of valour and delivered those whom they were aiding; and they were the first to be honoured by the State and laid to rest in this tomb.

Later on, when there was widespread war, and all the Greeks had marched against us and ravaged our country, most evilly requiting our city, and our men had defeated them by sea and had captured their Lacedaemonian leaders in Sphagia,[1] although they had it in their power to destroy them, yet they spared their lives and gave them back and made peace, since they deemed that against their fellow-Greeks it was right to wage war only up to the point of victory, and not to wreck the whole Greek community for the sake of a city's private grudge, but to wage war to the death against the barbarians. It is meet, indeed, that we should praise these men who were warriors in this war and now lie here, inasmuch as they demonstrated that if any contended that in the former war, against the barbarians, others were superior to the Athenians, their contention was false. This they now proved by their triumph in the war when the Greeks were at feud, and by their conquest of those who were the leaders of the rest of Greece, when, alone by themselves, they defeated that city by whose allied aid they had formerly defeated the barbarians.

This peace was followed by a third war, as formidable as it was unexpected, wherein many brave men lost their lives and now lie here. Many of these reared up numerous trophies of victory in Sicily,[2] fighting for the freedom of Leontini, to succour which

---

[2] The second Sicilian expedition took place in 413 B.C.

ὅρκους ἔπλευσαν εἰς ἐκείνους τοὺς τόπους, διὰ δὲ
μῆκος τοῦ πλοῦ εἰς ἀπορίαν τῆς πόλεως κατα-
στάσης καὶ οὐ δυναμένης αὐτοῖς ὑπηρετεῖν, τούτῳ
ἀπειπόντες ἐδυστύχησαν· ὧν οἱ ἐχθροὶ καὶ προσ-
πολεμήσαντες πλείω ἔπαινον ἔχουσι σωφροσύνης
καὶ ἀρετῆς ἢ τῶν ἄλλων οἱ φίλοι· πολλοὶ δ' ἐν
ταῖς ναυμαχίαις ταῖς καθ' Ἑλλήσποντον, μιᾷ μὲν
ἡμέρᾳ πάσας τὰς τῶν πολεμίων ἑλόντες ναῦς,
B πολλὰς δὲ καὶ ἄλλας νικήσαντες· ὃ δ' εἶπον δεινὸν
καὶ ἀνέλπιστον τοῦ πολέμου γενέσθαι, τόδε λέγω
τὸ εἰς τοσοῦτον φιλονεικίας ἐλθεῖν πρὸς τὴν πόλιν
τοὺς ἄλλους Ἕλληνας, ὥστε τολμῆσαι τῷ ἐχθίστῳ
ἐπικηρυκεύσασθαι βασιλεῖ, ὃν κοινῇ ἐξέβαλον μεθ'
ἡμῶν, ἰδίᾳ τοῦτον πάλιν ἐπάγεσθαι, βάρβαρον ἐφ'
Ἕλληνας, καὶ ξυναθροῖσαι ἐπὶ τὴν πόλιν πάντας
C Ἕλληνάς τε καὶ βαρβάρους. οὗ δὴ καὶ ἐκφανὴς
ἐγένετο ἡ τῆς πόλεως ῥώμη τε καὶ ἀρετή. οἰομένων γὰρ ἤδη αὐτὴν καταπεπολεμῆσθαι καὶ
ἀπειλημμένων ἐν Μυτιλήνῃ τῶν νεῶν, βοηθήσαντες
ἑξήκοντα ναυσίν, αὐτοὶ ἐμβάντες εἰς τὰς ναῦς, καὶ
ἄνδρες γενόμενοι ὁμολογουμένως ἄριστοι, νικήσαντες μὲν τοὺς πολεμίους, λυσάμενοι δὲ τοὺς
φιλίους, ἀναξίου τύχης τυχόντες, οὐκ ἀναιρεθέντες
ἐκ τῆς θαλάττης, κεῖνται ἐνθάδε. ὧν χρὴ ἀεὶ
D μεμνῆσθαί τε καὶ ἐπαινεῖν· τῇ μὲν γὰρ ἐκείνων

---

[1] This is an exaggeration if the occasion is that mentioned in Thucyd. viii. 9 ff., when ten empty ships were captured. But possibly the reference is to the victory at Cyzicus, B.C. 410, when sixty ships were taken or sunk.

[2] This refers to the Spartan treaty with Tissaphernes, B.C. 412, and the subsequent co-operation of the Persians against Athens.

MENEXENUS

city, and to honour their pledges, they sailed to those regions; but inasmuch as our city was in a helpless position and unable to reinforce them owing to the length of the voyage, fortune was against them and they renounced their design; yet for their prudence and their valour they have received more praise from their foes of the opposite army than the rest of men from their friends. Many others of them fought in the sea-fights in the Hellespont, where in one single day they captured all the enemy's ships,[1] besides winning many other engagements. But what I have termed the formidable and unexpected character of the war lay in this, that the rest of the Greeks had arrived at such a pitch of jealousy towards this city that they even brought themselves to solicit privately the aid of their deadliest foe, the very king whom they had publicly expelled with our assistance, inviting a barbarian as their ally against Greeks; and dared to range against our city the united forces of all the Greeks and barbarians.[2] And then it was that the strength and valour of our State shone out conspicuously. For when men fancied that she was already reduced by war, with her ships cut off at Mytilene, her citizens sent sixty ships to the rescue, manning the ships themselves and proving themselves indisputably to be men of valour by conquering their foes and setting free their friends;[3] albeit they met with undeserved misfortune, and were not recovered from the sea to find their burial here.[4] And for these reasons it behoves us to have them in remembrance and to praise them always; for it was owing to their

---

[3] The battle of Mytilene was fought in 407 B.C.
[4] At the battle of Arginusae, 406 B.C., twenty-five ships' crews were lost.

ἀρετῇ ἐνικήσαμεν οὐ μόνον τὴν τότε ναυμαχίαν ἀλλὰ καὶ τὸν ἄλλον πόλεμον· δόξαν γὰρ δι' αὐτοὺς ἡ πόλις ἔσχε μή ποτ' ἂν καταπολεμηθῆναι μηδ' ὑπὸ πάντων ἀνθρώπων· καὶ ἀληθῆ ἔδοξε. τῇ δὲ ἡμετέρᾳ αὐτῶν διαφορᾷ ἐκρατήθημεν, οὐχ ὑπὸ τῶν ἄλλων· ἀήττητοι γὰρ ἔτι καὶ νῦν ὑπό γε ἐκείνων ἐσμέν, ἡμεῖς δὲ αὐτοὶ ἡμᾶς αὐτοὺς καὶ ἐνικήσαμεν καὶ ἡττήθημεν.

Μετὰ δὲ ταῦτα ἡσυχίας γενομένης καὶ εἰρήνης πρὸς τοὺς ἄλλους, ὁ οἰκεῖος ἡμῖν πόλεμος οὕτως ἐπολεμήθη, ὥστε εἴπερ εἱμαρμένον εἴη ἀνθρώποις στασιάσαι, μὴ ἂν ἄλλως εὔξασθαι μηδένα πόλιν ἑαυτοῦ νοσῆσαι. ἔκ τε γὰρ τοῦ Πειραιέως καὶ τοῦ ἄστεος ὡς ἀσμένως καὶ οἰκείως ἀλλήλοις συνέμιξαν οἱ πολῖται καὶ παρ' ἐλπίδα τοῖς ἄλλοις Ἕλλησι, τόν τε πρὸς τοὺς Ἐλευσῖνι πόλεμον ὡς μετρίως ἔθεντο· καὶ τούτων ἁπάντων οὐδὲν ἄλλ' αἴτιον ἢ ἡ τῷ ὄντι ξυγγένεια, φιλίαν βέβαιον καὶ ὁμόφυλον οὐ λόγῳ ἀλλ' ἔργῳ παρεχομένη. χρὴ δὲ καὶ τῶν ἐν τούτῳ τῷ πολέμῳ τελευτησάντων ὑπ' ἀλλήλων μνείαν ἔχειν καὶ διαλλάττειν αὐτοὺς ᾧ δυνάμεθα, εὐχαῖς καὶ θυσίαις, ἐν τοῖς τοιοῖσδε, τοῖς κρατοῦσιν αὐτῶν εὐχομένους, ἐπειδὴ καὶ ἡμεῖς διηλλάγμεθα. οὐ γὰρ κακίᾳ ἀλλήλων ἥψαντο οὐδ' ἔχθρᾳ ἀλλὰ δυστυχίᾳ. μάρτυρες δὲ ἡμεῖς αὐτοί ἐσμεν τούτων οἱ ζῶντες· οἱ αὐτοὶ γὰρ ὄντες ἐκείνοις γένει συγ-

---

[1] *i.e.* the oligarchical party at Athens who held swa   r about eighteen months (404–403 B.C.) till ousted by the democrats under Thrasybulus.

MENEXENUS

valour that we were conquerors not only in the sea-fight on that day but in all the rest of the war; and it was due to them that men formed the conviction regarding our city (and it was a true conviction) that she could never be warred down, not even by all the world. And in truth it was by our own dissensions that we were brought down and not by the hands of other men; for by them we are still to this day undefeated, and it is we ourselves who have both defeated and been defeated by ourselves.

After these happenings, when we were at peace and amity with other States, our civil war at home was waged in such a way that—if men are fated to engage in civil strife—there is no man but would pray for his own State that its sickness might resemble ours. So kindly and so friendly was the way in which the citizens from the Peiraeus and from the city consorted with one another, and also—beyond men's hopes—with the other Greeks; and such moderation did they show in their settlement of the war against the men at Eleusis.[1] And the cause of all these actions was nothing else than that genuine kinship which produces, not in word only but in deed, a firm friendship founded on community of race. And of those who fell in this war also it is meet to make mention and to reconcile them by such means as we can under present conditions,—by prayer, that is, and by sacrifice,—praying for them to those that have them in their keeping, seeing that we ourselves also have been reconciled. For it was not through wickedness that they set upon one another, nor yet through hatred, but through misfortune. And to this we ourselves, who now live, can testify; for we who are of the same stock as they grant forgiveness to one

γνώμην ἀλλήλοις ἔχομεν ὧν τ' ἐποιήσαμεν ὧν τ' ἐπάθομεν.

15. Μετὰ δὲ τοῦτο παντελοῦς εἰρήνης ἡμῖν γενομένης, ἡσυχίαν ἦγεν ἡ πόλις, τοῖς μὲν βαρβάροις συγγιγνώσκουσα, ὅτι παθόντες ὑπ' αὐτῆς κακῶς [ἱκανῶς]¹ οὐκ ἐνδεῶς ἠμύναντο, τοῖς δὲ Ἕλλησιν ἀγανακτοῦσα, μεμνημένη ὡς εὖ παθόντες ὑπ' αὐτῆς οἵαν χάριν ἀπέδοσαν, κοινωσάμενοι τοῖς βαρβάροις, τάς τε ναῦς περιελόμενοι αἵ ποτ' ἐκείνους ἔσωσαν, καὶ τείχη καθελόντες ἀνθ' ὧν ἡμεῖς τἀκείνων ἐκωλύσαμεν πεσεῖν. διανοουμένη δὲ ἡ πόλις μὴ ἂν ἔτι ἀμῦναι μήτε Ἕλλησι πρὸς ἀλλήλων δουλουμένοις μήτε ὑπὸ βαρβάρων, οὕτως ᾤκει. ἡμῶν οὖν ἐν τοιαύτῃ διανοίᾳ ὄντων ἡγησάμενοι Λακεδαιμόνιοι τοὺς μὲν τῆς ἐλευθερίας ἐπικούρους πεπτωκέναι ἡμᾶς, σφέτερον δὲ ἤδη ἔργον εἶναι καταδουλοῦσθαι τοὺς ἄλλους, ταῦτ' ἔπραττον.

16. Καὶ μηκύνειν μὲν τί δεῖ; οὐ γὰρ πάλαι οὐδὲ παλαιῶν² ἀνθρώπων γεγονότα λέγοιμ' ἂν τὰ μετὰ ταῦτα· αὐτοὶ γὰρ ἴσμεν, ὡς ἐκπεπληγμένοι ἀφίκοντο εἰς χρείαν τῆς πόλεως τῶν τε Ἑλλήνων οἱ πρῶτοι, Ἀργεῖοι καὶ Βοιωτοὶ καὶ Κορίνθιοι, καὶ τό γε θειότατον πάντων, τὸ καὶ βασιλέα εἰς τοῦτο ἀπορίας ἀφικέσθαι ὥστε περιστῆναι αὐτῷ μηδαμόθεν ἄλλοθεν τὴν σωτηρίαν γενέσθαι ἀλλ' ἢ ἐκ ταύτης τῆς πόλεως, ἣν προθύμως ἀπώλλυ. καὶ δὴ καὶ εἴ τις βούλοιτο τῆς πόλεως κατηγορῆσαι δικαίως, τοῦτ' ἂν μόνον λέγων ὀρθῶς ἂν κατηγοροῖ,

---

¹ ἱκανῶς bracketed by Bekker.
² παλαιῶν one ms.: πολλῶν most mss., Zur.

MENEXENUS

another both for what we have done and what we have suffered.

After this, when peace was completely re-established, the city remained quiet, granting forgiveness to the barbarians for the vigorous defence they had offered when she had done them injury, but feeling aggrieved with the Greeks at the thought of the return they had made for the benefits she had done them, in that they joined themselves to the barbarians, and stripped her of those ships which had once been the means of their own salvation, and demolished her walls as a recompense for our saving their walls from ruin.[1] Our city, therefore, resolved that never again would she succour Greeks when in danger of enslavement either by one another or at the hands of barbarians; and in this mind she abode. Such then being our policy, the Lacedaemonians supposed that we, the champions of liberty, were laid low, and that it was now open to them to enslave the rest, and this they proceeded to do.

But why should I prolong the story? For what followed next is no tale of ancient history about men of long ago. Nay, we ourselves know how the Argives, the Boeotians and the Corinthians—the leading States of Greece—came to need our city, being stricken with terror, and how even the Persian king himself—most marvellous fact of all—was reduced to such a state of distress that eventually he could hope for salvation from no other quarter save this city of ours which he had been so eager to destroy. And in truth, if one desired to frame a just accusation against the city, the only true accusation one could

[1] These formed part of the terms exacted by the Spartans after the battle of Aegospotami, B.C. 405.

PLATO

ὡς ἀεὶ λίαν φιλοικτίρμων ἐστὶ καὶ τοῦ ἥττονος θεραπίς. καὶ δὴ καὶ ἐν τῷ τότε χρόνῳ οὐχ οἷά τε ἐγένετο καρτερῆσαι οὐδὲ διαφυλάξαι ἃ ἐδέδοκτο αὐτῇ, τὸ μηδενὶ δουλουμένῳ βοηθεῖν τῶν σφᾶς ἀδικησάντων, ἀλλὰ ἐκάμφθη καὶ ἐβοήθησε, καὶ τοὺς μὲν Ἕλληνας αὐτὴ βοηθήσασα ἀπελύσατο δουλείας, ὥστ' ἐλευθέρους εἶναι μέχρι οὗ πάλιν αὐτοὶ αὑτοὺς κατεδουλώσαντο, βασιλεῖ δὲ αὐτὴ μὲν οὐκ ἐτόλμησε βοηθῆσαι, αἰσχυνομένη τὰ τρόπαια τά τε Μαραθῶνι καὶ Σαλαμῖνι καὶ Πλαταιαῖς, φυγάδας δὲ καὶ ἐθελοντὰς ἐάσασα μόνον βοηθῆσαι ὁμολογουμένως ἔσωσε. τειχισαμένη δὲ καὶ ναυπηγησαμένη, ἐκδεξαμένη τὸν πόλεμον, ἐπειδὴ ἠναγκάσθη πολεμεῖν, ὑπὲρ Παρίων ἐπολέμει Λακεδαιμονίοις.

17. Φοβηθεὶς δὲ βασιλεὺς τὴν πόλιν, ἐπειδὴ ἑώρα Λακεδαιμονίους τῷ κατὰ θάλατταν πολέμῳ ἀπαγορεύοντας, ἀποστῆναι βουλόμενος ἐξῄτει τοὺς Ἕλληνας τοὺς ἐν τῇ ἠπείρῳ, οὕσπερ πρότερον Λακεδαιμόνιοι αὐτῷ ἐξέδοσαν, εἰ μέλλοι συμμαχήσειν ἡμῖν τε καὶ τοῖς ἄλλοις συμμάχοις, ἡγούμενος οὐκ ἐθελήσειν, ἵν' αὐτῷ πρόφασις εἴη τῆς ἀποστάσεως. καὶ τῶν μὲν ἄλλων ξυμμάχων ἐψεύσθη—ἠθέλησαν γὰρ αὐτῷ ἐκδιδόναι καὶ ξυνέθεντο καὶ ὤμοσαν Κορίνθιοι καὶ Ἀργεῖοι καὶ Βοιωτοὶ καὶ ἄλλοι σύμμαχοι, εἰ μέλλοι χρήματα παρέξειν, ἐκδώσειν τοὺς ἐν τῇ ἠπείρῳ Ἕλληνας—,

---

[1] *e.g.* the Athenian Conon became a Persian admiral and operated against the Spartans, 395–390 B.C.

[2] *i.e.* the Ionian Greeks, whom the Spartans offered to hand over to the Persians in 392 B.C.

MENEXENUS

bring would be this,—that she has always been compassionate to excess and the handmaid of the weak. And in fact, on that occasion, she proved unable to harden her heart and adhere firmly to her resolved policy of refusing to assist any in danger of enslavement against those who wronged them; on the contrary she gave way and lent assistance. The Greeks she aided herself and rescued them from slavery, so that they remained free until such time as they enslaved each other once more; but to the King she could not bring herself to lend official aid for fear of disgracing the trophies of Marathon, Salamis and Plataea, but she permitted exiles only and volunteers to assist him, and thereby, beyond a doubt, she saved him.[1] Having, then, restored her walls and rebuilt her navy, she entered upon the war, since war was forced upon her, and in defence of the Parians warred against the Lacedaemonians.

But the King, becoming afraid of our city when he saw that the Lacedaemonians were desisting from the naval struggle, wished to desert us; so he demanded the surrender of the Greeks in the Continent, whom the Lacedaemonians had formerly given over to him, as the price of his continuing his alliance with us and the other allies, thinking that we would refuse and thus furnish him with a pretext for his desertion. Now in the case of the rest of his allies he was mistaken; for they all—including the Corinthians, Argives, Boeotians, and the rest—consented to hand them over to him, making a sworn agreement that if he would supply them with money they would hand over the Greeks in the Continent [2]; but we, and we alone, could not bring

245 μόνοι δὲ ἡμεῖς οὐκ ἐτολμήσαμεν οὔτε ἐκδοῦναι οὔτε ὀμόσαι. οὕτω δή τοι τό γε τῆς πόλεως γενναῖον καὶ ἐλεύθερον βέβαιόν τε καὶ ὑγιές ἐστι καὶ
D φύσει μισοβάρβαρον, διὰ τὸ εἰλικρινῶς εἶναι Ἕλληνες καὶ ἀμιγεῖς βαρβάρων. οὐ γὰρ Πέλοπες οὐδὲ Κάδμοι οὐδὲ Αἴγυπτοί τε καὶ Δαναοὶ οὐδὲ ἄλλοι πολλοὶ φύσει μὲν βάρβαροι ὄντες, νόμῳ δὲ Ἕλληνες, συνοικοῦσιν ἡμῖν, ἀλλ' αὐτοὶ Ἕλληνες, οὐ μιξοβάρβαροι οἰκοῦμεν, ὅθεν καθαρὸν τὸ μῖσος ἐντέτηκε τῇ πόλει τῆς ἀλλοτρίας φύσεως. ὅμως δ' οὖν ἐμονώθημεν πάλιν διὰ τὸ μὴ ἐθέλειν αἰσχρὸν καὶ ἀνόσιον ἔργον ἐργάσασθαι Ἕλληνας βαρβάροις
E ἐκδόντες. ἐλθόντες οὖν εἰς ταὐτὰ ἐξ ὧν καὶ τὸ πρότερον κατεπολεμήθημεν, σὺν θεῷ ἄμεινον ἢ τότε ἐθέμεθα τὸν πόλεμον· καὶ γὰρ ναῦς καὶ τείχη ἔχοντες καὶ τὰς ἡμετέρας αὐτῶν ἀποικίας ἀπηλλάγημεν τοῦ πολέμου· οὕτως ἀγαπητῶς ἀπηλλάττοντο καὶ οἱ πολέμιοι. ἀνδρῶν μέντοι ἀγαθῶν καὶ ἐν τούτῳ τῷ πολέμῳ ἐστερήθημεν, τῶν τε ἐν Κορίνθῳ χρησαμένων δυσχωρίᾳ καὶ ἐν Λεχαίῳ
246 προδοσίᾳ. ἀγαθοὶ δὲ καὶ οἱ βασιλέα ἐλευθερώσαντες καὶ ἐκβαλόντες ἐκ τῆς θαλάττης Λακεδαιμονίους. ὧν ἐγὼ μὲν ὑμᾶς ἀναμιμνήσκω, ὑμᾶς δὲ πρέπει ξυνεπαινεῖν τε καὶ κοσμεῖν τοιούτους ἄνδρας.

18. Καὶ τὰ μὲν δὴ ἔργα ταῦτα τῶν ἀνδρῶν τῶν ἐνθάδε κειμένων καὶ τῶν ἄλλων, ὅσοι ὑπὲρ τῆς πόλεως τετελευτήκασι, πολλὰ μὲν τὰ εἰρημένα καὶ καλά, πολὺ δ' ἔτι πλείω καὶ καλλίω τὰ ὑπολειπό-

---

[1] This refers to "the King's Peace" (or Peace of Antalcidas) of 387-386 B.C.

MENEXENUS

ourselves either to hand them over or to join in the agreement. So firmly-rooted and so sound is the noble and liberal character of our city, and endowed also with such a hatred of the barbarian, because we are pure-blooded Greeks, unadulterated by barbarian stock. For there cohabit with us none of the type of Pelops, or Cadmus, or Aegyptus or Danaus, and numerous others of the kind, who are naturally barbarians though nominally Greeks; but our people are pure Greeks and not a barbarian blend; whence it comes that our city is imbued with a whole-hearted hatred of alien races. None the less, we were isolated once again because of our refusal to perform the dishonourable and unholy act of surrendering Greeks to barbarians. And thus we found ourselves in the same position which had previously led to our military overthrow; but, by the help of God, we brought the war to a more favourable conclusion [1] than on that occasion. For we still retained our ships, our walls, and our own colonies, when we ceased from the war, —so welcome to our enemies also was its cessation. Yet truly in this war also we suffered the loss of valiant men,—the men who had difficult ground to cope with at Corinth and treachery at Lechaeum [2]; valiant, too, were the men who rescued the King and drove the Lacedaemonians off the seas. These men I recall to your memory, and you it becomes to join in praising and celebrating men such as these.

And now we have related many of the noble deeds done by the men who are lying here, and by all the others who have died in defence of their city; yet far more numerous and more noble are those that

---

[2] The Corinthian oligarchs were supported by the Spartans, against whom the Athenians fought in 393–392.

B μενα· πολλαὶ γὰρ ἂν ἡμέραι καὶ νύκτες οὐχ ἱκαναὶ
γένοιντο τῷ τὰ πάντα μέλλοντι περαίνειν. τούτων
οὖν χρὴ μεμνημένους τοῖς τούτων ἐκγόνοις πάντ᾽
ἄνδρα παρακελεύεσθαι, ὥσπερ ἐν πολέμῳ, μὴ
λείπειν τὴν τάξιν τὴν τῶν προγόνων μηδ᾽ εἰς
τοὐπίσω ἀναχωρεῖν εἴκοντας κάκῃ. ἐγὼ μὲν οὖν
καὶ αὐτός, ὦ παῖδες ἀνδρῶν ἀγαθῶν, νῦν τε παρακελεύομαι καὶ ἐν τῷ λοιπῷ χρόνῳ, ὅπου ἄν τῳ
C ἐντυγχάνω ὑμῶν, καὶ ἀναμνήσω καὶ διακελεύσομαι
προθυμεῖσθαι εἶναι ὡς ἀρίστους· ἐν δὲ τῷ παρόντι
δίκαιός εἰμι εἰπεῖν ἃ οἱ πατέρες ἡμῖν ἐπέσκηπτον
ἀπαγγέλλειν τοῖς λειπομένοις, εἴ τι πάσχοιεν,
ἡνίκα κινδυνεύειν ἔμελλον. φράσω δὲ ὑμῖν ἅ τε
αὐτῶν ἤκουσα ἐκείνων καὶ οἷα νῦν ἡδέως ἂν εἴποιεν
ὑμῖν λαβόντες δύναμιν, τεκμαιρόμενος ἐξ ὧν τότε
ἔλεγον. ἀλλὰ νομίζειν χρὴ αὐτῶν ἀκούειν ἐκείνων
ἃ ἂν ἀπαγγέλλω. ἔλεγον δὲ τάδε.

D 19. Ὦ παῖδες, ὅτι μέν ἐστε πατέρων ἀγαθῶν,
αὐτὸ μηνύει τὸ νῦν παρόν· ἡμῖν δὲ ἐξὸν ζῆν μὴ
καλῶς, καλῶς αἱρούμεθα μᾶλλον τελευτᾶν, πρὶν
ὑμᾶς τε καὶ τοὺς ἔπειτα εἰς ὀνείδη καταστῆσαι
καὶ πρὶν τοὺς ἡμετέρους πατέρας καὶ πᾶν τὸ
πρόσθεν γένος αἰσχῦναι, ἡγούμενοι τῷ τοὺς αὐτοῦ
αἰσχύνοντι ἀβίωτον εἶναι καὶ τῷ τοιούτῳ οὔτε τινὰ
ἀνθρώπων οὔτε θεῶν φίλον εἶναι οὔτ᾽ ἐπὶ γῆς οὔθ᾽
ὑπὸ γῆς τελευτήσαντι. χρὴ οὖν μεμνημένους τῶν
E ἡμετέρων λόγων, ἐάν τι καὶ ἄλλο ἀσκῆτε, ἀσκεῖν

remain unmentioned; for many days and nights would not suffice were one to relate them all in full. Wherefore it is right that every man, bearing these men in mind, should exhort these men's children, just as in time of war, not to fall out of rank with their fathers nor to give way to cowardice and beat a retreat. And I myself for my own part, O ye children of valiant men, am now exhorting you and in the future, wheresoever I shall encounter any of you, I shall continue to remind you and admonish you to be zealous to show yourselves supremely valiant. But on this occasion it is my duty to record the message which your fathers, at the time when they were about to risk their lives, enjoined us, in case any ill befell them, to give to those who survived them. I will repeat to you both the words which I heard from their lips and those which they would now desire to say to you, if they had the power, judging from what they actually said on that occasion. You must, however, imagine that you are hearing from their own lips the message which I shall deliver. This, then, is what they said:

"O children, that ye are born of valiant sires is clearly shown by the facts now before you: we, who might have ignobly lived choose rather to die nobly, before we bring you and those after you to disgrace, and before we bring shame upon our own fathers and all our earlier forebears, since we deem that life is unworthy to be lived for the man who brings shame upon his own, and that such an one has no friend amongst gods or man, either here on earth, or under the earth when he is dead. Wherefore ye must bear in mind our words, and whatsoever else ye practise ye must practise it in union with valour,

μετ' ἀρετῆς, εἰδότας ὅτι τούτου λειπόμενα πάντα καὶ κτήματα καὶ ἐπιτηδεύματα αἰσχρὰ καὶ κακά. οὔτε γὰρ πλοῦτος κάλλος φέρει τῷ κεκτημένῳ μετ' ἀνανδρίας—ἄλλῳ γὰρ ὁ τοιοῦτος πλουτεῖ καὶ οὐχ ἑαυτῷ—, οὔτε σώματος κάλλος καὶ ἰσχὺς δειλῷ καὶ κακῷ ξυνοικοῦντα πρέποντα φαίνεται ἀλλ' ἀπρεπῆ, καὶ ἐπιφανέστερον ποιεῖ τὸν ἔχοντα καὶ ἐκφαίνει τὴν δειλίαν· πᾶσά τε ἐπιστήμη χωριζομένη δικαιοσύνης καὶ τῆς ἄλλης ἀρετῆς πανουργία, οὐ σοφία φαίνεται. ὧν ἕνεκα καὶ πρῶτον καὶ ὕστατον καὶ διὰ παντὸς πᾶσαν πάντως προθυμίαν πειρᾶσθε ἔχειν, ὅπως μάλιστα μὲν ὑπερβαλεῖσθε καὶ ἡμᾶς καὶ τοὺς πρόσθεν εὐκλείᾳ· εἰ δὲ μή, ἴστε ὡς ἡμῖν, ἂν μὲν νικῶμεν ὑμᾶς ἀρετῇ, ἡ νίκη αἰσχύνην φέρει, ἡ δὲ ἧττα, ἐὰν ἡττώμεθα, εὐδαιμονίαν. μάλιστα δ' ἂν νικώμεθα καὶ ὑμεῖς νικῷτε, εἰ παρασκευάσαισθε τῇ τῶν προγόνων δόξῃ μὴ καταχρησόμενοι μηδ' ἀναλώσοντες αὐτήν, γνόντες ὅτι ἀνδρὶ οἰομένῳ τι εἶναι οὐκ ἔστιν αἴσχιον οὐδὲν ἢ παρέχειν ἑαυτὸν τιμώμενον μὴ δι' ἑαυτὸν ἀλλὰ διὰ δόξαν προγόνων. εἶναι μὲν γὰρ τιμὰς γονέων ἐκγόνοις καλὸς θησαυρὸς καὶ μεγαλοπρεπής· χρῆσθαι δὲ καὶ χρημάτων καὶ τιμῶν θησαυρῷ, καὶ μὴ τοῖς ἐκγόνοις παραδιδόναι, αἰσχρὸν καὶ ἄνανδρον, ἀπορίᾳ ἰδίων αὐτοῦ κτημάτων τε καὶ εὐδοξιῶν. καὶ ἐὰν μὲν ταῦτα ἐπιτηδεύσητε, φίλοι παρὰ φίλους ἡμᾶς ἀφίξεσθε, ὅταν ὑμᾶς ἡ προσήκουσα μοῖρα κομίσῃ· ἀμελήσαντας

MENEXENUS

being well assured that when divorced from this all possessions and pursuits are base and ignoble. For neither does wealth bring honour to its possessor if combined with cowardice—for such an one is rich for another rather than for himself,—nor do beauty and strength appear comely, but rather uncomely, when they are attached to one that is cowardly and base, since they make their possessor more conspicuous and show up his cowardice; and every form of knowledge when sundered from justice and the rest of virtue is seen to be plain roguery rather than wisdom. For these reasons do ye make it your endeavour, first and last and always, in every way to show all zeal that ye may exceed, if possible, both us and those who went before us in renown; but if not, be ye well assured that if we vanquish you in virtue our victory brings us shame, whereas, if we are defeated, our defeat brings happiness. And most of all would we be the vanquished, you the victors, if ye are careful in your conduct not to trade upon the glory of your ancestors nor yet to squander it, believing that for a man who holds himself of some account there is nothing more shameful than to find himself held in honour not for his own sake but because of the glory of his ancestors. In the honours which belong to their parents, the children truly possess a noble and splendid treasure; but to use up one's treasure, whether of wealth or of honour, and bequeath none to one's children, is the base and unmanly act of one who lacks all wealth and distinctions of his own. And if ye practise these precepts ye will come unto us as friends unto friends whensoever the appointed doom shall convey you hither; but if ye neglect them and play the coward,

PLATO

δὲ ὑμᾶς καὶ κακισθέντας οὐδεὶς εὐμενῶς ὑποδέξεται. τοῖς μὲν οὖν παισὶ ταῦτ' εἰρήσθω.

20. Πατέρας δὲ ἡμῶν, οἷς εἰσί, καὶ μητέρας ἀεὶ χρὴ παραμυθεῖσθαι ὡς ῥᾷστα φέρειν τὴν ξυμφοράν, ἐὰν ἄρα ξυμβῇ γενέσθαι, καὶ μὴ ξυνοδύρεσθαι—οὐ γὰρ τοῦ λυπήσοντος προσδεήσονται· ἱκανὴ γὰρ ἔσται καὶ ἡ γενομένη τύχη τοῦτο πορίζειν—ἀλλ' ἰωμένους καὶ πραΰνοντας ἀναμιμνήσκειν αὐτοὺς ὅτι ὧν εὔχοντο τὰ μέγιστα αὐτοῖς οἱ θεοὶ ἐπήκοοι γεγόνασιν. οὐ γὰρ ἀθανάτους σφίσι παῖδας εὔχοντο γενέσθαι ἀλλ' ἀγαθοὺς καὶ εὐκλεεῖς· ὧν ἔτυχον, μεγίστων ἀγαθῶν ὄντων. πάντα δὲ οὐ ῥᾴδιον θνητῷ ἀνδρὶ κατὰ νοῦν ἐν τῷ ἑαυτοῦ βίῳ ἐκβαίνειν. καὶ φέροντες μὲν ἀνδρείως τὰς συμφορὰς δόξουσι τῷ ὄντι ἀνδρείων παίδων πατέρες εἶναι καὶ αὐτοὶ τοιοῦτοι· ὑπείκοντες δὲ ὑποψίαν παρέξουσιν ἢ μὴ ἡμέτεροι εἶναι ἢ ἡμῶν τοὺς ἐπαινοῦντας καταψεύδεσθαι. χρὴ δὲ οὐδέτερα τούτων, ἀλλ' ἐκείνους μάλιστα ἡμῶν ἐπαινέτας εἶναι ἔργῳ, παρέχοντας αὑτοὺς φαινομένους τῷ ὄντι πατέρας ὄντας ἄνδρας ἀνδρῶν. πάλαι γὰρ δὴ τὸ Μηδὲν ἄγαν λεγόμενον καλῶς δοκεῖ λέγεσθαι· τῷ γὰρ ὄντι εὖ λέγεται. ὅτῳ γὰρ ἀνδρὶ εἰς ἑαυτὸν ἀνήρτηται πάντα τὰ πρὸς εὐδαιμονίαν φέροντα ἢ ἐγγὺς τούτου, καὶ μὴ ἐν ἄλλοις ἀνθρώποις αἰωρεῖται ἐξ ὧν ἢ εὖ ἢ κακῶς πραξάντων πλανᾶσθαι ἠνάγκασται καὶ τὰ ἐκείνου, τούτῳ ἄριστα παρεσκεύασται ζῆν, οὗτός ἐστιν ὁ σώφρων καὶ οὗτος ὁ ἀνδρεῖος καὶ φρόνιμος· οὗτος γιγνομένων χρημά-

MENEXENUS

ye will be welcomed graciously by none." Let such, then, be the words we address to our children.

"Those of us who have fathers or mothers must counsel them always to bear their calamity—if so be that such has befallen them—as cheerfully as possible, and not join in their lamentations; for in sooth they will need no further cause of grief; the present misfortune will provide grief in plenty. Rather should we mollify and assuage their sorrow by reminding them that in the greatest matters the gods have already hearkened unto their prayers. For they prayed not that their sons should become immortal, but valiant and renowned; and these, which are the greatest of boons, they obtained. But that all things should turn out thus according to his mind, in respect of his own life, is for a mortal man no easy matter. Moreover, by bearing their calamities thus bravely they will clearly show that they are in truth the fathers of brave sons and of a like bravery themselves; whereas if they give way they will afford grounds for suspecting either that they are no fathers of ours or that we have been falsely belauded. But neither of these should they allow; rather should they belaud us most by their actions, showing themselves plainly to be in very truth the manly fathers of us men. That ancient saying, 'Nothing overmuch' is judged to be a noble saying; and in truth it is well said. For that man is best prepared for life who makes all that concerns his welfare depend upon himself, or nearly so, instead of hanging his hopes on other men, whereby with their rise or fall his own fortunes also inevitably sway up or down: he it is that is temperate, he it is that is courageous and wise; he it is that, when gaining

PLATO

τῶν καὶ παίδων καὶ διαφθειρομένων μάλιστα πείσεται τῇ παροιμίᾳ· οὔτε γὰρ χαίρων οὔτε λυπούμενος ἄγαν φανήσεται διὰ τὸ αὑτῷ πεποιθέναι. τοιούτους δὲ ἡμεῖς γε ἀξιοῦμεν καὶ τοὺς ἡμετέρους εἶναι καὶ βουλόμεθα καὶ φανῆναι,[1] καὶ ἡμᾶς αὐτοὺς νῦν παρέχομεν τοιούτους, οὐκ ἀγανακτοῦντας οὐδὲ φοβουμένους ἄγαν, εἰ δεῖ τελευτᾶν ἐν τῷ παρόντι. δεόμεθα δὴ καὶ πατέρων καὶ μητέρων τῇ αὐτῇ ταύτῃ διανοίᾳ χρωμένους τὸν ἐπίλοιπον βίον διάγειν, καὶ εἰδέναι ὅτι οὐ θρηνοῦντες οὐδὲ ὀλοφυρόμενοι ἡμᾶς ἡμῖν μάλιστα χαριοῦνται, ἀλλ᾿ εἴ τις ἔστι τοῖς τετελευτηκόσιν αἴσθησις τῶν ζώντων, οὕτως ἀχάριστοι εἶεν ἂν μάλιστα, ἑαυτούς τε κακοῦντες καὶ βαρέως φέροντες τὰς συμφοράς· κούφως δὲ καὶ μετρίως μάλιστ᾿ ἂν χαρίζοιντο. τὰ μὲν γὰρ ἡμέτερα τελευτὴν ἤδη ἕξει ἥπερ καλλίστη γίγνεται ἀνθρώποις, ὥστε πρέπει αὐτὰ μᾶλλον κοσμεῖν ἢ θρηνεῖν· γυναικῶν δὲ τῶν ἡμετέρων καὶ παίδων ἐπιμελούμενοι καὶ τρέφοντες καὶ ἐνταῦθα τὸν νοῦν τρέποντες τῆς τε τύχης μάλιστ᾿ ἂν εἶεν ἐν λήθῃ καὶ ζῷεν κάλλιον καὶ ὀρθότερον καὶ ἡμῖν προσφιλέστερον.

Ταῦτα δὴ ἱκανὰ τοῖς ἡμετέροις παρ᾿ ἡμῶν ἀγγέλλειν. τῇ δὲ πόλει παρακελευοίμεθ᾿ ἂν ὅπως ἡμῖν καὶ πατέρων καὶ υἱέων ἐπιμελήσονται, τοὺς μὲν παιδεύοντες κοσμίως, τοὺς δὲ γηροτροφοῦντες ἀξίως. νῦν δὲ ἴσμεν ὅτι, καὶ ἐὰν μὴ ἡμεῖς παρακελευώμεθα, ἱκανῶς ἐπιμελήσεται.

21. Ταῦτα οὖν, ὦ παῖδες καὶ γονεῖς τῶν τελευτησάντων, ἐκεῖνοί τε ἐπέσκηπτον ἡμῖν ἀπαγγέλλειν, καὶ ἐγὼ ὡς δύναμαι προθυμότατα ἀπαγγέλλω·

[1] φανῆναι] φαμέν mss., Zur.

MENEXENUS

or losing riches or children, will best exemplify the proverb; for, because he puts his trust in himself, he will neither be seen rejoicing nor yet grieving overmuch. Of such a character we request our friends to be, and desire them to appear, even as we now display ourselves as such, being neither aggrieved nor alarmed overmuch if so be that at this present crisis we must die. We beseech both fathers and mothers to pass the rest of their lives holding to this same conviction, and to be well assured that it is not by mourning and lamenting us that they will gratify us most; nay, if the dead have any perception of the living, it is thus that they would gratify us least, by debasing themselves and bearing their sorrows with a heavy heart; whereas by a light-hearted and temperate demeanour they would gratify us most. As for our own fortunes, they have already reached that climax which is the noblest of all for mortal men; wherefore it is more fitting to magnify than to mourn them. But to our wives and children let them give care and nurture and devote their mind to them; for thus they will best forget their ill fortune and live a life that is nobler and truer and more pleasing in our eyes.

"Let this, then, suffice as our message to our kinsfolk. To the City we would add an exhortation that on our behalf they care for our parents and our sons, bestowing on the latter an orderly training, and on the former the fitting nurture of old age; and, as it is, we are well assured that even without our exhortation the city will bestow upon them ample care."

Such is the message, O ye children and parents of the fallen, which they enjoined upon us to deliver, and which I, with all the earnestness in my power,

καὶ αὐτὸς δέομαι ὑπὲρ ἐκείνων, τῶν μὲν μιμεῖσθαι τοὺς αὑτῶν, τῶν δὲ θαρρεῖν ὑπὲρ αὑτῶν, ὡς ἡμῶν καὶ ἰδίᾳ καὶ δημοσίᾳ γηροτροφησόντων ὑμᾶς καὶ ἐπιμελησομένων, ὅπου ἂν ἕκαστος ἑκάστῳ ἐντυγχάνῃ ὁτῳοῦν τῶν ἐκείνων. τῆς δὲ πόλεως ἴστε που καὶ αὐτοὶ τὴν ἐπιμέλειαν, ὅτι νόμους θεμένη περὶ τοὺς τῶν ἐν τῷ πολέμῳ τελευτησάντων παῖδάς τε καὶ γεννήτορας ἐπιμελεῖται, καὶ διαφερόντως τῶν ἄλλων πολιτῶν προστέτακται φυλάττειν ἀρχῇ ἥπερ μεγίστη ἐστίν, ὅπως ἂν οἱ τούτων μὴ ἀδικῶνται πατέρες τε καὶ μητέρες· τοὺς δὲ παῖδας συνεκτρέφει αὐτή, προθυμουμένη ὅ τι μάλιστ᾿ ἄδηλον αὐτοῖς τὴν ὀρφανίαν γενέσθαι, ἐν πατρὸς σχήματι καταστᾶσα αὐτοῖς αὐτὴ ἔτι τε παισὶν οὖσι, καὶ ἐπειδὰν εἰς ἀνδρὸς τέλος ἴωσιν, ἀποπέμπει ἐπὶ τὰ σφέτερ᾿ αὐτῶν πανοπλίᾳ κοσμήσασα, ἐνδεικνυμένη καὶ ἀναμιμνῄσκουσα τὰ τοῦ πατρὸς ἐπιτηδεύματα ὄργανα τῆς πατρῴας ἀρετῆς διδοῦσα, καὶ ἅμα οἰωνοῦ χάριν ἄρχεσθαι ἰέναι ἐπὶ τὴν πατρῴαν ἑστίαν ἄρξοντα μετ᾿ ἰσχύος ὅπλοις κεκοσμημένον. αὐτοὺς δὲ τοὺς τελευτήσαντας τιμῶσα οὐδέποτε ἐκλείπει, καθ᾿ ἕκαστον ἐνιαυτὸν αὐτὴ τὰ νομιζόμενα ποιοῦσα κοινῇ πᾶσιν ἅπερ ἰδίᾳ ἑκάστῳ ἴδια γίγνεται, πρὸς δὲ τούτοις ἀγῶνας γυμνικοὺς καὶ ἱππικοὺς τιθεῖσα καὶ μουσικῆς πάσης, καὶ ἀτεχνῶς τῶν μὲν τελευτησάντων ἐν κληρονόμου καὶ ὑέος μοίρᾳ καθεστηκυῖα, τῶν δὲ ὑέων ἐν πατρός, γονέων δὲ καὶ τῶν τοιούτων ἐν ἐπιτρόπου,

MENEXENUS

have now delivered; and I myself, on their behalf, entreat the children to imitate their fathers, and the parents to have no fear for themselves, seeing that we, both privately and publicly, will give nurture to your age and bestow care upon you, wheresoever one of us meets with one of you. And as regards the care bestowed by the City, of your own selves ye know well that she has made laws regarding both the children and the begetters of those who have fallen in the war, to ensure their care; and that the highest authority in the State is instructed to watch over them beyond all other citizens, that the fathers and mothers of these men may suffer no wrong. And the City herself helps in the bringing up of their children, endeavouring to render them as little conscious as possible of their orphaned condition; while they are yet children she stands towards them as a father, and when they arrive at man's estate she presents them with full military equipment and sends them back to their own place, thereby exhibiting and putting them in mind of their fathers' profession by bestowing on each of them the instruments of his father's prowess, while at the same time desiring that he should be auspiciously equipped with arms on commencing his journey to his ancestral hearth, there to rule with power. Nor does the City ever omit to pay honour to the dead heroes themselves, seeing that she herself year by year performs publicly, on behalf of all, those customary rites which are privately performed for each; and moreover, she institutes contests in athletics and horse-racing and music of every kind. And thus, in simple fact, she stands towards the fallen in the position of son and heir, towards the sons in that of father, and towards the

249
πᾶσαν πάντων παρὰ πάντα τὸν χρόνον ἐπιμέλειαν ποιουμένη. ὧν χρὴ ἐνθυμουμένους πραότερον φέρειν τὴν ξυμφοράν· τοῖς τε γὰρ τελευτήσασι καὶ τοῖς ζῶσιν οὕτως ἂν προσφιλέστατοι εἴητε καὶ ῥᾷστοι θεραπεύειν τε καὶ θεραπεύεσθαι.

Νῦν δὲ ἤδη ὑμεῖς τε καὶ οἱ ἄλλοι πάντες κοινῇ κατὰ τὸν νόμον τοὺς τετελευτηκότας ἀπολοφυράμενοι ἄπιτε.

22. Οὗτός σοι ὁ λόγος, ὦ Μενέξενε, Ἀσπασίας D τῆς Μιλησίας ἐστίν.

ΜΕ. Νὴ Δία, ὦ Σώκρατες, μακαρίαν γε λέγεις τὴν Ἀσπασίαν, εἰ γυνὴ οὖσα τοιούτους λόγους οἵα τ' ἐστὶ συντιθέναι.

ΣΩ. Ἀλλ' εἰ μὴ πιστεύεις, ἀκολούθει μετ' ἐμοῦ, καὶ ἀκούσει αὐτῆς λεγούσης.

ΜΕ. Πολλάκις, ὦ Σώκρατες, ἐγὼ ἐντετύχηκα Ἀσπασίᾳ, καὶ οἶδα οἵα ἐστίν.

ΣΩ. Τί οὖν; οὐκ ἄγασαι αὐτὴν καὶ νῦν χάριν ἔχεις τοῦ λόγου αὐτῇ;

ΜΕ. Καὶ πολλήν γε, ὦ Σώκρατες, ἐγὼ χάριν ἔχω E τούτου τοῦ λόγου ἐκείνῃ ἢ ἐκείνῳ ὅστις σοι ὁ εἰπών ἐστιν αὐτόν· καὶ πρός[1] γε ἄλλων πολλῶν χάριν ἔχω τῷ εἰπόντι.

ΣΩ. Εὖ ἂν ἔχοι. ἀλλ' ὅπως μου μὴ κατερεῖς, ἵνα καὶ αὖθίς σοι πολλοὺς καὶ καλοὺς λόγους παρ' αὐτῆς πολιτικοὺς ἀπαγγέλλω.

ΜΕ. Θάρρει, οὐ κατερῶ· μόνον ἀπάγγελλε.

ΣΩ. Ἀλλὰ ταῦτ' ἔσται.

[1] πρός one ms.: πρό most mss., Zur.

MENEXENUS

parents of the dead in that of guardian, thus exercising towards all all manner of care throughout all time. Laying which to heart it behoves you to bear your sorrow with the greater calm; for thus will ye best content both the dead and the living, and tend and be tended with the greatest ease.

And now that ye and all the rest have already made public lamentation for the dead as the law ordains, go ye your ways.

There, Menexenus, you have the oration of Aspasia, the Milesian.

MEN. And by Zeus, Socrates, Aspasia, by your account, deserves to be congratulated if she is really capable of composing a speech like that, woman though she is.

SOC. Nay, then, if you are incredulous, come along with me and listen to a speech from her own lips.

MEN. I have met with Aspasia many a time, Socrates, and I know well what she is like.

SOC. Well, then, don't you admire her, and are you not grateful to her now for her oration?

MEN. Yes, I am exceedingly grateful, Socrates, for the oration to her or to him—whoever it was that repeated it to you; and what is more, I owe many other debts of gratitude to him that repeated it.

SOC. That will be fine! Only be careful not to give me away, so that I may report to you later on many other fine political speeches of hers.

MEN. Have no fear: I won't give you away; only do you report them.

SOC. Well, it shall be done.

EPISTLES

INTRODUCTION TO THE *EPISTLES*

In our manuscripts of Plato this collection of Thirteen Epistles is placed at the end, just before the spurious dialogues. This arrangement dates from the time of Thrasyllus, a contemporary of the Emperor Tiberius, who arranged the Platonic writings in "tetralogies," or groups of four, and placed the Epistles in his last tetralogy. But as we have them numbered in the manuscripts the Epistles do not seem to be arranged on any intelligible principle; and for purposes of study it would be better either to group them according to the persons to whom they were addressed, or to arrange them in chronological order. In chronological order the sequence should probably be this— xiii, ii, xi, x, iv, iii, vii, viii, vi (omitting i, v, ix, and xii as certainly spurious). While if we arrange them according to the persons to whom the Letters are addressed they will naturally fall into three main groups, viz. :—

(1) Letters to Dionysius—i, ii, iii, xiii.

(2) Letters to Dion and Dion's friends—iv, vii, viii, x.

(3) Letters to various Rulers and Statesmen—v, vi, ix, xi, xii.

It will be noticed that eight Letters, out of the thirteen, including the most important, are concerned with the Sicilian rulers, Dion and Dionysius; and

PLATO

it may be helpful to insert here a brief summary of the course of affairs in Sicily in so far as they affected Plato.

Dionysius the Elder was the tyrant of Syracuse when Plato, then about forty years of age, paid his first visit to Italy and Sicily (388–387 B.C.). At Syracuse Plato fell in with Dion, the brother-in-law of Dionysius, who was then about twenty years old, and was greatly struck with his character and abilities. A firm friendship, based on their enthusiasm for philosophy, sprang up between the two and lasted, without a break, till Dion's death some thirty-three years later. On his return to Athens from this expedition Plato founded the Academy. During the next twenty years, until his death in 367 B.C., Dionysius was active in extending and consolidating his empire in Sicily and S. Italy, and in making Syracuse a centre of light and learning as well as a military capital. Unfortunately his son and successor, Dionysius the Younger, when weighed in the balance of all-testing Time, was found wanting in all those qualities which go to make a successful governor. To practical incompetence and want of judgement he added vanity, jealousy, and mutability. During the early days of his reign he was guided in the government by his uncle Dion; and it was because of Dion's influence that Plato was persuaded to make his second visit to Syracuse (367 B.C.). By the aid of Dion Plato hoped to realize there his ideal of a state ruled by a philosopher-king. Unfortunately the reforms initiated by Dion raised enemies against him in the city; and the young Dionysius also became impatient of the Spartan life inculcated by Plato and by Dion, so that he lent only too ready an ear to the enemies

INTRODUCTION TO THE *EPISTLES*

who insinuated doubts of Dion's loyalty. So it came about that within a few months of Plato's arrival Dion, the champion of his political ideals, was driven into exile, and Dionysius, whom he had hoped to train up in the paths of philosophy, showed no desire to submit to the discipline proposed. Plato, therefore, after some efforts to secure the restoration of Dion and to bring Dionysius to a better mind, decided to leave Sicily.

During the next five years Plato, it would seem, continued to interest himself in Sicilian affairs, and helped, for one thing, to promote a friendly understanding between Archytas of Tarentum and Dionysius, which resulted in the latter aiding the Tarentines in their war against the Lucanians. Apparently Dionysius was anxious at this time to exploit his friendship with Plato for his own advantage, being ambitious to pose not only as a patron of letters but a man of science himself. So he surrounded himself with sophists or scientists before whom he could display his erudition as a full-fledged member of the great Academy of Athens, and his enthusiasm for the ideal philosophy. Meanwhile, however, he showed no disposition to recall Dion. Finally, his ambition to make of his court a centre of Hellenic culture roused him to make fresh overtures to Plato, so as to induce him to revisit Syracuse. Against his own better judgement (as he protests) Plato yielded to his entreaties, backed as it was by Dion's advice and that of other friends, hoping against hope that he might yet intercede successfully for Dion and convert Dionysius to a serious interest in philosophic politics.

But alas! his hopes were doomed to disappoint-

PLATO

ment. The third visit to Sicily (361–360 B.C.) proved an utter failure. In spite of his utmost efforts Plato could not induce the tyrant either to restore Dion or to undertake the serious study of philosophy and the prosecution of the political reforms which he professed to believe in. Consequently before long the relations between the two became strained, Plato being disgusted with the fickleness and jealousy of Dionysius, and Dionysius resenting Plato's loyalty to the interests of Dion and his own intellectual reputation. In the end Dionysius showed increased bitterness against Dion, whose property he confiscated, and deprived him even of his wife and child; while he treated Plato with marked disrespect, making him a virtual prisoner in the precincts of the palace, until he went actually in fear of his life from the violence of the soldiery. From this humiliating position Plato was only rescued by the intervention of his friend Archytas of Tarentum, to whom he had appealed for help.

On his return from this third visit to Sicily, in 360 B.C., Plato made a journey to Olympia, where he met Dion at the Festival, and discussed with him the situation of affairs at Syracuse. Rather against the wishes of Plato, Dion resolved on military operations against Dionysius, and summoned all his friends to help him. His attack on Syracuse in 357 was welcomed by the popular party, and he gained possession of the city without difficulty, although the tyrant continued to hold out in his palace in the island of Ortygia. But when Dion proceeded to put his political reforms into execution he lost the favour of the populace, who united against him under the leadership of Heracleides. After a series of dis-

INTRODUCTION TO THE *EPISTLES*

turbances Dion found himself obliged to execute Heracleides; but this strong measure instead of securing peace only increased his unpopularity and caused further trouble; until the climax of misfortune came with the assassination of Dion by a pair of Athenians who had posed as his friends, Callippus and his brother (353 B.C.).

After the death of their chief Dion's friends retired to Leontini, while Callippus held sway in Syracuse. But his hold on the city was far from secure, and after little more than a year the partisans of Dion, under his nephew Hipparinus, succeeded in capturing Syracuse. But though Dion's adherents thus held Syracuse and Leontini, the rest of Sicily remained a prey to civil war, and repeated incursions of foreign foes, Carthaginians and Italians, added to the general distress. Nor did the position of affairs at either Syracuse or Leontini continue satisfactory, from the point of view of Dion's friends, for any length of time. Hipparinus, the ruler of Syracuse, died in 350 B.C.; and his brother Nysaeus did not long survive him. At Leontini things took a still worse course; for Hicetas, who had assumed control as the nominee of Dion's party, fell out with them and even went so far as to destroy all the members of Dion's family upon whom he could lay hands.

It was thus that the last years of Plato, who died in 347 B.C., were saddened by the sight of the storm-clouds which rolled ever darkly over Sicily, the land for which his beloved Dion had spent his life, and spent it, alas! in vain.

Whatever view we take of the authenticity of the *Epistles* it is certain at least that some of them represent an ancient tradition, within a century of

Plato's time, regarding the historical and biographical details which they relate. Their interest and value consist mainly in the welcome light they throw on the practical activity of Plato and on the political influence of the Academy. They show us that Plato was really in earnest with the Ideal State which he had sketched in the *Republic*; and they put before us all the practical measures he took, with Dion as his colleague, to realize that Ideal. They also help us to understand how, in default of the Ideal, he was led to fall back upon the rule of Law, as a second-best; so that the *Laws*, rather than a new *Republic*, was the main work of his declining years.

The epistle as a special literary type was well established by the second and first centuries B.C. We have several letters, probably genuine, of Epicurus; and Isocrates, the contemporary of Plato, is credited with quite a number of letters, some of them addressed to Philip of Macedon, of which a few may be accepted as genuine. Some of these Isocratean letters are obviously not private but what we should term " open " letters, intended for publication. And in attempting to decide the question of authenticity it is important to bear in mind this distinction between the private and personal letter and the " open " letter, which is really a manifesto or essay under the guise of a letter and meant for the edification of a public far wider than the persons to whom it is nominally addressed. For it is evident that the chances of preservation, and therefore the chances of authenticity, are far greater in the case of the open " letter." It stands on much the same footing as the oration or the dialogue. In the case of the purely private letter, on the other hand, we have to assign

INTRODUCTION TO THE *EPISTLES*

a reason for its preservation. Did the writer retain a duplicate of his letters? Or did the recipient file and preserve them? And if a few have been thus preserved, why not all? Thus one is justified in doubting *a priori* the genuineness of the private letter unless there is overwhelming evidence in its favour. And this all the more because it is so easy to account for the origin of the spurious letter; since we have it on Galen's authority that good prices were paid by the libraries for letters signed with illustrious names. This put a premium upon forgeries, especially skilful forgeries; so that it was well worth while for an unscrupulous man of letters to study the style of a celebrated author such as Plato with a view to foisting on the learned world a plausibly fabricated epistle. In some cases, too, school-exercises composed by his scholars of the first or second generation might be mistaken at a later age for genuine compositions of the master himself. Or again, some teacher or propagandist, innocent of any real intention to deceive, might adopt the device of issuing his own views and theories under cover of the name of some earlier authority. And thus in one way or another spurious documents would come to be accepted at their face-value, and ascribed to the authors whose names they ostensibly bear.

If we turn now to the Platonic collection of thirteen Epistles we find that the only two which we can with any confidence regard as genuine are precisely those two " open " letters, the seventh and the eighth. Fortunately these are the longest, most important, and most informative of the thirteen. Of the other letters, iii, iv, and xiii are admitted to be genuine by a good many modern scholars; some also admit

vi, x, and xi, and, less confidently, ii ; while i, v, ix, and xii are generally acknowledged to be forgeries. The readiness of so many recent scholars to ascribe as many Epistles as possible to Plato is in marked contrast to the attitude of the great Platonic students of fifty or a hundred years ago (such as Ast and Zeller) who unhesitatingly pronounced the whole collection apocryphal. It would seem that the swing of the pendulum has now gone too far in the other direction ; and the most reasonable view is that the two great letters, seventh and eighth, are really from Plato's hand, while all the rest are fabrications after his manner. For further justification of this view the reader may be referred to the Prefatory Notes at the head of the individual Epistles.

The Greek text here printed is based on that of the Zurich edition, the chief deviations being indicated in the footnotes.

EPISTLE I

Prefatory Note.—In this letter the writer reproaches Dionysius for the ill-treatment he has received from him and his friends. In return for his public services he has been dismissed with ignominy, and has been offered a paltry sum to meet his travelling-expenses. He rejects the offer with scorn, and prophesies for the tyrant an evil future " bereft of friends," embellishing his tirade with quotations from tragic poets.

Such a letter, if written by Plato, could only belong to the year 360, immediately after his return from his third visit to Sicily and his final break with Dionysius. But Plato could never have described himself as the " sole Dictator " (αὐτοκράτωρ) of Syracuse ; nor does the account here given of his abrupt dismissal agree with what we are told in Epistle vii. Accordingly, it has been suggested (by Ficinus) that Dion was the writer, since he was " driven out with ignominy " (*Ep.* vii. 329 c). But neither is this a probable hypothesis, since Dion was never " Dictator," nor would he, a native of the place, have spoken of himself as sojourning in Syracuse, in the way the writer speaks in the opening sentence. The letter is of interest mainly for the poetical quotations it preserves ; and the very manner in which these are introduced is in itself sufficient disproof of Platonic authorship.

ΕΠΙΣΤΟΛΑΙ

Α

Πλάτων[1] Διονυσίῳ εὖ πράττειν.

Διατρίψας ἐγὼ παρ' ὑμῖν χρόνον τοσοῦτον καὶ διοικῶν τὴν ὑμετέραν ἀρχὴν πεπιστευμένος πάντων μάλιστα, τὰς ὠφελείας ὑμῶν λαμβανόντων, τὰς διαβολὰς δυσχερεῖς οὔσας ὑπέμενον· ᾔδειν γὰρ ὅτι τῶν ὠμοτέρων οὐδὲν ἐμοῦ συνεθέλοντος ὑμῖν δόξει πεπρᾶχθαι· πάντες γὰρ οἱ συμπολιτευόμενοι μεθ' ὑμῶν ὑπάρχουσί μοι μάρτυρες, ὧν ἐγὼ πολλοῖς συνηγωνισάμην, ἀπολύσας αὐτοὺς οὐ σμικρᾶς ζημίας. αὐτοκράτωρ δὲ πολλάκις τὴν ὑμετέραν πόλιν διαφυλάξας ἀπεπέμφθην ἀτιμότερον ἢ πτωχὸν ὑμῶν ἀποστελλόντων προσήκει καὶ κελευόντων ἐκπλεῦσαι, τοσοῦτον παρ' ὑμῖν διατρίψαντα χρόνον.

Ἐγὼ μὲν οὖν περὶ ἐμαυτοῦ βουλεύσομαι τὸ[2] λοιπὸν τρόπον ἀπανθρωπότερον, σὺ δὲ τοιοῦτος ὢν τύραννος οἰκήσεις μόνος. τὸ δὲ χρυσίον τὸ

[1] Πλάτων mss.: Δίων Ficinus, Zur.
[2] τὸ mss.: τὸν Zur.

---

[1] The Greek phrase εὖ πράττειν is purposely ambiguous, meaning either " act well " or " fare well " (*i.e.* " prosper "); *cf. Gorg.* 495 E, *Rep.* 353 E. It is the form of address regularly used in these Epistles, *cf. Epist.* iii. *ad init.*

EPISTLES

EPISTLE I

PLATO TO DIONYSIUS WISHES WELL-DOING [1]

AFTER I had spent so long a time with you and was trusted above all others in my administration of your government, while you were enjoying the benefits I was enduring the slanders, grievous as they were. For I knew that men would not believe that any of your more brutal acts were done with my consent, seeing that I have for my witnesses all those who take a part in your government, many of whom I have helped in their times of trial and saved them from no small damage. But after I had oftentimes kept guard over your City as sole Dictator, I was dismissed with more ignominy than a beggar would deserve who had stayed with you for so long a time, were you to pack him off and order him to sail away.

For the future, therefore, I for my part will consult my own interests in less philanthropic fashion, while you, "gross tyrant that you are, will dwell alone." [2]

---

[2] Apparently a "tag" from some tragedy. Note that "you" in this second paragraph refers to Dionysius alone, whereas in the first paragraph "you," in the plural, includes D.'s associates.

PLATO

C λαμπρόν, ὅπερ ἔδωκας εἰς ἀποστολήν, ἄγει σοι
Βακχεῖος ὁ τὴν ἐπιστολὴν φέρων· οὔτε γὰρ ἐφόδιον
ἐκεῖνό γ' ἦν ἱκανὸν οὔτε πρὸς τὸν ἄλλον βίον
ξυμφέρον, ἀδοξίαν δὲ πλείστην μὲν τῷ διδόντι σοὶ
παρασκευάζον, οὐ πολλῷ δὲ ἐλάττω κἀμοὶ λαμ-
βάνοντι, διόπερ οὐ λαμβάνω· σοὶ δ' οὐδὲν διαφέρει
δῆλον ὅτι καὶ λαβεῖν καὶ δοῦναι τοσοῦτον, ὥστε
κομισάμενος ἄλλον τινὰ τῶν ἑταίρων θεράπευσον
ὥσπερ ἐμέ· κἀγὼ γὰρ ἱκανῶς ὑπὸ σοῦ τεθερά-
D πευμαι· καί μοι τὸ τοῦ Εὐριπίδου κατὰ καιρὸν
ἔστιν εἰπεῖν, ὅτι σοὶ πραγμάτων ἄλλων ποτὲ
ξυμπεσόντων

εὔξει τοιοῦτον ἄνδρα σοι παρεστάναι.

ὑπομνῆσαι δέ σε βούλομαι διότι καὶ τῶν ἄλλων
τραγῳδοποιῶν οἱ πλεῖστοι, ὅταν ὑπό τινος ἀπο-
θνήσκοντα τύραννον εἰσάγωσιν, ἀναβοῶντα ποιοῦσι

310   φίλων ἔρημος, ὦ τάλας, ἀπόλλυμαι·

χρυσίου δὲ σπάνει ἀπολλύμενον οὐδεὶς πεποίηκε.
κἀκεῖνο δὲ τὸ ποίημα τοῖς νοῦν ἔχουσιν οὐ κακῶς
ἔχειν δοκεῖ·

οὐ χρυσὸς ἀγλαὸς σπανιώτατος ἐν θνατῶν δυσ-
    ελπίστῳ βίῳ,
οὐδ' ἀδάμας οὐδ' ἀργύρου κλῖναι πρὸς ἀνθρώπων
    δοκιμαζόμεν' ἀστράπτει πρὸς ὄψεις·
οὐδὲ γαίας εὐρυπέδου γόνιμοι βρίθοντες αὐτάρ-
    κεις γύαι,
ὡς ἀγαθῶν ἀνδρῶν ὁμοφράδμων νόησις.

B ἔρρωσο, καὶ γίγνωσκε τοσοῦτον ἡμῶν διημαρτηκὼς
ἵνα πρὸς τοὺς ἄλλους βέλτιον προσφέρῃ.

---
[1] Eurip. *Frag.* 956 (Nauck).

EPISTLE I

And as for the splendid sum of gold which you gave for my journey home, Baccheius, the bearer of this letter, is taking it back to you. For it was neither a sufficient sum for my journey nor was it otherwise useful for my support; and since it reflects the greatest disgrace on you who offer it, and not much less on me if I accept it, I therefore refuse to accept it. But evidently neither the giving nor the accepting of such an amount makes any difference to you; take it, then, and befriend therewith some other companion of yours as you did me; for I, in sooth, have had enough of your "befriending." Indeed, I may appropriately quote the verse of Euripides — that one day, when other fortunes befall you,

> Thou'lt pray for such a helper by thy side.[1]

And I desire to remind you that most of the other tragedians also, when they show a tyrant on the stage slaughtered by someone, represent him as crying out—

> Bereft of friends,—ah! woe is me,—I die.[2]

But not one of them has represented him as dying for lack of gold. This other poem also "to men of judgement seemeth not amiss"—

> In this our human life, with halting hopes,
> It is not glittering gold that rarest is:
> Not diamond nor couches silver-wrought
> Appear so brilliant in the eyes of men:
> Nor do the fertile fields of Earth's broad breast,
> Laden with crops, so all-sufficing seem
> As gallant men's unanimous resolve.[3]

Farewell; and may you learn how much you have lost in us, so that you may behave yourself better towards all others.

[2] *Trag. Gr. Frag. Adesp.* 347 (Nauck).
[3] *Lyr. Gr. Frag. Adesp.* 138 (Bergk).

EPISTLE II

Prefatory Note.—Plato begins by replying to Dionysius's complaint that he has recently been maligned by Plato's friends. He says that he is not responsible for his friends' conduct; yet, all the same, the reports Dionysius has heard about them are false. Next, he passes on to his main theme—the mutual relations which should subsist between himself and Dionysius, arguing that this is a matter of high importance because men everywhere love to discuss the mutual intercourse between potentates and sages, such as Cyrus and Solon. Nor is the verdict of future ages to be lightly esteemed by the wise. " And we, unlike the earlier examples of power in contact with wisdom, have it still in our power to amend our behaviour towards each other."

After this lesson on behaviour, Plato proceeds to deal with a philosophic problem which was puzzling Dionysius, namely, the nature of " the First " (principle). His explanation, he says, must be couched " in enigmas " lest the secrets of his doctrine should fall into the hands of the profane vulgar. Realities must be distinguished from qualities; and knowledge of the three grades of Being can only be acquired by long study and travail of soul. The doubts and difficulties felt by Dionysius are common to all students of the subject; and Plato commends him

EPISTLE II

for consulting him about them, and advises him to go on doing so. Only he must never write down these doctrines; just as Plato himself has never put them in writing—the writings ascribed to him being really those of "Socrates rejuvenated." Even this letter must be burned, once it has been committed to memory.

The letter concludes with some personal observations and requests.

As regards the *authenticity* of this letter, it may be taken as fairly certain that it is not by Plato. The following considerations, amongst others, tell strongly against it.

The Olympic Games mentioned in the opening paragraph cannot well be those of 360 B.C. (as some have supposed), since the general tone of the letter shows that it must be earlier than Plato's return from his third visit to Syracuse in that year. The reference, then, must be to the Games of 364 B.C.; and if so, the Syracusan visit alluded to in 312 A can only be the second visit of 367-366 B.C. But the account there given of the failure of that visit owing to the suspicious attitude of Dionysius plainly contradicts what we are told of Dionysius's hospitable treatment of Plato in *Ep.* vii. (329 D ff.). Moreover, there is no other evidence that Plato visited Olympia in 364 B.C.; although we are told (*Ep.* vii. 350 B) that he did so in 360 B.C.

In addition to these historical inconsistencies there is much to arouse suspicion in the tone and matter of the letter. Can we imagine the real Plato saying that his object in visiting Syracuse was "to make philosophy honoured by the multitude"? Or denying that he ever wrote serious books on philosophy?

PLATO

Or bidding his correspondent burn a letter because it contained a doctrine which nobody else (as he says) could possibly understand since it was written " in riddles " ? Or trotting out a long list of sages and potentates to suggest his own magnanimity and the magnificence of Dionysius ? Moreover, the numerous echoes of passages in other letters and dialogues (see the footnotes) are such as to confirm the suspicion that the writer of this letter was a rhetor of a later age and of Pythagorean leanings, who thought to glorify the memory of Plato by composing this epistolary exercise on the theme—" How ought Plato and Dionysius to behave towards each other ? "

A word may be added regarding the " secret doctrine " which professes to explain " the nature of the First " in the passage 312 E ff. Among the interpretations offered, some identify the Three (principles) with the Idea of Good, Reason ($νοῦς$) and the Soul; others with God, the Ideas, the World-Soul; others with the Ideas, the World of Sense, Matter; others with the Idea of Good (=Reason), the World as known by Science ($διάνοια$), and the World as known by Sense. It has also been suggested that the writer had *Ep.* vii. 342 A ·ff. in mind and meant by his three the Name, the Definition, and the Idea. Now it is, perhaps, futile to attempt to interpret what seems a piece of wilful mystification. If, however, we are to venture on the thankless task and credit the writer with at least the minimum of sense, we must suppose him to be referring to three grades of existence in each of which there is a plurality surrounding, or related to, a central Unity or common principle. Now what is here said of " the King of All " is closely parallel to

EPISTLE II

the description given of the Idea of Good in *Republic* 509 B, D, 517 C; so it is natural to equate the First Principle and the first grade of Being with the Idea of Good and the other Ideas, or with the Divine Reason (*cf. Phileb.* 30 D) and the realm of rational objects (νοητά). If so, the Second (principle) may probably be the form of intelligence which comes next to reason in the *Republic*, *i.e.* διάνοια, " related to " which are " the Mathematicals " (τὰ μεταξύ); while in the Third place will come Sense-perception (αἴσθησις) and the world of sense-objects.

But it is impossible to be sure that this was what the writer really intended to convey in his oracular exposition, which, for its vagueness, may be compared with the cryptic allusion to two Divine Powers at the end of the sixth letter.

B

Πλάτων Διονυσίῳ εὖ πράττειν.

Ἤκουσα Ἀρχεδήμου ὅτι σὺ ἡγεῖ χρῆναι περὶ σοῦ μὴ μόνον ἐμὲ ἡσυχίαν ἄγειν ἀλλὰ καὶ τοὺς ἐμοὺς ἐπιτηδείους τοῦ φλαῦρόν τι ποιεῖν ἢ λέγειν περὶ σοῦ· Δίωνα δὲ μόνον ἐξαίρετον ποιεῖ. οὗτος δὲ ὁ λόγος σημαίνει, τὸ Δίωνα ἐξαίρετον εἶναι, ὅτι οὐκ ἄρχω ἐγὼ τῶν ἐμῶν ἐπιτηδείων· εἰ γὰρ ἦρχον ἐγὼ οὕτω τῶν τε ἄλλων καὶ σοῦ καὶ Δίωνος, πλείω ἂν ἦν ἡμῖν¹ τε πᾶσιν ἀγαθὰ τοῖς τε ἄλλοις Ἕλλησιν, ὡς ἐγώ φημι. νῦν δὲ μέγας ἐγώ εἰμι ἐμαυτὸν παρέχων τῷ ἐμῷ λόγῳ ἑπόμενον. καὶ ταῦτα λέγω ὡς οὐχ ὑγιές τι Κρατιστόλου καὶ Πολυξένου πρὸς σὲ εἰρηκότων, ὧν φασὶ λέγειν τὸν ἕτερον ὅτι ἀκούοι Ὀλυμπίασι πολλῶν τινῶν τῶν μετ' ἐμοῦ σε κακηγορούντων. ἴσως γὰρ ὀξύτερον ἐμοῦ ἀκούει· ἐγὼ μὲν γὰρ οὐκ ἤκουσα. χρὴ δέ, ὡς ἐμοὶ δοκεῖ, οὕτωσί σε ποιεῖν τοῦ λοιποῦ, ὅταν τι τοιοῦτον λέγῃ τις περὶ ἡμῶν τινός, γράμματα πέμψαντα ἐμὲ ἐρέσθαι· ἐγὼ γὰρ τἀληθῆ λέγειν οὔτε ὀκνήσω οὔτε αἰσχυνοῦμαι.

¹ ἡμῖν MSS.: ὑμῖν Zur.

---

[1] A disciple of Archytas of Tarentum, the Pythagorean scientist; cf. *Epp.* iii. 319 A; vii. 339 A, 349 D.
[2] Cf. *Ep.* vii. 347 C.
[3] This closely resembles *Laws* 835 C (with μόνος for μέγας).

EPISTLE II

PLATO TO DIONYSIUS WISHES WELL-DOING

I HEAR from Archedemus [1] that you think that not only I myself should keep quiet but my friends also from doing or saying anything bad about you; and that "you except Dion only." [2] Now your saying this, that Dion is excepted, implies that I have no control over my friends; for had I had this control over you and Dion, as well as the rest, more blessings would have come to us all and to the rest of the Greeks also, as I affirm. But as it is, my greatness consists in making myself follow my own instructions.[3] However, I do not say this as though what Cratistolus and Polyxenus [4] have told you is to be trusted; for it is said that one of these men declares that at Olympia [5] he heard quite a number of my companions maligning you. No doubt his hearing is more acute than mine; for I certainly heard no such thing. For the future, whenever anyone makes such a statement about any of us, what you ought, I think, to do is to send me a letter of inquiry; for I shall tell the truth without scruple or shame.

---

[4] Polyxenus was a Sophist and a disciple of Bryson of Megara, cf. 314 D, and *Ep.* xiii. 360 c. Of Cratistolus nothing further is known.

[5] Probably the Olympic Festival of 364 B.C. (not 360 B.C. as in *Ep.* vii. 350 B); see the Prefatory Note.

PLATO

Ἐμοὶ δὲ δὴ καὶ σοὶ τὰ πρὸς ἀλλήλους οὑτωσὶ τυγχάνει ἔχοντα· οὔτε αὐτοὶ ἀγνῶτές ἐσμεν οὐδενὶ Ἑλλήνων ὡς ἔπος εἰπεῖν, οὔτε ἡ συνουσία ἡμῶν σιγᾶται. μὴ λανθανέτω δέ σε ὅτι οὐδ' εἰς τὸν ἔπειτα χρόνον σιγηθήσεται· τοσοῦτοι[1] οἱ παραδεδεγμένοι εἰσὶν αὐτήν, ἅτε οὐκ ὀλίγην γεγενημένην οὐδ' ἠρέμα. τί οὖν δὴ λέγω νυνί; ἐρῶ ἄνωθεν ἀρξάμενος. πέφυκε ξυνιέναι εἰς ταὐτὸ φρόνησίς τε καὶ δύναμις μεγάλη, καὶ ταῦτ' ἀλλήλ' ἀεὶ διώκει καὶ ζητεῖ καὶ ξυγγίγνεται· ἔπειτα καὶ οἱ ἄνθρωποι χαίρουσι περὶ τούτων αὐτοί τε διαλεγόμενοι καὶ ἄλλων ἀκούοντες ἔν τε ἰδίαις ξυνουσίαις καὶ ἐν ταῖς ποιήσεσιν· οἷον καὶ περὶ Ἱέρωνος ὅταν διαλέγωνται ἄνθρωποι καὶ Παυσανίου τοῦ Λακεδαιμονίου, χαίρουσι τὴν Σιμωνίδου ξυνουσίαν παραφέροντες, ἅ τε ἔπραξε καὶ εἶπε πρὸς αὐτούς· καὶ Περίανδρον τὸν Κορίνθιον καὶ Θαλῆν τὸν Μιλήσιον ὑμνεῖν εἰώθασιν ἅμα, καὶ Περικλέα καὶ Ἀναξαγόραν, καὶ Κροῖσον αὖ καὶ Σόλωνα ὡς σοφοὺς καὶ Κῦρον ὡς δυνάστην. καὶ δὴ ταῦτα μιμούμενοι οἱ ποιηταὶ Κρέοντα μὲν καὶ Τειρεσίαν συνάγουσι, Πολύειδον δὲ καὶ Μίνω, Ἀγαμέμνονα δὲ καὶ Νέστορα καὶ Ὀδυσσέα καὶ Παλαμήδη. ὡς δ' ἐμοὶ δοκεῖ, καὶ Προμηθέα Διὶ ταύτῃ πῃ συνῆγον οἱ πρῶτοι ἄνθρωποι. τούτων δὲ τοὺς μὲν εἰς διαφοράν, τοὺς δ' εἰς φιλίαν ἀλλήλοις ἰόντας, τοὺς

---

[1] τοσοῦτοι H. Richards: τοιοῦτοι mss., Zur.

---

[1] Hiero, the elder, was tyrant of Gela and Syracuse 485–467 B.C. Pausanias defeated the Persians at Plataea 479 B.C. Simonides of Ceos was a famous lyric poet.

[2] Periander was tyrant of Corinth; Thales the first of the Ionian philosophers; Pericles the famous Athenian statesman; Anaxagoras, of Clazomenae, the philosopher;

EPISTLE II

Now as for you and me, the relation in which we stand towards each other is really this. There is not a single Greek, one may say, to whom we are unknown, and our intercourse is a matter of common talk; and you may be sure of this, that it will be common talk also in days to come, because so many have heard tell of it owing to its duration and its publicity. What, now, is the point of this remark? I will go back to the beginning and tell you. It is natural for wisdom and great power to come together, and they are for ever pursuing and seeking each other and consorting together. Moreover, these are qualities which people delight in discussing themselves in private conversation and hearing others discuss in their poems. For example, when men talk about Hiero[1] or about Pausanias the Lacedaemonian they delight to bring in their meeting with Simonides and what he did and said to them; and they are wont to harp on Periander of Corinth and Thales of Miletus, and on Pericles and Anaxagoras, and on Croesus also and Solon as wise men with Cyrus as potentate.[2] The poets, too, follow their example, and bring together Creon and Tiresias, Polyeidus and Minos, Agamemnon and Nestor, Odysseus and Palamedes[3]; and so it was, I suppose, that the earliest men also brought together Prometheus and Zeus. And of these some were—as the poets tell—at feud with each other, and others were

Croesus, king of Lydia, famed for his wealth; Solon, the Athenian legislator; Cyrus, the Persian king, who overthrew Croesus.

[3] Creon and Tiresias are characters in Sophocles' *Oed. Tyr.* and *Antig.*; Polyeidus and Minos in Eurip. *Polyeidus*; the rest in Homer; Aeschylus, in *Prom. Vinct.*, tells us about Zeus and Prometheus.

δὲ τοτὲ μὲν εἰς φιλίαν, τοτὲ δ' εἰς διαφοράν, καὶ τὰ μὲν ὁμονοοῦντας, τὰ δὲ διαφερομένους ᾄδουσι.

Πάντα δὴ ταῦτα λέγω τόδε βουλόμενος ἐνδείξασθαι, ὅτι οὐκ ἐπειδὰν ἡμεῖς τελευτήσωμεν καὶ οἱ λόγοι οἱ περὶ ἡμῶν αὐτῶν σεσιγήσονται, ὥστ' ἐπιμελητέον αὐτῶν ἐστίν. ἀνάγκη γάρ, ὡς ἔοικε, μέλειν ἡμῖν καὶ τοῦ ἔπειτα χρόνου, ἐπειδὴ καὶ τυγχάνουσι κατά τινα φύσιν οἱ μὲν ἀνδραποδωδέστατοι οὐδὲν φροντίζοντες αὐτοῦ, οἱ δ' ἐπιεικέστατοι πᾶν ποιοῦντες ὅπως ἂν εἰς τὸν ἔπειτα χρόνον εὖ ἀκούσωσιν. ὃ δὴ καὶ ἐγὼ τεκμήριον ποιοῦμαι ὅτι ἔστι τις αἴσθησις τοῖς τεθνεῶσι τῶν ἐνθάδε· αἱ γὰρ βέλτισται ψυχαὶ μαντεύονται ταῦτα οὕτως ἔχειν, αἱ δὲ μοχθηρόταται οὔ φασι, κυριώτερα δὲ τὰ τῶν θείων ἀνδρῶν μαντεύματα ἢ τὰ τῶν μή.

Οἶμαι δ' ἔγωγε τοῖς ἔμπροσθεν, περὶ ὧν λέγω, εἰ ἐξείη αὐτοῖς ἐπανορθώσασθαι τὰς αὑτῶν συνουσίας, πάνυ ἂν σπουδάσαι ὥστε βελτίω λέγεσθαι περὶ αὐτῶν ἢ νῦν. τοῦτο οὖν ἡμῖν ἔτι σὺν θεῷ εἰπεῖν ἔξεστιν, εἴ τι ἄρα μὴ καλῶς πέπρακται κατὰ τὴν ἔμπροσθεν συνουσίαν, ἐπανορθώσασθαι καὶ ἔργῳ καὶ λόγῳ· περὶ γὰρ φιλοσοφίαν φημὶ ἐγὼ τὴν ἀληθινὴν δόξαν ἔσεσθαι καὶ λόγον[1] ἡμῶν μὲν ὄντων ἐπιεικῶν βελτίω, φαύλων δὲ τοὐναντίον. καί τοι περὶ τούτου ἡμεῖς ἐπιμελούμενοι οὐδὲν ἂν εὐσεβέστερον πράττοιμεν, οὐδ' ἀμελοῦντες ἀσεβέστερον.

Ὡς δὴ δεῖ γίγνεσθαι, καὶ τὸ δίκαιον ᾗ ἔχει, ἐγὼ φράσω. ἦλθον ἐγὼ εἰς Σικελίαν δόξαν ἔχων πολὺ

[1] καὶ λόγον MSS.: om. Zur.

---

[1] This question is also alluded to in *Menex.* 248 c, *Apol.* 40 c ff.

friends; while others again were now friends and now foes, and partly in agreement and partly in disagreement.

Now my object in saying all this is to make it clear, that when we ourselves die men's talk about us will not likewise be silenced; so that we must be careful about it. We must necessarily, it seems, have a care also for the future, seeing that, by some law of nature, the most slavish men pay no regard to it, whereas the most upright do all they can to ensure that they shall be well spoken of in the future. Now I count this as a proof that the dead have some perception of things here on earth;[1] for the best souls divine that this is so, while the worst deny it; and the divinings of men who are godlike are of more authority than those of men who are not.

I certainly think that, had it been in their power to rectify what was wrong in their intercourse, those men of the past whom I have mentioned would have striven to the utmost to ensure a better report of themselves than they now have.[2] In our case, then—if God so grant—it still remains possible to put right whatever has been amiss in word or deed during our intercourse in the past. For I maintain that, as regards the true philosophy, men will think and speak well of it if we ourselves are upright, and ill if we are base. And in truth we could do nothing more pious than to give attention to this matter, nothing more impious than to disregard it.

How this result should be brought about, and what is the just course to pursue, I will now explain. I came to Sicily with the reputation of being by far

---

[1] On the subject of posthumous fame *cf. Sympos.* 208 c ff.

PLATO

311 τῶν ἐν φιλοσοφίᾳ διαφέρειν, βουλόμενος δὲ ἐλθὼν
312 εἰς Συρακούσας συμμάρτυρα λαβεῖν σέ, ἵνα δή
μοι τιμῷτο φιλοσοφία καὶ παρὰ τῷ πλήθει.
τοῦτο δ' οὐκ εὐαγές μοι ἀπέβη. τὸ δ' αἴτιον οὐ
λέγω ὅπερ ἂν πολλοὶ εἴποιεν, ἀλλ' ὅτι ἐφαίνου οὐ
πάνυ ἐμοὶ πιστεύειν σύ, ἀλλ' ἐμὲ μέν πως ἀπο-
πέμψασθαι ἐθέλειν, ἑτέρους δὲ μεταπέμψασθαι, καὶ
ζητεῖν τὸ πρᾶγμα τί τὸ ἐμόν ἐστιν, ἀπιστῶν, ὡς
ἐμοὶ δοκεῖ. καὶ οἱ ἐπὶ τούτοις βοῶντες πολλοὶ
ἦσαν, λέγοντες ὡς σὺ ἐμοῦ μὲν καταπεφρόνηκας,
B ἄλλα δὲ ἐσπούδακας. ταῦτα δὴ διαβεβόηται.

Ὁ δὲ μετὰ ταῦτα δίκαιόν ἐστι ποιεῖν, ἄκουε, ἵνα
σοι καὶ ἀποκρίνωμαι ὃ σὺ ἐρωτᾷς, πῶς χρὴ ἔχειν
ἐμὲ καὶ σὲ πρὸς ἀλλήλους. εἰ μὲν ὅλως φιλοσοφίας
καταπεφρόνηκας, ἐᾶν χαίρειν· εἰ δὲ παρ' ἑτέρου
ἀκήκοας ἢ αὐτὸς βελτίονα εὕρηκας τῶν παρ' ἐμοί,
ἐκεῖνα τίμα· εἰ δ' ἄρα τὰ παρ' ἡμῶν σοι ἀρέσκει,
τιμητέον καὶ ἐμὲ μάλιστα. νῦν οὖν, ὥσπερ καὶ
ἐξ ἀρχῆς, σὺ καθηγοῦ, ἕψομαι δὲ ἐγώ· τιμώμενος
C μὲν γὰρ ὑπὸ σοῦ τιμήσω σέ, μὴ τιμώμενος δὲ
ἡσυχίαν ἄξω. ἔτι δὲ σὺ μὲν ἐμὲ τιμῶν καὶ τούτου
καθηγούμενος φιλοσοφίαν δόξεις τιμᾶν, καὶ αὐτὸ
τοῦτο ὅτι διεσκόπεις καὶ ἄλλους[1] πρὸς πολλῶν
εὐδοξίαν σοι οἴσει ὡς φιλοσόφῳ ὄντι· ἐγὼ δὲ σὲ
τιμῶν μὴ τιμῶντα πλοῦτον δόξω θαυμάζειν τε
καὶ διώκειν, τοῦτο δ' ἴσμεν ὅτι παρὰ πᾶσιν ὄνομα

---

[1] ἄλλους MSS.: ἄλλως Zur.

---

[1] A most un-Platonic sentiment: contrast *Rep.* 493 E ff., and 314 A below.
[2] For Dionysius as a philosopher *cf. Ep.* vii. 345 B; and for the discussion of honour and dishonour as between D. and P. *cf. Ep.* vii. 345 C, 350 C.

EPISTLE II

the most eminent of those engaged in philosophy; and I desired, on my arrival in Syracuse, to gain your testimony as well, in order that I might get philosophy held in honour even by the multitude.[1] In this, however, I was disappointed. But the reason I give for this is not that which is commonly given; rather it was because you showed that you did not fully trust me but wished rather to get rid of me somehow and invite others in my place; and owing, as I believe, to your distrust of me, you showed yourself inquisitive as to what my business was. Thereupon it was proclaimed aloud by many that you utterly despised me and were devoted to other affairs. This certainly was the story noised abroad.

And now I will tell you what it is right to do after this, that so I may reply also to your question how you and I ought to behave towards each other. If you altogether despise philosophy, leave it alone. If, again, you have been taught by someone else or have yourself invented better doctrines than mine, hold them in honour.[2] But if you are contented with my doctrines, then you should hold me also in special honour. So now, just as at the beginning, do you lead the way and I will follow. If I am honoured by you, I will honour you; but if I am not honoured I will keep to myself. Moreover, if you honour me and take the lead in so doing, you will be thought to be honouring philosophy; and the very fact that you have studied other systems as well will gain you the credit, in the eyes of many, of being a philosopher yourself. But if I honour you, while you do not honour me, I shall be deemed to be a man who worships and pursues after wealth; and to such conduct everyone, we know, gives a bad

PLATO

οὐ καλὸν ἔχει. ὡς δ' ἐν κεφαλαίῳ εἰπεῖν, σοῦ μὲν τιμῶντος ἀμφοτέροις κόσμος, ἐμοῦ δὲ ὄνειδος ἀμφοῖν. περὶ μὲν οὖν τούτων ταῦτα.

Τὸ δὲ σφαιρίον οὐκ ὀρθῶς ἔχει· δηλώσει δέ σοι Ἀρχέδημος, ἐπειδὰν ἔλθῃ. καὶ δὴ καὶ περὶ τοῦδε, ὃ τούτου τιμιώτερόν τ' ἐστὶ καὶ θειότερον, καὶ μάλα σφόδρ' αὐτῷ δηλωτέον, ὑπὲρ οὗ σὺ πέπομφας ἀπορούμενος. φῇς γὰρ δὴ κατὰ τὸν ἐκείνου λόγον οὐχ ἱκανῶς ἀποδεδεῖχθαί σοι περὶ τῆς τοῦ πρώτου φύσεως. φραστέον δή σοι δι' αἰνιγμῶν, ἵν' ἄν τι ἡ δέλτος ἢ πόντου ἢ γῆς ἐν πτυχαῖς πάθῃ, ὁ ἀναγνοὺς μὴ γνῷ. ὧδε γὰρ ἔχει. περὶ τὸν πάντων βασιλέα 'πάντ' ἐστὶ καὶ ἐκείνου ἕνεκα πάντα, καὶ ἐκεῖνο αἴτιον ἁπάντων τῶν καλῶν. δεύτερον δὲ πέρι τὰ δεύτερα, καὶ τρίτον πέρι τὰ τρίτα. ἡ οὖν ἀνθρωπίνη ψυχὴ περὶ αὐτὰ ὀρέγεται μαθεῖν ποῖ' ἄττα ἐστί, βλέπουσα εἰς τὰ αὑτῆς συγγενῆ, ὧν οὐδὲν ἱκανῶς ἔχει. τοῦ δὴ βασιλέως πέρι καὶ ὧν εἶπον, οὐδέν ἐστι τοιοῦτον. τὸ δὴ μετὰ τοῦτο ἡ ψυχή φησιν· ἀλλὰ ποῖόν τι μήν; τοῦτ' ἐστίν, ὦ παῖ Διονυσίου καὶ Δωρίδος, τὸ ἐρώτημα ὃ πάντων αἴτιόν ἐστι κακῶν, μᾶλλον δὲ ἡ περὶ τούτου ὠδὶς ἐν τῇ ψυχῇ ἐγγιγνομένη, ἣν εἰ μή τις ἐξαιρεθήσεται, τῆς ἀληθείας ὄντως οὐ μή ποτε τύχῃ.

Σὺ δὲ τοῦτο πρὸς ἐμὲ ἐν τῷ κήπῳ ὑπὸ ταῖς δάφναις αὐτὸς ἔφησθα ἐννενοηκέναι καὶ εἶναι σὸν

---

[1] Apparently some form of orrery, devised to illustrate the motions of the heavenly bodies; *cf.* Cicero, *De Rep.* i. 14; *De nat. deor.* ii. 34.

[2] For this phrase *cf. Laws* 886 c. The explanation of "the Three" (principles) which follows is a piece of wanton mystification, of which it is impossible to suppose that Plato

EPISTLE II

name. So, to sum it all up, if you pay the honour, it will be a credit to both of us, but if I pay it a disgrace to both. So much, then, about this subject.

As to the globe,[1] there is something wrong with it; and Archedemus will point it out to you when he arrives. There is also another matter—much more valuable and divine than the globe—which he most certainly must explain, as you were puzzled about it when you sent him. For, according to his report, you say that you have not had a sufficient demonstration of the doctrine concerning the nature of " the First."[2] Now I must expound it to you in a riddling way in order that, should the tablet come to any harm " in folds of ocean or of earth," he that readeth may not understand.

The matter stands thus: Related to the King of All are all things, and for his sake they are, and of all things fair He is the cause. And related to the Second are the second things; and related to the Third the third. About these, then, the human soul strives to learn, looking to the things that are akin to itself, whereof none is fully perfect. But as to the King and the objects I have mentioned, they are of quite different quality. In the next place the soul inquires—" Well then, what quality have they ? " But the cause of all the mischief, O son of Dionysius and Doris, lies in this very question, or rather in the travail which this question creates in the soul; and unless a man delivers himself from this he will never really attain the truth.

You, however, declared to me in the garden, under the laurels, that you had formed this notion yourself

could ever have been guilty. For attempts to solve " the riddle " see Prefatory Note.

PLATO

313
B εὕρημα. καὶ ἐγὼ εἶπον ὅτι τοῦτο εἰ φαίνοιτό σοι οὕτως ἔχειν, πολλῶν ἂν εἴης λόγων ἐμὲ ἀπολελυκώς. οὐ μὴν ἄλλῳ γέ ποτ' ἔφην ἐντετυχηκέναι τοῦθ' εὑρηκότι, ἀλλὰ ἡ πολλή μοι πραγματεία περὶ τοῦτ' εἴη. σὺ δὲ ἴσως μὲν ἀκούσας του, τάχα δ' ἂν θείᾳ μοίρᾳ κατὰ τοῦθ' ὁρμήσας, ἔπειτα αὐτοῦ τὰς ἀποδείξεις ὡς ἔχων βεβαίως οὐ κατέδησας, ἀλλ' ᾄττει σοι[1] τοτὲ μὲν οὕτω, τοτὲ δὲ ἄλλως περὶ τὸ φανταζόμενον, τὸ δὲ οὐδέν ἐστι τοιοῦτον. καὶ
C τοῦτο οὐ σοὶ μόνῳ γέγονεν, ἀλλ' εὖ ἴσθι μηδένα πώποτέ μου τὸ πρῶτον ἀκούσαντα ἔχειν ἄλλως πως ἢ οὕτω κατ' ἀρχάς, καὶ ὁ μὲν πλείω ἔχων πράγματα, ὁ δὲ ἐλάττω μόγις ἀπαλλάττονται, σχεδὸν δὲ οὐδεὶς ὀλίγα.

Τούτων δὴ γεγονότων καὶ ἐχόντων οὕτω, σχεδὸν κατὰ τὴν ἐμὴν δόξαν εὑρήκαμεν ὃ σὺ ἐπέστειλας, ὅπως δεῖ πρὸς ἀλλήλους ἡμᾶς ἔχειν. ἐπεὶ γὰρ βασανίζεις αὐτὰ ξυγγιγνόμενός τε ἄλλοις καὶ
D παραθεώμενος παρὰ τὰ τῶν ἄλλων καὶ αὐτὰ καθ' αὑτά, νῦν σοι ταῦτά τε, εἰ ἀληθὴς ἡ βάσανος, προσφύσεται, καὶ οἰκεῖος τούτοις τε καὶ ἡμῖν ἔσει.

Πῶς οὖν αὐτά τ'[2] ἔσται καὶ πάντα ἃ εἰρήκαμεν; τὸν Ἀρχέδημον νῦν τε ὀρθῶς ἐποίησας πέμψας, καὶ τὸ λοιπόν, ἐπειδὰν ἔλθῃ πρὸς σὲ καὶ ἀπαγγείλῃ τὰ παρ' ἐμοῦ, μετὰ ταῦτα ἴσως ἄλλαι σε ἀπορίαι

---

[1] ᾄττει σοι Burnet : ἀττισοι MSS. : ἄττεις Zur.
[2] αὐτά τ' some MSS. : οὐ ταῦτ' best MSS., Zur.

---

[1] This phrase echoes *Theaet.* 188 c.
[2] There are echoes here of *Meno* 97 E ff., 100 A, and *Theaet.* 151 A ff. *Cf.* also *Ep.* vii. 340 B, 343 C, 344 B.

EPISTLE II

and that it was a discovery of your own; and I made answer that if it was plain to you that this was so, you would have saved me from a long discourse.[1] I said, however, that I had never met with any other person who had made this discovery; on the contrary most of the trouble I had was about this very problem. So then, after you had either, as is probable, got the true solution from someone else, or had possibly (by Heaven's favour) hit on it yourself, you fancied you had a firm grip on the proofs of it, and so you omitted to make them fast; thus your view of the truth sways now this way, now that, round about the apparent object; whereas the true object is wholly different.[2] Nor are you alone in this experience; on the contrary, there has never yet been anyone, I assure you, who has not suffered the same confusion at the beginning, when he first learnt this doctrine from me; and they all overcome it with difficulty, one man having more trouble and another less, but scarcely a single one of them escapes with but little.

So now that this has occurred, and things are in this state, we have pretty well found an answer, as I think, to the question how we ought to behave towards each other. For seeing that you are testing my doctrines both by attending the lectures of other teachers and by examining my teaching side by side with theirs, as well as by itself, then, if the test you make is a true one, not only will these doctrines implant themselves now in your mind, but you also will be devoted both to them and to us.

How, then, will this, and all that I have said, be brought to pass? You have done right now in sending Archedemus; and in the future also, after he returns to you and reports my answer, you will probably be

PLATO

313 λήψονται. πέμψεις οὖν αὖθις, ἂν ὀρθῶς βουλεύῃ, παρ' ἐμὲ τὸν Ἀρχέδημον, ὁ δ' ἐμπορευσάμενος ἥξει
E πάλιν· καὶ τοῦτο ἐὰν δὶς ἢ τρὶς ποιήσῃς καὶ βασανίσῃς τὰ παρ' ἐμοῦ πεμφθέντα ἱκανῶς, θαυμάζοιμ' ἂν εἰ μὴ τὰ νῦν[1] ἀπορούμενα πολύ σοι διοίσει ἢ νῦν.[2] θαρροῦντες οὖν ποιεῖτε οὕτως· οὐ μὴ γάρ ποτε τῆς ἐμπορίας ταύτης οὔτε σὺ στείλῃς οὔτε Ἀρχέδημος ἐμπορεύσεται καλλίω καὶ θεοφιλεστέραν.

314 Εὐλαβοῦ μέντοι μή ποτε ἐκπέσῃ ταῦτα εἰς ἀνθρώπους ἀπαιδεύτους· σχεδὸν γάρ, ὡς ἐμοὶ δοκεῖ, οὐκ ἔστι τούτων πρὸς τοὺς πολλοὺς καταγελαστότερα ἀκούσματα, οὐδ' αὖ πρὸς τοὺς εὐφυεῖς θαυμαστότερά τε καὶ ἐνθουσιαστικώτερα. πολλάκις δὲ λεγόμενα καὶ ἀεὶ ἀκουόμενα καὶ πολλὰ ἔτη μόγις, ὥσπερ χρυσός, ἐκκαθαίρεται μετὰ πολλῆς πραγματείας. ὃ δὲ θαυμαστὸν αὐτοῦ γέγονεν, ἄκουσον. εἰσὶ γὰρ ἄνθρωποι ταῦτα
B ἀκηκοότες καὶ πλείους, δυνατοὶ μὲν μαθεῖν, δυνατοὶ δὲ μνημονεῦσαι καὶ βασανίσαντες πάντῃ πάντως κρῖναι, γέροντες ἤδη καὶ οὐκ ἐλάττω τριάκοντα ἐτῶν ἀκηκοότες, οἳ νῦν ἄρτι σφίσι φασὶ τὰ μὲν τότε ἀπιστότατα δόξαντα εἶναι νῦν πιστότατα καὶ ἐναργέστατα φαίνεσθαι, ἃ δὲ τότε πιστότατα, νῦν τοὐναντίον. πρὸς ταῦτ' οὖν σκοπῶν εὐλαβοῦ μή ποτέ σοι μεταμελήσῃ τῶν νῦν ἀναξίως ἐκπεσόντων. μεγίστη δὲ φυλακὴ τὸ μὴ γράφειν ἀλλ' ἐκμαν-

---

[1] νῦν mss.: πρὶν Zur.
[2] νῦν some mss.: τὰ νῦν other mss., Zur.

---

[1] A Pythagorean touch, *cf.* Horace's "odi profanum volgus et arceo."
[2] This would make Plato's teaching go back to 393 B.C.,

EPISTLE II

beset later on with fresh perplexities. Then, if you are rightly advised, you will send Archedemus back to me, and he with his cargo will return to you again. And if you do this twice or thrice, and fully test the doctrines I send you, I shall be surprised if your present difficulties do not assume quite a new aspect. Do you, therefore, act so, and with confidence; for there is no merchandise more fair than this or dearer to Heaven which you can ever dispatch or Archedemus transport.

Beware, however, lest these doctrines be ever divulged to uneducated people.[1] For there are hardly any doctrines, I believe, which sound more absurd than these to the vulgar, or, on the other hand, more admirable and inspired to men of fine disposition. For it is through being repeated and listened to frequently for many years that these doctrines are refined at length, like gold, with prolonged labour. But listen now to the most remarkable result of all. Quite a number of men there are who have listened to these doctrines—men capable of learning and capable also of holding them in mind and judging them by all sorts of tests—and who have been hearers of mine for no less than thirty years [2] and are now quite old; and these men now declare that the doctrines that they once held to be most incredible appear to them now the most credible, and what they then held most credible now appears the opposite. So, bearing this in mind, have a care lest one day you should repent of what has now been divulged improperly. The greatest safeguard is to avoid writing and to learn by heart; for it is not

e. five or six years before he founded the Academy—which seems improbable.

C θάνειν· οὐ γὰρ ἔστι τὰ γραφέντα μὴ οὐκ ἐκπεσεῖν. διὰ ταῦτα οὐδὲν πώποτ' ἐγὼ περὶ τούτων γέγραφα, οὐδ' ἔστι σύγγραμμα Πλάτωνος οὐδὲν οὐδ' ἔσται, τὰ δὲ νῦν λεγόμενα Σωκράτους ἐστὶ καλοῦ καὶ νέου γεγονότος. ἔρρωσο καὶ πείθου, καὶ τὴν ἐπιστολὴν ταύτην νῦν πρῶτον πολλάκις ἀναγνοὺς κατάκαυσον.

Ταῦτα μὲν ταύτῃ. περὶ δὲ Πολυξένου ἐθαύμασας ὅτι [οὐ] πέμψαιμί σοι. ἐγὼ δὲ καὶ περὶ Λυκόφρονος
D καὶ τῶν ἄλλων τῶν παρὰ σοὶ ὄντων λέγω καὶ πάλαι καὶ νῦν τὸν αὐτὸν λόγον, ὅτι πρὸς τὸ διαλεχθῆναι καὶ φύσει καὶ τῇ μεθόδῳ τῶν λόγων πάμπολυ διαφέρεις αὐτῶν, καὶ οὐδεὶς αὐτῶν ἑκὼν ἐξελέγχεται, ὥς τινες ὑπολαμβάνουσιν, ἀλλ' ἄκοντες. καὶ δοκεῖς μέντοι πάνυ μετρίως κεχρῆσθαί τε αὐτοῖς καὶ δεδωρῆσθαι. ταῦτα μὲν περὶ τούτων πολλὰ ὡς περὶ τοιούτων.

Φιλιστίωνι δέ, εἰ μὲν αὐτὸς χρῇ, σφόδρα χρῶ,
E εἰ δὲ οἷόν τε, Σπευσίππῳ χρῆσον καὶ ἀπόπεμψον· δεῖται δὲ σοῦ καὶ Σπεύσιππος· ὑπέσχετο δέ μοι καὶ Φιλιστίων, εἰ σὺ ἀφείης¹ αὐτόν, ἥξειν προθύμως Ἀθήναζε. τὸν ἐκ τῶν λιθοτομιῶν εὖ ἐποίησας ἀφείς, ἐλαφρὰ δὲ ἡ δέησις καὶ περὶ τῶν οἰκετῶν αὐτοῦ καὶ περὶ Ἡγησίππου τοῦ Ἀρίστωνος· ἐπέστειλας γάρ μοι, ἄν τις ἀδικῇ ἢ τοῦτον ἢ

---

¹ ἀφείης Hermann : ἀφίης mss., Zur.

---

[1] This curious statement seems based on *Ep.* vii. 341 c combined perhaps with an allusion to the *Parmenides*.
[2] A contemporary Sophist.
[3] A physician at the court of Dionysius.

EPISTLE II

possible that what is written down should not get divulged. For this reason I myself have never yet written anything on these subjects, and no treatise by Plato exists or will exist, but those which now bear his name belong to a Socrates become fair and young.[1] Fare thee well, and give me credence; and now, to begin with, read this letter over repeatedly and then burn it up.

So much, then, for that. You were surprised at my sending Polyxenus to you; but now as of old I repeat the same statement about Lycophron[2] also and the others you have with you, that, as respects dialectic, you are far superior to them all both in natural intelligence and in argumentative ability; and I maintain that if any of them is beaten in argument, this defeat is not voluntary, as some imagine, but involuntary. All the same, it appears that you treat them with the greatest consideration and make them presents. So much, then, about these men: too much, indeed, about such as they!

As for Philistion,[3] if you are making use of him yourself by all means do so; but if not, lend him if possible to Speusippus[4] and send him home. Speusippus, too, begs you to do so; and Philistion also promised me, that, if you would release him, he would gladly come to Athens. Many thanks for releasing the man in the stone-quarries; and my request with regard to his household and Hegesippus, the son of Ariston,[5] is no hard matter; for in your

[4] Plato's nephew, who succeeded him as head of the Academy. If, as seems probable, S. was unknown to D. until he went to Sicily with Plato in 361 B.C., this request seems strange.

[5] Nothing further is known of any of the persons here mentioned.

417

PLATO

ἐκείνους καὶ σὺ αἴσθῃ, μὴ ἐπιτρέψειν. καὶ περὶ Λυσικλείδου τἀληθὲς εἰπεῖν ἄξιον· μόνος γὰρ τῶν ἐκ Σικελίας Ἀθήναζε ἀφικομένων οὐδὲν μετεβάλετο περὶ τῆς σῆς καὶ ἐμῆς συνουσίας, ἀλλ' ἀεί τι ἀγαθὸν καὶ ἐπὶ τὰ βελτίω λέγων περὶ τῶν γεγονότων διατελεῖ.

EPISTLE II

letter you said that should anyone wrong him or them and you come to know of it you would not allow it. It is proper for me also to say what is true about Lysicleides; for of all those who have come to Athens from Sicily he is the only one who has not misrepresented your association with me; on the contrary, he always speaks nicely about past events and puts the best construction on them.

EPISTLE III

Prefatory Note.—This letter begins with a piece of moralizing on the propriety or otherwise of commencing a letter by "wishing joy" to the person addressed; Plato's custom rather being to "wish well-doing" to his friends. He then proceeds to complain that Dionysius has been spreading slanders about him by saying that it was Plato who had prevented him from carrying out the reforms he had always had in mind—namely, converting the tyranny at Syracuse into a constitutional monarchy and planting settlers in the Greek cities of Sicily. Other slanders, also, have been going about to the effect that all that was wrong in the administration of Dionysius was due to the mischievous interference and influence of Plato. The twofold charge calls for a twofold defence. This takes the form of a summary of Plato's objects and activities in connexion with Syracusan affairs. In the first instance he went to Syracuse because of his interest in Dion, and when Dion was exiled he ceased to take part in Sicilian politics, except for attempting to reconcile Dionysius and Dion, and returned to Athens. The reason for his next visit was that Dionysius had made it a condition of his coming to terms with Dion, and that all his friends urged him to go to Syracuse. He did his duty as a friend in going, but the journey turned

PLATO

out as fruitless as he himself expected. Though he did all he could in the interests of Dion, Dionysius kept constantly putting him off with promises he had no intention of keeping, while he was making away with Dion's property in Sicily. At first by his promises he induced Plato to prolong his stay, and then he turned round and accused him of making plots against the tyranny, so as to scare him away. Thus the result of Plato's loyalty to Dion was that he incurred the enmity of the tyrant. This narrative of events is sufficient to dispose of the charge that Plato was responsible for the political acts of Dionysius.

The other charge, that Plato had stood in the way of the reforms proposed by Dionysius, is refuted by recalling a conversation they had in the presence of witnesses. And Plato concludes by affirming that these still form part of the policy he recommends, and bidding Dionysius take back his false charge and "sing a palinode."

It is clear that this letter, if written by Plato, belongs to a time subsequent to his third Syracusan visit, *i.e.* after 360 B.C.; and the tone of it seems rather to point to a date after Dion's successful attempt to establish himself in Sicily in 357 B.C. It reads like an "open" rather than a private letter, in fact a political manifesto intended to support Dion in his policy of political reform.

But it may well be doubted whether this letter is authentic. The preamble, to begin with, is suspiciously like the work of a sophistical rhetor of a later age. There are significant resemblances between the plan of the letter, with its twofold defence against earlier and more recent accusations, and the

EPISTLE III

plan of the *Apology* in which Socrates replies to two sets of accusers. Further, there are many obvious parallels between this letter and the seventh letter which suggest that the writer of the former was borrowing from the latter; for if either letter is really Plato's, it is more likely to be the seventh than the third.

Γ

"Πλάτων Διονυσίῳ χαίρειν" ἐπιστείλας ἆρ' ὀρθῶς ἂν τυγχάνοιμι τῆς βελτίστης προσρήσεως; ἢ μᾶλλον κατὰ τὴν ἐμὴν συνήθειαν γράφων "εὖ πράττειν," ὥσπερ εἴωθα ἐν ταῖς ἐπιστολαῖς τοὺς φίλους προσαγορεύειν; σὺ μὲν γὰρ δὴ καὶ τὸν θεόν, ὡς ἤγγειλαν οἱ τότε θεωροῦντες, προσεῖπες ἐν Δελφοῖς αὐτῷ τούτῳ θωπεύσας τῷ ῥήματι, καὶ γέγραφας, ὥς φασί,

χαῖρε καὶ ἡδόμενον βίοτον διάσωζε τυράννου·

ἐγὼ δὲ οὐδὲ ἀνθρώπῳ κλήσει, μήτι[1] δὴ θεῷ, παρακελευσαίμην ἂν δρᾶν τοῦτο, θεῷ μὲν ὅτι παρὰ φύσιν προστάττοιμ' ἄν, πόρρω γὰρ ἡδονῆς ἵδρυται καὶ λύπης τὸ θεῖον, ἀνθρώπῳ δὲ ὅτι τὰ πολλὰ βλάβην ἡδονὴ καὶ λύπη γεννᾷ, δυσμαθίαν καὶ λήθην καὶ ἀφροσύνην καὶ ὕβριν τίκτουσα ἐν τῇ ψυχῇ. καὶ ταῦτα μὲν οὕτως εἰρήσθω παρ' ἐμοῦ περὶ τῆς προσρήσεως· σὺ δ' ἀναγνοὺς αὐτά, ὅπῃ βούλει δέξασθαι, ταύτῃ δέχου.

Φασὶ δ' οὐκ ὀλίγοι λέγειν σε πρός τινας τῶν παρὰ σὲ πρεσβευόντων, ὡς ἄρα σοῦ ποτε λέγοντος

---

[1] μήτι Burnet : οὔτι (om. δὴ) mss. : μὴ ὅτι Zur.

---

[1] This discussion of the proper form of address is suspiciously like *Charm.* 164 D.

EPISTLE III

"PLATO to Dionysius wishes Joy!" If I wrote thus, should I be hitting on the best mode of address? Or rather, by writing, according to my custom, "Wishes well-doing," this being my usual mode of address, in my letters to my friends? You, indeed,—as was reported by the spectators then present—addressed even the God himself at Delphi in this same flattering phrase, and wrote, as they say, this verse—

> I wish thee joy! And may'st thou always keep
> The tyrant's life a life of pleasantness.

But as for me, I would not call upon a man, and much less a god, and bid him enjoy himself—a god, because I would be imposing a task contrary to his nature (since the Deity has his abode far beyond pleasure or pain),—nor yet a man, because pleasure and pain generate mischief for the most part, since they breed in the soul mental sloth and forgetfulness and witlessness and insolence.[1] Let such, then, be my declaration regarding the mode of address; and you, when you read it, accept it in what sense you please.

It is stated by not a few that you related to some of the ambassadors at your Court, that upon one occasion I heard you speaking of your intention to

ἀκούσας ἐγὼ μέλλοντος τάς τε Ἑλληνίδας πόλεις
ἐν Σικελίᾳ οἰκίζειν καὶ Συρακουσίους ἐπικουφίσαι,
τὴν ἀρχὴν ἀντὶ τυραννίδος εἰς βασιλείαν μεταστήσαντα, ταῦτ' ἄρα σὲ μὲν τότε διεκώλυσα, ὡς
σὺ φῄς, σοῦ σφόδρα προθυμουμένου, νῦν δὲ Δίωνα
διδάσκοιμι δρᾶν αὐτὰ ταῦτα, καὶ τοῖς διανοήμασι
τοῖς σοῖς τὴν σὴν ἀρχὴν ἀφαιρούμεθά σε. σὺ δ'
Ε εἰ μέν τι διὰ τοὺς λόγους τούτους ὠφελεῖ, γιγνώσκεις αὐτός, ἀδικεῖς δ' οὖν ἐμὲ τἀναντία τῶν
γενομένων λέγων. ἅδην γὰρ ὑπὸ Φιλιστίδου καὶ
ἄλλων πολλῶν πρὸς τοὺς μισθοφόρους καὶ εἰς
τὸ Συρακουσίων πλῆθος διεβλήθην διὰ τὸ μένειν
ἐν ἀκροπόλει, τοὺς δ' ἔξωθεν, εἴ τι γίγνοιτο
ἁμάρτημα, πᾶν εἰς ἐμὲ τρέπειν, σὲ φάσκοντας
πάντα ἐμοὶ πείθεσθαι. σὺ δ' αὐτὸς οἶσθα σαφέ-
316 στατα τῶν πολιτικῶν ἐμὲ σοὶ κοινῇ πραγματευ-
σάμενον ἑκόντα ὀλίγα δὴ κατ' ἀρχάς, ὅτε τι[1] πλέον
ποιεῖν ἂν ᾠήθην, ἄλλα τε βραχέα ἄττα καὶ τὰ
περὶ[2] τῶν νόμων προοίμια σπουδάσαντα μετρίως,
χωρὶς ὧν σὺ προσέγραψας ἤ τις ἕτερος· ἀκούω
γὰρ ὕστερον ὑμῶν τινας αὐτὰ διασκευωρεῖν, δῆλα
μὴν ἑκάτερα ἔσται τοῖς τὸ ἐμὸν ἦθος δυναμένοις
κρίνειν.

Ἀλλ' οὖν, ὅπερ ἀρτίως εἶπον, οὐ διαβολῆς
προσδέομαι πρός τε Συρακουσίους καὶ εἰ δή τινας
ἑτέρους πείθεις λέγων αὐτά, ἀλλὰ πολὺ μᾶλλον
Β ἀπολογίας πρός τε τὴν προτέραν γενομένην
διαβολὴν καὶ τὴν νῦν μετ' ἐκείνην μείζω φυομένην

[1] ὅτε τι MS. corr.: ὅπῃ mss., Zur.
[2] τὰ περὶ MSS.: περὶ τὰ Zur.

[1] To be identified, possibly, with the Sicilian historian

EPISTLE III

occupy the Greek cities in Italy and to relieve the Syracusans by changing the government to a monarchy instead of a tyranny, and at that time (as you assert) I stopped you from doing so, although you were most eager to do it, whereas now I am urging Dion to do precisely the same thing; and thus we are robbing you of your empire by means of your own plans. Whether you derive any benefit from this talk you know best yourself, but you certainly wrong me by saying what is contrary to the fact. For of false accusation I have had enough from Philistides [1] and many others who accused me to the mercenaries and to the Syracusan populace because I stayed in the acropolis; and the people outside, whenever a mistake occurred, ascribed it entirely to me, alleging that you obeyed me in all things. But you yourself know for certain that I willingly took part in some few of your political acts at the first, when I thought that I was doing some good by it; and that I gave a fair amount of attention to the Preludes of the laws,[2] besides other small matters, apart from the additions in writing made by you or anyone else—for I am told that some of you afterwards revised my Preludes; but no doubt the several contributions will be evident to those who are competent to appreciate my style.

Well then, as I said just now, what I need is not any further accusation to the Syracusans, or any others there may be who believe your story, but much rather a defence not only against the previous false accusations, but also against the graver and more violent accusation which is now being concocted

---

[1] Philistus, exiled by Dionysius I. and subsequently restored to favour (*cf.* Plutarch, *Dion*, c. 19).   [2] *Cf. Laws* 722 D ff.

καὶ σφοδροτέραν. πρὸς δύο δή μοι διττὰς ἀναγκαῖον ποιήσασθαι τὰς ἀπολογίας, πρῶτον μὲν ὡς εἰκότως σοι ἔφυγον κοινωνεῖν περὶ τὰ τῆς πόλεως πράγματα, τὸ δὲ δεύτερον ὡς οὐκ ἐμὴν ταύτην εἴρηκας συμβουλὴν οὐδὲ διακώλυσιν, μέλλοντί σοι κατοικίζειν Ἑλληνίδας πόλεις ἐμποδὼν ἐμὲ γεγενῆσθαι. τὴν οὖν ἀρχὴν ὧν εἶπον περὶ προτέρων ἄκουε πρότερον.

Ἦλθον καλούμενος εἰς Συρακούσας ὑπό τε σοῦ καὶ Δίωνος, τοῦ μὲν δεδοκιμασμένου παρ' ἐμοὶ καὶ ξένου πάλαι γεγονότος, ἐν ἡλικίᾳ δὲ ὄντος μέσῃ τε καὶ καθεστηκυίᾳ, ὧν δὴ παντάπασι χρεία τοῖς νοῦν καὶ σμικρὸν κεκτημένοις μέλλουσι περὶ τοσούτων ὅσα ἦν τότε τὰ σὰ βουλεύεσθαι, σοῦ δὲ ὄντος μὲν σφόδρα νέου, πολλῆς δὲ ἀπειρίας οὔσης περί σε τούτων ὧν ἔμπειρον ἔδει γεγονέναι, καὶ σφόδρα ἀγνῶτος ἐμοί. τὸ μετὰ τοῦτο εἴτ' ἄνθρωπος εἴτε θεὸς εἴτε τύχη τις μετὰ σοῦ Δίωνα ἐξέβαλε, καὶ ἐλείφθης μόνος. ἆρ' οὖν οἴει μοι τότε πολιτικῶν εἶναι κοινωνίαν πρὸς σέ, τὸν μὲν ἔμφρονα κοινωνὸν ἀπολωλεκότι, τὸν δὲ ἄφρονα ὁρῶντι μετὰ πονηρῶν καὶ πολλῶν ἀνθρώπων καταλελειμμένον, οὐκ ἄρχοντα, οἰόμενον δ' ἄρχειν, ὑπὸ δὲ τοιούτων ἀνθρώπων ἀρχόμενον; ἐν οἷς τί χρῆν ποιεῖν ἐμέ; μῶν οὐχ ὅπερ ἐποίουν ἀναγκαῖον ἐκ τῶν λοιπῶν τὰ μὲν πολιτικὰ χαίρειν ἐᾶν, εὐλαβούμενον τὰς ἐκ τῶν φθόνων διαβολάς, ὑμᾶς δὲ πάντως, καίπερ ἀλλήλων χωρὶς γεγονότας καὶ διαφόρους ὄντας, πειρᾶσθαι φίλους ἀλλήλοις ὅ τι μάλιστα ποιεῖν; τούτων δὴ καὶ σὺ μάρτυς, ὅτι

EPISTLE III

to follow it. Against the two accusations I must necessarily make a twofold defence—stating, firstly, that I reasonably avoided sharing in your political transactions; and, secondly, that neither the advice was mine, nor yet the hindrance you alleged,—when you said that I had stopped you when you proposed to plant colonists in the Greek cities. So, listen first to the origin of the first of the accusations I have mentioned.

It was on your invitation and Dion's that I came to Syracuse. Dion was a tried comrade of mine and a guest-friend of old standing, and he was a man of staid middle age,—qualities that are specially required by men who possess even a particle of sense when they intend to advise concerning affairs so important as yours then were. You, on the other hand, were extremely young, and in your case I was quite without experience of those points regarding which experience was required, as I was totally unacquainted with you. Thereafter, some man or god or chance, with your assistance, cast out Dion, and you were left alone. Do you suppose, then, that I took any part with you in your political acts, when I had lost my wise partner and saw the unwise one left behind in the company of a crowd of evil men, not ruling himself, but being ruled by men of that sort, while fancying himself the ruler? In these circumstances what ought I to have done? Was I not bound to do as I did,—to bid farewell for the future to politics, shunning the slanders which proceed from envy, and to use every endeavour to make you and Dion as friendly to each other as possible, separated though you were and at variance with each other? Yea, you yourself also are a witness of

τοῦτο αὐτὸ ξυντείνων ᾿οὐκ ἀνῆκα πώποτε· καὶ μόγις μέν, ὅμως δ᾽ ὡμολογήθη νῷν πλεῦσαι· μὲν οἴκαδε ἐμέ, ἐπειδὴ πόλεμος ὑμᾶς κατεῖχεν, εἰρήνης δ᾽ αὖ γενομένης ἐλθεῖν ἐμέ τε καὶ Δίωνα εἰς Συρακούσας, σὲ δὲ καλεῖν ἡμᾶς. καὶ ταῦτα μὲν οὕτως ἐγένετο τῆς ἐμῆς εἰς Συρακούσας ἀποδημίας πέρι τῆς πρώτης καὶ τῆς πάλιν οἴκαδε σωτηρίας.

Τὸ δὲ δεύτερον εἰρήνης γενομένης ἐκάλεις με οὐ κατὰ τὰς ὁμολογίας, ἀλλὰ μόνον ἥκειν ἐπέστελλες, Δίωνα δ᾽ εἰσαῦθις ἔφησθα μεταπέμψεσθαι. διὰ ταῦτα οὐκ ἦλθον, ἀλλὰ καὶ Δίωνι τότ᾽ ἀπηχθόμην· ᾤετο γὰρ εἶναι βέλτιον ἐλθεῖν ἐμὲ καὶ ὑπακοῦσαί σοι. τὸ δὲ μετὰ ταῦτα ὕστερον ἐνιαυτῷ τριήρης ἀφίκετο καὶ ἐπιστολαὶ παρὰ σοῦ, τῶν δ᾽ ἐν ταῖς ἐπιστολαῖς γραμμάτων ἦρχεν ὡς ἂν ἀφίκωμαι τὰ Δίωνός μοι γενήσοιτο πράγματα πάντα κατὰ νοῦν τὸν ἐμόν, μὴ ἀφικομένου δὲ τἀναντία. αἰσχύνομαι δὴ λέγειν ὅσαι τότε ἐπιστολαὶ παρὰ σοῦ καὶ παρ᾽ ἄλλων ἦλθον διὰ σὲ ἐξ Ἰταλίας καὶ Σικελίας, καὶ παρ᾽ ὅσους τῶν ἐμῶν οἰκείων καὶ τῶν γνωρίμων, καὶ πᾶσαι διακελευόμεναί μοι ἰέναι καὶ δεόμεναι σοὶ πάντως ἐμὲ πείθεσθαι. ἐδόκει δὴ πᾶσιν, ἀρξαμένοις ἀπὸ Δίωνος, δεῖν ἐμὲ πλεῦσαι καὶ μὴ μαλθακίζεσθαι. καί τοι τήν θ᾽ ἡλικίαν αὐτοῖς προὐτεινόμην καὶ περὶ σοῦ διισχυριζόμην ὡς οὐχ οἷός τ᾽ ἔσοιο ἀνταρκέσαι τοῖς διαβάλλουσιν ἡμᾶς καὶ βουλομένοις εἰς ἔχθραν ἐλθεῖν· ἑώρων γὰρ καὶ

---

[1] Probably the war against the Lucanians.
[2] For the events of Plato's first visit *cf. Ep.* vii. 327 c ff.,

EPISTLE III

this, that I have never yet ceased to strive for this very object. And it was agreed between us—although with difficulty—that I should sail home, since you were engaged in war,[1] and that, when peace was restored, Dion and I should go to Syracuse and that you should invite us. And that was how things took place as regards my first sojourn at Syracuse[2] and my safe return home again.

But on the second occasion, when peace was restored, you did not keep to our agreement in the invitation you gave me but wrote that I should come alone, and stated that you would send for Dion later on. On this account I did not go; and, moreover, I was vexed also with Dion; for he was of opinion that it was better for me to go and to yield to your wishes. Subsequently, after a year's interval, a trireme arrived with letters from you, and the first words written in the letters were to the effect that if I came I should find that Dion's affairs would all proceed as I desired, but the opposite if I failed to come. And indeed I am ashamed to say how many letters came at that time from Italy and Sicily from you and from others on your account, or to how many of my friends and acquaintances they were addressed, all enjoining me to go and beseeching me to trust you entirely. It was the firm opinion of everyone, beginning with Dion, that it was my duty to make the voyage and not be faint-hearted. But I always made my age[3] an excuse; and as for you, I kept assuring them that you would not be able to withstand those who slander us and desire that we should quarrel; for I saw then, as I see now, that,

338 A, B; for those of the second visit, *Ep.* vii. 338 B ff., 345 c ff.   [3] In 361 B.C. Plato was about 67.

317

τότε καὶ νῦν ὁρῶ τὰς μεγάλας οὐσίας καὶ ὑπερόγκους τῶν τε ἰδιωτῶν καὶ τῶν μονάρχων σχεδόν,
D ὅσωπερ ἂν μείζους ὦσι, τοσούτῳ πλείους καὶ μείζους τοὺς διαβάλλοντας καὶ πρὸς ἡδονὴν μετὰ αἰσχρᾶς βλάβης ὁμιλοῦντας τρεφούσας, οὗ κακὸν οὐδὲν μεῖζον γεννᾷ πλοῦτός τε καὶ ἡ τῆς ἄλλης ἐξουσίας δύναμις. ὅμως δ' οὖν πάντα ταῦτα χαίρειν ἐάσας ἦλθον, διανοηθεὶς ὡς οὐδένα δεῖ τῶν ἐμῶν φίλων ἐμὲ αἰτιᾶσθαι ὡς διὰ τὴν ἐμὴν ῥαθυμίαν τὰ σφέτερα πάντα ἐξὸν μὴ ἀπολέσθαι διώλετο.

E Ἐλθὼν δέ, οἶσθα γὰρ δὴ σὺ πάντα τἀντεῦθεν ἤδη γενόμενα, ἐγὼ μὲν ἠξίουν δή που κατὰ τὴν ὁμολογίαν τῶν ἐπιστολῶν πρῶτον μὲν κατάγειν Δίωνα οἰκειωσάμενον, φράζων τὴν οἰκειότητα, ἣν εἰ ἐμοὶ τότε ἐπείθου, τάχ' ἂν βέλτιον τῶν νῦν γεγονότων ἔσχε καὶ σοὶ καὶ Συρακούσαις καὶ τοῖς ἄλλοις Ἕλλησιν, ὡς ἡ ἐμὴ δόξα μαντεύεται· ἔπειτα τὰ Δίωνος τοὺς οἰκείους ἔχειν ἠξίουν
318 καὶ μὴ διανείμασθαι τοὺς διανειμαμένους, οὓς οἶσθα σύ· πρὸς δὲ τούτοις ᾤμην δεῖν τὰ κατ' ἐνιαυτὸν ἕκαστον εἰωθότα αὐτῷ κομίζεσθαι καὶ μᾶλλον ἔτι καὶ οὐχ ἧττον ἐμοῦ παραγενομένου πέμπεσθαι. τούτων οὐδενὸς τυγχάνων ἠξίουν ἀπιέναι. τὸ μετὰ ταῦτα ἔπειθές με μεῖναι τὸν ἐνιαυτόν, φάσκων τὴν Δίωνος ἀποδόμενος οὐσίαν πᾶσαν τὰ μὲν ἡμίσεα ἀποπέμψειν εἰς Κόρινθον, τὰ δ' ἄλλα τῷ παιδὶ καταλείψειν αὐτοῦ. πολλὰ
B ἔχων εἰπεῖν, ὧν ὑποσχόμενος οὐδὲν ἐποίησας, διὰ τὸ πλῆθος αὐτῶν συντέμνω. τὰ γὰρ δὴ χρήματα

EPISTLE III

as a rule, when great and exorbitant wealth is in the hands either of private citizens or of monarchs, the greater it is, the greater and more numerous are the slanderers it breeds and the hordes of parasites and wastrels—than which there is no greater evil generated by wealth or by the other privileges of power. Notwithstanding, I put aside all these considerations and went, resolving that none of my friends should lay it to my charge that owing to my lack of energy all their fortunes were ruined when they might have been saved from ruin.

On my arrival—for you know, to be sure, all that subsequently took place—I, of course, requested, in accordance with the agreement in your letters, that you should, in the first place, recall Dion on terms of friendship—which terms I mentioned; and if you had then yielded to this request, things would probably have turned out better than they have done now both for you and Syracuse and for the rest of Greece —that, at least, is my own intuitive belief. Next, I requested that Dion's family should have possession of his property, instead of the distributors, whom you wot of, having the distribution of it. And further, I deemed it right that the revenue which was usually paid over to him year by year should be forwarded to him all the more, rather than all the less, because of my presence. None of these requests being granted, I asked leave to depart. Thereupon you kept urging me to stop for the year, declaring that you would sell all Dion's property and send one half of the proceeds to Corinth and retain the other half for his son. And I could mention many other promises none of which you fulfilled; but the number of them is so great that I cut it short. For when

PLATO

πάντα ἀποδόμενος, οὐ πείσας Δίωνα, φάσκων οὐ
πωλήσειν ἄνευ τοῦ πείθειν, τὸν κολοφῶνα, ὦ
θαυμάσιε, ταῖς ὑποσχέσεσιν ἀπάσαις νεανικώτατον
ἐπέθηκας· μηχανὴν γὰρ οὔτε καλὴν οὔτε κομψὴν
οὔτε δικαίαν οὔτε ξυμφέρουσαν εὗρες, ἐμὲ ἐκφοβεῖν,
ὡς ἀγνοοῦντα τὰ τότε γιγνόμενα, ἵνα μηδὲ ἐγὼ
C ζητοίην τὰ χρήματα ἀποπέμπεσθαι. ἡνίκα γὰρ
Ἡρακλείδην ἐξέβαλλες,[1] οὔτε Συρακοσίοις δοκοῦν
δικαίως οὔτ' ἐμοί, διότι μετὰ Θεοδότου καὶ
Εὐρυβίου συνεδεήθην σου μὴ ποιεῖν ταῦτα, ταύτην
λαβὼν ὡς ἱκανὴν πρόφασιν εἶπες ὅτι καὶ πάλαι
σοι δῆλος εἴην σοῦ μὲν οὐδὲν φροντίζων, Δίωνος
δὲ καὶ τῶν Δίωνος φίλων καὶ οἰκείων, καὶ ἐπειδὴ
νῦν Θεοδότης καὶ Ἡρακλείδης ἐν διαβολαῖς εἶεν
οἰκεῖοι Δίωνος ὄντες, πᾶν μηχανώμην ὅπως οὗτοι
μὴ δώσουσι δίκην.

D Καὶ ταῦτα μὲν ταύτῃ περὶ τὰ πολιτικὰ κοινωνίας
τῆς ἐμῆς καὶ σῆς· καὶ εἴ τινα ἑτέραν ἀλλοτριότητα
ἐνεῖδες ἐν ἐμοὶ πρὸς σέ, εἰκότως οἴου[2] ταύτῃ πάντα
ταῦτα γεγονέναι. καὶ μὴ θαύμαζε· κακὸς γὰρ ἂν
ἔχοντί γε νοῦν ἀνδρὶ φαινοίμην ἐνδίκως πεισθεὶς
ὑπὸ τοῦ μεγέθους τῆς σῆς ἀρχῆς τὸν μὲν παλαιὸν
φίλον καὶ ξένον, κακῶς πράττοντα διὰ σέ, μηδὲν
σοῦ χείρω, ἵνα οὕτως εἴπω, τοῦτον μὲν προδοῦναι,
E σὲ δὲ τὸν ἀδικοῦντα ἑλέσθαι καὶ πᾶν δρᾶν ὅπῃ σὺ
προσέταττες, ἕνεκα χρημάτων δῆλον ὅτι· οὐδὲν
γὰρ ἂν ἕτερον ἔφησεν αἴτιόν τις εἶναι τῆς ἐμῆς
μεταβολῆς, εἰ μετεβαλόμην. ἀλλὰ ταῦτα μὲν

---
[1] ἐξέβαλλες R. Hackforth: ἐξέβαλες mss., Zur.
[2] οἴου H. Richards: οἴει mss., Zur.

---

[1] A leading Syracusan noble, supporter of Dion; *cf. Ep.*
iv. 320 E, vii. 348 B; Theodotes was a connexion of H.

EPISTLE III

you had sold all the goods, without Dion's consent—though you had declared that without his consent you would not dispose of them—you put the coping-stone on all your promises, my admirable friend, in a most outrageous way: you invented a plan that was neither noble nor ingenious nor just nor profitable—namely, to scare me off from so much as seeking for the dispatch of the money, as being in ignorance of the events then going on. For when you sought to expel Heracleides [1]—unjustly, as it seemed to the Syracusans as well as to myself—because I had joined with Theodotes and Eurybius in entreating you not to do so, you took this as an ample excuse, and asserted that it had long been plain to you that I paid no regard to you, but only to Dion and Dion's friends and connexions, and now that Theodotes and Heracleides, who were Dion's connexions, were the subjects of accusations, I was using every means to prevent their paying the just penalty.

Such, then, was the course of events as regards our association in political affairs. And if you perceived any other estrangement in my attitude towards you, you may reasonably suppose that that was the way in which all these things took place. Nor need you be surprised; for I should justly be accounted base by any man of sense had I been influenced by the greatness of your power to betray my old and intimate guest-friend—a man, to say the least, in no wise inferior to you—when, because of you, he was in distress, and to prefer you, the man who did the wrong, and to do everything just as you bade me—for filthy lucre's sake, obviously; for to this, and nothing else, men would have ascribed this change of front in me, if I had changed. Well, then,

PLATO

ταύτῃ γενόμενα τὴν ἐμὴν καὶ σὴν λυκοφιλίαν καὶ ἀκοινωνίαν διὰ σὲ ἀπειργάσατο.

Σχεδὸν δ' εἰς λόγον ὁ λόγος ἥκει μοι ξυνεχὴς τῷ νῦν δὴ γενομένῳ, περὶ οὗ μοι τὸ δεύτερον ἀπολογητέον ἔφην εἶναι. σκόπει δὴ καὶ πρόσεχε πάντως, ἄν τί σοι ψεύδεσθαι δόξω καὶ μὴ τἀληθῆ λέγειν. φημὶ γάρ σε Ἀρχεδήμου παρόντος ἐν τῷ κήπῳ καὶ Ἀριστοκρίτου, σχεδὸν ἡμέραις πρότερον εἴκοσι τῆς ἐμῆς ἐκ Συρακουσῶν οἴκαδ' ἀποδημίας, ἃ νῦν δὴ λέγεις ⟨εἰπεῖν⟩[1] ἐμοὶ μεμφόμενον, ὡς Ἡρακλείδου τέ μοι καὶ τῶν ἄλλων πάντων μᾶλλον ἢ σοῦ μέλοι. καί με τούτων ἐναντίον διηρώτησας εἰ μνημονεύω, κατ' ἀρχὰς ὅτ' ἦλθον, κελεύων σε τὰς πόλεις τὰς Ἑλληνίδας κατοικίζειν· ἐγὼ δὲ συνεχώρουν μεμνῆσθαι καὶ ἔτι νῦν μοι δοκεῖν ταῦτ' εἶναι βέλτιστα. ῥητέον δέ, ὦ Διονύσιε, καὶ τοὐπὶ τούτῳ τότε λεχθέν. ἠρόμην γὰρ δή σε πότερον αὐτὸ τοῦτό σοι ξυμβουλεύσαιμι μόνον ἢ τι καὶ ἄλλο πρὸς τούτῳ· σὺ δὲ καὶ μάλα ἀπεκρίνω μεμηνιμένως καὶ ὑβριστικῶς εἰς ἐμέ, ὡς ᾤου— διὸ τὸ τότε σοι ὕβρισμα νῦν ὕπαρ ἀντ' ὀνείρατος γέγονεν—, εἶπες δὲ καὶ μάλα πλάστως[2] γελῶν, εἰ μέμνημαι, ὡς παιδευθέντα με ἐκέλευες ποιεῖν πάντα ταῦτα ἢ μὴ ποιεῖν. ἔφην ἐγὼ κάλλιστα μνημονεῦσαί σε. Οὐκοῦν παιδευθέντα, ἔφησθα, γεωμετρεῖν; ἢ πῶς; κἀγὼ τὸ μετὰ ταῦτα ὃ ἐπῄει

---

[1] εἰπεῖν I add.
[2] μάλα πλάστως one MS.: μάλ' ἀπλάστως most MSS., Zur.

---

[1] i.e. quarrelling. Cf. Rep. 566 A; Phaedr. 241 C, D; Laws 906 D.   [2] Cf. Ep. ii. 310 B.
[3] Cf. Ep. xiii. 363 D. For "in the garden" cf. Ep. ii. 313 A.

EPISTLE III

it was the fact that things took this course, owing to you, which produced this wolf-love [1] and want of fellowship between you and me.

Practically continuous with the statement made just now there comes, I find, that other statement against which, as I said, I have to make my second defence. Consider now and pay the closest attention, in case I seem to you to be lying at all and not speaking the truth. I affirm that when Archedemus [2] and Aristocritus [3] were with us in the garden, some twenty days before I departed home from Syracuse, you made the same complaint against me that you are making now—that I cared more for Heracleides and for all the rest than for you. And in the presence of those men you asked me whether I remembered bidding you, when I first arrived, to plant settlers in the Greek cities. I granted you that I did remember, and that I still believed that this was the best policy. But, Dionysius, I must also repeat, the next observation that was made on this occasion. For I asked you whether this and this only was what I advised, or something else besides; and you made answer to me in a most indignant and most mocking tone, as you supposed—and consequently the object of your mockery then has now turned out a reality instead of a dream [4]; for you said with a very artificial laugh, if my memory serves me—" You bade me be educated before I did all these things or else not do them." I replied that your memory was excellent. You then said—" Did you mean educated in land-measuring or what?" But I refrained from making

[4] This seems to mean that Plato's scheme of education, scoffed at by Dionysius, was the secret of Dion's success—the "dream" of the "philosopher-king" being realized in his person.

319 μοι εἰπεῖν οὐκ εἶπον, φοβούμενος μὴ σμικροῦ
ῥήματος ἕνεκα τὸν ἔκπλουν ὃν προσεδόκων, μή
μοι στενὸς γίγνοιτο ἀντ' εὐρυχωρίας.

Ἀλλ' οὖν ὧν ἕνεκα πάντ' εἴρηται, ταῦτ' ἐστί·
μή με διάβαλλε λέγων ὡς οὐκ εἴων ἐγώ σε πόλεις
Ἑλληνίδας ἐρρούσας ὑπὸ βαρβάρων οἰκίζειν, οὐδὲ
D Συρακουσίους ἐπικουφίσαι βασιλείαν ἀντὶ τυραννίδος μεταστήσαντα. τούτων γὰρ οὔθ' ἧττον ἐμοὶ
πρέποντα ἔχοις ἄν ποτε λέγων μου καταψεύσασθαι,
πρὸς δὲ τούτοις ἔτι σαφεστέρους τούτων εἰς
ἔλεγχον λόγους ἐγὼ δοίην ἄν, εἴ τις ἱκανή που
φαίνοιτο κρίσις, ὡς ἐγὼ μὲν ἐκέλευον, σὺ δ' οὐκ
ἤθελες πράττειν αὐτά· καὶ μὴν οὐ χαλεπὸν εἰπεῖν
ἐναργῶς ὡς ἦν ταῦτα ἄριστα πραχθέντα καὶ σοὶ
καὶ Συρακοσίοις καὶ Σικελιώταις πᾶσιν. ἀλλ' ὦ
E τᾶν, εἰ μὲν μὴ φῇς εἰρηκέναι εἰρηκὼς ταῦτα, ἔχω τὴν
δίκην· εἰ δ' ὁμολογεῖς, τὸ μετὰ τοῦτο ἡγησάμενος
εἶναι σοφὸν τὸν Στησίχορον, τὴν παλινῳδίαν αὐτοῦ
μιμησάμενος, ἐκ τοῦ ψεύδους εἰς τὸν ἀληθῆ λόγον
μετάστηθι.[1]

[1] μετάστηθι ms. corr.: μεταστήσει Zur. (-ση mss.).

EPISTLE III

the retort which it occurred to me to make, for I was alarmed about the homeward voyage I was hoping for, lest instead of having an open road I should find it shut, and all because of a short saying.

Well then, the purpose of all I have said is this: do not slander me by declaring that I was hindering you from colonizing the Greek cities that were ruined by the barbarians, and from relieving the Syracusans by substituting a monarchy for a tyranny. For you could never bring any false accusation against me that was less appropriate than these; and, moreover, in refutation of them I could bring still clearer statements—if any competent tribunal were anywhere to be seen—showing that it was I who was urging you, and you who were refusing, to execute these plans. And, verily, it is easy to affirm frankly that these plans, if they had been executed, were the best both for you and the Syracusans, and for all the Siceliots. But, my friend, if you deny having said this, when you have said it, I am justified; while if you confess it, you should further agree that Stesichorus [1] was a wise man, and imitate his palinode, and renounce the false for the true tale.

---

[1] A lyric poet, *circa* 600 B.C., said to have been struck blind for his attacks on the reputation of Helen of Troy, which he subsequently withdrew in his recantation ("palinode"); *cf. Phaedr.* 243 A, B.

EPISTLE IV

Prefatory Note.—This letter belongs, ostensibly, to the same period as the Third, when Dion had already achieved some measure of success in his struggle for the mastery of Syracuse. The writer expresses his satisfaction at the progress of Dion's cause; admonishes him and his followers that they must show by their deeds that they pay more than lip-service to virtue; warns Dion, in particular, that the eyes of the world are upon him and that he must not become a prey to jealousy or arrogance. He also expresses a desire for reliable news of Sicilian affairs.

This letter again, if it be Plato's, may well be regarded as an open letter, intended to encourage Dion's party, with a private postscript meant only for Dion's eyes. But it contains nothing to convince us that it really was written by Plato. On the contrary, we should hardly expect Plato to write in this style to Dion, with a mixture of flattery and reproof, and quoting for his admonition the ancient instances of Cyrus and Lycurgus, and comparing himself to a child clapping an actor, and urging Dion to become a men-pleaser and cultivate the arts of a popular demagogue. Moreover, Plato's attitude, in thus publicly supporting Dion's military efforts, is hardly compatible with the attitude he professes to adopt in *Ep.* vii. (350 c), even allowing for the interval

EPISTLE IV

which separates, we may suppose, the two letters. Nor does it decrease our doubts when we find a close parallel to the sentence about actors being stimulated by the applause of children in the *Evagoras* of Isocrates (xxxii. 3).

On the whole, then, it would seem that this is another epistolary composition of a later age falsely ascribed to Plato.

Δ

Πλάτων Δίωνι Συρακοσίῳ εὖ πράττειν.

Οἶμαι μὲν φανερὰν εἶναι διὰ παντὸς τοῦ χρόνου τὴν ἐμὴν προθυμίαν περὶ τὰς συμβεβηκυίας πράξεις, καὶ ὅτι πολλὴν εἶχον περὶ αὐτῶν σπουδὴν εἰς τὸ ξυμπερανθῆναι, οὐκ ἄλλου τινὸς ἕνεκα μᾶλλον ἢ τῆς ἐπὶ τοῖς καλοῖς φιλοτιμίας· νομίζω γὰρ δίκαιον εἶναι τοὺς ὄντας τῇ ἀληθείᾳ ἐπιεικεῖς καὶ πράττοντας τοιαῦτα τυγχάνειν δόξης τῆς προσηκούσης. τὰ μὲν οὖν εἰς τὸ παρὸν σὺν θεῷ εἰπεῖν ἔχει καλῶς, τὰ δὲ περὶ τῶν μελλόντων ὁ μέγιστός ἐστιν ἀγών. ἀνδρείᾳ μὲν γὰρ καὶ τάχει καὶ ῥώμῃ διενεγκεῖν δόξειεν ἂν καὶ ἑτέρων εἶναί τινων, ἀληθείᾳ δὲ καὶ δικαιοσύνῃ καὶ μεγαλοπρεπείᾳ καὶ τῇ περὶ πάντα ταῦτα εὐσχημοσύνῃ, ξυμφαίη τις ἂν τοὺς ἀντιποιουμένους τὰ τοιαῦτα τιμᾶν εἰκότως τῶν ἄλλων διαφέρειν.

Νῦν οὖν δῆλον μέν ἐστιν ὃ λέγω, ἀναμιμνήσκειν δὲ ὅμως δεῖ ἡμᾶς αὐτοὺς ὅτι προσήκει πλέον ἢ παίδων τῶν ἄλλων ἀνθρώπων διαφέρειν τοὺς οἶσθα δή που. φανεροὺς οὖν δεῖ ἡμᾶς γενέσθαι ὅτι ἐσμὲν

---

[1] This refers to Dion's military operations in Sicily in 357 B.C., and perhaps later.
[2] The reference is to Dion's plans for the political reformation of Sicily.

EPISTLE IV

PLATO TO DION OF SYRACUSE WISHES WELL DOING

It has been plain, I believe, all along that I took a keen interest in the operations [1] that have been carried out, and that I was most anxious to see them finally completed. In this I was mainly prompted by my jealous regard for what is noble [2]; for I esteem it just that those who are truly virtuous, and who act accordingly, should achieve the reputation they deserve. Now for the present (God willing) affairs are going well; but it is in the future that the chief struggle lies. For while it might be thought that excellence in courage and speed and strength might belong to various other men, everyone would agree that surpassing excellence in truth, justice, generosity and the outward exhibition of all these virtues naturally belongs to those who profess to hold them in honour.

Now the point of this remark is plain; but none the less it is right that we should remind ourselves that it behoves certain persons (who these are of course you know) [3] to surpass the rest of mankind as if they were less than children.[4] It is, therefore,

[3] The persons meant are Plato's own pupils and Dion's political supporters.

[4] For this (perhaps proverbial) phrase (*cf.* "no better than a child") *cf. Phaedr.* 279 A.

443

PLATO

320 τοιοῦτοι οἷοίπερ φαμέν, ἄλλως τε καὶ ἐπειδὴ σὺν θεῷ εἰπεῖν ῥᾴδιον ἔσται. τοῖς μὲν γὰρ ἄλλοις συμβέβηκεν ἀναγκαῖον εἶναι πλανηθῆναι πολὺν D τόπον, εἰ μέλλουσι γνωσθῆναι· τὸ δὲ νῦν ὑπάρχον περὶ σὲ τοιοῦτόν ἐστιν ὥστε τοὺς ἐξ ἁπάσης τῆς οἰκουμένης, εἰ καὶ νεανικώτερόν ἐστιν εἰπεῖν, εἰς ἕνα τόπον ἀποβλέπειν, καὶ ἐν τούτῳ μάλιστα πρὸς σέ. ὡς οὖν ὑπὸ πάντων ὁρώμενος παρασκευάζου τόν τε Λυκοῦργον ἐκεῖνον ἀρχαῖον ἀποδείξων καὶ τὸν Κῦρον καὶ εἴ τις ἄλλος πώποτε ἔδοξεν ἤθει καὶ πολιτείᾳ διενεγκεῖν, ἄλλως τε καὶ ἐπειδὴ E πολλοὶ καὶ σχεδὸν ἅπαντες οἱ τῇδε λέγουσιν ὡς πολλὴ ἐστιν ἐλπὶς ἀναιρεθέντος Διονυσίου διαφθαρῆναι τὰ πράγματα διὰ τὴν σὴν τε καὶ Ἡρακλείδου καὶ Θεοδότου καὶ τῶν ἄλλων γνωρίμων φιλοτιμίαν. μάλιστα μὲν οὖν μηδεὶς εἴη τοιοῦτος· ἐὰν δ' ἄρα καὶ γίγνηταί τις, σὺ φαίνου ἰατρεύων, καὶ πρὸς τὸ βέλτιστον ἔλθοιτ' ἄν.

321 Ταῦτα δὲ ἴσως γελοῖόν σοι φαίνεται εἶναι τὸ ἐμὲ λέγειν· διότι καὶ αὐτὸς οὐκ ἀγνοεῖς· ἐγὼ δὲ καὶ ἐν τοῖς θεάτροις ὁρῶ τοὺς ἀγωνιστὰς ὑπὸ τῶν παίδων παροξυνομένους, μή τι δὴ ὑπό γε τῶν φίλων, οὕς ἄν τις οἴηται μετὰ σπουδῆς κατ' εὔνοιαν παρακελεύεσθαι. νῦν οὖν αὐτοί τε ἀγωνίζεσθε καὶ ἡμῖν εἴ του δεῖ ἐπιστέλλετε.

Τὰ δ' ἐνθάδε παραπλησίως ἔχει καθάπερ καὶ ὑμῶν παρόντων. ἐπιστέλλετε δὲ καὶ ὅ τι πέ-

---

[1] For Lycurgus, the Spartan lawgiver, *cf. Ep.* viii. 354 B; for Cyrus *cf. Ep.* ii. 311 A, *Laws* 693 D ff.
[2] See *Ep.* iii. 318 c note.
[3] *Cf.* Isocr. *Evag.* 32. 3.

EPISTLE IV

incumbent upon us to show plainly that we are the sort of men we claim to be, and that all the more because (God willing) it will be an easy task. For whereas all other men find it necessary to wander far afield if they mean to get themselves known, you are in such a position now that people all the world over—bold though it be to say so—have their eyes fixed on one place only, and in that place upon you above all men. Seeing, then, that you have the eyes of all upon you, prepare yourself to play the part of that ancient worthy Lycurgus and of Cyrus [1] and of all those others who have been famed hitherto for their excellence of character and of statesmanship; and that all the more because there are many, including nearly all the people here, who keep saying that, now that Dionysius is overthrown, there is every prospect that things will go to ruin owing to the jealous rivalry of yourself, and Heracleides and Theodotes [2] and the other notables. I pray, then, that no one, if possible, may suffer from this complaint; but in case anyone should, after all, do so, you must play the part of a physician; and so things will turn out best for you all.

Probably it strikes you as ridiculous that I should say this, seeing that you yourself also know it quite well; but I notice how even in the theatres the players are spurred on by the plaudits of the children —not to speak of their own friends—whenever a player believes them to be genuine and well-meaning in their encouragement.[3] So do you also play your parts now; and if you have need of anything send us word.

Affairs with us are in much the same state as when you were here. Send us word also about what you

PLATO

πρακται ὑμῖν ἢ πράττοντες τυγχάνετε, ὡς ἡμεῖς πολλὰ ἀκούοντες οὐδὲν ἴσμεν. καὶ νῦν ἐπιστολαὶ παρὰ μὲν Θεοδότου καὶ Ἡρακλείδου ἥκουσιν εἰς Λακεδαίμονα καὶ Αἴγιναν, ἡμεῖς δέ, καθάπερ εἴρηται, πολλὰ ἀκούοντες παρὰ[1] τῶν τῇδε οὐδὲν ἴσμεν. ἐνθυμοῦ δὲ καὶ ὅτι δοκεῖς τισὶν ἐνδεεστέρως τοῦ προσήκοντος θεραπευτικὸς εἶναι· μὴ οὖν λανθανέτω σε ὅτι διὰ τοῦ ἀρέσκειν τοῖς ἀνθρώποις καὶ τὸ πράττειν ἐστίν, ἡ δ' αὐθάδεια ἐρημίᾳ ξύνοικος. Εὐτύχει.

[1] παρὰ W.-Möllendorff : περὶ mss., Zur.

EPISTLE IV

have already done or happen to be doing now, since we know nothing although we hear many reports.

Even at this moment letters have come to Lacedaemon and Aegina from Theodotes and Heracleides; but we, as I said, know nothing, although we hear many reports from the people here. And, Dion, do you also bear in mind that you are thought by some to be unduly wanting in affability; so do not forget that successful action depends on pleasing people, whereas arrogance is next neighbour to isolation.

Good-luck attend thee!

EPISTLE V

Prefatory Note.—This is a letter of "counsel" to Perdiccas the Third, king of Macedon, recommending him to avail himself of the counsels of Euphraeus. As every form of government has a "voice" peculiar to itself, it requires study to distinguish these "voices"; and Euphraeus will be of help in this study. If it be objected that Plato himself took no part in public life and thus gave no proof of his understanding the "voice" of democracy, the answer is that the Athenian democracy in his time was incapable of profiting by his "counsel." But Perdiccas is open to advice and not incurable.

This letter, if genuine, must fall within the limits of the reign of Perdiccas, 365 and 360 B.C. But its genuineness is defended by few scholars. The curious passage (321 D) about the "voices" of polities is obviously borrowed from *Rep.* 493 A, B. The defence of Plato's abstention from politics is not pertinent to the real subject of the letter; for it is the competence of Euphraeus, not Plato, that is in question. And the language here used about "counsel" looks very much like the work of a later composer who had *Ep.* vii. 330 c ff. before him. Unless the writer were himself a monarchist, the ascription of this letter to Plato may have been due (as has been suggested) to a malicious desire to paint Plato as a supporter of Macedon and its tyrants.

E

Πλάτων Περδίκκᾳ εὖ πράττειν.

Εὐφραίῳ μὲν συνεβούλευσα, καθάπερ ἐπέστελλες, τῶν σῶν ἐπιμελούμενον περὶ ταῦτα διατρίβειν. δίκαιος δ' εἰμὶ καὶ σοὶ ξενικὴν καὶ ἱερὰν ξυμβουλὴν λεγομένην συμβουλεύειν περί τε τῶν ἄλλων ὧν ἂν φράζῃς καὶ ὡς Εὐφραίῳ δεῖ τὰ νῦν χρῆσθαι. πολλὰ μὲν γὰρ ὁ ἀνὴρ χρήσιμος, μέγιστον δὲ οὗ καὶ σὺ νῦν ἐνδεὴς εἶ διά τε τὴν ἡλικίαν καὶ διὰ τὸ μὴ πολλοὺς αὐτοῦ πέρι ξυμβούλους εἶναι τοῖς νέοις. ἔστι γὰρ δή τις φωνὴ τῶν πολιτειῶν ἑκάστης καθαπερεί τινων ζώων, ἄλλη μὲν δημοκρατίας, ἄλλη δ' ὀλιγαρχίας, ἡ δ' αὖ μοναρχίας. ταύτας φαῖεν μὲν ἂν ἐπίστασθαι πάμπολλοι, πλεῖστον δ' ἀπολείπονται τοῦ κατανοεῖν αὐτὰς πλὴν ὀλίγων δή τινων. ἥτις μὲν ἂν οὖν τῶν πολιτειῶν τὴν αὑτῆς φθέγγηται φωνὴν πρός τε θεοὺς καὶ πρὸς ἀνθρώπους, καὶ τῇ φωνῇ τὰς πράξεις ἑπομένας ἀποδιδῷ, θάλλει τε ἀεὶ καὶ σώζεται, μιμουμένη δ' ἄλλην φθείρεται. πρὸς ταῦτ' οὖν Εὐφραῖός σοι γίγνοιτ' οὐχ ἥκιστα ἂν χρήσιμος, καίπερ καὶ πρὸς ἄλλα ὢν ἀνδρεῖος· τοὺς γὰρ τῆς μοναρχίας λόγους

---
[1] Perdiccas was king of Macedon 365–360 B.C.
[2] A native of Euboea and pupil of Plato.
[3] *Cf. Rep.* 493 A–C.

EPISTLE V

PLATO TO PERDICCAS [1] WISHES WELL-DOING

I COUNSELLED Euphraeus,[2] in accordance with your message, to devote his time to the task of caring for your interests; and I feel myself bound also to give you friendly, and what is called " sacred," counsel both about the other matters you mention and as to how you ought now to make use of Euphraeus. For the man is useful for many things, the most important being that in which you yourself are deficient owing to your youth, and also because it is a matter about which there are not many counsellors available for the young. For forms of government, like animals, have each their own kind of language,[3] one for democracy, another for oligarchy, and a third kind for monarchy; and though a vast number of people would assert that they understand these languages, yet all but a few of them are very far indeed from discerning them. Now each of these polities, if it speaks its own language both to gods and to men, and renders its actions conformable to its language, remains always flourishing and secure; but if it imitates another it becomes corrupted. It is for this study, then, that Euphraeus will be specially useful to you, although there are also other studies in which he is competent. For he, I hope, will help you to

PLATO

322 οὐχ ἥκιστ᾽ αὐτὸν ἐλπίζω ξυνεξευρήσειν τῶν περὶ τὴν σὴν διατριβὴν ὄντων. εἰς ταῦτ᾽ οὖν αὐτῷ χρώμενος ὀνήσει τε αὐτὸς καὶ ἐκεῖνον πλεῖστα ὠφελήσεις.

Ἐὰν δέ τις ἀκούσας ταῦτα εἴπῃ, Πλάτων, ὡς ἔοικε, προσποιεῖται μὲν τὰ δημοκρατίᾳ ξυμφέροντα εἰδέναι, ἐξὸν δ᾽ ἐν τῷ δήμῳ λέγειν καὶ συμβουλεύειν αὐτῷ τὰ βέλτιστα οὐ πώποτε ἀναστὰς ἐφθέγξατο, πρὸς ταῦτ᾽ εἰπεῖν ἔστιν ὅτι Πλάτων ὀψὲ ἐν τῇ πατρίδι γέγονε καὶ τὸν δῆμον B κατέλαβεν ἤδη πρεσβύτερον καὶ εἰθισμένον ὑπὸ τῶν ἔμπροσθεν πολλὰ καὶ ἀνόμοια τῇ ἐκείνου ξυμβουλῇ πράττειν· ἐπεὶ πάντων ἂν ἥδιστα καθάπερ πατρὶ συνεβούλευεν αὐτῷ, εἰ μὴ μάτην μὲν κινδυνεύσειν ᾤετο, πλέον δ᾽ οὐδὲν ποιήσειν. ταὐτὸν δὴ οἶμαι δρᾶσαι ἂν καὶ τὴν ἐμὴν ξυμβουλήν. εἰ γὰρ δόξαιμεν ἀνιάτως ἔχειν, πολλὰ ἂν χαίρειν ἡμῖν εἰπὼν ἐκτὸς ἂν γίγνοιτο τῆς περὶ ἐμὲ καὶ τὰ ἐμὰ C ξυμβουλῆς. Εὐτύχει.

EPISTLE V

explore the speech of monarchy as well as any of the persons you employ. So if you make use of him for this purpose you will not only benefit yourself but will also be helping him immensely.

Suppose, however, that on hearing this someone were to say: "Plato, as it seems, is claiming to know what is of advantage to democracy; yet when he has had it in his power to speak before the *demos* and to counsel it for the best he has never yet stood up and made a speech"—to this you may reply that "Plato was born late in the history of his country, and he found the *demos* already old and habituated by the previous statesmen to do many things at variance with his own counsel.[1] For he would have given counsel to it, as to his father, with the greatest possible pleasure, had he not supposed that he would be running risks in vain, and would do no good. And I suppose that he would do the same as regards counselling me. For if he deemed us to be in an incurable state, he would bid us a long farewell and leave off giving counsel about me or my affairs." Good-luck be thine!

[1] *Cf. Ep.* vii. 325 A, c ff.; and, for a theory of "counsel," *Ep.* vii. 330 c ff.

EPISTLE VI

Prefatory Note.—The purpose of this letter is to urge Hermeias, the tyrant of Atarneus, to cultivate friendly relations with Erastus and Coriscus who lived in the neighbouring town of Scepsis. This, argues the writer, will be for the benefit of both parties by joining the practical efficiency of the one to the theoretical ability of the other. And to cement their union this letter, read over many times in common, should serve as a bond; and they should swear fidelity by the All-ruling God and his Lord and Father.

This letter was condemned long ago on the ground that what it says (in 322 E) of Plato's personal ignorance of Hermeias is in conflict with the account of Hermeias given by Strabo. Possibly Strabo was misinformed; but even granting this, there are various other objections to the letter which make it difficult to believe in its authenticity. In its tone and style it has many points of resemblance to *Ep.* ii. In both the writer dwells on the value of combining " wisdom " with " power " (vi. 322 D ff., and ii. 310 E); in both he advises that, in case of dispute, reference should be made to himself (vi. 323 B and ii. 310 D); in both he charges his correspondents to read over his letter frequently (vi. 323 c, and ii. 314 c); and in both he makes obscure

EPISTLE VI

references to Divine Principles. It seems, therefore, fairly safe to conclude that *Ep.* vi. is by the same author as *Ep.* ii.; and if further evidence were needed that the latter epistle is un-Platonic it is sufficiently provided by the former with its astounding conclusion that these three gentlemen are to swear by strange divinities, and, moreover (all-important though the matter is) not to be wholly in earnest about their oath but to take it " with a blend of earnestness and jest " (an ill-timed reminiscence of some passages in the *Laws*). The ostensible date of this letter is *circa* 350 B.C.

ς

Πλάτων Ἑρμείᾳ καὶ Ἐράστῳ καὶ Κορίσκῳ
εὖ πράττειν.

Ἐμοὶ φαίνεται θεῶν τις ὑμῖν τύχην ἀγαθήν, ἂν εὖ δέξησθε, εὐμενῶς καὶ ἱκανῶς παρασκευάζειν. οἰκεῖτε γὰρ δὴ γείτονές τε ὑμῖν αὐτοῖς καὶ χρείαν ἔχοντες ὥστε ἀλλήλους εἰς τὰ μέγιστα ὠφελεῖν. Ἑρμείᾳ μὲν γὰρ οὔτε ἵππων πλῆθος οὔτε ἄλλης πολεμικῆς συμμαχίας οὐδ' αὖ χρυσοῦ προσγενομένου γένοιτ' ἂν μείζων εἰς τὰ πάντα δύναμις, ἢ φίλων βεβαίων τε καὶ ἦθος ἐχόντων ὑγιές· Ἐράστῳ δὲ καὶ Κορίσκῳ, πρὸς τῇ τῶν εἰδῶν σοφίᾳ τῇ καλῇ ταύτῃ φήμ' ἐγώ, καίπερ γέρων ὤν, προσδεῖν σοφίας τῆς περὶ τοὺς πονηροὺς καὶ ἀδίκους φυλακτικῆς καί τινος ἀμυντικῆς δυνάμεως. ἄπειροι γάρ εἰσι διὰ τὸ μεθ' ἡμῶν μετρίων ὄντων καὶ οὐ κακῶν συχνὸν διατετριφέναι τοῦ βίου. διὸ δὴ τούτων προσδεῖν εἶπον, ἵνα μὴ ἀναγκάζωνται τῆς ἀληθινῆς μὲν ἀμελεῖν σοφίας, τῆς δὲ ἀνθρωπίνης τε καὶ ἀναγκαίας ἐπιμελεῖσθαι μειζόνως ἢ δεῖ.

---

[1] Hermeias was tyrant of Atarneus, *circa* 351 B.C. Erastus and Coriscus were pupils of Plato who lived at Scepsis, near Atarneus.

[2] Plato would be about 77 in 351–350 B.C. The point of

EPISTLE VI

PLATO TO HERMEIAS AND ERASTUS AND CORISCUS [1]
WISHES WELL-DOING

SOME God, as it seems plain to me, is preparing for you good fortune in a gracious and bountiful way, if only you accept it with grace. For you dwell near together as neighbours in close association so that you can help one another in the things of greatest importance. For Hermeias will find in his multitude of horses or of other military equipment, or even in the gaining of gold itself, no greater source of power for all purposes than in the gaining of steadfast friends possessed of a sound character; while Erastus and Coriscus, in addition to this fair Science of Ideas, need also—as I, old though I am,[2] assert—the science which is a safeguard in dealing with the wicked and unjust, and a kind of self-defensive power. For they lack experience owing to the fact that they have spent a large part of their lives in company with us who are men of moderation and free from vice; and for this reason, as I have said, they need these additional qualities, so that they may not be compelled to neglect the true Science, and to pay more attention than is right to that which is human

this allusion to his age may be that old men ought rather to cultivate other-worldliness.

ταύτην δ' αὖ τὴν δύναμιν Ἑρμείας μοι φαίνεται φύσει τε, ὅσα μήπω ξυγγεγονότι, καὶ τέχνῃ δι' ἐμπειρίας εἰληφέναι.

Τί οὖν δὴ λέγω; σοὶ μέν, Ἑρμεία, πεπειραμένος Ἐράστου καὶ Κορίσκου πλέονα ἢ σύ, φημὶ καὶ μηνύω καὶ μαρτυρῶ μὴ ῥᾳδίως εὑρήσειν σε ἀξιοπιστότερα ἤθη τούτων τῶν γειτόνων· ἔχεσθαι δὴ παντὶ ξυμβουλεύω δικαίῳ τρόπῳ τούτων τῶν ἀνδρῶν, μὴ πάρεργον ἡγουμένῳ· Κορίσκῳ δὲ καὶ Ἐράστῳ πάλιν Ἑρμείου ἀντέχεσθαι ξύμβουλός εἰμι καὶ πειρᾶσθαι ταῖς ἀνθέξεσιν ἀλλήλων εἰς μίαν ἀφικέσθαι φιλίας ξυμπλοκήν. ἂν δέ τις ὑμῶν ἄρα ταύτην πῃ λύειν δοκῇ, τὸ γὰρ ἀνθρώπινον οὐ παντάπασι βέβαιον, δεῦρο παρ' ἐμὲ καὶ τοὺς ἐμοὺς πέμπετε μομφῆς κατήγορον ἐπιστολήν· οἶμαι γὰρ δίκῃ τε καὶ αἰδοῖ τοὺς παρ' ἡμῶν ἐντεῦθεν ἐλθόντας λόγους, εἰ μή τι τὸ λυθὲν μέγα τύχοι γενόμενον, ἐπῳδῆς ἡστινοσοῦν μᾶλλον ἂν συμφῦσαι καὶ συνδῆσαι πάλιν εἰς τὴν προϋπάρχουσαν φιλότητά τε καὶ κοινωνίαν, ἣν ἂν[1] μὲν φιλοσοφῶμεν ἅπαντες ἡμεῖς τε καὶ ὑμεῖς, ὅσον ἂν δυνώμεθα καὶ ἑκάστῳ παρείκῃ, κύρια τὰ νῦν κεχρησμῳδημένα ἔσται· τὸ δὲ ἂν μὴ δρῶμεν ταῦτα οὐκ ἐρῶ· φήμην γὰρ ἀγαθὴν μαντεύομαι, καὶ φημὶ δὴ ταῦθ' ἡμᾶς πάντ' ἀγαθὰ ποιήσειν, ἂν θεὸς ἐθέλῃ.

Ταύτην τὴν ἐπιστολὴν πάντας ὑμᾶς τρεῖς ὄντας

---

[1] ἂν MSS. corr.: ὅταν MSS., Zur.

---

[1] For the language here *cf.* *Symp.* 192 E, 215 C.

EPISTLE VI

and necessitated. Now Hermeias, on the other hand, seems to me—so far as I can judge without having met him as yet—to possess this practical ability both by nature and also through the skill bred of experience.

What, then, do I suggest? To you, Hermeias, I, who have made trial of Erastus and Coriscus more fully than you, affirm and proclaim and testify that you will not easily discover more trustworthy characters than these your neighbours; and I counsel you to hold fast to these men by every righteous means, and regard this as a duty of no secondary importance. To Coriscus and Erastus the counsel I give is this—that they in turn should hold fast to Hermeias, and endeavour by thus holding to one another to become united in the bonds of friendship. But in case any one of you should be thought to be breaking up this union in any way—for what is human is not altogether durable—send a letter here to me and my friends stating the grounds of complaint; for I believe that—unless the disruption should happen to be serious—the arguments sent you from here by us, based on justice and reverence, will serve better than any incantation to weld you and bind you together [1] once again into your former state of friendship and fellowship. If, then, all of us —both we and you—practise this philosophy, as each is able, to the utmost of our power, the prophecy I have now made will come true; but if we fail to do this, I keep silence as to the consequence; for the prophecy I am making is one of good omen, and I declare that we shall, God willing, do all these things well.

All you three must read this letter, all together if

ἀναγνῶναι χρή, μάλιστα μὲν ἀθρόους, εἰ δὲ μή, κατὰ δύο, κοινῇ κατὰ δύναμιν ὡς οἷόν τ' ἐστὶ πλειστάκις, καὶ χρῆσθαι συνθήκῃ καὶ νόμῳ κυρίῳ, D ὅ ἐστι δίκαιον, ἐπομνύντας σπουδῇ τε ἅμα μὴ ἀμούσῳ καὶ τῇ τῆς σπουδῆς ἀδελφῇ παιδιᾷ, καὶ τὸν τῶν πάντων θεὸν ἡγεμόνα τῶν τε ὄντων καὶ τῶν μελλόντων, τοῦ τε ἡγεμόνος καὶ αἰτίου πατέρα κύριον ἐπομνύντας, ὅν, ἂν ὄντως φιλοσοφῶμεν, εἰσόμεθα πάντες σαφῶς εἰς δύναμιν ἀνθρώπων εὐδαιμόνων.

---

[1] For similar expressions cf. *Laws* 761 D, 803 C.
[2] The divine " Ruler " and his " Father " may perhaps be identified with the World-Soul and Demiurge of the

EPISTLE VI

possible, or if not by twos; and as often as you possibly can read it in common, and use it as a form of covenant and a binding law, as is right; and with an earnestness that is not out of tune combined with the playfulness that is sister to earnestness,[1] swear by the God that is Ruler of all that is and that shall be, and swear by the Lord and Father of the Ruler and Cause,[2] Whom, if we are real philosophers, we shall all know truly so far as men well-fortuned[3] can.

Timaeus; or else with the Sun and the Idea of Good in the *Republic* (508 A, 516 B, c, 517 c). *Cf.* also *Ep.* ii. 312 E ff.

[3] εὐδαίμων, in Platonic usage, implies nobility of spirit as well as felicity; *cf. Ep.* viii. 354 c, 355 c.

EPISTLE VII

Prefatory Note.—This is the longest and most important of the Platonic Epistles, and has the best claims to authenticity. From internal evidence we may infer that it was written after the murder of Dion (in 353 B.C.) and before the overthrow of the usurper Callippus in the following year.

While the letter purports to be a message of "counsel" to Dion's friends it really contains a description and a defence of the whole course of Plato's participation in the political affairs of Sicily, and thus constitutes an elaborate *Apologia pro vita sua*.

The letter is so long and so full of digressions that a brief summary may be found useful.

323 D–326 B. Plato's policy the same as Dion's. History of Plato's early life, and how he came to form his political creed, and to stand aloof from public life at Athens owing to its corrupt state.

326 B–328 D. Plato's *first visit* to Sicily. His view of its evil social and political conditions. The friendship he formed with Dion, who came to share his ethical and political creed. How he was urged by Dion, after the death of Dionysius the Elder, to revisit Syracuse, and aid him in effecting a political reformation by training up the young Dionysius to become a philosopher-king.

328 D–330 B. Plato's *second visit* to Sicily. How he

PLATO

was induced to go by the fear of seeming to prove false both to his friendship for Dion and to the cause of philosophy. But his visit proved a failure. Hostile factions slandered Dion and secured his banishment by the young Dionysius, while Plato himself was treated with suspicion. None the less, he kept doing his utmost to influence Dionysius aright.

330 c–331 d. Now Plato must turn to the main purpose of his letter, which is to give *counsel to Dion's friends.* But a counsellor, like a doctor, can only prescribe for those who are willing to act on his advice. And it is a mistake to force the unwilling or to use violent means to rectify the conduct of a father or a fatherland.

331 d–334 c. So Plato's present advice will be similar to that formerly given by him and Dion to Dionysius. They urged him to cultivate self-control and to make loyal friends, warning him by the unhappy example of his father; and they advised him as to his policy. But slander and treachery again prevailed and Dion was exiled. But he returned and by deeds instead of words taught Dionysius in severer fashion, until treachery and slander again attacked him with fatal results. He was accused of seeking to make himself a despot, and two false friends did him to death. It is true that his murderers were Athenians; but no slur should be cast on Athens on that account; for was not his best friend also an Athenian?

334 c–337 e. This account of the advice he gave Dionysius and its sequel is intended as an admonition also to those who consult him now. The policy advised was the abolition of despotism in

EPISTLE VII

Sicily and the establishment of constitutional government, with just laws in all the cities. Dionysius, because he rejected this advice, now lives an ignoble life ; Dion, because he followed it, has met a noble death. If only Dion had been successful he would have secured for all Sicily the blessings which can only come from the reign of Law. Let his friends, therefore, follow now in his footsteps ; let them cease from party-strife and reprisals ; let them practise moderation and self-control in the hour of victory ; and, seeing that the Ideal State under a philosopher-king is now impracticable, let them form a constitution in which Law is King.

337 E–340 B. Plato, having thus concluded his " counsel " to Dion's friends, proceeds with the narrative of his relations with Dionysius. His second visit to Sicily had been ended by the outbreak of war, but he had promised to return after the war on condition that Dion was recalled from exile. But when the tyrant wished to defer the recall of Dion Plato was reluctant to return. Finally, however, he yielded to the urgent entreaties of Dionysius, backed up by the advice of Dion, his Athenian friends, and his friend Archytas of Tarentum. It was reported that Dionysius had recovered his enthusiasm for philosophy ; and Plato felt that, if this were true, he dare not miss the possible chance of seeing his dreams and Dion's fulfilled.

340 B–341 A. On this his *third visit* to Sicily Plato decided to begin by putting to the test the tyrant's interest in philosophy. The test was made by explaining the toil and time it involved owing to the length of the necessary propaedeutic. Only those

PLATO

who can face the ordeal of " plain living and high thinking " survive this test.

341 B–345 C. A long *digression* is here made, dealing with Plato's views on philosophy and its teaching. Dionysius, he says, was an unsatisfactory pupil, since he claimed to be already an expert in philosophy. Later on, it is said, he wrote a treatise on metaphysics himself which he claimed to be superior to Plato's lectures. But he and all others who make such claims are impostors. The deepest doctrines do not admit of written expression, and can only be the fruit of lifelong study; hence, says Plato, I have never written them down myself, nor would the attempt be anything but harmful (342 A).

Why the ultimate realities are thus incommunicable is shown by an analysis of philosophic apprehension and expression. *Knowledge*, and the *Real* which is its object, are approached through sense-perception and verbal description. The elements of this last are the *Name* and the *Definition*; while what the senses perceive is the phenomenon or *Image*. And we must apprehend Name, Definition, Image and Knowledge (" the first Four ") before we attain to the Real (" the Fifth ") (342 E).

For we must be clear as to how the Real differs from the Sensible and its expression. The Name and the Definition give us *quality*, not *essence*. And Name and Definition, like the sensible Image, are never fixed but always shifting and relative. So because the nature of " the Four " is thus defective, the student who seeks to apprehend through them " the Fifth " (Ideal Reality) is filled with confusion; for in seeking the *essence* he finds the *quality* always

EPISTLE VII

intruding. And it is only by searching scrutiny of the " First Four " concomitants of apprehension that the student can hope to win through to a vision of the Real—and then only if he be of his own nature akin to that Ideal Object of reason: and that vision comes, when it does come, by a sudden flash, " as it were a light from heaven " (344 B).

The approach to Philosophy being thus arduous, no " serious " teacher would ever try to teach these " serious " themes in public or write them down. So that if Dionysius has written on metaphysics it only shows that he misunderstands the subject, and that his motive is to gain a cheap reputation for culture. For he received one lesson only on metaphysics from Plato. But whatever be the tyrant's views regarding philosophy and his own philosophic competence, it is monstrous that he should have shown such disrespect as he did to Plato, the acknowledged Master-philosopher (345 c).

345 c-350 B. The narrative of the *third Sicilian visit* is now *resumed*, after the philosophic digression; and it is a narrative of the insults heaped on Plato by Dionysius. To begin with, the latter broke his agreement by refusing to allow the revenues of the exiled Dion to be sent to him. In anger at this Plato said he would return home; but on the tyrant's proposing easier terms for Dion, he consented to remain until they got a reply from Dion. Presently, however, Dionysius turned round and said that he would have all Dion's property sold, keep a half of it for Dion's son, and let Plato take the balance to Dion (347 D).

As this happened when the sailing-season

PLATO

(summer of 361 B.C.) was already over, it was useless to expostulate further; and Plato, caged like a bird, was intent only on escaping as soon as possible. Soon after this a mutiny arose among the mercenary force at Syracuse, owing to the attempt of Dionysius to cut down their pay. For this the blame was thrown on Heracleides, the democratic leader, and his arrest was ordered. Theodotes, however, pleaded for his life, and Dionysius agreed to let him leave the country unharmed. But this agreement he broke the next day, in spite of the renewed intervention of Theodotes and Plato, by sending out soldiers to hunt for Heracleides and seize him. Luckily, however, he made his escape (349 c).

Dionysius's next piece of disrespect was to turn Plato out of the Acropolis and give him a lodging near the soldiers' quarters. And he found a new pretext for quarrelling in the visits paid by Plato to Theodotes, the friend of Dion and Heracleides, which he regarded as a slight to himself. Plato found himself threatened also with violence at the hands of the soldiers amongst whom he lived; so he appealed for help to Archytas of Tarentum, and a Tarentine vessel was dispatched with a request to Dionysius that he should allow Plato to leave Sicily; which request was granted (350 B).

350 B–351 E. On his return from this third Sicilian expedition Plato visited Olympia where he met Dion. Dion was eager to begin military operations against Dionysius, but Plato refused his support on the ground that he had been the guest of the tyrant and was averse to fomenting discord, foreseeing the evils that would inevitably result from

EPISTLE VII

civil war. But his counsels of moderation went unheeded. Yet Dion did not seek power for his own sake but for the sake of the public good; he was not selfish or avaricious or vengeful, but strove to establish the reign of Justice in the State by just means. Prudent though he was, his fall was due to an error of judgement: he failed to gauge accurately the depth of the wickedness of the men with whom he had to deal. The dagger that slew Dion in the hour of his success plunged deep into the very heart of Sicily (351 E).

351 E–352 A. *Conclusion.*—The " counsel " based on the experiences now related has been already set forth. And the narrative of the third visit has been given at this length in order to refute distorted accounts in which Plato's acts and motives had been misrepresented. If it serves to fulfil this purpose and convince his readers that he was justified in what he said and did, he will be well content.

From this summary it will be seen that the letter is, in the main, autobiographical. Its professed object, to offer " counsel " to Dion's friends, is obviously not its chief object, since only one page (336 E–337 E) out of nearly thirty is devoted to the actual statement of that " counsel." The chief object can only be that of pleading justification for the part played by Plato in the internal affairs of Sicily and in the struggle between the rival leaders Dionysius and Dion. The main points of the argument, as derived from the personal experiences narrated, would seem to be these: First, a strong re-assertion of his political creed, namely, that it is only under the rule of the philosopher-king, or, failing

that, under the rule of just laws in a constitutional republic, that any State can hope to flourish. Plato's conviction of this was the outcome of his early experiences in Athens, and all that he saw later, both at home and abroad, only served to confirm it. Next, he wished to make it clear that this conviction, this political philosophy, was one of the main principles which had governed all his actions in regard to Sicilian affairs. He felt himself forced, as he puts it, to have dealings with Dionysius " lest he should be betraying Philosophy." When Providence seemed to be offering a splendid opportunity of realizing the philosopher's dream of the Ideal State, he felt it incumbent upon him to seize that opportunity: his conscience compelled him. Another reason for his actions which is strongly emphasized throughout the letter was his close friendship with Dion, a friendship based on community of conviction. Dion was a convert to Plato's ethical system and shared his political creed. Therefore Dion's cause and the cause of Philosophy were inextricably intertwined; and the claims of friendship came to reinforce the claims of creed.

These are the main points pressed as supplying a justification of Plato's actions and their motives. But his actions, however well-intentioned, were not successful. Therefore much of the narrative, and of the underlying argument, is framed with the view of explaining this ill-success. The main cause lay in the character of Dionysius, who was fickle, treacherous and vain. Others who should share the blame are Dion's enemies at the Court of Syracuse, who set the tyrant against him. Instance after instance is given of the suspicion and the treachery of Diony-

EPISTLE VII

sius in his dealings with Dion and with Plato, and of the prevalence of calumny at the Court of Syracuse. Nor was Dion himself wholly blameless, for it was against Plato's advice that he set out on the final enterprise against Dionysius which cost him his life.

These, then, are the main points—apart from the philosophical digression—which emerge from this lengthy, and somewhat confused, narrative. And from a consideration of these points we may gather something of the reasons which moved Plato to write this letter of self-justification. Evidently he is trying to meet hostile criticism; and we may fairly suppose that the main points of the attack corresponded to the main points of his defence. After Dion's failure and death in 353 B.C. no doubt his supporters were ready enough to throw the blame on someone, and Plato, as his most influential adviser, was the most obvious person to blame. He, like the murderer Callippus (they would say), was an Athenian; he, very likely, had helped to embroil Dion with Dionysius; all his pretended influence at the Court of Syracuse had only proved mischievous, judged by results; and, in fact, if only this Athenian had not come meddling with Sicilian affairs everything might have turned out much better. Possibly also they accused Plato of fraud in connexion with Dion's property.

It is easy enough to understand how such attacks might be made at such a crisis on the probity and good sense and consistency of Plato, and how he might have felt himself driven to defend himself against such baseless charges. But it is rather more difficult to see the relevance of what is known as

PLATO

" the philosophical digression "—a passage which some critics have condemned (not unnaturally) as a spurious insertion. It may be suggested that Plato's purpose in stressing the abstruse and difficult nature of philosophy is to rebut the charge that he had failed to convert Dionysius to views shared by himself and Dion. We may also conjecture that his exposition of the nature of Reality, on which he bases his denial that metaphysics can be explained in writing, is inserted with the object of exploding the notion that Dionysius, or any of his other teachers, were philosophers at all in any true sense of the word. For it appears that Dionysius claimed to be a competent exponent of Idealism, and that many were inclined to accept his claims; and doing so, they might be tempted to ascribe Plato's quarrel with the tyrant to professional jealousy. Or else they might argue that if Dionysius could master the subject so easily and quickly, what need can there be for the prolonged course of training prescribed by Plato? And it is to correct such ignorance of the true nature of philosophy, and to expose the hollowness of the claims of philosophic impostors, and thereby to justify his own attitude towards Dionysius, that Plato writes at such length on the subject. He writes, also, with something like passion, because he feels that the criticisms levelled at him are levelled at Philosophy herself, and that her honour is at stake.

As regards the philosophical exposition itself, there is little or nothing that is not either expressed or implied in the statements of Idealism contained in Plato's Dialogues. Two points only need here be indicated, to supplement the paraphrase already given in our summary, and the references in the

footnotes. For one thing, the use of the term "knowledge" is somewhat confusing, since it sometimes seems to be equated with intellectual apprehension in general, and at other times with pure cognition by the reason. As applied to Reality, or the Ideas, it can, of course, only be used in this latter sense of "scientific knowledge."

The other point of technical interest is that here Ideas are postulated of artificial as well as natural objects, contrary to what Aristotle says about the Platonic theory, as well as to some well-known recent expositions of "the later Platonism." Without entering upon this controverted subject, it is enough to say here that, whether or not Plato ever adopted a later theory of the kind described, the Idealism propounded in this letter is, in all essentials, the same as "the earlier theory" of the *Phaedo* and *Republic*. The Idea is the inexpressible and incommunicable Real which lies behind all existence, objective or subjective.

Now while the apologetic character of this letter is sufficiently clear, doubts have been raised as regards its historical setting. Is it really likely that Dion's followers, whether at Syracuse or at Leontini, would have written to Plato for advice, and put in writing also the criticisms and charges implied in this written answer? And can we easily imagine Plato penning this long narrative of events in Sicily for the benefit of people who must have been perfectly familiar with Sicilian history for years past? Moreover it is difficult to suppose that the tyrant Callippus would allow the dispatch of any non-official communications between Syracuse and foreign ports. These considerations seem to render it more

probable that not only is this letter an "open" letter addressed rather to the general public than to the parties named in the superscription, but that superscription itself is merely a literary device. The letter was never meant to be sent to Sicily at all. And, this being so, the natural corollary is that the hypothetical letter from Dion's party asking for advice is equally imaginary. So that what Plato is doing in this letter is to indulge in a literary fiction which enables him to publish in epistolary form what is at once a history, an apology and a manifesto. For what public, then, was this intended, if we rule out the Sicilians? There can be little doubt as to the answer: it was the public opinion of his own countrymen which Plato was chiefly concerned to influence: the ignorant gossip, the malicious rumours, the damaging misrepresentations current at Athens, were what annoyed him most and what he was most anxious to disprove.

It may be noticed, further, that this view of the letter is supported by the points of contact it has with the *Antidosis* of Isocrates, a speech contemporary with the letter and, like it, largely autobiographical and apologetic. The way in which Isocrates there criticizes Plato and tries to belittle his work as a writer and teacher is sufficient to show the kind of misrepresentation and professional jealousy against which Plato had to contend at home. And in the defence contained in this letter there is probably much of pointed reference to those domestic critics —pseudo-philosophers of the Dionysian type, sophistical quibblers, and rhetors and writers the dupes of unstable words.

Lastly, the severity with which Sicilian luxury is

EPISTLE VII

condemned, combined with the care taken to exculpate Athens from any complicity in the murder of Dion, helps to confirm the view that this seventh letter was published, in the first instance at least, for circulation in Athens and not in Syracuse.

Z

Πλάτων τοῖς Δίωνος οἰκείοις τε καὶ ἑταίροις εὖ πράττειν.

Ἐπεστείλατέ μοι νομίζειν δεῖν τὴν διάνοιαν ὑμῶν εἶναι τὴν αὐτὴν ἣν εἶχε καὶ Δίων, καὶ δὴ καὶ κοινωνεῖν διεκελεύεσθέ μοι, καθ᾽ ὅσον οἷός τ᾽ εἰμὶ ἔργῳ καὶ λόγῳ. ἐγὼ δέ, εἰ μὲν δόξαν καὶ ἐπιθυμίαν τὴν αὐτὴν ἔχετε ἐκείνῳ, ξύμφημι κοινωνήσειν, εἰ δὲ μή, βουλεύσεσθαι πολλάκις. τίς δ᾽ ἦν ἡ ἐκείνου διάνοια καὶ ἐπιθυμία, σχεδὸν οὐκ εἰκάζων ἀλλ᾽ ὡς εἰδὼς σαφῶς εἴποιμ᾽ ἄν. ὅτε γὰρ κατ᾽ ἀρχὰς εἰς Συρακούσας ἐγὼ ἀφικόμην σχεδὸν ἔτη τετταράκοντα γεγονώς, Δίων εἶχε τὴν ἡλικίαν ἣν τὰ νῦν Ἱππαρῖνος γέγονε, καὶ ἣν ἔσχε τότε δόξαν, ταύτην καὶ διετέλεσεν ἔχων, Συρακοσίους οἴεσθαι δεῖν ἐλευθέρους εἶναι, κατὰ νόμους τοὺς ἀρίστους οἰκοῦντας· ὥστε οὐδὲν θαυμαστὸν εἴ τις θεῶν καὶ τοῦτον εἰς τὴν αὐτὴν δόξαν περὶ πολιτείας ἐκείνῳ γενέσθαι σύμφρονα ποιήσειε. τίς δ᾽ ἦν ὁ τρόπος τῆς γενέσεως αὐτῆς, οὐκ ἀπάξιον ἀκοῦσαι νέῳ καὶ μὴ νέῳ, πειράσομαι δὲ ἐξ ἀρχῆς αὐτὴν ἐγὼ πρὸς ὑμᾶς διεξελθεῖν· ἔχει γὰρ καιρὸν τὰ νῦν.

---

[1] Dion was about twenty in 388–387 B.C., the date of Plato's first visit to Syracuse; so if this letter was written in 353 B.C. the birth of Hipparinus (probably Dion's son, not his nephew) should be put at about 373 B.C. *Cf. Ep.* viii. Prefatory Note and 355 E.

EPISTLE VII

PLATO TO DION'S ASSOCIATES AND FRIENDS WISHES WELL-DOING

You wrote to me that I ought to consider that your policy was the same as that which Dion had; and moreover you charged me to support it, so far as I can, both by deed and word. Now if you really hold the same views and aims as he, I consent to support them, but if not, I will ponder the matter many times over. And what was his policy and his aim I will tell you, and that, as I may say, not from mere conjecture but from certain knowledge. For when I originally arrived at Syracuse, being about forty years old, Dion was of the age which Hipparinus has now reached,[1] and the views which he had then come to hold he continued to hold unchanged; for he believed that the Syracusans ought to be free and dwell under the best laws. Consequently, it is no matter of surprise if some Deity has made Hipparinus also come to share his views about government and be of the same mind. Now the manner in which these views originated is a story well worth hearing for young and old alike, and I shall endeavour to narrate it to you from the beginning; for at the present moment it is opportune.

PLATO

324   Νέος ἐγώ ποτε ὢν πολλοῖς δὴ ταὐτὸν ἔπαθον·
ᾠήθην, εἰ θᾶττον ἐμαυτοῦ γενοίμην κύριος, ἐπὶ τὰ
C κοινὰ τῆς πόλεως εὐθὺς ἰέναι. καί μοι τύχαι τινὲς
τῶν τῆς πόλεως πραγμάτων τοιαίδε παρέπεσον.

Ὑπὸ πολλῶν γὰρ τῆς τότε πολιτείας λοιδορου-
μένης μεταβολὴ γίγνεται, καὶ τῆς μεταβολῆς εἷς
καὶ πεντήκοντά τινες ἄνδρες προῦστησαν ἄρχοντες,
ἕνδεκα μὲν ἐν ἄστει, δέκα δ' ἐν Πειραιεῖ, περί τε
ἀγορὰν ἑκάτεροι τούτων ὅσα τ' ἐν τοῖς ἄστεσι
διοικεῖν ἔδει, τριάκοντα δὲ πάντων ἄρχοντες κατ-
D έστησαν αὐτοκράτορες. τούτων δή τινες οἰκεῖοί τε
ὄντες καὶ γνώριμοι ἐτύγχανον ἐμοί, καὶ δὴ καὶ
παρεκάλουν εὐθὺς ὡς ἐπὶ προσήκοντα πράγματά
με. καὶ ἐγὼ θαυμαστὸν οὐδὲν ἔπαθον ὑπὸ νεό-
τητος· ᾠήθην γὰρ αὐτοὺς ἔκ τινος ἀδίκου βίου ἐπὶ
δίκαιον τρόπον ἄγοντας διοικήσειν δὴ τὴν πόλιν,
ὥστε αὐτοῖς σφόδρα προσεῖχον τὸν νοῦν, τί
πράξοιεν. καὶ ἑώρων[1] δή που τοὺς ἄνδρας ἐν
χρόνῳ ὀλίγῳ χρυσῆν ἀποδείξαντας τὴν ἔμπροσθεν
πολιτείαν, τά τε ἄλλα καὶ φίλον ἄνδρα ἐμοὶ
E πρεσβύτερον Σωκράτη, ὃν ἐγὼ σχεδὸν οὐκ ἂν
αἰσχυνοίμην εἰπὼν δικαιότατον εἶναι τῶν τότε, ἐπί
τινα τῶν πολιτῶν μεθ' ἑτέρων ἔπεμπον, βίᾳ ἄξοντα
325 ὡς ἀποθανούμενον, ἵνα δὴ μετέχοι τῶν πραγμάτων
αὐτοῖς, εἴτε βούλοιτο εἴτε μή· ὁ δ' οὐκ ἐπείθετο,
πᾶν δὲ παρεκινδύνευσε παθεῖν πρὶν ἀνοσίων αὐτοῖς
ἔργων γενέσθαι κοινωνός· ἃ δὴ πάντα καθορῶν καὶ

---

[1] ἑώρων : ὁρῶν mss., Zur.

---

[1] Plato's uncle Charmides and his cousin Critias were among the leaders of "the Thirty."
[2] For this episode see *Apol. Socr.* 32 c.

EPISTLE VII

In the days of my youth my experience was the same as that of many others. I thought that as soon as I should become my own master I would immediately enter into public life. But it so happened, I found, that the following changes occurred in the political situation.

In the government then existing, reviled as it was by many, a revolution took place; and the revolution was headed by fifty-one leaders, of whom eleven were in the City and ten in the Piraeus—each of these sections dealing with the market and with all municipal matters requiring management — and Thirty were established as irresponsible rulers of all. Now of these some were actually connexions and acquaintances of mine [1]; and indeed they invited me at once to join their administration, thinking it would be congenial. The feelings I then experienced, owing to my youth, were in no way surprising: for I imagined that they would administer the State by leading it out of an unjust way of life into a just way, and consequently I gave my mind to them very diligently, to see what they would do. And indeed I saw how these men within a short time caused men to look back on the former government as a golden age; and above all how they treated my aged friend Socrates, whom I would hardly scruple to call the most just of men then living, when they tried to send him, along with others, after one of the citizens, to fetch him by force that he might be put to death —their object being that Socrates, whether he wished or no, might be made to share in their political actions; he, however, refused to obey and risked the uttermost penalties rather than be a partaker in their unholy deeds.[2] So when I beheld all these actions and

PLATO

325
εἴ τιν' ἄλλα τοιαῦτα οὐ σμικρά, ἐδυσχέρανά τε καὶ
ἐμαυτὸν ἐπανήγαγον ἀπὸ τῶν τότε κακῶν. χρόνῳ
δὲ οὐ πολλῷ μετέπεσε τὰ τῶν τριάκοντά τε καὶ
πᾶσα ἡ τότε πολιτεία. πάλιν δὲ βραδύτερον μέν,
εἷλκε δέ με ὅμως ἡ περὶ τὸ πράττειν τὰ κοινὰ καὶ
B πολιτικὰ ἐπιθυμία. ἦν οὖν καὶ ἐν ἐκείνοις, ἅτε
τεταραγμένοις, πολλὰ γιγνόμενα ἅ τις ἂν δυσχε-
ράνειε, καὶ οὐδέν τι θαυμαστὸν ἦν τιμωρίας ἐχθρῶν
γίγνεσθαί τινων τισι μείζους ἐν μεταβολαῖς· καίτοι πολλῇ γε ἐχρήσαντο οἱ τότε κατελθόντες
ἐπιεικείᾳ. κατὰ δέ τινα τύχην αὖ τὸν ἑταῖρον
ἡμῶν Σωκράτη τοῦτον δυναστεύοντές τινες εἰσ-
άγουσιν εἰς δικαστήριον, ἀνοσιωτάτην αἰτίαν ἐπι-
βάλλοντες καὶ πάντων ἥκιστα Σωκράτει προσ-
C ήκουσαν· ὡς ἀσεβῆ γὰρ οἱ μὲν εἰσήγαγον, οἱ δὲ
κατεψηφίσαντο καὶ ἀπέκτειναν τὸν τότε τῆς
ἀνοσίου ἀγωγῆς οὐκ ἐθελήσαντα μετασχεῖν περὶ
ἕνα τῶν τότε φευγόντων φίλων, ὅτε φεύγοντες
ἐδυστύχουν αὐτοί.

Σκοποῦντι δή μοι ταῦτά τε καὶ τοὺς ἀνθρώπους
τοὺς πράττοντας τὰ πολιτικά, καὶ τοὺς νόμους γε
καὶ ἔθη, ὅσῳ μᾶλλον διεσκόπουν ἡλικίας τε εἰς τὸ
πρόσθε προὔβαινον, τοσούτῳ χαλεπώτερον ἐφαίνετο
D ὀρθῶς εἶναί μοι τὰ πολιτικὰ διοικεῖν. οὔτε γὰρ
ἄνευ φίλων ἀνδρῶν καὶ ἑταίρων πιστῶν οἷόν τ'
εἶναι πράττειν,—οὓς οὔθ' ὑπάρχοντας ἦν εὑρεῖν
εὐπετές, οὐ γὰρ ἔτι ἐν τοῖς τῶν πατέρων ἤθεσι καὶ
ἐπιτηδεύμασιν ἡ πόλις ἡμῶν διῳκεῖτο, καινούς τε

---

[1] Possibly an allusion to the execution of Theramenes by Critias.

[2] *i.e.* the democrats under Thrasybulus and Thrasyllus.

[3] Meletus and Anytus, the accusers of Socrates; see the *Apology*.

EPISTLE VII

others of a similar grave kind,[1] I was indignant, and I withdrew myself from the evil practices then going on. But in no long time the power of the Thirty was overthrown together with the whole of the government which then existed. Then once again I was really, though less urgently, impelled with a desire to take part in public and political affairs. Many deplorable events, however, were still happening in those times, troublous as they were, and it was not surprising that in some instances, during these revolutions, men were avenging themselves on their foes too fiercely; yet, notwithstanding, the exiles who then returned [2] exercised no little moderation. But, as ill-luck would have it, certain men of authority [3] summoned our comrade Socrates before the law-courts, laying a charge against him which was most unholy, and which Socrates of all men least deserved; for it was on the charge of impiety that those men summoned him and the rest condemned and slew him—the very man who on the former occasion, when they themselves had the misfortune to be in exile, had refused to take part in the unholy arrest of one of the friends of the men then exiled.

When, therefore, I considered all this, and the type of men who were administering the affairs of State, with their laws too and their customs, the more I considered them and the more I advanced in years myself, the more difficult appeared to me the task of managing affairs of State rightly. For it was impossible to take action without friends and trusty companions; and these it was not easy to find ready to hand, since our State was no longer managed according to the principles and institutions of our forefathers; while to acquire other new friends

PLATO

325 ἄλλους ἀδύνατον ἦν κτᾶσθαι μετά τινος ῥαστώνης, —τά τε τῶν νόμων γράμματα καὶ ἔθη διεφθείρετο καὶ ἐπεδίδου θαυμαστὸν ὅσον, ὥστε με, τὸ πρῶτον
E πολλῆς μεστὸν ὄντα ὁρμῆς ἐπὶ τὸ πράττειν τὰ κοινά, βλέποντα εἰς ταῦτα καὶ φερόμενα ὁρῶντα πάντῃ πάντως, τελευτῶντα ἰλιγγιᾶν, καὶ τοῦ μὲν σκοπεῖν μὴ ἀποστῆναι πῇ ποτε ἄμεινον ἂν γίγνοιτο περί τε αὐτὰ ταῦτα καὶ δὴ καὶ περὶ τὴν πᾶσαν
326 πολιτείαν, τοῦ δὲ πράττειν αὖ περιμένειν ἀεὶ καιρούς, τελευτῶντα δὲ νοῆσαι περὶ πασῶν τῶν νῦν πόλεων ὅτι κακῶς ξύμπασαι πολιτεύονται· τὰ γὰρ τῶν νόμων αὐταῖς σχεδὸν ἀνιάτως ἔχοντά ἐστιν ἄνευ παρασκευῆς θαυμαστῆς τινὸς μετὰ τύχης· λέγειν τε ἠναγκάσθην, ἐπαινῶν τὴν ὀρθὴν φιλοσοφίαν, ὡς ἐκ ταύτης ἐστὶ τά τε πολιτικὰ δίκαια καὶ τὰ τῶν ἰδιωτῶν πάντα κατιδεῖν· κακῶν οὖν οὐ λήξειν τὰ ἀνθρώπινα γένη, πρὶν ἂν ἢ τὸ τῶν
B φιλοσοφούντων ὀρθῶς γε καὶ ἀληθῶς γένος εἰς ἀρχὰς ἔλθῃ τὰς πολιτικὰς ἢ τὸ τῶν δυναστευόντων ἐν ταῖς πόλεσιν ἔκ τινος μοίρας θείας ὄντως φιλοσοφήσῃ.

Ταύτην δὴ τὴν διάνοιαν ἔχων εἰς Ἰταλίαν τε καὶ Σικελίαν ἦλθον, ὅτε πρῶτον ἀφικόμην. ἐλθόντα δέ με ὁ ταύτῃ λεγόμενος αὖ βίος εὐδαίμων, Ἰταλιωτικῶν τε καὶ Συρακουσίων τραπεζῶν πλήρης, οὐδαμῇ οὐδαμῶς ἤρεσε, δίς τε τῆς ἡμέρας ἐμπιπλάμενον ζῆν καὶ μηδέποτε κοιμώμενον μόνον
C νύκτωρ, καὶ ὅσα τούτῳ ἐπιτηδεύματα ξυνέπεται τῷ βίῳ· ἐκ γὰρ τούτων τῶν ἐθῶν οὔτ᾽ ἂν φρόνιμος

---

[1] An obvious reference to *Rep.* 473 D, 501 E.
[2] This echoes the famous passage in *Rep.* v. 473 D; *cf.* 328 A *infra*.
[3] *Cf. Rep.* 404 D.

EPISTLE VII

with any facility was a thing impossible. Moreover, both the written laws and the customs were being corrupted, and that with surprising rapidity. Consequently, although at first I was filled with an ardent desire to engage in public affairs, when I considered all this and saw how things were shifting about anyhow in all directions, I finally became dizzy; and although I continued to consider by what means some betterment could be brought about not only in these matters but also in the government as a whole, yet as regards political action I kept constantly waiting for an opportune moment; until, finally, looking at all the States which now exist, I perceived that one and all they are badly governed; for the state of their laws is such as to be almost incurable without some marvellous overhauling and good-luck to boot. So in my praise of the right philosophy I was compelled to declare [1] that by it one is enabled to discern all forms of justice both political and individual. Wherefore the classes of mankind (I said) will have no cessation from evils until either the class of those who are right and true philosophers attains political supremacy, or else the class of those who hold power in the States becomes, by some dispensation of Heaven, really philosophic.[2]

This was the view I held when I came to Italy and Sicily, at the time of my first arrival. And when I came I was in no wise pleased at all with "the blissful life," as it is there termed, replete as it is with Italian and Syracusan banquetings[3]; for thus one's existence is spent in gorging food twice a day and never sleeping alone at night, and all the practices which accompany this mode of living. For not a single man of all who live beneath the heavens

PLATO

326
οὐδείς ποτε γενέσθαι τῶν ὑπὸ τὸν οὐρανὸν ἀνθρώπων ἐκ νέου ἐπιτηδεύων δύναιτο,—οὐχ οὕτω θαυ-
μαστῇ φύσει κραθήσεται,—σώφρων δὲ οὐδ' ἂν
μελλῆσαί ποτε γενέσθαι, καὶ δὴ καὶ περὶ τῆς ἄλλης
ἀρετῆς ὁ αὐτὸς λόγος ἂν εἴη. πόλις τε οὐδεμία ἂν
ἠρεμῆσαι κατὰ νόμους οὐδ' οὑστινασοῦν ἀνδρῶν
οἰομένων ἀναλίσκειν μὲν δεῖν πάντα ἐς ὑπερβολάς,
D ἀργῶν δὲ εἰς ἅπαντα ἡγουμένων αὖ δεῖν γίγνεσθαι
πλὴν εἰς εὐωχίας καὶ πότους καὶ ἀφροδισίων
σπουδὰς διαπονουμένας· ἀναγκαῖον δὲ εἶναι ταύτας
τὰς πόλεις εἰς τυραννίδας τε καὶ ὀλιγαρχίας καὶ
δημοκρατίας μεταβαλλούσας μηδέποτε λήγειν,
δικαίου δὲ καὶ ἰσονόμου πολιτείας τοὺς ἐν αὐταῖς
δυναστεύοντας μηδ' ὄνομα ἀκούοντας ἀνέχεσθαι.
ταῦτα δὴ πρὸς τοῖς πρόσθε διανοούμενος εἰς
Συρακούσας διεπορεύθην, ἴσως μὲν κατὰ τύχην,
E ἔοικε μὴν τότε μηχανωμένῳ τινὶ τῶν κρειττόνων
ἀρχὴν βαλέσθαι τῶν νῦν γεγονότων πραγμάτων
περὶ Δίωνα καὶ τῶν περὶ Συρακούσας· δέος δὲ μὴ
καὶ πλειόνων ἔτι, ἐὰν μὴ νῦν ὑμεῖς ἐμοὶ πείθησθε
τὸ δεύτερον συμβουλεύοντι.

Πῶς οὖν δὴ λέγω πάντων ἀρχὴν γεγονέναι τὴν
327 τότε εἰς Σικελίαν ἐμὴν ἄφιξιν; ἐγὼ συγγενό-
μενος Δίωνι τότε νέῳ κινδυνεύω, τὰ δοκοῦντα
ἐμοὶ βέλτιστα ἀνθρώποις εἶναι μηνύων διὰ λόγων
καὶ πράττειν αὐτὰ ξυμβουλεύων, ἀγνοεῖν ὅτι
τυραννίδος τινὰ τρόπον κατάλυσιν ἐσομένην μηχανώμενος ἐλάνθανον ἐμαυτόν. Δίων μὲν γὰρ δὴ

---

[1] These are the three defective forms of government, contrasting with the three correct forms, monarchy, aristocracy, and constitutional republic; see *Politicus* 291 D ff., 302 B ff.

EPISTLE VII

could ever become wise if these were his practices from his youth, since none will be found to possess a nature so admirably compounded; nor would he ever be likely to become temperate; and the same may truly be said of all other forms of virtue. And no State would remain stable under laws of any kind, if its citizens, while supposing that they ought to spend everywhere to excess, yet believed that they ought to cease from all exertion except feastings and drinkings and the vigorous pursuit of their amours. Of necessity these States never cease changing into tyrannies, oligarchies, and democracies,[1] and the men who hold power in them cannot endure so much as the mention of the name of a just government with equal laws. Holding these views, then, as well as those previously formed, I travelled through to Syracuse—possibly as luck would have it, though it seems likely that one of the Superior Powers was contriving at that time to lay the foundation of the events which have now taken place in regard to Dion and in regard to Syracuse; and of still more events, as is to be feared, unless you now hearken to the counsel I offer you now, for the second time.[2]

What, then, do I mean by saying that my arrival in Sicily on that occasion was the foundation of everything? When I associated with Dion, who was then a youth, instructing him verbally in what I believed was best for mankind and counselling him to realize it in action, it seems that I was not aware that I was, in a way, unwittingly contriving the future overthrow of the tyranny. For Dion in truth, being

[2] The first occasion being at Olympia in 360 B.C.; cf. 350 B ff.

μάλ' εὐμαθὴς ὢν πρός τε τἆλλα καὶ πρὸς τοὺς τότε ὑπ' ἐμοῦ λόγους λεγομένους οὕτως ὀξέως ὑπήκουσε καὶ σφόδρα, ὡς οὐδεὶς πώποτε ὧν ἐγὼ προσέτυχον νέων, καὶ τὸν ἐπίλοιπον βίον ζῆν ἠθέλησε διαφερόντως τῶν πολλῶν Ἰταλιωτῶν τε καὶ Σικελιωτῶν, ἀρετὴν περὶ πλείονος ἡδονῆς τῆς τε ἄλλης τρυφῆς ἠγαπηκώς. ὅθεν ἐπαχθέστερον τοῖς περὶ τὰ τυραννικὰ νόμιμα ζῶσιν ἐβίω μέχρι τοῦ θανάτου τοῦ περὶ Διονύσιον γενομένου.

Μετὰ δὲ τοῦτο διενοήθη μὴ μόνον ἐν αὑτῷ ποτ' ἂν γενέσθαι ταύτην τὴν διάνοιαν, ἣν αὐτὸς ὑπὸ τῶν ὀρθῶν λόγων ἔσχεν, ἐγγιγνομένην δ' αὐτὴν καὶ ἐν ἄλλοις ὁρῶν κατενόει, πολλοῖς μὲν οὔ, γιγνομένην δ' οὖν ἔν τισιν, ὧν καὶ Διονύσιον ἡγήσατο ἕνα γενέσθαι τάχ' ἂν ξυλλαμβανόντων θεῶν, γενομένου δ' αὖ τοῦ τοιούτου τόν τε αὐτοῦ βίον καὶ τὸν τῶν ἄλλων Συρακουσίων ἀμήχανον ἂν μακαριότητι ξυμβῆναι γενόμενον. πρὸς δὴ τούτοις ᾠήθη δεῖν ἐκ παντὸς τρόπου εἰς Συρακούσας ὅ τι τάχιστα ἐλθεῖν ἐμὲ κοινωνὸν τούτων, μεμνημένος τήν τε αὑτοῦ καὶ ἐμὴν συνουσίαν, ὡς εὐπετῶς ἐξειργάσατο εἰς ἐπιθυμίαν ἐλθεῖν αὐτὸν τοῦ καλλίστου τε καὶ ἀρίστου βίου· ὃ δὴ καὶ νῦν εἰ διαπράξαιτο ἐν Διονυσίῳ ὡς ἐπεχείρησε, μεγάλας ἐλπίδας εἶχεν ἄνευ σφαγῶν καὶ θανάτων καὶ τῶν νῦν γεγονότων κακῶν βίον ἂν εὐδαίμονα καὶ ἀληθινὸν ἐν πάσῃ τῇ χώρᾳ κατασκευάσαι.

Ταῦτα Δίων ὀρθῶς διανοηθεὶς ἔπεισε μεταπέμπεσθαι Διονύσιον ἐμέ, καὶ αὐτὸς ἐδεῖτο πέμπων ἥκειν ὅ τι τάχιστα ἐκ παντὸς τρόπου, πρίν τινας

---

[1] Dionysius the Elder died in 367 B.C.

EPISTLE VII

quick-witted, both in other respects and in grasping the arguments I then put forward, hearkened to me with a keenness and ardour that I have never yet found in any of the youth whom I have met ; and he determined to live the rest of his life in a different manner from the majority of the Italians and Sicilians, counting virtue worthy of more devotion than pleasure and all other kinds of luxury. In consequence, his way of life was in ill-odour with those who were conforming to the customary practices of the tyranny, until the death of Dionysius [1] occurred.

After this event, he came to the belief that this belief, which he himself had acquired through right instruction, would not always be confined to himself ; and in fact he saw it being implanted in others also— not in many, it is true, but yet implanted in some ; and of these he thought that Dionysius (with Heaven's help) might become one, and that, if he did become a man of this mind, both his own life and that of all the rest of the Syracusans would, in consequence, be a life of immeasurable felicity. Moreover, Dion considered that I ought, by all means, to come to Syracuse with all speed to be his partner in this task, since he bore in mind our intercourse with one another and how happily it had wrought on him to acquire a longing for the noblest and best life ; and if now, in like manner, he could effect this result in Dionysius, as he was trying to do, he had great hopes of establishing the blissful and true life throughout all the land without massacres and murders and the evils which have now come about.

Holding these right views, Dion persuaded Dionysius to summon me ; and he himself also sent a request that I should by all means come with all

PLATO

327

Ε ἄλλους ἐντυχόντας Διονυσίῳ ἐπ' ἄλλον βίον αὐτὸν
τοῦ βελτίστου παρατρέψαι. λέγων δὲ τάδε ἐδεῖτο,
εἰ καὶ μακρότερα εἰπεῖν. τίνας γὰρ καιρούς, ἔφη,
μείζους περιμενοῦμεν τῶν νῦν παραγεγονότων θείᾳ
τινὶ τύχῃ; καταλέγων δὲ τήν τε ἀρχὴν τῆς
328 Ἰταλίας καὶ Σικελίας καὶ τὴν αὐτοῦ δύναμιν ἐν
αὐτῇ, καὶ τὴν νεότητα καὶ τὴν ἐπιθυμίαν τὴν
Διονυσίου, φιλοσοφίας τε καὶ παιδείας ὡς ἔχοι
σφόδρα, λέγων, τούς τε αὐτοῦ ἀδελφιδοῦς καὶ τοὺς
οἰκείους ὡς εὐπαράκλητοι εἶεν πρὸς τὸν ὑπ' ἐμοῦ
λεγόμενον ἀεὶ λόγον καὶ βίον, ἱκανώτατοί τε
Διονύσιον συμπαρακαλεῖν, ὥστε, εἴπερ ποτέ, καὶ
νῦν ἐλπὶς πᾶσα ἀποτελεσθήσεται τοῦ τοὺς αὐτοὺς
φιλοσόφους τε καὶ πόλεων ἄρχοντας μεγάλων
ξυμβῆναι γενομένους.

Β Τὰ μὲν δὴ παρακελεύματα ἦν ταῦτά τε καὶ
τοιαῦτα ἕτερα πάμπολλα, τὴν δ' ἐμὴν δόξαν, τὸ
μὲν περὶ τῶν νέων ὅπῃ ποτὲ γενήσοιτο, εἶχε φόβος
—αἱ γὰρ ἐπιθυμίαι τῶν τοιούτων ταχεῖαι καὶ πολλάκις ἑαυταῖς ἐναντίαι φερόμεναι—, τὸ δὲ Δίωνος
ἠπιστάμην τῆς ψυχῆς πέρι φύσει τε ἐμβριθὲς ὂν
ἡλικίας τε ἤδη μετρίως ἔχον. ὅθεν μοι σκοπουμένῳ
καὶ διστάζοντι πότερον εἴη πορευτέον καὶ ὑπακουστέον[1] ἢ πῶς, ὅμως ἔρρεψε δεῖν, εἴ ποτέ τις τὰ
C διανοηθέντα περὶ νόμων τε καὶ πολιτείας ἀποτελεῖν
ἐγχειρήσοι, καὶ νῦν πειρατέον εἶναι· πείσας γὰρ

---

[1] καὶ ὑπακουστέον mss. corr.: om. mss., Zur.

---

[1] Among the philosophers and sophists who are said to
have been entertained by Dionysius were Aristippus the
Cyrenaic, Aeschines the Socratic, Polyxenus (*cf. Ep.* ii.
310 c), and Philistus (*cf. Ep.* iii. 315 E).

EPISTLE VII

speed, before that any others [1] should encounter Dionysius and turn him aside to some way of life other than the best. And these were the terms—long though they are to repeat—in which his request was couched: "What opportunities (he asked) are we to wait for that could be better than those that have now been presented by a stroke of divine good fortune?" And he dwelt in detail on the extent of the empire in Italy and Sicily and his own power therein, and the youth of Dionysius, mentioning also how great a desire he had for philosophy and education, and he spoke of his own nephews [2] and connexions, and how they would be not only easily converted themselves to the doctrines and the life I always taught, but also most useful in helping to influence Dionysius; so that now, if ever (he concluded), all our hopes will be fulfilled of seeing the same persons at once philosophers and rulers of mighty States.

By these and a vast number of other like arguments Dion kept exhorting me; but as regards my own opinion, I was afraid how matters would turn out so far as the young people were concerned—for the desires of such as they change quickly, and frequently in a contrary direction; although, as regards Dion's own character, I knew that it was stable by nature and already sufficiently mature. Wherefore as I pondered the matter and was in doubt whether I should make the journey and take his advice, or what, I ultimately inclined to the view that if we were ever to attempt to realize our theories concerning laws and government, now was the time to undertake it; for should I succeed in convincing one single person sufficiently

[2] Probably sisters' sons of Dion, and not including Hipparinus (who would be too young at this date).

PLATO

328 ἕνα μόνον ἱκανῶς πάντα ἐξειργασμένος ἐσοίμην ἀγαθά. ταύτῃ μὲν δὴ τῇ διανοίᾳ τε καὶ τόλμῃ ἀπῆρα οἴκοθεν, οὐχ ᾗ τινὲς ἐδόξαζον, ἀλλ' αἰσχυνόμενος μὲν ἐμαυτὸν τὸ μέγιστον, μὴ δόξαιμί ποτε ἐμαυτῷ παντάπασι λόγος μόνον ἀτεχνῶς εἶναί τις, ἔργου δὲ οὐδενὸς ἄν ποτε ἑκὼν ἀνθάψασθαι, κινδυνεύσειν δὲ προδοῦναι πρῶτον μὲν τὴν Δίωνος D ξενίαν τε καὶ ἑταιρείαν ἐν κινδύνοις ὄντως γεγονότος οὐ σμικροῖς. εἴτ' οὖν πάθοι τι, εἴτ' ἐκπεσὼν ὑπὸ Διονυσίου καὶ τῶν ἄλλων ἐχθρῶν ἔλθοι παρ' ἡμᾶς φεύγων καὶ ἀνέροιτο εἰπών Ὦ Πλάτων, ἥκω σοι φυγὰς οὐχ ὁπλιτῶν δεόμενος οὐδὲ ἱππέων ἐνδεὴς γενόμενος τοῦ ἀμύνασθαι τοὺς ἐχθρούς, ἀλλὰ λόγων καὶ πειθοῦς, ᾗ σὲ μάλιστα ἠπιστάμην ἐγὼ δυνάμενον ἀνθρώπους νέους ἐπὶ τὰ ἀγαθὰ καὶ τὰ δίκαια προτρέποντα εἰς φιλίαν τε καὶ ἑταιρείαν E ἀλλήλοις καθιστάναι ἑκάστοτε· ὧν ἐνδείᾳ κατὰ τὸ σὸν μέρος νῦν ἐγὼ καταλιπὼν Συρακούσας ἐνθάδε πάρειμι. καὶ τὸ μὲν ἐμὸν ἔλαττον ὄνειδός σοι φέρει· φιλοσοφία δέ, ἣν ἐγκωμιάζεις ἀεὶ καὶ ἀτίμως φῇς ὑπὸ τῶν λοιπῶν ἀνθρώπων φέρεσθαι, πῶς οὐ προδέδοται τὰ νῦν μετ' ἐμοῦ μέρος ὅσον 329 ἐπὶ σοὶ γέγονε; καὶ Μεγαροῖ μὲν εἰ κατοικοῦντες ἐτυγχάνομεν, ἦλθες δή που ἄν μοι βοηθὸς ἐφ' ἅ σε παρεκάλουν, ἢ πάντων ἂν φαυλότατον ἡγοῦ σαυτόν· νῦν δ' ἄρα τὸ μῆκος τῆς πορείας καὶ τὸ μέγεθος δὴ

---

[1] The *second* danger was of "proving false to Philosophy," see 328 E *infra*.

[2] A town close to Athens, to which the disciples of Socrates retreated after his death.

EPISTLE VII

I should have brought to pass all manner of good. Holding this view and in this spirit of adventure it was that I set out from home,—not in the spirit which some have supposed, but dreading self-reproach most of all, lest haply I should seem to myself to be utterly and absolutely nothing more than a mere voice and never to undertake willingly any action, and now to be in danger of proving false, in the first [1] instance, to my friendship and association with Dion, when he is actually involved in no little danger. Suppose, then, that some evil fate should befall him, or that he should be banished by Dionysius and his other foes and then come to us as an exile and question us in these words—" O Plato, I come to you as an exile not to beg for foot-soldiers, nor because I lack horse-soldiers to ward off mine enemies, but to beg for arguments and persuasion, whereby you above all, as I know, are able to convert young men to what is good and just and thereby to bring them always into a state of mutual friendliness and comradeship. And it is because you have left me destitute of these that I have now quitted Syracuse and come hither. My condition, however, casts a lesser reproach on you; but as for Philosophy, which you are always belauding, and saying that she is treated with ignominy by the rest of mankind, surely, so far as it depends on you, she too is now betrayed as well as I. Now if we had happened to be living at Megara,[2] you would no doubt have come to assist me in the cause for which I summoned you, on pain of deeming yourself of all men the most base; and now, forsooth, do you imagine that when you plead in excuse the length of the journey and the great strain of the voyage and of the labour involved

τοῦ πλοῦ καὶ τοῦ πόνου ἐπαιτιώμενος οἴει δόξαν κακίας ἀποφευξεῖσθαί ποτε; πολλοῦ γε καὶ δεήσει.

Λεχθέντων δὲ τούτων τίς ἂν ἦν μοι πρὸς ταῦτα εὐσχήμων ἀπόκρισις; οὐκ ἔστιν. ἀλλ' ἦλθον μὲν κατὰ λόγον ἐν δίκῃ τε, ὡς οἷόν τε ἄνθρωπον μάλιστα, διὰ τὰ τοιαῦτα καταλιπὼν τὰς ἐμαυτοῦ διατριβάς, οὔσας οὐκ ἀσχήμονας, ὑπὸ τυραννίδα δοκοῦσαν οὐ πρέπειν τοῖς ἐμοῖς λόγοις οὐδὲ ἐμοί· ἐλθών τε ἐμαυτὸν ἠλευθέρωσα Διὸς ξενίου καὶ τῆς φιλοσόφου ἀνέγκλητον μοίρας παρέσχον, ἐπονειδίστου γενομένης ἂν εἴ τι καταμαλθακισθεὶς καὶ ἀποδειλιῶν αἰσχύνης μετέσχον κακῆς.

Ἐλθὼν δέ, οὐ γὰρ δεῖ μηκύνειν, εὗρον στάσεως τὰ περὶ Διονύσιον μεστὰ ξύμπαντα καὶ διαβολῶν πρὸς τὴν τυραννίδα Δίωνος πέρι. ἤμυνον μὲν οὖν καθ' ὅσον ἠδυνάμην, σμικρὰ δ' οἷός τ' ἦν, μηνὶ δὲ σχεδὸν ἴσως τετάρτῳ Δίωνα Διονύσιος αἰτιώμενος ἐπιβουλεύειν τῇ τυραννίδι, σμικρὸν εἰς πλοῖον ἐμβιβάσας, ἐξέβαλεν ἀτίμως. οἱ δὴ Δίωνος τὸ μετὰ τοῦτο πάντες φίλοι ἐφοβούμεθα μή τινα ἐπαιτιώμενος τιμωροῖτο ὡς συναίτιον τῆς Δίωνος ἐπιβουλῆς· περὶ δ' ἐμοῦ καὶ διῆλθε λόγος τις ἐν Συρακούσαις, ὡς τεθνεὼς εἴην ὑπὸ Διονυσίου ὡς τούτων ἁπάντων τῶν τότε γεγονότων αἴτιος. ὁ δὲ αἰσθανόμενος πάντας ἡμᾶς οὕτω διατεθέντας, φοβούμενος μὴ μεῖζον ἐκ τῶν φόβων γένοιτό τι, φιλοφρόνως πάντας ἀνελάμβανε, καὶ δὴ καὶ τὸν ἐμὲ παρεμυθεῖτό τε καὶ θαρρεῖν διεκελεύετο καὶ

---

[1] Zeus " the Guardian of guests " is mentioned because Plato was a guest-friend of Dion.

EPISTLE VII

you can possibly be acquitted of the charge of cowardice? Far from it, indeed."

If he had spoken thus, what plausible answer should I have had to such pleadings? There is none. Well then, I came for good and just reasons so far as it is possible for men to do so; and it was because of such motives that I left my own occupations, which were anything but ignoble, to go under a tyranny which ill became, as it seemed, both my teaching and myself. And by my coming I freed myself from guilt in the eyes of Zeus Xenios [1] and cleared myself from reproach on the part of Philosophy, seeing that she would have been calumniated if I, through poorness of spirit and timidity, had incurred the shame of cowardice.

On my arrival—I must not be tedious—I found Dionysius's kingdom all full of civil strife and of slanderous stories brought to the court concerning Dion. So I defended him, so far as I was able, though it was little I could do; but about three months later, charging Dion with plotting against the tyranny, Dionysius set him aboard a small vessel and drove him out with ignominy. After that all of us who were Dion's friends were in alarm lest he should punish any of us on a charge of being accomplices in Dion's plot; and regarding me a report actually went abroad in Syracuse that I had been put to death by Dionysius as being responsible for all the events of that time. But when Dionysius perceived us all in this state of mind, he was alarmed lest our fears should bring about some worse result; so he was for receiving us all back in a friendly manner; and, moreover, he kept consoling me and bidding me be of good courage and begging me by all means

329
ἐδεῖτο πάντως μένειν· ἐγίγνετο γάρ οἱ τὸ μὲν ἐμὲ
φυγεῖν ἀπ' αὐτοῦ καλὸν οὐδέν, τὸ δὲ μένειν, διὸ
δὴ καὶ σφόδρα προσεποιεῖτο δεῖσθαι· τὰς δὲ τῶν
τυράννων δεήσεις ἴσμεν, ὅτι μεμιγμέναι ἀνάγκαις
Ε εἰσίν· ὃ δὴ μηχανώμενος διεκώλυέ μου τὸν ἔκπλουν,
εἰς ἀκρόπολιν ἀγαγὼν καὶ κατοικίσας ὅθεν οὐδ' ἂν
εἷς ἔτι με ναύκληρος μὴ ὅτι κωλύοντος ἐξήγαγε
Διονυσίου, ἀλλ' οὐδ' εἰ μὴ πέμπων αὐτὸς τὸν
κελεύοντα ἐξαγαγεῖν ἐπέστελλεν, οὔτ' ἂν ἔμπορος
οὔτε τῶν ἐν ταῖς τῆς χώρας ἐξόδοις ἀρχόντων οὐδ'
ἂν εἷς περιεῖδέ με μόνον ἐκπορευόμενον, ὃς οὐχ ἂν
συλλαβὼν εὐθέως παρὰ Διονύσιον πάλιν ἀπήγαγεν,
330 ἄλλως τε καὶ διηγγελμένον ἤδη ποτὲ τοὐναντίον ἢ
τὸ πρότερον πάλιν, ὡς Πλάτωνα Διονύσιος θαυμαστῶς ὡς ἀσπάζεται. τὸ δ' εἶχε δὴ πῶς; τὸ γὰρ
ἀληθὲς δεῖ φράζειν. ἠσπάζετο μὲν ἀεὶ προϊόντος
τοῦ χρόνου μᾶλλον κατὰ τὴν τοῦ τρόπου τε καὶ
ἤθους συνουσίαν, ἑαυτὸν δὲ ἐπαινεῖν μᾶλλον ἢ
Δίωνα ἐβούλετό με καὶ φίλον ἡγεῖσθαι διαφερόντως
μᾶλλον ἢ 'κεῖνον, καὶ θαυμαστῶς ἐφιλονείκει πρὸς
τὸ τοιοῦτον. ᾗ δ' ἂν οὕτως ἐγένετο, εἴπερ ἐγίγνετο,
Β κάλλιστα, ὤκνει ὡς δὴ μανθάνων καὶ ἀκούων τῶν
περὶ φιλοσοφίαν λόγων οἰκειοῦσθαι καὶ ἐμοὶ συγ-
γίγνεσθαι, φοβούμενος τοὺς τῶν διαβαλλόντων
λόγους, μή πῃ παραποδισθείη καὶ Δίων δὴ πάντα
εἴη διαπεπραγμένος. ἐγὼ δὲ πάντα ὑπέμενον, τὴν
πρώτην διάνοιαν φυλάττων ᾗπερ ἀφικόμην, εἴ πως

---

[1] The citadel of Syracuse, where Plato was housed during both his visits, the tyrant thus having him under his eye.
[2] Philistus and the anti-reform party alleged that Dion was plotting against the tyrant, aided and abetted by Plato, *cf.* 333 E *infra*.

EPISTLE VII

to remain. For my fleeing away from him would have brought him no credit, but rather my remaining; and that was why he pretended to beg it of me so urgently. But the requests of tyrants are coupled, as we know, with compulsory powers. So in order to further this plan he kept hindering my departure; for he brought me into the Acropolis[1] and housed me in a place from which no skipper would have brought me off, and that not merely if prevented by Dionysius but also if he failed to send them a messenger charging them to take me off. Nor would any trader nor any single one of the officers at the ports of the country have let me pass out by myself, without arresting me on the spot and bringing me back again to Dionysius, especially as it had already been proclaimed abroad, contrary to the former report, that "Dionysius is wonderfully devoted to Plato." But what were the facts? For the truth must be told. He became indeed more and more devoted as time advanced, according as he grew familiar with my disposition and character, but he was desirous that I should praise him more than Dion and regard him rather than Dion as my special friend, and this triumph he was marvellously anxious to achieve. But the best way to achieve this, if it was to be achieved—namely, by occupying himself in learning and in listening to discourses on philosophy and by associating with me—this he always shirked owing to his dread of the talk of slanderers, lest he might be hampered in some measure and Dion might accomplish all his designs.[2] I, however, put up with all this, holding fast the original purpose with which I had come, in the hope

PLATO

εἰς ἐπιθυμίαν ἔλθοι τῆς φιλοσόφου ζωῆς. ὁ δ' ἐνίκησεν ἀντιτείνων.

Καὶ ὁ πρῶτος δὴ χρόνος τῆς εἰς Σικελίαν ἐμῆς ἐπιδημήσεώς τε καὶ διατριβῆς διὰ πάντα ταῦτα ξυνέβη γενόμενος. μετὰ δὲ τοῦτο ἀπεδήμησά τε καὶ πάλιν ἀφικόμην πάσῃ σπουδῇ μεταπεμπομένου Διονυσίου· ὧν δὲ ἕνεκα καὶ ὅσα ἔπραξα, ὡς εἰκότα τε καὶ δίκαια, ὑμῖν πρῶτον μὲν ξυμβουλεύσας ἃ χρὴ ποιεῖν ἐκ τῶν νῦν γεγονότων, ὕστερον τὰ περὶ ταῦτα διέξειμι, τῶν ἐπανερωτώντων ἕνεκα τί δὴ βουλόμενος ἦλθον τὸ δεύτερον, ἵνα μὴ τὰ πάρεργα ὡς ἔργα μοι ξυμβαίνῃ λεγόμενα. λέγω δὴ τάδε ἐγώ.[1]

Τὸν συμβουλεύοντα ἀνδρὶ κάμνοντι καὶ δίαιταν διαιτωμένῳ μοχθηρὰν πρὸς ὑγίειαν ἄλλο τι χρὴ πρῶτον μὲν αὐτὸν μεταβάλλειν τὸν βίον, καὶ ἐθέλοντι μὲν πείθεσθαι καὶ τἆλλα ἤδη παραινεῖν· μὴ ἐθέλοντι δέ, φεύγοντα ἀπὸ τῆς τοῦ τοιούτου ξυμβουλῆς ἄνδρα τε ἡγοίμην ἂν καὶ ἰατρικόν, τὸν δὲ ὑπομένοντα τοὐναντίον ἄνανδρόν τε καὶ ἄτεχνον. ταὐτὸν δὴ καὶ πόλει, εἴτε αὐτῆς εἷς εἴη κύριος εἴτε καὶ πλείους, εἰ μὲν κατὰ τρόπον ὀρθῇ πορευομένης ὁδῷ τῆς πολιτείας ξυμβουλεύοιτό τι τῶν προσφόρων, νοῦν ἔχοντος τὸ τοῖς τοιούτοις ξυμβουλεύειν· τοῖς δ' ἔξω τὸ παράπαν βαίνουσι τῆς ὀρθῆς πολιτείας καὶ μηδαμῇ ἐθέλουσιν αὐτῆς εἰς ἴχνος ἰέναι, προαγορεύουσι δὲ τῷ ξυμβούλῳ τὴν μὲν πολιτείαν ἐᾶν καὶ μὴ κινεῖν, ὡς ἀποθανουμένῳ ἐὰν

[1] τάδε ἐγώ] τάδε. Ἐγώ Zur.

[1] *i.e.* "first place" must be given to what is (ostensibly) the main object of the letter, viz. the advising of Dion's friends; see further the Prefatory Note.

EPISTLE VII

that he might possibly gain a desire for the philosophic life; but he, with his resistance, won the day.

These, then, were the causes which brought about my visit to Sicily and my sojourn there, on the first occasion. After this I went away, and I returned again on receiving a most urgent summons from Dionysius. That my motives for doing so and all my actions were reasonable and just, all this I will try to explain later on, for the benefit of those who ask what object I had in going the second time. But first I must counsel you as to the course you ought to adopt in view of the present circumstances, so as not to give the first place to matters of secondary importance.[1] What I have to say, then, is this:

Ought not the doctor that is giving counsel to a sick man who is indulging in a mode of life that is bad for his health to try first of all to change his life, and only proceed with the rest of his advice if the patient is willing to obey? But should he prove unwilling, then I would esteem him both manly and a true doctor if he withdraws from advising a patient of that description, and contrariwise unmanly and unskilled if he continues to advise.[2] So too with a State, whether it has one ruler or many, if so be that it asks for some salutary advice when its government is duly proceeding by the right road, then it is the act of a judicious man to give advice to such people. But in the case of those who altogether exceed the bounds of right government and wholly refuse to proceed in its tracks, and who warn their counsellor to leave the government alone and not

[2] For the comparison of the political adviser to a physician *cf.* *Rep.* 425 E ff., *Laws* 720 A ff.

PLATO

331 κινῇ, τὰ[1] δὲ βουλήσεσι καὶ ἐπιθυμίαις αὐτῶν ὑπηρετοῦντα ξυμβουλεύειν κελεύουσι τίνα τρόπον γίγνοιτ' ἂν ῥᾷστά τε καὶ τάχιστα εἰς τὸν ἀεὶ χρόνον, τὸν μὲν ὑπομένοντα ξυμβουλὰς τοιαύτας ἡγοίμην ἂν ἄνανδρον, τὸν δ' οὐχ ὑπομένοντα ἄνδρα.

Ταύτην δὴ τὴν διάνοιαν ἐγὼ κεκτημένος, ὅταν τίς μοι ξυμβουλεύηται περί τινος τῶν μεγίστων περὶ τὸν αὑτοῦ βίον, οἷον περὶ χρημάτων κτήσεως
B ἢ περὶ σώματος ἢ ψυχῆς ἐπιμελείας, ἂν μέν μοι τὸ καθ' ἡμέραν ἔν τινι τρόπῳ δοκῇ ζῆν ἢ συμβουλεύσαντος ἂν ἐθέλειν πείθεσθαι περὶ ὧν ἀνακοινοῦται, προθύμως ξυμβουλεύω καὶ οὐκ ἀφοσιωσάμενος μόνον ἐπαυσάμην· ἐὰν δὲ μὴ ξυμβουλεύηταί μοι τὸ παράπαν ἢ συμβουλεύοντι δῆλος ᾖ μηδαμῇ πεισόμενος, αὐτόκλητος ἐπὶ τὸν τοιοῦτον οὐκ ἔρχομαι ξυμβουλεύσων, βιασόμενος δὲ οὐδ' ἂν υἱὸς ᾖ μου· δούλῳ δὲ ξυμβουλεύσαιμ' ἂν καὶ μὴ ἐθέλοντά γε προσβιαζοίμην. πατέρα δὲ ἢ μητέρα
C οὐχ ὅσιον ἡγοῦμαι προσβιάζεσθαι μὴ νόσῳ παραφροσύνης ἐχομένους· ἐὰν δέ τινα καθεστῶτα ζῶσι βίον, ἑαυτοῖς ἀρέσκοντα, ἐμοὶ δὲ μή, μήτε ἀπεχθάνεσθαι μάτην νουθετοῦντα μήτε δὴ κολακεύοντά γε ὑπηρετεῖν αὐτοῖς, πληρώσεις ἐπιθυμιῶν ἐκπορίζοντα ἃς αὐτὸς ἀσπαζόμενος οὐκ ἂν ἐθέλοιμι ζῆν. ταὐτὸν δὴ καὶ περὶ πόλεως αὐτοῦ διανοούμενον χρὴ ζῆν τὸν ἔμφρονα· λέγειν μέν, εἰ μὴ καλῶς αὐ-
D τῷ φαίνοιτο πολιτεύεσθαι, εἰ μέλλοι μήτε ματαίως

---

[1] τὰ] ταῖς MSS., Zur.

---

[1] On the subject of filial piety *cf. Crito* 51 c, *Laws* 717 B ff.

EPISTLE VII

disturb it, on pain of death if he does disturb it, while ordering him to advise as to how all that contributes to their desires and appetites may most easily and quickly be secured for ever and ever—then, in such a case, I should esteem unmanly the man who continued to engage in counsels of this kind, and the man who refused to continue manly.

This, then, being the view I hold, whenever anyone consults me concerning any very important affair relating to his life—the acquisition of wealth, for instance, or the care of his body or his soul,—if I believe that he is carrying on his daily life in a proper way, or that he will be willing to obey my advice in regard to the matters disclosed, then I give counsel readily and do not confine myself to some merely cursory reply. But if he does not ask my advice at all or plainly shows that he will in no wise obey his adviser, I do not of my own instance come forward to advise such an one, nor yet to compel him, not even were he my own son. To a slave, however, I would give advice, and if he refused it I would use compulsion. But to a father or mother I deem it impious to apply compulsion,[1] unless they are in the grip of the disease of insanity; but if they are living a settled life which is pleasing to them, though not to me, I would neither irritate them with vain exhortations nor yet minister to them with flatteries by providing them with means to satisfy appetites of a sort such that I, were I addicted to them, would refuse to live. So likewise it behoves the man of sense to hold, while he lives, the same view concerning his own State: if it appears to him to be ill governed he ought to speak, if so be that his speech is not likely to prove fruitless nor to cause his

PLATO

ἐρεῖν μήτε ἀποθανεῖσθαι λέγων, βίαν δὲ πατρίδι πολιτείας μεταβολῆς μὴ προσφέρειν, ὅταν ἄνευ φυγῶν καὶ σφαγῆς ἀνδρῶν μὴ δυνατὸν ᾖ γίγνεσθαι τὴν ἀρίστην, ἡσυχίαν δὲ ἄγοντα εὔχεσθαι τὰ ἀγαθὰ αὑτῷ τε καὶ τῇ πόλει.

Κατὰ δὴ τοῦτον τὸν τρόπον ἐγὼ ὑμῖν τ' ἂν ξυμβουλεύοιμι, ξυνεβούλευον δὲ καὶ Διονυσίῳ μετὰ Δίωνος, ζῆν μέντοι τὸ καθ' ἡμέραν πρῶτον, ὅπως ἐγκρατὴς αὐτὸς αὑτοῦ ὅ τι μάλιστα ἔσεσθαι μέλλοι καὶ πιστοὺς φίλους τε καὶ ἑταίρους κτήσεσθαι, ὅπως μὴ πάθοι ἅπερ ὁ πατὴρ αὐτοῦ, ὃς παραλαβὼν Σικελίας πολλὰς καὶ μεγάλας πόλεις ὑπὸ τῶν βαρβάρων ἐκπεπορθημένας, οὐχ οἷός τ' ἦν κατοικίσας πολιτείας ἐν ἑκάσταις καταστήσασθαι πιστὰς ἑταίρων ἀνδρῶν, οὔτε ἄλλων δή ποθεν ὀθνείων οὔτε ἀδελφῶν, οὓς ἔθρεψέ τε αὐτὸς νεωτέρους ὄντας ἔκ τε ἰδιωτῶν ἄρχοντας καὶ ἐκ πενήτων πλουσίους ἐπεποιήκει διαφερόντως. τούτων κοινωνὸν τῆς ἀρχῆς οὐδένα οἷός τ' ἦν πειθοῖ καὶ διδαχῇ καὶ εὐεργεσίαις καὶ ξυγγενείαις ἀπεργασάμενος ποιήσασθαι, Δαρείου δὲ ἑπταπλασίῳ φαυλότερος ἐγένετο, ὃς οὐκ ἀδελφοῖς πιστεύσας οὐδ' ὑφ' αὑτοῦ τραφεῖσι, κοινωνοῖς δὲ μόνον τῆς τοῦ Μήδου τε καὶ εὐνούχου χειρώσεως, διένειμέ τε μέρη μείζω ἕκαστα Σικελίας πάσης ἑπτὰ καὶ πιστοῖς ἐχρήσατο τοῖς κοινωνοῖς καὶ οὐκ ἐπιτιθεμένοις οὔτε αὐτῷ

---

[1] Cf. *Ep.* v. 322 B.
[2] The reference is to the two brothers of Dionysius the Elder, Leptines and Thearidas.

EPISTLE VII

death[1]; but he ought not to apply violence to his fatherland in the form of a political revolution, whenever it is impossible to establish the best kind of polity without banishing and slaughtering citizens, but rather he ought to keep quiet and pray for what is good both for himself and for his State.

This, then, is the way in which I would counsel you,—even as Dion and I together used to counsel Dionysius that he should, in the first place, so order his daily life as to gain the greatest possible mastery over himself, and to win for himself trusty friends and companions; that so he might avoid the evils suffered by his father. For he, when he had recovered many great cities of Sicily which had been laid waste by the barbarians, was unable, when he settled them, to establish in each a loyal government composed of true comrades,—whether strangers from abroad or men of his own kin[2] whom he himself had reared up in their youth and had raised from a private position to one of authority and from a state of poverty to surpassing wealth. Neither by persuasion nor instruction, neither by benefits nor by ties of kindred, was he able to make any one of them worthy of a share in his government. Thus he was seven times more unhappy than Darius[3] who trusted men who neither were his brothers nor reared up by himself but merely colleagues who had helped him to crush the Mede and the Eunuch; and he divided amongst them seven provinces, each greater than the whole of Sicily; and these colleagues he found loyal, neither did they make any attack either

[3] Darius wrested the kingdom of Persia from the usurper Pseudo-Smerdis by the aid of six other Persian nobles, *cf. Laws* 695 B ff. For the numerical computation of comparative happiness *cf. Rep.* 587 B ff.

PLATO

332 οὔτε ἀλλήλοις, ἔδειξέ τε παράδειγμα οἷον χρὴ τὸν νομοθέτην καὶ βασιλέα τὸν ἀγαθὸν γίγνεσθαι· νόμους γὰρ κατασκευάσας ἔτι καὶ νῦν διασέσωκε τὴν Περσῶν ἀρχήν. ἔτι δὲ Ἀθηναῖοι πρὸς τούτοις, οὐκ αὐτοὶ κατοικίσαντες πολλὰς τῶν Ἑλλήνων πόλεις ὑπὸ βαρβάρων ἐκβεβλημένας [ἀλλ' οἰκουμένας]¹ παραλαβόντες, ὅμως ἑβδομήκοντα ἔτη
C διεφύλαξαν τὴν ἀρχὴν ἄνδρας φίλους ἐν ταῖς πόλεσιν ἑκάσταις κεκτημένοι. Διονύσιος δὲ εἰς μίαν πόλιν ἀθροίσας πᾶσαν Σικελίαν ὑπὸ σοφίας πιστεύων οὐδενὶ μόγις ἐσώθη· πένης γὰρ ἦν ἀνδρῶν φίλων καὶ πιστῶν, οὗ μεῖζον σημεῖον εἰς ἀρετὴν καὶ κακίαν οὐκ ἔστιν οὐδέν, τοῦ ἔρημον ἢ μὴ τοιούτων ἀνδρῶν εἶναι.

Ἃ δὴ καὶ Διονυσίῳ ξυνεβουλεύομεν ἐγὼ καὶ Δίων ἐπειδὴ τὰ παρὰ πατρὸς αὐτῷ ξυνεβεβήκει οὕτως
D ἀνομιλήτῳ μὲν παιδείας, ἀνομιλήτῳ δὲ συνουσιῶν τῶν προσηκουσῶν γεγονέναι, πρῶτον ἐπὶ ταῦτα ὁρμήσαντα φίλους ἄλλους αὐτῷ τῶν οἰκείων ἅμα καὶ ἡλικιωτῶν καὶ συμφώνους πρὸς ἀρετὴν κτήσασθαι, μάλιστα δ' αὐτὸν αὑτῷ, τούτου γὰρ αὐτὸν θαυμαστῶς ἐνδεᾶ γεγονέναι, λέγοντες οὐκ ἐναργῶς οὕτως, οὐ γὰρ ἦν ἀσφαλές, αἰνιττόμενοι δὲ καὶ διαμαχόμενοι τοῖς λόγοις ὡς οὕτω μὲν πᾶς ἀνὴρ αὑτόν τε καὶ ἐκείνους ὧν ἂν ἡγεμὼν γίγνηται σώσει, μὴ ταύτῃ δὲ τραπόμενος τἀναντία πάντα
E ἀποτελεῖ· πορευθεὶς δὲ ὡς λέγομεν, καὶ ἑαυτὸν ἔμφρονά τε καὶ σώφρονα ἀπεργασάμενος, εἰ τὰς

¹ ἀλλ' οἰκουμένας om. best ms.

---

[1] The maritime empire of the Athenians lasted for some seventy years after Salamis (480 B.C.).

EPISTLE VII

on himself or on one another. And thus he left an example of the character which should belong to the good lawgiver and king; for by the laws he framed he has preserved the empire of the Persians even until this day. Moreover, the Athenians also, after taking over many of the Greek cities which had fallen into the hands of the barbarians, though they had not colonized them themselves yet held their sway over them securely for seventy years because they possessed citizens who were their friends in each of those cities.[1] But Dionysius, though he amalgamated the whole of Sicily into one City-State, because in his wisdom he distrusted everyone, barely achieved safety; for he was poor in men who were loyal friends, and there exists no surer sign of a man's virtue or vice than whether he is or is not destitute of men of that kind.

Such, then, was the counsel which Dion and I always gave to Dionysius. Inasmuch as the result of his father's conduct was to leave him unprovided with education and unprovided with suitable intercourse, he should, in the first place, make it his aim to acquire other friends for himself from among his kindred and contemporaries who were in harmony about virtue; and to acquire, above all else, this harmony within himself, since in this he was surprisingly deficient. Not that we expressed this openly, for it would not have been safe; but we put it in veiled terms and maintained by argument that this is how every man will save both himself and all those under his leadership, whereas if he does not adopt this course he will bring about entirely opposite results. And if he pursued the course we describe, and made himself right-minded and sober-minded,

PLATO

332 ἐξηρημωμένας Σικελίας πόλεις κατοικίσειε νόμοις τε ξυνδήσειε καὶ πολιτείαις, ὥστε αὑτῷ τε οἰκείας καὶ ἀλλήλαις εἶναι πρὸς τὰς τῶν βαρβάρων βοηθείας, οὐ διπλασίαν τὴν πατρῴαν ἀρχὴν μόνον 333 ποιήσοι, πολλαπλασίαν δὲ ὄντως· ἕτοιμον γὰρ εἶναι τούτων γενομένων πολὺ μᾶλλον δουλώσασθαι Καρχηδονίους τῆς ἐπὶ Γέλωνος αὐτοῖς γενομένης δουλείας, ἀλλ' οὐχ ὥσπερ νῦν τοὐναντίον ὁ πατὴρ αὐτοῦ φόρον ἐτάξατο φέρειν τοῖς βαρβάροις.

Ταῦτα ἦν τὰ λεγόμενα καὶ παρακελευόμενα ὑφ' ἡμῶν τῶν ἐπιβουλευόντων Διονυσίῳ, ὡς πολλαχόθεν ἐχώρουν οἱ τοιοῦτοι λόγοι, οἳ δὴ καὶ κρατήσαντες παρὰ Διονυσίῳ ἐξέβαλον μὲν Δίωνα, ἡμᾶς B δ' εἰς φόβον κατέβαλον. ἵνα δ' ἐκπεράνωμεν οὐκ ὀλίγα πράγματα [τὰ]¹ ἐν ὀλίγῳ χρόνῳ, ἐλθὼν ἐκ Πελοποννήσου καὶ Ἀθηνῶν Δίων ἔργῳ τὸν Διονύσιον ἐνουθέτησεν. ἐπειδὴ δ' οὖν ἠλευθέρωσέ τε καὶ ἀπέδωκεν αὐτοῖς δὶς τὴν πόλιν, ταὐτὸν πρὸς Δίωνα Συρακόσιοι τότε ἔπαθον, ὅπερ καὶ Διονύσιος, ὅτε αὐτὸν ἐπεχείρει παιδεύσας καὶ θρέψας βασιλέα τῆς ἀρχῆς ἄξιον οὕτω κοινωνεῖν αὐτῷ τοῦ βίου C παντός, ὁ δὲ τοῖς διαβάλλουσιν ⟨ὑπήκουσεν⟩² καὶ λέγουσιν ὡς ἐπιβουλεύων τῇ τυραννίδι Δίων πράττοι πάντα ὅσα ἔπραττεν ἐν τῷ τότε χρόνῳ, ἵνα ὁ μὲν παιδείᾳ δὴ τὸν νοῦν κηληθεὶς ἀμελοῖ τῆς ἀρχῆς ἐπιτρέψας ἐκείνῳ, ὁ δὲ σφετερίσαιτο καὶ Διονύσιον

¹ τὰ bracketed by Hermann.
² ὑπήκουσεν I add (ἐπίστευε add. Cornarius).

---

[1] Gelon succeeded Hippocrates as tyrant of Gela about 490 B.C., and then captured Syracuse and made it his capital.

EPISTLE VII

then, if he were to re-people the devastated cities of Sicily and bind them together by laws and constitutions so that they should be leagued both with himself and with one another against barbarian reinforcements, he would thus not merely double the empire of his father but actually multiply it many times over; for if this came to pass, it would be an easy task to enslave the Carthaginians far more than they had been enslaved in the time of Gelon,[1] whereas now, on the contrary, his father had contracted to pay tribute to the barbarians.

Such was the advice and exhortation given to Dionysius by us, who were plotting against him, as statements pouring in from many quarters alleged; which statements in fact so prevailed with Dionysius that they caused Dion's expulsion and threw us into a state of alarm. Then—to cut a long story short—Dion came from the Peloponnesus and from Athens and admonished Dionysius by deed.[2] When, however, Dion had delivered the Syracusans and given them back their city twice, they showed the same feeling towards him as Dionysius had done. For when Dion was trying to train and rear him up to be a king worthy of the throne, that so he might share with him in all his life, Dionysius listened to the slanderers who said that Dion was plotting against the tyranny in all that he was then doing, his scheme being that Dionysius, with his mind infatuated with education, should neglect his empire and entrust it to Dion, who should then seize on it

---

His defeat of the Carthaginians at Himera, 480 B.C., was celebrated by the poet Simonides.

[2] *i.e.* by a military campaign ("deed" as opposed to "word") in 357 B.C.

ἐκβάλοι ἐκ τῆς ἀρχῆς δόλῳ. ταῦτα τότε ἐνίκησε καὶ τὸ δεύτερον ἐν Συρακοσίοις λεγόμενα, καὶ μάλα ἀτόπῳ τε καὶ αἰσχρᾷ νίκῃ τοῖς τῆς νίκης αἰτίοις.

Οἷον γὰρ γέγονεν, ἀκοῦσαι χρὴ τοὺς ἐμὲ παρακαλοῦντας πρὸς τὰ νῦν πράγματα. ἦλθον Ἀθηναῖος ἀνὴρ ἐγώ, ἑταῖρος Δίωνος, σύμμαχος αὐτῷ, πρὸς τὸν τύραννον, ὅπως ἀντὶ πολέμου φιλίαν ποιήσαιμι· διαμαχόμενος δὲ τοῖς διαβάλλουσιν ἡττήθην. πείθοντος δὲ Διονυσίου τιμαῖς καὶ χρήμασι γενέσθαι μετ᾽ αὐτοῦ ἐμέ, μάρτυρά τε καὶ φίλον πρὸς τὴν εὐπρέπειαν τῆς ἐκβολῆς τῆς Δίωνος αὐτῷ γίγνεσθαι, τούτων δὴ τὸ πᾶν διήμαρτεν. ὕστερον δὲ δὴ κατιὼν οἴκαδε Δίων ἀδελφὼ δύο προσλαμβάνει Ἀθήνηθεν, οὐκ ἐκ φιλοσοφίας γεγονότε φίλω ἀλλ᾽ ἐκ τῆς περιτρεχούσης ἑταιρείας ταύτης τῆς τῶν πλείστων φίλων, ἣν ἐκ τοῦ ξενίζειν τε καὶ μυεῖν καὶ ἐποπτεύειν πραγματεύονται. καὶ δὴ καὶ τούτω τὼ ξυγκαταγαγόντε αὐτὸν φίλω ἐκ τούτων τε καὶ ἐκ τῆς πρὸς τὴν κάθοδον ὑπηρεσίας ἐγενέσθην ἑταίρω. ἐλθόντες δὲ εἰς Σικελίαν, ἐπειδὴ Δίωνα ᾔσθοντο διαβεβλημένον εἰς τοὺς ἐλευθερωθέντας ὑπ᾽ αὐτοῦ Σικελιώτας ὡς ἐπιβουλεύοντα γενέσθαι τύραννον, οὐ μόνον τὸν ἑταῖρον καὶ ξένον προὔδοσαν, ἀλλ᾽ οἷον τοῦ φόνου αὐτόχειρες ἐγένοντο, ὅπλα ἔχοντες ἐν ταῖς χερσὶν αὐτοὶ τοῖς φονεῦσι παρεστῶτες ἐπίκουροι. καὶ τὸ μὲν αἰσχρὸν καὶ ἀνόσιον οὔτε παρίεμαι ἔγωγε οὔτε τι

---

[1] Callippus and Philostratus; *cf.* Plutarch, *Dion*, cc. 54 ff.
[2] After the Little Mysteries of Eleusis the initiated became a μυστής, after the Great Mysteries an ἐπόπτης.

EPISTLE VII

for himself and expel Dionysius from his kingship by craft. And then, for the second time, these slanderous statements triumphed with the Syracusans, and that with a triumph that was most monstrous and shameful for the authors of the triumph.

Those who are urging me to address myself to the affairs of to-day ought to hear what then took place. I, a citizen of Athens, a companion of Dion, an ally of his own, went to the tyrant in order that I might bring about friendship instead of war; but in my struggle with the slanderers I was worsted. But when Dionysius tried to persuade me by means of honours and gifts of money to side with him so that I should bear witness, as his friend, to the propriety of his expulsion of Dion, in this design he failed utterly. And later on, while returning home from exile, Dion attached to himself two brothers from Athens,[1] men whose friendship was not derived from philosophy, but from the ordinary companionship out of which most friendships spring, and which comes from mutual entertaining and sharing in religion and mystic ceremonies.[2] So, too, in the case of these two friends who accompanied him home; it was for these reasons and because of their assistance in his homeward voyage that they became his companions. But on their arrival in Sicily, when they perceived that Dion was slanderously charged before the Siceliots whom he had set free with plotting to become tyrant, they not only betrayed their companion and host but became themselves, so to say, the authors of his murder, since they stood beside the murderers, ready to assist, with arms in their hands. For my own part, I neither slur over the shamefulness and sinfulness of their action nor do I

λέγω· πολλοῖς γὰρ καὶ ἄλλοις ὑμνεῖν ταῦτα ἐπιμελὲς καὶ εἰς τὸν ἔπειτα μελήσει χρόνον· τὸ δὲ Ἀθηναίων πέρι λεγόμενον, ὡς αἰσχύνην οὗτοι περιῆψαν τῇ πόλει, ἐξαιροῦμαι· φημὶ γὰρ κἀκεῖνον Ἀθηναῖον εἶναι ὃς οὐ προὔδωκε τὸν αὐτὸν τοῦτον, ἐξὸν χρήματα καὶ ἄλλας τιμὰς πολλὰς λαμβάνειν. οὐ γὰρ διὰ βαναύσου φιλότητος ἐγεγόνει φίλος, διὰ δὲ ἐλευθέρας παιδείας κοινωνίαν, ᾗ μόνῃ χρὴ πιστεύειν τὸν νοῦν κεκτημένον μᾶλλον ἢ ξυγγενείᾳ ψυχῶν καὶ σωμάτων. ὥστε οὐκ ἀξίω ὀνείδους γεγόνατον τῇ πόλει τὼ Δίωνα ἀποκτείναντε, ὡς ἐλλογίμω πώποτε ἄνδρε γενομένω.

Ταῦτα εἴρηταί πάντα τῆς ξυμβουλῆς ἕνεκα τῶν Διωνείων φίλων καὶ ξυγγενῶν. ξυμβουλεύω δὲ δή τι πρὸς τούτοις τὴν αὐτὴν ξυμβουλὴν καὶ λόγον τὸν αὐτὸν λέγων ἤδη τρίτον τρίτοις ὑμῖν· μὴ δουλοῦσθαι Σικελίαν ὑπ' ἀνθρώποις δεσπόταις, μηδὲ ἄλλην πόλιν, ὅ γ' ἐμὸς λόγος, ἀλλ' ὑπὸ νόμοις· οὔτε γὰρ τοῖς δουλουμένοις οὔτε τοῖς δουλωθεῖσιν ἄμεινον, αὐτοῖς καὶ παισὶ παίδων τε ἐκγόνοις, ἀλλ' ὀλέθριος πάντως ἡ πεῖρα, σμικρὰ δὲ καὶ ἀνελεύθερα ψυχῶν ἤθη τὰ τοιαῦτα ἁρπάζειν κέρδη φιλεῖ, οὐδὲν τῶν εἰς τὸν ἔπειτα καὶ εἰς τὸν παρόντα καιρὸν ἀγαθῶν καὶ δικαίων εἰδότα θείων τε καὶ ἀνθρωπίνων. ταῦτα πρῶτον μὲν Δίωνα ἐγὼ ἐπεχείρησα πείθειν, δεύτερον δὲ Διονύσιον, τρίτους δὲ ὑμᾶς νῦν. καί μοι πείθεσθε Διὸς τρίτου σωτῆρος χάριν, εἶτα εἰς Διονύσιον βλέψαντες καὶ

---

[1] *Cf.* 336 D, *Laws* 961 A ff.
[2] An allusion to the custom of offering the third (and last) cup at banquets as a libation to Zeus Soter; *cf. Rep.* 583 B, *Charm.* 167 B.

EPISTLE VII

dwell on it, since there are many others who make it their care to recount these doings and will continue to do so in time to come. But I do take exception to what is said about the Athenians, that these men covered their city with shame; for I assert that it was also an Athenian who refused to betray the very same man when, by doing so, he might have gained wealth and many other honours. For he had become his friend not in the bonds of a venal friendship but owing to association in liberal education; since it is in this alone that the judicious man should put his trust, rather than in kinship of soul or of body. Consequently, the two murderers of Dion are not important enough to cast a reproach upon our city,[1] as though they had ever yet shown themselves men of mark.

All this has been said by way of counsel to Dion's friends and relatives. And one piece of counsel I add, as I repeat now for the third time to you in the third place the same counsel as before, and the same doctrine. Neither Sicily, nor yet any other State—such is my doctrine—should be enslaved to human despots but rather to laws; for such slavery is good neither for those who enslave nor those who are enslaved—themselves, their children and their children's children; rather is such an attempt wholly ruinous, and the dispositions that are wont to grasp gains such as these are petty and illiberal, with no knowledge of what belongs to goodness and justice, divine or human, either in the present or in the future. Of this I attempted to persuade Dion in the first place, secondly Dionysius, and now, in the third place, you. Be ye, then, persuaded for the sake of Zeus, Third Saviour,[2] and considering also the case of Dionysius and of Dion, of whom the former was

PLATO

334
Δίωνα, ὧν ὁ μὲν μὴ πειθόμενος ζῇ τὰ νῦν οὐ
E καλῶς, ὁ δὲ πειθόμενος τέθνηκε καλῶς· τὸ γὰρ
τῶν καλλίστων ἐφιέμενον αὑτῷ τε καὶ πόλει
πάσχειν ὅ τι ἂν πάσχῃ πᾶν ὀρθὸν καὶ καλόν. οὔτε
γὰρ πέφυκεν ἀθάνατος ἡμῶν οὐδείς, οὔτ' εἴ τῳ
ξυμβαίη, γένοιτο ἂν εὐδαίμων, ὡς δοκεῖ τοῖς
πολλοῖς. κακὸν γὰρ καὶ ἀγαθὸν οὐδὲν λόγου ἄξιόν
335 ἐστι τοῖς ἀψύχοις, ἀλλ' ἢ μετὰ σώματος οὔσῃ
ψυχῇ τοῦτο ξυμβήσεται ἑκάστῃ ἢ κεχωρισμένῃ.
πείθεσθαι δὲ ὄντως ἀεὶ χρὴ τοῖς παλαιοῖς τε καὶ
ἱεροῖς λόγοις, οἳ δὴ μηνύουσιν ἡμῖν ἀθάνατον
ψυχὴν εἶναι δικαστάς τε ἴσχειν καὶ τίνειν τὰς
μεγίστας τιμωρίας, ὅταν τις ἀπαλλαχθῇ τοῦ
σώματος. διὸ καὶ τὰ μεγάλα ἁμαρτήματα καὶ
ἀδικήματα σμικρότερον εἶναι χρὴ νομίζειν κακὸν
πάσχειν ἢ δρᾶσαι. ὧν ὁ φιλοχρήματος πένης τε
B ἀνὴρ τὴν ψυχὴν οὔτε ἀκούει, ἐάν τε ἀκούσῃ, κατα-
γελῶν, ὡς οἴεται, πανταχόθεν ἀναιδῶς ἁρπάζει
πᾶν ὅ τί περ ἂν οἴηται, καθάπερ θηρίον, φαγεῖν
ἢ πιεῖν ἢ περὶ τὴν ἀνδραποδώδη καὶ ἀχάριστον,
ἀφροδίσιον λεγομένην οὐκ ὀρθῶς, ἡδονὴν ἐκποριεῖν
αὑτῷ τοὐμπίπλασθαι,[1] τυφλὸς ὢν καὶ οὐχ ὁρῶν
οἷα[2] ξυνέπεται [τῶν ἁρπαγμάτων][3] ἀνοσιουργία,
κακὸν ἡλίκον, ἀεὶ μετ' ἀδικήματος ἑκάστου, ἣν
ἀναγκαῖον τῷ ἀδικήσαντι συνεφέλκειν ἐπί τε γῇ
C στρεφομένῳ καὶ ὑπὸ γῆς νοστήσαντι πορείαν
ἄτιμόν τε καὶ ἀθλίαν πάντως πανταχῇ.

Δίωνα δὴ ἐγὼ λέγων ταῦτά τε καὶ ἄλλα τοιαῦτα

[1] τοὐμπίπλασθαι Hermann : τῷ (MSS. τοῦ) μὴ πίμπλασθαι Zur.
[2] οἷα] οἷς MSS., Zur.
[3] τῶν ἁρπαγμάτων I bracket (v.l. τῶν πραγμάτων).

EPISTLE VII

unpersuaded and is living now no noble life, while the latter was persuaded and has nobly died. For whatsoever suffering a man undergoes when striving after what is noblest both for himself and for his State is always right and noble. For by nature none of us is immortal, and if any man should come to be so he would not be happy, as the vulgar believe; for no evil nor good worthy of account belongs to what is soulless, but they befall the soul whether it be united with a body or separated therefrom. But we ought always truly to believe the ancient and holy doctrines which declare to us that the soul is immortal and that it has judges and pays the greatest penalties, whensoever a man is released from his body; wherefore also one should account it a lesser evil to suffer than to perform the great iniquities and injustices.[1] But to these doctrines the man who is fond of riches but poor in soul listens not, or if he listens he laughs them (as he thinks) to scorn, while he shamelessly plunders from all quarters everything which he thinks likely to provide himself, like a beast, with food or drink or the satiating himself with the slavish and graceless pleasure which is miscalled by the name of the Goddess of Love[2]; for he is blind and fails to see what a burden of sin—how grave an evil—ever accompanies each wrong-doing; which burden the wrong-doer must of necessity drag after him both while he moves about on earth and when he has gone beneath the earth again on a journey that is unhonoured and in all ways utterly miserable.

Of these and other like doctrines I tried to per-

---

[1] This theme is to be found also in the *Gorgias* and *Republic*; cf. also *Lysis* 217 B.

[2] Cf. *Gorg.* 493 E, *Phaedo* 81 B, *Phileb.* 12 B.

ἔπειθον, καὶ τοῖς ἀποκτείνασιν ἐκεῖνον δικαιότατ᾽
ἂν ὀργιζοίμην ἐγὼ τρόπον τινὰ ὁμοιότατα καὶ
Διονυσίῳ· ἀμφότεροι γὰρ ἐμὲ καὶ τοὺς ἄλλους, ὡς
ἔπος εἰπεῖν, ἅπαντας τὰ μέγιστα ἔβλαψαν ἀνθρώπους, οἱ μὲν τὸν βουλόμενον δικαιοσύνῃ χρῆσθαι
διαφθείραντες, ὁ δὲ οὐδὲν ἐθελήσας χρήσασθαι
D δικαιοσύνῃ διὰ πάσης τῆς ἀρχῆς, μεγίστην δύναμιν
ἔχων, ἐν ᾗ γενομένη φιλοσοφία τε καὶ δύναμις
ὄντως ἐν ταὐτῷ διὰ πάντων ἀνθρώπων Ἑλλήνων
τε καὶ βαρβάρων λάμψασ᾽ ἄν[1] ἱκανῶς δόξαν
παρέστησε πᾶσι τὴν ἀληθῆ, ὡς οὐκ ἄν ποτε
γένοιτο εὐδαίμων οὔτε πόλις οὔτ᾽ ἀνὴρ οὐδεὶς ὃς
ἂν μὴ μετὰ φρονήσεως ὑπὸ δικαιοσύνῃ διαγάγῃ
τὸν βίον, ἤτοι ἐν αὑτῷ κεκτημένος ἢ ὁσίων
ἀνδρῶν ἀρχόντων ἐν ἤθεσι τραφείς τε καὶ παι-
E δευθεὶς ἐνδίκως. ταῦτα μὲν Διονύσιος ἔβλαψε· τὰ
δὲ ἄλλα σμικρὰ ἂν εἴη πρὸς ταῦτά μοι βλάβη.
ὁ δὲ Δίωνα ἀποκτείνας οὐκ οἶδε ταὐτὸν ἐξειργασμένος τούτῳ. Δίωνα γὰρ ἐγὼ σαφῶς οἶδα,
ὡς οἷόν τε περὶ ἀνθρώπων ἄνθρωπον διισχυρίζε-
σθαι, ὅτι τὴν ἀρχὴν εἰ κατέσχεν, ὡς οὐκ ἄν
ποτε ἐπ᾽ ἄλλο γε σχῆμα ἀρχῆς ἐτράπετο ἢ ἐπὶ τὸ
336 Συρακούσας μὲν πρῶτον τὴν πατρίδα τὴν ἑαυτοῦ,
ἐπεὶ τὴν δουλείαν αὐτῆς ἀπήλλαξε [καὶ] φαιδρύνας
ἐλευθερίῳ δ᾽ ἐν σχήματι κατέστησε, τὸ μετὰ τοῦτ᾽
ἂν πάσῃ μηχανῇ ἐκόσμησε νόμοις τοῖς προσήκουσί
τε καὶ ἀρίστοις τοὺς πολίτας, τό τε ἐφεξῆς τούτοις
προὐθυμεῖτ᾽ ἂν πρᾶξαι, πᾶσαν Σικελίαν κατοικίζειν

---

[1] λάμψασ᾽ ἂν Schneider : λάμψασαν Zur.

EPISTLE VII

suade Dion, and I have the best of rights to be angry with the men who slew him, very much as I have to be angry also with Dionysius; for both they and he have done the greatest of injuries both to me, and, one may say, to all the rest of mankind—they by destroying the man who purposed to practise justice, and he by utterly refusing to practise justice, when he had supreme power, throughout all his empire; although if, in that empire, philosophy and power had really been united in the same person the radiance thereof would have shone through the whole world of Greeks and barbarians, and fully imbued them with the true conviction that no State nor any individual man can ever become happy unless he passes his life in subjection to justice combined with wisdom, whether it be that he possesses these virtues within himself or as the result of being reared and trained righteously under holy rulers in their ways. Such were the injuries committed by Dionysius; and, compared to these, the rest of the injuries he did I would count but small. And the murderer of Dion is not aware that he has brought about the same result as Dionysius. For as to Dion, I know clearly—in so far as it is possible for a man to speak with assurance about men—that, if he had gained possession of the kingdom, he would never have adopted for his rule any other principle than this: when he had first brought gladness to Syracuse, his own fatherland, by delivering her from bondage, and had established her in a position of freedom, he would have endeavoured next, by every possible means, to set the citizens in order by suitable laws of the best kind; and as the next step after this, he would have done his utmost to colonize the whole of Sicily and

PLATO

καὶ ἐλευθέραν ἀπὸ τῶν βαρβάρων ποιεῖν, τοὺς μὲν ἐκβάλλων, τοὺς δὲ χειρούμενος ῥᾶον Ἱέρωνος· τούτων δ' αὖ γενομένων δι' ἀνδρὸς δικαίου τε καὶ ἀνδρείου καὶ σώφρονος καὶ φιλοσόφου τὴν αὐτὴν ἀρετῆς ἂν πέρι γενέσθαι δόξαν τοῖς πολλοῖς ἥπερ ἄν, εἰ Διονύσιος ἐπείσθη, ⟨πάντα⟩¹ παρὰ πᾶσιν ἂν ὡς ἔπος εἰπεῖν ἀνθρώποις ἀπέσωσε γενομένη. νῦν δὲ ἥ πού τις δαίμων ἤ τις ἀλιτήριος ἐμπεσὼν ἀνομίᾳ καὶ ἀθεότητι καὶ τὸ μέγιστον τόλμαις ἀμαθίας, ἐξ ἧς πάντα κακὰ πᾶσιν ἐρρίζωται καὶ βλαστάνει καὶ εἰς ὕστερον ἀποτελεῖ καρπὸν τοῖς γεννήσασι πικρότατον, αὕτη πάντα τὸ δεύτερον ἀνέτρεψέ τε καὶ ἀπώλεσε.

Νῦν δὲ δὴ εὐφημῶμεν χάριν οἰωνοῦ τὸ τρίτον. ὅμως δὲ μιμεῖσθαι μὲν συμβουλεύω Δίωνα ὑμῖν τοῖς φίλοις τήν τε τῆς πατρίδος εὔνοιαν καὶ τὴν τῆς τροφῆς σώφρονα δίαιταν, ἐπὶ λῳόνων² δὲ ὀρνίθων τὰς ἐκείνου βουλήσεις πειρᾶσθαι ἀποτελεῖν· αἱ δὲ ἦσαν, ἀκηκόατε παρ' ἐμοῦ σαφῶς· τὸν δὲ μὴ δυνάμενον ὑμῶν Δωριστὶ ζῆν κατὰ τὰ πάτρια, διώκοντα δὲ τόν τε τῶν Δίωνος σφαγέων καὶ τὸν Σικελικὸν βίον, μήτε παρακαλεῖν μήτε οἴεσθαι πιστὸν ἄν τι καὶ ὑγιὲς πρᾶξαί ποτε· τοὺς δὲ ἄλλους παρακαλεῖν ἐπὶ πάσης Σικελίας κατοικισμόν τε καὶ ἰσονομίαν ἔκ τε αὐτῆς Σικελίας καὶ ἐκ Πελοποννήσου ξυμπάσης, φοβεῖσθαι δὲ μηδὲ Ἀθήνας·

---
¹ πάντα I add.
² λῳόνων Schneider: λῷον ὡς Zur. (λ. ὧν mss.).

---
[1] Hiero, tyrant of Syracuse (478–466), waged successful war against the Carthaginians.

EPISTLE VII

to make it free from the barbarians, by driving out some of them and subduing others more easily than did Hiero.[1] And if all this had been done by a man who was just and courageous and temperate and wisdom-loving, the most of men would have formed the same opinion of virtue which would have prevailed, one may say, throughout the whole world, if Dionysius had been persuaded by me, and which would have saved all. But as it is, the onset of some deity or some avenging spirit, by means of lawlessness and godlessness and, above all, by the rash acts of ignorance [2]—that ignorance which is the root whence all evils for all men spring and which will bear hereafter most bitter fruit for those who have planted it—this it is which for the second time has wrecked and ruined all.

But now, for the third time, let us speak good words, for the omen's sake. Nevertheless, I counsel you, his friends, to imitate Dion in his devotion to his fatherland and in his temperate mode of life; and to endeavour to carry out his designs, though under better auspices; and what those designs were you have learnt from me clearly. But if any amongst you is unable to live in the Dorian fashion of his forefathers and follows after the Sicilian way of life and that of Dion's murderers, him you should neither call to your aid nor imagine that he could ever perform a loyal or sound action; but all others you should call to aid you in repeopling all Sicily and giving it equal laws, calling them both from Sicily itself and from the whole of the Peloponnese, not fearing even Athens itself; for there too there are

[2] For the calamitous effects of " ignorance " (or " folly ") *cf. Laws* 688 c ff., 863 c ff.

εἰσὶ γὰρ καὶ ἐκεῖ πάντων ἀνθρώπων διαφέροντες πρὸς ἀρετὴν ξενοφόνων τε ἀνδρῶν μισοῦντες τόλμας. εἰ δ' οὖν ταῦτα μὲν ὕστερα γένοιτ' ἄν, κατεπείγουσι δὲ ὑμᾶς αἱ τῶν στάσεων πολλαὶ καὶ παντοδαπαὶ φυόμεναι ἑκάστης ἡμέρας διαφοραί, εἰδέναι μέν που χρὴ πάντα τινὰ ἄνδρα, ᾧ καὶ βραχὺ δόξης ὀρθῆς μετέδωκε θεία τις τύχη, ὡς οὐκ ἔστι παῦλα κακῶν τοῖς στασιάσασι, πρὶν ἂν οἱ κρατήσαντες μάχαις καὶ ἐκβολαῖς ἀνθρώπων καὶ σφαγαῖς μνησικακοῦντες καὶ ἐπὶ τιμωρίας παύσωνται τρεπόμενοι τῶν ἐχθρῶν, ἐγκρατεῖς δὲ ὄντες αὑτῶν, θέμενοι νόμους κοινοὺς μηδὲν μᾶλλον πρὸς ἡδονὴν αὑτοῖς ἢ τοῖς ἡττηθεῖσι κειμένους, ἀναγκάσωσιν αὐτοὺς χρῆσθαι τοῖς νόμοις διτταῖς οὔσαις ἀνάγκαις, αἰδοῖ καὶ φόβῳ, φόβῳ μὲν διὰ τὸ κρείττους αὐτῶν εἶναι δεικνύντες τὴν βίαν, αἰδοῖ δὲ αὖ διὰ τὸ κρείττους φαίνεσθαι περί τε τὰς ἡδονὰς καὶ τοῖς νόμοις μᾶλλον ἐθέλοντές τε καὶ δυνάμενοι δουλεύειν. ἄλλως δὲ οὐκ ἔστιν ὡς ἄν ποτε κακῶν λήξαι πόλις ἐν αὑτῇ στασιάσασα, ἀλλὰ στάσεις καὶ ἔχθραι καὶ μίση καὶ ἀπιστίαι ταῖς οὕτω διατεθείσαις πόλεσιν αὐταῖς πρὸς αὑτὰς ἀεὶ γίγνεσθαι φιλεῖ.

Τοὺς δὴ κρατήσαντας ἀεὶ χρή, ὅτανπερ ἐπιθυμήσωσι σωτηρίας, αὐτοὺς ἐν αὑτοῖς ἄνδρας προκρῖναι τῶν Ἑλλήνων οὓς ἂν πυνθάνωνται ἀρίστους ὄντας, πρῶτον μὲν γέροντας, καὶ παῖδας καὶ γυναῖκας κεκτημένους οἴκοι καὶ προγόνους αὐτῶν ὅ τι μάλιστα πολλούς τε καὶ ὀνομαστοὺς καὶ

---

[1] *Cf. Laws* 646 E ff., 671 D.   [2] *Cf. Laws* 715 A ff.

EPISTLE VII

those who surpass all men in virtue, and who detest the enormities of men who slay their hosts. But—though these results may come about later,—if for the present you are beset by the constant quarrels of every kind which spring up daily between the factions, then every single man on whom the grace of Heaven has bestowed even a small measure of right opinion must surely be aware that there is no cessation of evils for the warring factions until those who have won the mastery cease from perpetuating feuds by assaults and expulsions and executions, and cease from seeking to wreak vengeance on their foes; and, exercising mastery over themselves, lay down impartial laws which are framed to satisfy the vanquished no less than themselves; and compel the vanquished to make use of these laws by means of two compelling forces, namely, Reverence and Fear [1] —Fear, inasmuch as they make it plain that they are superior to them in force; and Reverence, because they show themselves superior both in their attitude to pleasures and in their greater readiness and ability to subject themselves to the laws. In no other way is it possible for a city at strife within itself to cease from evils, but strife and enmity and hatred and suspicion are wont to keep for ever recurring in cities when their inner state is of this kind.[2]

Now those who have gained the mastery, whenever they become desirous of safety, ought always to choose out among themselves such men of Greek origin as they know by inquiry to be most excellent —men who are, in the first place, old, and who have wives and children at home, and forefathers as numerous and good and famous as possible, and who

PLATO

κτῆσιν κεκτημένους πάντας ἱκανήν· ἀριθμὸν δὲ εἶναι μυριάνδρῳ πόλει πεντήκοντα ἱκανοὶ τοιοῦτοι. τούτους δὲ δεήσεσι καὶ τιμαῖς ὅ τι μεγίσταις οἴκοθεν μεταπέμπεσθαι, μεταπεμψαμένους δὲ ὁμόσαντας δεῖσθαι καὶ κελεύειν θεῖναι νόμους, μήτε νικήσασι μήτε νικηθεῖσι νέμειν πλέον, τὸ δὲ ἴσον καὶ κοινὸν πάσῃ τῇ πόλει. τεθέντων δὲ τῶν νόμων ἐν τούτῳ δὴ τὰ πάντα ἐστίν. ἂν μὲν γὰρ οἱ νενικηκότες ἥττους αὑτοὺς τῶν νόμων μᾶλλον τῶν νενικημένων παρέχωνται, πάντ' ἔσται σωτηρίας τε καὶ εὐδαιμονίας μεστὰ καὶ πάντων κακῶν ἀποφυγή· εἰ δὲ μή, μήτ' ἐμὲ μήτ' ἄλλον κοινωνὸν παρακαλεῖν ἐπὶ τὸν μὴ πειθόμενον τοῖς νῦν ἐπεσταλμένοις. ταῦτα γάρ ἐστιν ἀδελφὰ ὧν τε Δίωνι ὧν τ' ἐγὼ ἐπεχειρήσαμεν Συρακούσαις εὖ φρονοῦντες συμπρᾶξαι, δεύτερα μήν. πρῶτα δ' ἦν ἃ τὸ πρῶτον ἐπεχειρήθη μετ' αὐτοῦ Διονυσίου πραχθῆναι πᾶσι κοινὰ ἀγαθά, τύχη δέ τις ἀνθρώπων κρείττων διεφόρησε. τὰ δὲ νῦν ὑμεῖς πειρᾶσθε εὐτυχέστερον αὐτὰ ἀγαθῇ πρᾶξαι μοίρᾳ καὶ θείᾳ τινὶ τύχῃ.

Ξυμβουλὴ μὲν δὴ καὶ ἐπιστολὴ εἰρήσθω καὶ ἡ παρὰ Διονύσιον ἐμὴ προτέρα ἄφιξις· ἡ δὲ δὴ ὑστέρα πορεία τε καὶ πλοῦς ὡς εἰκότως τε ἅμα καὶ ἐμμελῶς γέγονεν, ᾧ μέλει ἀκούειν ἔξεστι τὸ μετὰ τοῦτο. ὁ μὲν γὰρ δὴ πρῶτος χρόνος τῆς ἐν

---

[1] For this scheme *cf. Laws* 752 D ff.; and for the qualifications of the law-givers *cf. Laws* 765 D.

[2] For the Law-governed State as the second-best, after the Ideal Republic, *cf. Polit.* 297 D ff.

[3] Alluding to the attempt then being made by Dion's party at Leontini, under Hipparinus (his nephew), to overthrow Callippus.

EPISTLE VII

are all in possession of ample property; and for a city of ten thousand citizens, fifty such men would be a sufficient number.[1] These men they should fetch from their homes by means of entreaties and the greatest possible honours; and when they have fetched them they should entreat and enjoin them to frame laws, under oath that they will give no advantage either to conquerors or conquered, but equal rights in common to the whole city. And when the laws have been laid down, then everything depends upon the following condition. On the one hand, if the victors prove themselves subservient to the laws more than the vanquished, then all things will abound in safety and happiness, and all evils will be avoided; but should it prove otherwise, neither I nor anyone else should be called in to take part in helping the man who refuses to obey our present injunctions. For this course of action is closely akin to that which Dion and I together, in our plans for the welfare of Syracuse, attempted to carry out, although it is but the second-best[2]; for the first was that which we first attempted to carry out with the aid of Dionysius himself—a plan which would have benefited all alike, had it not been that some Chance, mightier than men, scattered it to the winds. Now, however, it is for you to endeavour to carry out our policy with happier results by the aid of Heaven's blessing and divine good-fortune.[3]

Let this, then, suffice as my counsel and my charge, and the story of my former visit to the court of Dionysius. In the next place, he that cares to listen may hear the story of my later journey by sea, and how naturally and reasonably it came about. For (as I said) I had completed my account of the first

PLATO

338 Σικελίᾳ διατριβῆς μοι διεπεράνθη, καθάπερ εἶπον, πρὶν συμβουλεύειν τοῖς οἰκείοις καὶ ἑταίροις τοῖς περὶ Δίωνα· τὸ μετ' ἐκεῖνα δ' οὖν ἔπεισα ὅπῃ δή ποτ' ἐδυνάμην Διονύσιον ἀφεῖναί με, εἰρήνης δὲ γενομένης, ἦν γὰρ τότε πόλεμος ἐν Σικελίᾳ ξυνωμολογήσαμεν ἀμφότεροι, Διονύσιος μὲν [ἔφη] μεταπέμψεσθαι[1] Δίωνα καὶ ἐμὲ πάλιν καταστησάμενος τὰ περὶ τὴν ἀρχὴν ἀσφαλέστερον ἑαυτῷ, Δίωνα δὲ ἠξίου διανοεῖσθαι μὴ φυγὴν αὑτῷ γε-
B γονέναι τότε, μετάστασιν δέ· ἐγὼ δ' ἥξειν ὡμολόγησα ἐπὶ τούτοις τοῖς λόγοις. γενομένης δὲ εἰρήνης μετεπέμπετ' ἐμέ, Δίωνα δὲ ἐπισχεῖν ἔτι ἐνιαυτὸν ἐδεῖτο, ἐμὲ δὲ ἥκειν ἐκ παντὸς τρόπου ἠξίου. Δίων μὲν οὖν ἐκέλευέ τέ με πλεῖν καὶ ἐδεῖτο· καὶ γὰρ δὴ λόγος ἐχώρει πολὺς ἐκ Σικελίας ὡς Διονύσιος θαυμαστῶς φιλοσοφίας ἐν ἐπιθυμίᾳ πάλιν εἴη γεγονὼς τὰ νῦν, ὅθεν ὁ Δίων συντεταμένως ἐδεῖτο ἡμῶν τῇ μεταπέμψει μὴ ἀπειθεῖν.
C ἐγὼ δὲ ἤδη μέν που κατὰ τὴν φιλοσοφίαν τοῖς νέοις πολλὰ τοιαῦτα γιγνόμενα, ὅμως δ' οὖν ἀσφαλέστερόν μοι ἔδοξε χαίρειν τότε γε πολλὰ καὶ Δίωνα καὶ Διονύσιον ἐᾶν, καὶ ἀπηχθόμην ἀμφοῖν ἀποκρινάμενος ὅτι γέρων τε εἴην καὶ κατὰ τὰς ὁμολογίας οὐδὲν γίγνοιτο τῶν τὰ νῦν πραττομένων.

Ἔοικε δὴ τὸ μετὰ τοῦτο Ἀρχύτης τε παρὰ Διονύσιον ἀφικέσθαι—, ἐγὼ γὰρ πρὶν ἀπιέναι ξενίαν καὶ φιλίαν Ἀρχύτῃ καὶ τοῖς ἐν Τάραντι καὶ

---

[1] ἔφη bracketed by H. Richards: μεταπέμψασθαι MSS., Zur.

---

[1] This refers back to 330 C, D, just before he begins his "counsel" to Dion's friends.
[2] Cf. *Ep.* iii. 317 A.

EPISTLE VII

period of my stay in Sicily [1] before I gave my counsel to the intimates and companions of Dion. What happened next was this: I urged Dionysius by all means possible to let me go, and we both made a compact that when peace was concluded (for at that time there was war in Sicily [2]) Dionysius, for his part, should invite Dion and me back again, as soon as he had made his own power more secure; and he asked Dion to regard the position he was now in not as a form of exile but rather as a change of abode; and I gave a promise that upon these conditions I would return. When peace was made he kept sending for me; but he asked Dion to wait still another year, although he kept demanding most insistently that I should come. Dion, then, kept urging and entreating me to make the voyage; for in truth constant accounts were pouring in from Sicily how Dionysius was now once more marvellously enamoured of philosophy; and for this reason Dion was strenuously urging me not to disobey his summons. I was of course well aware that such things often happen to the young in regard to philosophy; but none the less I deemed it safer, at least for the time, to give a wide berth both to Dion and Dionysius, and I angered them both by replying that I was an old man and that none of the steps which were now being taken were in accordance with our compact.

Now it seems that after this Archytas [3] arrived at the court of Dionysius; for when I sailed away, I had, before my departure, effected a friendly alliance between Archytas and the Tarentines and Dionysius;

---

[3] A famous scientist and statesman of Tarentum; *cf.* 350 A *infra*, *Ep.* xiii. 360 c.

PLATO

D Διονυσίῳ ποιήσας ἀπέπλεον—, ἄλλοι τέ τινες ἐν Συρακούσαις ἦσαν Δίωνός τε ἄττα διακηκοότες καὶ τούτων τινὲς ἄλλοι, παρακουσμάτων τινῶν ἔμμεστοι τῶν κατὰ φιλοσοφίαν· οἳ δοκοῦσί μοι Διονυσίῳ πειρᾶσθαι διαλέγεσθαι πως[1] περὶ τὰ τοιαῦτα, ὡς Διονυσίου πάντα διακηκοότος ὅσα διενοούμην ἐγώ. ὁ δὲ οὔτε ἄλλως ἐστὶν ἀφυὴς πρὸς τὴν τοῦ μανθάνειν δύναμιν φιλότιμός τε θαυμαστῶς· ἤρεσκέ τε οὖν ἴσως αὐτῷ τὰ λεγόμενα ᾐσχύνετό τε φανερὸς γιγνόμενος οὐδὲν ἀκηκοὼς
E ὅτ' ἐπεδήμουν ἐγώ, ὅθεν ἅμα μὲν εἰς ἐπιθυμίαν ᾔει τοῦ διακοῦσαι ἐναργέστερον, ἅμα δ' ἡ φιλοτιμία κατήπειγεν αὐτόν· δι' ἃ δὲ οὐκ ἤκουσεν ἐν τῇ πρόσθεν ἐπιδημίᾳ, διεξήλθομεν ἐν τοῖς ἄνω ῥηθεῖσι νῦν δὴ λόγοις. ἐπειδὴ οὖν οἴκαδέ τε ἐσώθην καὶ καλοῦντος τὸ δεύτερον ἀπηρνήθην, καθάπερ εἶπον νῦν δή, δοκεῖ μοι Διονύσιος παντάπασι φιλοτιμηθῆναι μή ποτέ τισι δόξαιμι κατα-
339 φρονῶν αὐτοῦ τῆς φύσεώς τε καὶ ἕξεως ἅμα καὶ τῆς διαίτης ἔμπειρος γεγονώς, οὐκέτ' ἐθέλειν δυσχεραίνων παρ' αὐτὸν ἀφικνεῖσθαι.

Δίκαιος δὴ λέγειν εἰμὶ τἀληθὲς καὶ ὑπομένειν, εἴ τις ἄρα τὰ γεγονότα ἀκούσας καταφρονήσει τῆς ἐμῆς φιλοσοφίας, τὸν τύραννον δὲ ἡγήσεται νοῦν ἔχειν. ἔπεμψε μὲν γὰρ δὴ Διονύσιος τρίτον ἐπ' ἐμὲ τριήρη ῥᾳστώνης ἕνεκα τῆς πορείας, ἔπεμψε δὲ Ἀρχέδημον, ὃν ἡγεῖτό με τῶν ἐν Σικελίᾳ περὶ πλείστου ποιεῖσθαι, τῶν Ἀρχύτῃ ξυγγεγονότων

---

[1] πως] τῶν mss., Zur.

---

[1] Cf. 330 B.
[2] Plato had refused the second time; see 338 E.

522

EPISTLE VII

and there were certain others in Syracuse who had had some teaching from Dion, and others again who had been taught by these, men who were stuffed with some borrowed philosophical doctrines. These men, I believe, tried to discuss these subjects with Dionysius, on the assumption that Dionysius was thoroughly instructed in all my system of thought. Now besides being naturally gifted otherwise with a capacity for learning Dionysius has an extraordinary love of glory. Probably, then, he was pleased with what was said and was ashamed of having it known that he had no lessons while I was in the country; and in consequence of this he was seized with a desire to hear my doctrines more explicitly, while at the same time he was spurred on by his love of glory: and we have already explained, in the account we gave a moment ago,[1] the reasons why he had not been a hearer of mine during my previous sojourn. So when I had got safely home and had refused his second summons, as I said just now, Dionysius was greatly afraid, I believe, because of his love of glory, lest any should suppose that it was owing to my contempt for his nature and disposition, together with my experience of his mode of life, that I was ungracious and was no longer willing to come to his court.

Now I am bound to tell the truth, and to put up with it should anyone, after hearing what took place, come to despise, after all, my philosophy and consider that the tyrant showed intelligence. For, in fact, Dionysius, on this third occasion,[2] sent a trireme to fetch me, in order to secure my comfort on the voyage; and he sent Archedemus, one of the associates of Archytas, believing that I esteemed him above

PLATO

Β ἕνα, καὶ ἄλλους γνωρίμους τῶν ἐν Σικελίᾳ· οὗτοι δὲ ἡμῖν ἤγγελλον πάντες τὸν αὐτὸν λόγον, ὡς θαυμαστὸν ὅσον Διονύσιος ἐπιδεδωκὼς εἴη πρὸς φιλοσοφίαν. ἔπεμψε δὲ ἐπιστολὴν πάνυ μακράν, εἰδὼς ὡς πρὸς Δίωνα διεκείμην καὶ τὴν αὖ Δίωνος προθυμίαν τοῦ ἐμὲ πλεῖν καὶ εἰς Συρακούσας ἐλθεῖν· πρὸς γὰρ δὴ πάντα ταῦτα ἦν παρεσκευασμένη τὴν ἀρχὴν ἔχουσα ἡ ἐπιστολή, τῇδέ πῃ φράζουσαν[1]· Διονύσιος Πλάτωνι· τὰ νόμιμα
C ἐπὶ τούτοις εἰπὼν οὐδὲν τὸ μετὰ τοῦτο εἶπε πρότερον ἢ ὡς "Ἂν εἰς Σικελίαν πεισθεὶς ὑφ' ἡμῶν ἔλθῃς τὰ νῦν, πρῶτον μέν σοι τὰ περὶ Δίωνα ὑπάρξει ταύτῃ γιγνόμενα ὅπηπερ ἂν αὐτὸς ἐθέλῃς· θελήσεις δὲ οἶδ' ὅτι τὰ μέτρια, καὶ ἐγὼ συγχωρήσομαι· εἰ δὲ μή, οὐδέν σοι τῶν περὶ Δίωνα ἕξει πραγμάτων οὔτε περὶ τἆλλα οὔτε περὶ αὐτὸν κατὰ νοῦν γιγνόμενα. ταῦθ' οὕτως εἶπε, τἆλλα δὲ
D μακρὰ ἂν εἴη καὶ ἄνευ καιροῦ λεγόμενα. ἐπιστολαὶ δὲ ἄλλαι ἐφοίτων παρά τε Ἀρχύτου καὶ τῶν ἐν Τάραντι, τήν τε φιλοσοφίαν ἐγκωμιάζουσαι τὴν Διονυσίου καὶ ὅτι, ἂν μὴ ἀφίκωμαι νῦν, τὴν πρὸς Διονύσιον αὐτοῖς γενομένην φιλίαν δι' ἐμοῦ οὐ σμικρὰν οὖσαν πρὸς τὰ πολιτικὰ παντάπασι διαβαλοίην. ταύτης δὴ τοιαύτης γενομένης ἐν τῷ τότε χρόνῳ τῆς μεταπέμψεως, τῶν μὲν ἐκ Σικελίας τε καὶ Ἰταλίας ἑλκόντων, τῶν δὲ Ἀθήνηθεν ἀτεχνῶς μετὰ δεήσεως οἷον ἐξωθούντων με, καὶ
Ε πάλιν ὁ λόγος ἧκεν ὁ αὐτός, τὸ μὴ δεῖν προδοῦναι Δίωνα μηδὲ τοὺς ἐν Τάραντι ξένους τε καὶ ἑταίρους.

[1] φράζουσαν Müller: φράζουσα mss., Zur.

EPISTLE VII

all others in Sicily, and other Sicilians of my acquaintance ; and all these were giving me the same account, how that Dionysius had made marvellous progress in philosophy. And he sent an exceedingly long letter, since he knew how I was disposed towards Dion and also Dion's eagerness that I should make the voyage[1] and come to Syracuse ; for his letter was framed to deal with all these circumstances, having its commencement couched in some such terms as these—" Dionysius to Plato," followed by the customary greetings ; after which, without further preliminary—" If you are persuaded by us and come now to Sicily, in the first place you will find Dion's affairs proceeding in whatever way you yourself may desire—and you will desire, as I know, what is reasonable, and I will consent thereto ; but otherwise none of Dion's affairs, whether they concern himself or anything else, will proceed to your satisfaction." Such were his words on this subject, but the rest it were tedious and inopportune to repeat. And other letters kept coming both from Archytas and from the men in Tarentum, eulogizing the philosophy of Dionysius, and saying that unless I come now I should utterly dissolve their friendship with Dionysius which I had brought about, and which was of no small political importance. Such then being the nature of the summons which I then received,—when on the one hand the Sicilians and Italians were pulling me in and the Athenians, on the other, were literally pushing me out, so to say, by their entreaties,—once again the same argument recurred, namely, that it was my duty not to betray Dion, nor yet my hosts and comrades in Tarentum. And I felt also myself

[1] *Cf. Ep.* iii. 317.

αὐτῷ δέ μοι ὑπῆν ὡς οὐδὲν θαυμαστὸν νέον ἄνθρωπον παρακούοντα ἀξίων λόγου πραγμάτων, εὐμαθῆ, πρὸς ἔρωτα ἐλθεῖν τοῦ βελτίστου βίου· δεῖν οὖν αὐτὸ ἐξελέγξαι σαφῶς, ὁποτέρως ποτὲ ἄρα σχοίη, καὶ τοῦτ' αὐτὸ μηδαμῇ προδοῦναι μηδ' ἐμὲ τὸν αἴτιον γενέσθαι τηλικούτου ἀληθῶς ὀνείδους, εἴπερ ὄντως εἴη τῳ ταῦτα λελεγμένα.

Πορεύομαι δὴ τῷ λογισμῷ τούτῳ κατακαλυψάμενος, πολλὰ δεδιὼς μαντευόμενός τε οὐ πάνυ καλῶς, ὡς ἔοικεν. ἐλθὼν δ' οὖν τὸ τρίτον τῷ σωτῆρι τοῦτό γε οὖν ἔπραξα ὄντως· ἐσώθην γάρ τοι πάλιν[1] εὐτυχῶς, καὶ τούτων γε μετὰ θεὸν Διονυσίῳ χάριν εἰδέναι χρεών, ὅτι πολλῶν βουληθέντων ἀπολέσαι με διεκώλυσε καὶ ἔδωκέ τι μέρος αἰδοῖ τῶν περὶ ἐμὲ πραγμάτων. ἐπειδὴ δὲ ἀφικόμην, ᾤμην τούτου πρῶτον ἔλεγχον δεῖν λαβεῖν, πότερον ὄντως εἴη Διονύσιος ἐξημμένος ὑπὸ φιλοσοφίας ὥσπερ πυρός, ἢ μάτην ὁ πολὺς οὗτος ἔλθοι λόγος Ἀθήναζε. ἔστι δή τις τρόπος τοῦ περὶ τὰ τοιαῦτα πεῖραν λαμβάνειν οὐκ ἀγεννὴς ἀλλ' ὄντως τυράννοις πρέπων, ἄλλως τε καὶ τοῖς τῶν παρακουσμάτων μεστοῖς, ὃ δὴ κἀγὼ Διονύσιον εὐθὺς ἐλθὼν ᾐσθόμην καὶ μάλα πεπονθότα. δεικνύναι δὴ δεῖ τοῖς τοιούτοις ὅ τι ἔστι πᾶν τὸ πρᾶγμα οἷόν τε καὶ δι' ὅσων πραγμάτων καὶ ὅσον πόνον ἔχει. ὁ γὰρ ἀκούσας, ἐὰν μὲν ὄντως ᾖ φιλόσοφος οἰκεῖός τε καὶ ἄξιος τοῦ πράγματος θεῖος ὤν, ὁδόν τε ἡγεῖται θαυμαστὴν ἀκηκοέναι ξυν-

[1] τοι πάλιν Hermann: τὸ πάλαι mss., Zur.

[1] *Cf.* 334 D.

EPISTLE VII

that there would be nothing surprising in a young man, who was apt at learning, attaining to a love of the best life through hearing lectures on subjects of importance. So it seemed to be my duty to determine clearly in which way the matter really stood, and in no wise to prove false to this duty, nor to leave myself open to a reproach that would be truly serious, if so be that any of these reports were true.

So having blindfolded myself with this argumentation I made the journey, although, naturally, with many fears and none too happy forebodings. However, when I arrived the third time, I certainly did find it really a case of "the Third to the Saviour"[1]: for happily I did get safely back again; and for this I ought to give thanks, after God, to Dionysius, seeing that, when many had planned to destroy me, he prevented them and paid some regard to reverence in his dealings with me. And when I arrived, I deemed that I ought first of all to gain proof of this point,—whether Dionysius was really inflamed by philosophy, as it were by fire, or all this persistent account which had come to Athens was empty rumour. Now there is a method of testing such matters which is not ignoble but really suitable in the case of tyrants, and especially such as are crammed with borrowed doctrines; and this was certainly what had happened to Dionysius, as I perceived as soon as I arrived. To such persons one must point out what the subject is as a whole, and what its character, and how many preliminary subjects it entails and how much labour. For on hearing this, if the pupil be truly philosophic, in sympathy with the subject and worthy of it, because divinely gifted, he believes that he has been shown

τατέον τε εἶναι νῦν καὶ οὐ βιωτὸν ἄλλως ποιοῦντι· μετὰ τοῦτο δὴ ξυντείνας αὐτόν τε καὶ τὸν ἡγούμενον τὴν ὁδὸν οὐκ ἀνίησι πρὶν ἂν ἢ τέλος ἐπιθῇ πᾶσι ἢ λάβῃ δύναμιν ὥστε αὐτὸς αὑτὸν χωρὶς τοῦ δείξαντος μὴ ἀδύνατος εἶναι ποδηγεῖν. ταύτῃ καὶ κατὰ ταῦτα διανοηθεὶς ὁ τοιοῦτος ζῇ, πράττων μὲν ἐν αἷς τισὶν ἂν ᾖ πράξεσι, παρὰ πάντα δὲ ἀεὶ φιλοσοφίας ἐχόμενος καὶ τροφῆς τῆς καθ' ἡμέραν, ἥτις ἂν αὐτὸν μάλιστα εὐμαθῆ τε καὶ μνήμονα καὶ λογίζεσθαι δυνατὸν ἐν αὑτῷ νήφοντα ἀπεργάζηται· τὴν δὲ ἐναντίαν ταύτῃ μισῶν διατελεῖ. οἱ δὲ ὄντως μὲν μὴ φιλόσοφοι, δόξαις δ' ἐπικεχρωσμένοι, καθάπερ οἱ τὰ σώματα ὑπὸ τοῦ ἡλίου ἐπικεκαυμένοι, ἰδόντες τε ὅσα μαθήματά ἐστι καὶ ὁ πόνος ἡλίκος καὶ δίαιτα ἡ καθ' ἡμέραν ὡς πρέπουσα ἡ κοσμία τῷ πράγματι, χαλεπὸν ἡγησάμενοι καὶ ἀδύνατον αὑτοῖς, οὔτε δὴ ἐπιτηδεύειν δυνατοὶ γίγνονται· ἔνιοι δὲ αὐτῶν πείθουσιν αὑτοὺς ὡς ἱκανῶς ἀκηκοότες εἰσὶ τὸ ὅλον καὶ οὐδὲν ἔτι δέονταί τινων πραγμάτων.

Ἡ μὲν δὴ πεῖρα αὕτη γίγνεται ἡ σαφής τε καὶ ἀσφαλεστάτη πρὸς τοὺς τρυφῶντάς τε καὶ ἀδυνάτους διαπονεῖν, ὡς μηδέποτε βαλεῖν ἐν αἰτίᾳ τὸν δεικνύντα ἀλλ' αὐτὸν αὑτόν, μὴ δυνάμενον πάντα τὰ πρόσφορα ἐπιτηδεύειν τῷ πράγματι.

Οὕτω δὴ καὶ Διονυσίῳ τότ' ἐρρήθη τὰ ῥηθέντα.

---

[1] Cf. Rep. 531 D.

EPISTLE VII

a marvellous pathway and that he must brace himself at once to follow it, and that life will not be worth living if he does otherwise. After this he braces both himself and him who is guiding him on the path, nor does he desist until either he has reached the goal of all his studies, or else has gained such power as to be capable of directing his own steps without the aid of the instructor. It is thus, and in this mind, that such a student lives, occupied indeed in whatever occupations he may find himself, but always beyond all else cleaving fast to philosophy and to that mode of daily life which will best make him apt to learn and of retentive mind and able to reason within himself soberly; but the mode of life which is opposite to this he continually abhors. Those, on the other hand, who are in reality not philosophic, but superficially tinged by opinions,— like men whose bodies are sunburnt on the surface —when they see how many studies are required and how great labour,[1] and how the orderly mode of daily life is that which befits the subject, they deem it difficult or impossible for themselves, and thus they become in fact incapable of pursuing it; while some of them persuade themselves that they have been sufficiently instructed in the whole subject and no longer require any further effort.

Now this test proves the clearest and most infallible in dealing with those who are luxurious and incapable of enduring labour, since it prevents any of them from ever casting the blame on his instructor instead of on himself and his own inability to pursue all the studies which are accessory to his subject.

This, then, was the purport of what I said to

PLATO

341

πάντα μὲν οὖν οὔτ' ἐγὼ διεξῆλθον οὔτε Διονύσιος
B ἐδεῖτο· πολλὰ γὰρ αὐτὸς καὶ τὰ μέγιστα εἰδέναι
τε καὶ ἱκανῶς ἔχειν προσεποιεῖτο διὰ τὰς ὑπὸ τῶν
ἄλλων παρακοάς. ὕστερον δὲ καὶ ἀκούω γεγρα-
φέναι αὐτὸν περὶ ὧν τότε ἤκουσε, συνθέντα ὡς
αὐτοῦ τέχνην, οὐδὲν τῶν αὐτῶν ὧν ἀκούοι· οἶδα
δὲ οὐδὲν τούτων· ἄλλους μέν τινας οἶδα γεγρα-
φότας περὶ τῶν αὐτῶν τούτων· οἵτινες δέ, οὐδ'
αὐτοὶ αὑτούς. τοσόνδε γε μὴν περὶ πάντων ἔχω
C φράζειν τῶν γεγραφότων καὶ γραψόντων, ὅσοι
φασὶν εἰδέναι περὶ ὧν ἐγὼ σπουδάζω, εἴτ' ἐμοῦ
ἀκηκοότες εἴτ' ἄλλων εἴθ' ὡς εὑρόντες αὐτοί·
τούτους οὐκ ἔστι κατά γε τὴν ἐμὴν δόξαν περὶ
τοῦ πράγματος ἐπαΐειν οὐδέν. οὔκουν ἐμόν γε
περὶ αὐτῶν ἔστι σύγγραμμα οὐδὲ μήποτε γένηται·
ῥητὸν γὰρ οὐδαμῶς ἐστιν ὡς ἄλλα μαθήματα, ἀλλ'
ἐκ πολλῆς συνουσίας γιγνομένης περὶ τὸ πρᾶγμα
αὐτὸ καὶ τοῦ συζῆν ἐξαίφνης, οἷον ἀπὸ πυρὸς
D πηδήσαντος ἐξαφθὲν φῶς, ἐν τῇ ψυχῇ γενόμενον
αὐτὸ ἑαυτὸ ἤδη τρέφει. καίτοι τοσόνδε γε οἶδα,
ὅτι γραφέντα ἢ λεχθέντα ὑπ' ἐμοῦ βέλτιστ' ἂν
λεχθείη· καὶ μὴν ὅτι γεγραμμένα κακῶς οὐχ
ἥκιστ' ἂν ἐμὲ λυποῖ. εἰ δέ μοι ἐφαίνετο γραπτέα
θ' ἱκανῶς εἶναι πρὸς τοὺς πολλοὺς καὶ ῥητά, τί

---

[1] Probably an allusion to the proverbial maxim "Know thyself."

[2] *Cf. Sympos.* 210 E for the "suddenness" of the mystic vision of the Idea.

[3] On the danger of writing such doctrines *cf. Ep.* ii. 314 c ff.; and for philosophy as possible only for "the few" *cf. Rep.* 494 A.

EPISTLE VII

Dionysius on that occasion. I did not, however, expound the matter fully, nor did Dionysius ask me to do so; for he claimed that he himself knew many of the most important doctrines and was sufficiently informed owing to the versions he had heard from his other teachers. And I am even told that later on he himself wrote a treatise on the subjects in which I then instructed him, composing it as though it were something of his own invention and quite different from what he had heard; but of all this I know nothing. I know indeed that certain others have written about these same subjects; but what manner of men they are not even themselves know.[1] But thus much I can certainly declare concerning all these writers, or prospective writers, who claim to know the subjects which I seriously study, whether as hearers of mine or of other teachers, or from their own discoveries; it is impossible, in my judgement at least, that these men should understand anything about this subject. There does not exist, nor will there ever exist, any treatise of mine dealing therewith. For it does not at all admit of verbal expression like other studies, but, as a result of continued application to the subject itself and communion therewith, it is brought to birth in the soul on a sudden,[2] as light that is kindled by a leaping spark, and thereafter it nourishes itself. Notwithstanding, of thus much I am certain, that the best statement of these doctrines in writing or in speech would be my own statement; and further, that if they should be badly stated in writing, it is I who would be the person most deeply pained. And if I had thought that these subjects ought to be fully stated in writing or in speech to the public,[3] what nobler action could

PLATO

341 τούτου κάλλιον ἐπέπρακτ' ἂν ἡμῖν ἐν τῷ βίῳ
ἢ τοῖς τε ἀνθρώποισι μέγα ὄφελος γράψαι καὶ
E τὴν φύσιν εἰς φῶς πᾶσι προαγαγεῖν; ἀλλ' οὔτε
ἀνθρώποις ἡγοῦμαι τὴν ἐπιχείρησιν περὶ αὐτῶν
γενομένην[1] ἀγαθόν, εἰ μή τισιν ὀλίγοις, ὁπόσοι
δυνατοὶ ἀνευρεῖν αὐτοὶ διὰ σμικρᾶς ἐνδείξεως·
τῶν τε δὴ ἄλλων τοὺς μὲν καταφρονήσεως οὐκ
ὀρθῆς ἐμπλήσειεν ἂν οὐδαμῇ ἐμμελῶς,[2] τοὺς δὲ
ὑψηλῆς καὶ χαύνης ἐλπίδος, ὡς σέμν' ἄττα μεμαθηκότας.

342 Ἔτι δὲ μακρότερα περὶ αὐτῶν ἐν νῷ μοι γέγονεν
εἰπεῖν· τάχα γὰρ ἂν περὶ ὧν λέγω σαφέστερον
ἂν εἴη τι λεχθέντων αὐτῶν. ἔστι γάρ τις λόγος
ἀληθὴς ἐναντίος τῷ τολμήσαντι γράφειν τῶν τοιούτων καὶ ὁτιοῦν, πολλάκις μὲν ὑπ' ἐμοῦ καὶ πρόσθεν
ῥηθείς, ἔοικε δ' οὖν εἶναι καὶ νῦν λεκτέος.

Ἔστι τῶν ὄντων ἑκάστῳ, δι' ὧν τὴν ἐπιστήμην
ἀνάγκη παραγίγνεσθαι, τρία· τέταρτον δ' αὐτή·
πέμπτον δ' αὐτὸ τιθέναι δεῖ ὃ δὴ γνωστόν τε καὶ
B ἀληθές ἐστιν· ὧν ἓν μὲν ὄνομα, δεύτερον δὲ λόγος,
τὸ δὲ τρίτον εἴδωλον, τέταρτον δὲ ἐπιστήμη. περὶ
ἓν οὖν λαβὲ βουλόμενος μαθεῖν τὸ νῦν λεγόμενον,
καὶ πάντων οὕτω πέρι νόησον. κύκλος ἐστί τι
λεγόμενον, ᾧ τοῦτ' αὐτό ἐστιν ὄνομα ὃ νῦν ἐφθέγμεθα. λόγος δ' αὐτοῦ τὸ δεύτερον, ἐξ ὀνομάτων
καὶ ῥημάτων συγκείμενος· τὸ γὰρ ἐκ τῶν ἐσχάτων
ἐπὶ τὸ μέσον ἴσον ἀπέχον πάντῃ, λόγος ἂν εἴη

---

[1] γενομένην Bonitz: λεγομένην mss., Zur.
[2] ὀρθῆς . . . ἐμμελῶς mss.: ὀρθῶς . . . ἐμμελοῦς Zur.

---

[1] Cf. 341 c.
[2] Cf. *Laws* 895 D, where Essence, Definition, and Name are enumerated; also *Parm.* 142 A.

EPISTLE VII

I have performed in my life than that of writing what is of great benefit to mankind and bringing forth to the light for all men the nature of reality? But were I to undertake this task it would not, as I think, prove a good thing for men, save for some few who are able to discover the truth themselves with but little instruction; for as to the rest, some it would most unseasonably fill with a mistaken contempt, and others with an overweening and empty aspiration, as though they had learnt some sublime mysteries.

But concerning these studies I am minded to speak still more at length; since the subject with which I am dealing [1] will perhaps be clearer when I have thus spoken. For there is a certain true argument which confronts the man who ventures to write anything at all of these matters,—an argument which, although I have frequently stated it in the past, seems to require statement also at the present time.

Every existing object has three things [2] which are the necessary means by which knowledge of that object is acquired; and the knowledge itself is a fourth thing; and as a fifth one must postulate the object itself which is cognizable and true. First of these comes the name; secondly the definition; thirdly the image; fourthly the knowledge. If you wish, then, to understand what I am now saying, take a single example and learn from it what applies to all. There is an object called a circle, which has for its *name* the word we have just mentioned; and, secondly, it has a *definition*, composed of names and verbs; for " that which is everywhere equidistant from the extremities to the centre " will be the definition of that object which has for its name

PLATO

ἐκείνου ᾧπερ στρογγύλον καὶ περιφερὲς ὄνομα καὶ κύκλος. τρίτον δὲ τὸ ζωγραφούμενόν τε καὶ ἐξαλειφόμενον καὶ τορνευόμενον καὶ ἀπολλύμενον· ὧν αὐτὸς ὁ κύκλος, ὃν πέρι πάντ᾽ ἐστὶ ταῦτα, οὐδὲν πάσχει τούτων ὡς ἕτερον ὄν. τέταρτον δὲ ἐπιστήμη καὶ νοῦς ἀληθής τε δόξα περὶ ταῦτ᾽ ἐστίν· ὡς δὲ ἓν τοῦτο αὖ πᾶν θετέον, οὐκ ἐν φωναῖς οὐδ᾽ ἐν σωμάτων σχήμασιν ἀλλ᾽ ἐν ψυχαῖς ἐνόν, ᾧ δῆλον ἕτερόν τε ὂν αὐτοῦ τοῦ κύκλου τῆς φύσεως τῶν τε ἔμπροσθεν λεχθέντων τριῶν. τούτων δὲ ἐγγύτατα μὲν ξυγγενείᾳ καὶ ὁμοιότητι τοῦ πέμπτου νοῦς πεπλησίακε, τἆλλα δὲ πλέον ἀπέχει.

Ταὐτὸν δὴ περί τε εὐθέος ἅμα καὶ περιφεροῦς σχήματος καὶ χρόας, περί τε ἀγαθοῦ καὶ καλοῦ καὶ δικαίου, καὶ περὶ σώματος ἅπαντος σκευαστοῦ τε καὶ κατὰ φύσιν γεγονότος, πυρὸς ὕδατός τε καὶ τῶν τοιούτων πάντων, καὶ ζώου ξύμπαντος πέρι καὶ ἐν ψυχαῖς ἤθους [καὶ]¹ περὶ ποιήματα καὶ παθήματα ξύμπαντα· οὐ γὰρ ἂν τούτων μή τις τὰ τέτταρα λάβῃ ἁμῶς γέ πως, οὔποτε τελέως ἐπιστήμης τοῦ πέμπτου μέτοχος ἔσται. πρὸς γὰρ τούτοις ταῦτα οὐχ ἧττον ἐπιχειρεῖ τὸ ποῖόν τι περὶ ἕκαστον δηλοῦν ἢ τὸ ὂν ἑκάστου διὰ τὸ τῶν λόγων ἀσθενές· ὧν ἕνεκα νοῦν ἔχων οὐδεὶς τολμήσει ποτὲ εἰς αὐτὸ τιθέναι τὰ νενοημένα, καὶ ταῦτα εἰς ἀμετακίνητον, ὃ δὴ πάσχει τὰ γεγραμμένα τύποις.

Τοῦτο δὲ πάλιν αὖ τὸ νῦν λεγόμενον δεῖ μαθεῖν. κύκλος ἕκαστος τῶν ἐν ταῖς πράξεσι γραφομένων

---

¹ καὶ bracketed by W.-Möllendorff.

---

[1] For the definition of "circle" cf. *Tim.* 33 B, *Parm.* 137 E.
[2] This echoes the language of *Rep.* 490 B.
[3] Cf. *Cratyl.* 438 D, E.

"round" and "spherical" and "circle."[1] And in the third place there is that object which is in course of being portrayed and obliterated, or of being shaped with a lathe, and falling into decay; but none of these affections is suffered by the circle itself, whereto all these others are related inasmuch as it is distinct therefrom. Fourth comes *knowledge* and intelligence and true opinion regarding these objects; and these we must assume to form a single whole, which does not exist in vocal utterance or in bodily forms but in souls; whereby it is plain that it differs both from the nature of the circle itself and from the three previously mentioned. And of those four intelligence approaches most nearly in kinship and similarity to the fifth,[2] and the rest are further removed.

The same is true alike of the straight and of the spherical form, and of colour, and of the good and the fair and the just, and of all bodies whether manufactured or naturally produced (such as fire and water and all such substances), and of all living creatures, and of all moral actions or passions in souls. For unless a man somehow or other grasps the four of these, he will never perfectly acquire knowledge of the fifth. Moreover, these four attempt to express the quality of each object no less than its real essence, owing to the weakness inherent in language[3]; and for this reason, no man of intelligence will ever venture to commit to it the concepts of his reason, especially when it is unalterable —as is the case with what is formulated in writing.

But here again you must learn further the meaning of this last statement. Every one of the circles which are drawn in geometric exercises or are turned by

PLATO

ᾗ καὶ τορνευθέντων μεστὸς τοῦ ἐναντίου ἐστὶ τῷ πέμπτῳ· τοῦ γὰρ εὐθέος ἐφάπτεται πάντη· αὐτὸς δέ, φαμέν, ὁ κύκλος οὔτε τι σμικρότερον οὔτε μεῖζον τῆς ἐναντίας ἔχει ἐν αὐτῷ φύσεως. ὄνομά τε αὐτῶν φαμὲν οὐδὲν οὐδενὶ βέβαιον εἶναι, κωλύειν δ' οὐδὲν τὰ νῦν στρογγύλα καλούμενα εὐθέα κεκλῆσθαι τά τε εὐθέα δὴ στρογγύλα, καὶ οὐδὲν ἧττον βεβαίως ἕξειν τοῖς μεταθεμένοις καὶ ἐναντίως καλοῦσι. καὶ μὴν περὶ λόγου γε ὁ αὐτὸς λόγος, εἴπερ ἐξ ὀνομάτων καὶ ῥημάτων σύγκειται, μηδὲν ἱκανῶς βεβαίως εἶναι βέβαιον. μυρίος δὲ λόγος αὖ περὶ ἑκάστου τῶν τεττάρων, ὡς ἀσαφές· τὸ δὲ μέγιστον, ὅπερ εἴπομεν ὀλίγον ἔμπροσθεν, ὅτι δυοῖν ὄντοιν, τοῦ τε ὄντος καὶ τοῦ ποιοῦ τινός, οὐ τὸ ποιόν τι, τὸ δὲ τί ζητούσης εἰδέναι τῆς ψυχῆς, τὸ μὴ ζητούμενον ἕκαστον τῶν τεττάρων προτεῖνον[1] τῇ ψυχῇ λόγῳ τε καὶ κατ' ἔργα, αἰσθήσεσιν εὐέλεγκτον τό τε λεγόμενον καὶ δεικνύμενον ἀεὶ παρεχόμενον ἕκαστον, ἀπορίας τε καὶ ἀσαφείας ἐμπίπλησι πάσης ὡς ἔπος εἰπεῖν πάντ' ἄνδρα. ἐν οἷσι μὲν οὖν μηδ' εἰθισμένοι τὸ ἀληθὲς ζητεῖν ἐσμὲν ὑπὸ πονηρᾶς τροφῆς, ἐξαρκεῖ δὲ τὸ προταθὲν τῶν εἰδώλων, οὐ καταγέλαστοι γιγνόμεθα ὑπ' ἀλλήλων, οἱ ἐρωτώμενοι ὑπὸ τῶν ἐρωτώντων, δυναμένων δὲ τὰ τέτταρα διαρρίπτειν τε καὶ ἐλέγχειν· ἐν οἷς δ' ἂν τὸ πέμπτον ἀπο-

---

[1] προτεῖνον MSS.: πρότερον MSS. corr., Zur.

---

[1] *i.e.* any number of straight tangents to a circle may be drawn; or, a circle, like a straight line, is composed of points, therefore the circular is full of the elements of the straight.

EPISTLE VII

the lathe is full of what is opposite to the fifth, since it is in contact with the straight everywhere [1]; whereas the circle itself, as we affirm, contains within itself no share greater or less of the opposite nature. And none of the objects, we affirm, has any fixed name, nor is there anything to prevent forms which are now called " round " from being called " straight," and the " straight " " round " [2]; and men will find the names no less firmly fixed when they have shifted them and apply them in an opposite sense. Moreover, the same account holds good of the Definition also, that, inasmuch as it is compounded of names and verbs, it is in no case fixed with sufficient firmness.[3] And so with each of the Four, their inaccuracy is an endless topic; but, as we mentioned a moment ago, the main point is this, that while there are two separate things, the real essence and the quality, and the soul seeks to know not the quality but the essence, each of the Four proffers to the soul either in word or in concrete form that which is not sought; and by thus causing each object which is described or exhibited to be always easy of refutation by the senses, it fills practically all men with all manner of perplexity and uncertainty. In respect, however, of those other objects the truth of which, owing to our bad training, we usually do not so much as seek—being content with such of the images as are proffered,—those of us who answer are not made to look ridiculous by those who question, we being capable of analysing and convicting the Four. But in all cases where we

[2] *Cf. Cratyl.* 384 D, E for the view that names are not natural but conventional fixities.

[3] *Cf. Theaet.* 208 B ff. for the instability of Definitions.

PLATO

κρίνασθαι καὶ δηλοῦν ἀναγκάζωμεν, ὁ βουλόμενος τῶν δυναμένων ἀνατρέπειν κρατεῖ, καὶ ποιεῖ τὸν ἐξηγούμενον ἐν λόγοις ἢ γράμμασιν ἢ ἀποκρίσεσι τοῖς πολλοῖς τῶν ἀκουόντων δοκεῖν μηδὲν γιγνώσκειν ὧν ἂν ἐπιχειρῇ γράφειν ἢ λέγειν, ἀγνοούντων ἐνίοτε ὡς οὐχ ἡ ψυχὴ τοῦ γράψαντος ἢ λέξαντος ἐλέγχεται, ἀλλ' ἡ τῶν τεττάρων φύσις ἑκάστου, πεφυκυῖα φαύλως. ἡ δὲ διὰ πάντων αὐτῶν διαγωγή, ἄνω καὶ κάτω μεταβαίνουσα ἐφ' ἕκαστον, μόγις ἐπιστήμην ἐνέτεκεν εὖ πεφυκότος εὖ πεφυκότι· κακῶς δὲ ἂν φυῇ, ὡς ἡ τῶν πολλῶν ἕξις τῆς ψυχῆς εἴς τε τὸ μαθεῖν εἴς τε τὰ λεγόμενα ἤθη πέφυκε, τὰ δὲ διέφθαρται, οὐδ' ἂν ὁ Λυγκεὺς ἰδεῖν ποιήσειε τοὺς τοιούτους. ἑνὶ δὲ λόγῳ, τὸν μὴ ξυγγενῆ τοῦ πράγματος οὔτ' ἂν εὐμαθία ⟨μαθεῖν⟩[1] ποιήσειέ ποτε οὔτε μνήμη· τὴν ἀρχὴν γὰρ ἐν ἀλλοτρίαις ἕξεσιν οὐκ ἐγγίγνεται· ὥστε ὁπόσοι τῶν δικαίων τε καὶ τῶν ἄλλων ὅσα καλὰ μὴ προσφυεῖς εἰσι καὶ ξυγγενεῖς, ἄλλοι δὲ ἄλλων εὐμαθεῖς ἅμα καὶ μνήμονες, οὐδ' ὅσοι ξυγγενεῖς, δυσμαθεῖς δὲ καὶ ἀμνήμονες, οὐδένες τούτων μήποτε μάθωσιν ἀλήθειαν ἀρετῆς εἰς τὸ δυνατὸν οὐδὲ κακίας. ἅμα γὰρ αὐτὰ ἀνάγκη μανθάνειν καὶ τὸ ψεῦδος ἅμα καὶ ἀληθὲς τῆς ὅλης οὐσίας, μετὰ τριβῆς πάσης καὶ χρόνου πολλοῦ, ὅπερ ἐν ἀρχαῖς εἶπον· μόγις

---

[1] μαθεῖν I add.

[1] An Argonaut, noted for his keenness of sight; here, by a playful hyperbole, he is supposed to be also a producer of sight in others; cf. Aristoph. Plut. 210.
[2] Cf. Laws 816 D.
[3] Cf. 341 C.

EPISTLE VII

compel a man to give the Fifth as his answer and to explain it, anyone who is able and willing to upset the argument gains the day, and makes the person who is expounding his view by speech or writing or answers appear to most of his hearers to be wholly ignorant of the subjects about which he is attempting to write or speak; for they are ignorant sometimes of the fact that it is not the soul of the writer or speaker that is being convicted but the nature of each of the Four, which is essentially defective. But it is the methodical study of all these stages, passing in turn from one to another, up and down, which with difficulty implants knowledge, when the man himself, like his object, is of a fine nature; but if his nature is bad—and, in fact, the condition of most men's souls in respect of learning and of what are termed "morals" is either naturally bad or else corrupted,—then not even Lynceus [1] himself could make such folk see. In one word, neither receptivity nor memory will ever produce knowledge in him who has no affinity with the object, since it does not germinate to start with in alien states of mind; consequently neither those who have no natural connexion or affinity with things just, and all else that is fair, although they are both receptive and retentive in various ways of other things, nor yet those who possess such affinity but are unreceptive and unretentive—none, I say, of these will ever learn to the utmost possible extent the truth of virtue nor yet of vice. For in learning these objects it is necessary to learn at the same time both what is false and what is true of the whole of Existence,[2] and that through the most diligent and prolonged investigation, as I said at the commencement [3]; and it

PLATO

δὲ τριβόμενα πρὸς ἄλληλα αὐτῶν ἕκαστα, ὀνόματα
καὶ λόγοι ὄψεις τε καὶ αἰσθήσεις, ἐν εὐμενέσιν
ἐλέγχοις ἐλεγχόμενα καὶ ἄνευ φθόνων ἐρωτήσεσι
καὶ ἀποκρίσεσι χρωμένων, ἐξέλαμψε φρόνησις περὶ
ἕκαστον καὶ νοῦς, συντείνοντι[1] ὅ τι μάλιστ' εἰς
δύναμιν ἀνθρωπίνην.

Διὸ δὴ πᾶς ἀνὴρ σπουδαῖος τῶν ὄντως[2] σπου-
δαίων πέρι πολλοῦ δεῖ μὴ γράψας ποτὲ ἐν ἀνθρώποις εἰς φθόνον καὶ ἀπορίαν καταβάλῃ· ἑνὶ δὴ
ἐκ τούτων δεῖ γιγνώσκειν λόγῳ, ὅταν ἴδῃ τίς του
συγγράμματα γεγραμμένα εἴτε ἐν νόμοις νομοθέτου
εἴτε ἐν ἄλλοις τισὶν ἅττ' οὖν, ὡς οὐκ ἦν τούτῳ
ταῦτα σπουδαιότατα, εἴπερ ἔστ' αὐτὸς σπουδαῖος,
κεῖται δέ που ἐν χώρᾳ τῇ καλλίστῃ τῶν τούτου·
εἰ δὲ ὄντως αὐτῷ ταῦτ' ἐσπουδασμένα ἐν γράμμασιν ἐτέθη, Ἐξ ἄρα δή οἱ ἔπειτα, θεοὶ μὲν οὔ,
βροτοὶ δὲ φρένας ὤλεσαν αὐτοί.

Τούτῳ δὴ τῷ μύθῳ τε καὶ πλάνῳ ὁ ξυνεπισπό-
μενος εὖ εἴσεται, εἴτ' οὖν Διονύσιος ἔγραψέ τι τῶν
περὶ φύσεως ἄκρων καὶ πρώτων εἴτε τις ἐλάττων
εἴτε μείζων, ὡς οὐδὲν ἀκηκοὼς οὐδὲ μεμαθηκὼς
ἦν ὑγιὲς ὧν ἔγραψε κατὰ τὸν ἐμὸν λόγον· ὁμοίως
γὰρ ἂν αὐτὰ ἐσέβετο ἐμοί, καὶ οὐκ ἂν αὐτὰ ἐτόλ-
μησεν εἰς ἀναρμοστίαν καὶ ἀπρέπειαν ἐκβάλλειν.
οὔτε γὰρ ὑπομνημάτων χάριν αὐτὰ ἔγραψεν· οὐδὲν
γὰρ δεινὸν μή τις αὐτὸ ἐπιλάθηται, ἐὰν ἅπαξ τῇ

---

[1] συντείνοντι E. Sachs : συντείνων mss., Zur.
[2] ὄντως W.-Möllendorff : ὄντων mss., Zur.

---

[1] For legislation as not a " serious " subject but " playful "
see *Laws* 769 A; *cf. Polit.* 294 A.
[2] *i.e.* in his head, the abode of unexpressed thoughts; *cf.
Tim.* 44 D.
[3] Homer, *Il.* vii. 360, xii. 234.

EPISTLE VII

is by means of the examination of each of these objects, comparing one with another—names and definitions, visions and sense-perceptions,—proving them by kindly proofs and employing questionings and answerings that are void of envy—it is by such means, and hardly so, that there bursts out the light of intelligence and reason regarding each object in the mind of him who uses every effort of which mankind is capable.

And this is the reason why every serious man in dealing with really serious subjects [1] carefully avoids writing, lest thereby he may possibly cast them as a prey to the envy and stupidity of the public. In one word, then, our conclusion must be that whenever one sees a man's written compositions—whether they be the laws of a legislator or anything else in any other form,—these are not his most serious works, if so be that the writer himself is serious: rather those works abide in the fairest region he possesses.[2] If, however, these really are his serious efforts, and put into writing, it is not "the gods" but mortal men who "Then of a truth themselves have utterly ruined his senses." [3]

Whosoever, then, has accompanied me in this story and this wandering of mine will know full well that, whether it be Dionysius or any lesser or greater man who has written something about the highest and first truths of Nature, nothing of what he has written, as my argument shows, is based on sound teaching or study. Otherwise he would have reverenced these truths as I do, and would not have dared to expose them to unseemly and degrading treatment. For the writings of Dionysius were not meant as aids to memory, since there is no fear lest anyone should forget the truth if once he grasps it with his

PLATO

344
ψυχῇ περιλάβῃ, πάντων γὰρ ἐν βραχυτάτοις κεῖται· φιλοτιμίας δὲ αἰσχρᾶς, εἴπερ, ἕνεκα, εἴθ' ὡς αὑτοῦ τιθέμενος εἴθ' ὡς παιδείας δὴ μέτοχος ὤν, ἧς οὐκ ἄξιος ἦν ἀγαπῶν δόξαν τὴν τῆς μετοχῆς γενομένην.
345 εἰ μὲν οὖν ἐκ τῆς μιᾶς συνουσίας Διονυσίῳ τοῦτο γέγονε, τάχ' ἂν εἴη· γέγονε δ' οὖν ὅπως, ἴττω Ζεύς, φησὶν ὁ Θηβαῖος· διεξῆλθον μὲν γὰρ ὡς εἶπόν τότε[1] ἐγὼ καὶ ἅπαξ μόνον, ὕστερον δὲ οὐ πώποτε ἔτι.

Ἐννοεῖν δὴ δεῖ τὸ μετὰ τοῦτο, ὅτῳ μέλει τὸ περὶ αὐτὰ γεγονὸς εὑρεῖν, ὅπῃ ποτὲ γέγονε, τίνι ποτ' αἰτίᾳ τὸ δεύτερον καὶ τὸ τρίτον πλεονάκις τε οὐ διεξῇμεν· πότερον Διονύσιος ἀκούσας μόνον
B ἅπαξ οὕτως εἰδέναι τε οἴεται καὶ ἱκανῶς οἶδεν, εἴτε αὐτὸς εὑρὼν ἢ καὶ μαθὼν ἔμπροσθεν παρ' ἑτέρων, ἢ φαῦλα εἶναι τὰ λεχθέντα, ἢ τὸ τρίτον οὐ καθ' αὑτόν, μείζονα δέ, καὶ ὄντως οὐκ ἂν δυνατὸς εἶναι φρονήσεώς τε καὶ ἀρετῆς ζῆν ἐπιμελούμενος. εἰ μὲν γὰρ φαῦλα, πολλοῖς μάρτυσι μαχεῖται τὰ ἐναντία λέγουσιν, οἳ περὶ τῶν τοιούτων πάμπολυ Διονυσίου κυριώτεροι ἂν εἶεν κριταί. εἰ δὲ εὑρηκέναι ἢ μεμαθηκέναι, ἄξια δ' οὖν εἶναι
C πρὸς παιδείαν ψυχῆς ἐλευθέραν, πῶς ἂν μὴ θαυμαστὸς ὢν ἄνθρωπος τὸν ἡγεμόνα τούτων καὶ

[1] τότε W.-Möllendorff: τε mss., Zur.

---

[1] *Cf. Phaedr.* 275 D, 278 A.
[2] *Cf. Phaedo* 62 A, B; the allusion is to the Theban dialect (ἴττω for ἴστω) used by Cebes.
[3] *Cf.* 341 A.
[4] *Cf. Ep.* ii. 314 A ff.
[5] *i.e.* Plato himself.

EPISTLE VII

soul, seeing that it occupies the smallest possible space [1]; rather, if he wrote at all, it was to gratify his base love of glory, either by giving out the doctrines as his own discoveries, or else by showing, forsooth, that he shared a culture which he by no means deserved because of his lust for the fame accruing from its possession. Well, then, if such was the effect produced on Dionysius by our one conversation, perhaps it was so; but how this effect was produced "God troweth," as the Theban says [2]; for as I said,[3] I explained my doctrine to him then on one occasion only, and never again since then.

And if anyone is concerned to discover how it was that things actually happened as they did in regard to this matter, he ought to consider next the reason why we did not explain our doctrine a second time, or a third time, or still more often. Does Dionysius fancy that he possesses knowledge, and is his knowledge adequate, as a result of hearing me once only, or as the result of his own researches, or of previous instruction from other teachers? Or does he regard my doctrines as worthless? Or, thirdly, does he believe them to be beyond and above his capacity, and that he himself would be really incapable of living a life devoted to wisdom and virtue? For if he deems them worthless he will be in conflict with many witnesses who maintain the opposite, men who should be vastly more competent judges of such matters than Dionysius.[4] While if he claims that he has found out these truths by research or by instruction, and if he admits their value for the liberal education of the soul, how could he possibly (unless he is a most extraordinary person) have treated the leading authority [5] on this subject with such ready

PLATO

345
κύριον οὕτως εὐχερῶς ἠτίμασέ ποτ' ἄν; πῶς δ'
ἠτίμασεν ἐγὼ φράζοιμ' ἄν.

Οὐ πολὺν χρόνον διαλιπὼν τὸ μετὰ τοῦτο, ἐν
τῷ πρόσθεν Δίωνα ἐῶν τὰ ἑαυτοῦ κεκτῆσθαι
καὶ καρποῦσθαι χρήματα, τότε οὐκέτ' εἴα τοὺς
ἐπιτρόπους αὐτοῦ πέμπειν εἰς Πελοπόννησον,
καθάπερ ἐπιλελησμένος τῆς ἐπιστολῆς παντάπασιν·
εἶναι γὰρ αὐτὰ οὐ Δίωνος ἀλλὰ τοῦ υἱέος, ὄντος
D μὲν ἀδελφιδοῦ αὐτοῦ κατὰ νόμους ἐπιτροπεύοντος.
τὰ μὲν δὴ πεπραγμένα μέχρι τούτου ταῦτ' ἦν ἐν
τῷ τότε χρόνῳ, τούτων δὲ οὕτω γενομένων ἑωρά-
κειν τε ἐγὼ ἀκριβῶς τὴν ἐπιθυμίαν τὴν¹ Διονυσίου
φιλοσοφίας, ἀγανακτεῖν τε ἐξῆν εἴτε βουλοίμην
εἴτε μή. ἦν γὰρ θέρος ἤδη τότε καὶ ἔκπλοι τῶν
νεῶν. ἐδόκει δὴ χαλεπαίνειν μὲν οὐ δεῖν ἐμὲ
Διονυσίῳ μᾶλλον ἢ ἐμαυτῷ τε καὶ τοῖς βιασαμένοις
ἐλθεῖν ἐμὲ τὸ τρίτον εἰς τὸν πορθμὸν τὸν περὶ τὴν
Σκύλλαν,

E ὄφρ' ἔτι τὴν ὀλοὴν ἀναμετρήσαιμι Χάρυβδιν,

λέγειν δὲ πρὸς Διονύσιον ὅτι μοι μένειν ἀδύνατον
εἴη Δίωνος οὕτω προπεπηλακισμένου. ὁ δὲ παρεμυθεῖτό τε καὶ ἐδεῖτο μένειν, οὐκ οἰόμενός οἱ
καλῶς ἔχειν ἐμὲ ἄγγελον αὐτὸν τῶν τοιούτων
ἐλθεῖν ὅ τι τάχος· οὐ πείθων δὲ αὐτός μοι πομπὴν
346 παρασκευάσειν ἔφη. ἐγὼ γὰρ ἐν τοῖς ἀποστόλοις
πλοίοις ἐμβὰς διενοούμην πλεῖν, τεθυμωμένος
πάσχειν τε οἰόμενος δεῖν, εἰ διακωλυοίμην, ὁτιοῦν,
ἐπειδὴ περιφανῶς ἠδίκουν μὲν οὐδέν, ἠδικούμην
δέ. ὁ δὲ οὐδέν με τοῦ καταμένειν προσιέμενος

---

¹ τὴν Burnet: τῆς mss., Zur.

¹ Homer, *Od.* xii. 428.

EPISTLE VII

disrespect? And how he showed this disrespect I will now relate.

It happened next, after no long interval, that whereas Dionysius had previously allowed Dion to remain in possession of his own property and to enjoy the income, he now ceased to permit Dion's trustees to remit it to the Peloponnese, just as though he had entirely forgotten the terms of his letter, claiming that the property belonged not to Dion but to his son, his own nephew, of whom he was the legal trustee. Such were his actions during this period up to this point; and when matters had turned out thus, I perceived clearly what kind of love Dionysius had for philosophy; and, moreover, I had good reason to be annoyed, whether I wished it or not. For by then it was already summer and the season for ships to sail. Still I judged that I had no right to be more angry with Dionysius than with myself and those who had forced me to come the third time to the straits adjoining Scylla—" There yet again to traverse the length of deadly Charybdis "[1]; rather I should inform Dionysius that it was impossible for me to remain now that Dion was so insultingly treated. He, however, tried to talk me over and entreated me to remain, as he thought it would not be to his own credit that I should hurry away in person to convey such tidings; and when he failed to persuade me he promised to provide a passage for me himself. For I was proposing to embark and sail in the trading-vessels; because I was enraged and thought that I ought to stop at nothing, in case I were hindered, seeing that I was manifestly doing no wrong but suffering wrong. But when he saw that I had no inclination to remain he devised a

346

ὁρῶν μηχανὴν τοῦ μεῖναι τὸν τότε ἔκπλουν μηχανᾶται τοιάνδε τινά. τῇ μετὰ ταῦτα ἐλθὼν ἡμέρᾳ λέγει πρός με πιθανὸν λόγον· Ἐμοὶ καὶ σοὶ Δίων, ἔφη, καὶ τὰ Δίωνος ἐκποδὼν ἀπ-
B αλλαχθήτω τοῦ περὶ αὐτὰ πολλάκις διαφέρεσθαι. ποιήσω γὰρ διὰ σέ, ἔφη, Δίωνι τάδε. ἀξιῶ ἐκεῖνον ἀπολαβόντα τὰ ἑαυτοῦ οἰκεῖν μὲν ἐν Πελοποννήσῳ, μὴ ὡς φυγάδα δέ, ἀλλ' ὡς αὐτῷ καὶ δεῦρο ἐξὸν ἀποδημεῖν, ὅταν ἐκείνῳ τε καὶ ἐμοὶ καὶ ὑμῖν τοῖς φίλοις κοινῇ ξυνδοκῇ· ταῦτα δ' εἶναι μὴ ἐπιβουλεύοντος ἐμοί· τούτων δὲ ἐγγυητὰς γίγνεσθαι σέ τε καὶ τοὺς σοὺς οἰκείους καὶ τοὺς ἐνθάδε Δίωνος, ὑμῖν δὲ τὸ βέβαιον
C ἐκεῖνος παρεχέτω. τὰ χρήματα δὲ ἃ ἂν λάβῃ, κατὰ Πελοπόννησον μὲν καὶ Ἀθήνας κείσθω παρ' οἷς τισὶν ἂν ὑμῖν δοκῇ, καρπούσθω δὲ Δίων, μὴ κύριος δὲ ἄνευ ὑμῶν γιγνέσθω ἀνελέσθαι. ἐγὼ γὰρ ἐκείνῳ μὲν οὐ σφόδρα πιστεύω τούτοις χρώμενον ἂν τοῖς χρήμασι δίκαιον γίγνεσθαι περὶ ἐμέ· οὐ γὰρ ὀλίγα ἔσται· σοὶ δὲ καὶ τοῖς σοῖς μᾶλλον πεπίστευκα. ὅρα δὴ ταῦτα εἴ σοι ἀρέσκει, καὶ μένε ἐπὶ τούτοις τὸν ἐνιαυτὸν τοῦτον, εἰς δὲ ὥρας ἄπιθι λαβὼν τὰ χρήματα ταῦτα· καὶ Δίων
D εὖ οἶδ' ὅτι πολλὴν χάριν ἕξει σοι διαπραξαμένῳ ταῦτα ὑπὲρ ἐκείνου.

Τοῦτον δὴ ἐγὼ τὸν λόγον ἀκούσας ἐδυσχέραινον μέν, ὅμως δὲ βουλευσάμενος ἔφην εἰς τὴν ὑστεραίαν αὐτῷ περὶ τούτων τὰ δόξαντα ἀπαγγελεῖν. ταῦτα ξυνεθέμεθα τότε. ἐβουλευόμην δὴ τὸ μετὰ ταῦτα κατ' ἐμαυτὸν γενόμενος, μάλα συγκεχυμένος·

[1] Amongst Plato's companions on this visit to Sicily were Speusippus and Xenocrates.

EPISTLE VII

scheme of the following kind to secure my remaining over that sailing-season. On the following day he came and addressed me in these plausible terms: "You and I," he said, "must get Dion and Dion's affairs cleared out of the way, to stop our frequent disputes about them. And this," said he, "is what I will do for Dion for your sake. I require that he shall remove his property and reside in the Peloponnese, not, however, as an exile but possessing the right to visit this country also whenever it is mutually agreed by him and by me and by you his friends. But this is on condition that he does not conspire against me; and you and your associates[1] and Dion's here in Sicily shall be the guarantors of these terms, and he shall furnish you with his security. And all the property he shall take shall be deposited in the Peloponnese and Athens with such persons as you shall think fit; and he shall enjoy the income from it but shall not be authorized to remove it without your consent. For I do not altogether trust him to act justly towards me if he had the use of these funds—for they will be by no means small; and I put more trust in you and your friends. So consider whether this arrangement contents you, and remain on these terms for the present year, and when next season arrives depart and take with you these funds of Dion. And I am well assured that Dion will be most grateful to you for having effected this arrangement on his behalf."

And I, when I heard this speech, was annoyed, but none the less I replied that I would think it over and let him know next day my decision about the matter; and to this we both then agreed. So after this, when I was by myself, I was thinking it over, very

PLATO

346 πρῶτος δ' ἦν μοι τῆς βουλῆς ἡγούμενος ὅδε λόγος·
Φέρε, εἰ διανοεῖται τούτων μηδὲν ποιεῖν Διονύσιος
E ὧν φησίν, ἀπελθόντος δ' ἐμοῦ ἐὰν ἐπιστέλλῃ
Δίωνι πιθανῶς αὐτός τε καὶ ἄλλοις πολλοῖς[1] τῶν
αὑτοῦ διακελευόμενος, ἃ νῦν πρὸς ἐμὲ λέγει, ὡς
αὐτοῦ μὲν ἐθέλοντος, ἐμοῦ δὲ οὐκ ἐθελήσαντος ἃ
προὐκαλεῖτό με δρᾶν, ἀλλ' ὀλιγωρήσαντος τῶν
ἐκείνου τὸ παράπαν πραγμάτων, πρὸς δὲ καὶ
τούτοισιν ἔτι μηδ' ἐθέλῃ με ἐκπέμπειν αὐτὸς τῶν
ναυκλήρων μηδενὶ προστάττων, ἐνδείξηται δὲ πᾶσι
347 ῥᾳδίως ὡς ἀβουλῶν ἐμὲ ἐκπλεῖν, ἆρά τις ἐθελήσει
με ἄγειν ναύτην[2] ὁρμώμενον ἐκ τῆς Διονυσίου
οἰκίας; (ᾤκουν γὰρ δὴ πρὸς τοῖς ἄλλοισι κακοῖς
ἐν τῷ κήπῳ τῷ περὶ τὴν οἰκίαν, ὅθεν οὐδ' ἂν ὁ
θυρωρὸς ἤθελέ με ἀφεῖναι μὴ πεμφθείσης αὐτῷ
τινος ἐντολῆς παρὰ Διονυσίου.) ἂν δὲ περιμείνω
τὸν ἐνιαυτόν, ἕξω μὲν Δίωνι ταῦτα ἐπιστέλλειν,
ἐν οἷς τ' αὖτ' εἰμὶ καὶ ἃ πράττω. καὶ ἐὰν μὲν δὴ
ποιῇ τι Διονύσιος ὧν φησίν, οὐ παντάπασιν ἔσται
B μοι καταγελάστως πεπραγμένα· τάλαντα γὰρ ἴσως
ἐστὶν οὐκ ἔλαττον, ἂν ἐκτιμᾷ τις ὀρθῶς, ἑκατὸν
ἡ Δίωνος οὐσία· ἂν δ' οὖν γίγνηται τὰ νῦν ὑπο-
φαίνοντα, οἷα εἰκὸς αὐτὰ γίγνεσθαι, ἀπορῶ μὲν ὅ
τι χρήσομαι ἐμαυτῷ, ὅμως δὲ ἀναγκαῖον ἴσως
ἐνιαυτόν γ' ἔτι πονῆσαι καὶ ἔργοις ἐλέγξαι πειρᾶσθαι τὰς Διονυσίου μηχανάς.

Ταῦτά μοι δόξαντα εἰς τὴν ὑστεραίαν εἶπον πρὸς

---

[1] ἄλλοις πολλοῖς MSS.: ἄλλοι πολλοὶ Zur.
[2] ναύτην MSS.: ναύτης Zur.

---

[1] For this use of the word ναύτης *cf.* Soph. *Philoct.* 901.

much perturbed. And in my deliberation the first and foremost reflexion was this—" Come now, suppose that Dionysius has no intention of performing any of his promises, and suppose that on my departure he sends a plausible note to Dion—both writing himself and charging many of his friends also to do so—stating the proposal he is now making to me, and how in spite of his wish I had refused to do what he had invited me to do, and had taken no interest at all in Dion's affairs; and beyond all this, suppose that he is no longer willing to send me away by giving his own personal order to one of the shipmasters, but makes it plain to them all that he has no wish for me to sail away in comfort—in this case would any of them consent to convey me as a passenger,[1] starting off from the residence of Dionysius?" For, in addition to my other misfortunes, I was lodging in the garden adjoining his residence, and out of this not even the doorkeeper would have allowed me to pass without a permit sent him from Dionysius. "On the other hand, if I stay on for the year I shall be able to write and tell Dion the position in which I am placed and what I am doing; and if Dionysius should actually perform any of his promises, I shall have accomplished something not altogether contemptible—for Dion's property, if it is rightly valued, amounts probably to as much as a hundred talents; whereas if the events now dimly threatening come to pass in the way that seems likely, I am at a loss to know what I shall do with myself. Notwithstanding, I am obliged, it appears, to endure another year of toil and endeavour to test by actual experience the devices of Dionysius."

When I had come to this decision, I said to

PLATO

Διονύσιον ὅτι Δέδοκταί μοι μένειν. ἀξιῶ μήν, ἔφην, μὴ κύριον ἡγεῖσθαί σε Δίωνος ἐμέ, πέμπειν δὲ μετ' ἐμοῦ σὲ παρ' αὐτὸν γράμματα τὰ νῦν δεδογμένα δηλοῦντα, καὶ ἐρωτᾶν εἴτε ἀρκεῖ ταῦτα αὐτῷ, καὶ εἰ μή, βούλεται δὲ ἄλλ' ἄττα καὶ ἀξιοῖ, καὶ ταῦτά ἐπιστέλλειν ὅ τι τάχιστα, σὲ δὲ νεωτερίζειν μηδέν πω τῶν περὶ ἐκεῖνον. ταῦτ' ἐρρήθη, ταῦτα ξυνωμολογήσαμεν, ὡς νῦν εἴρηται σχεδόν.

Ἐξέπλευσε δὴ τὰ πλοῖα μετὰ τοῦτο, καὶ οὐκέτι μοι δυνατὸν ἦν πλεῖν, ὅτε δή μοι καὶ Διονύσιος ἐμνήσθη λέγων ὅτι τὴν ἡμίσειαν τῆς οὐσίας εἶναι δέοι Δίωνος, τὴν δ' ἡμίσειαν τοῦ υἱέος· ἔφη δὴ πωλήσειν αὐτήν, πραθείσης δὲ τὰ μὲν ἡμίσεα ἐμοὶ δώσειν ἄγειν, τὰ δ' ἡμίσεα τῷ παιδὶ καταλείψειν αὐτοῦ· τὸ γὰρ δὴ δικαιότατον οὕτως ἔχειν. πληγεὶς δ' ἐγὼ τῷ λεχθέντι πάνυ μὲν ᾤμην γελοῖον εἶναι ἀντιλέγειν[1] ἔτι, ὅμως δ' εἶπον ὅτι χρείη τὴν παρὰ Δίωνος ἐπιστολὴν περιμένειν ἡμᾶς καὶ ταῦτα πάλιν αὐτὰ ἐπιστέλλειν. ὁ δὲ ἑξῆς τούτοις πάνυ νεανικῶς ἐπώλει τὴν οὐσίαν αὐτοῦ πᾶσαν, ὅπῃ τε καὶ ὅπως ἤθελε καὶ οἷς τισί, πρὸς ἐμὲ δὲ οὐδὲν ὅλως ἐφθέγγετο περὶ αὐτῶν, καὶ μὴν ὡσαύτως ἐγὼ πρὸς ἐκεῖνον αὖ περὶ τῶν Δίωνος πραγμάτων οὐδὲν ἔτι διελεγόμην· οὐδὲν γὰρ ἔτι πλέον ᾤμην ποιεῖν.

Μέχρι μὲν δὴ τούτων ταύτῃ μοι βεβοηθημένον ἐγεγόνει φιλοσοφίᾳ καὶ φίλοις· τὸ δὲ μετὰ ταῦτα

---

[1] ἀντιλέγειν Hermann: τι (ὅτι mss.) λέγειν Zur.

---

[1] For this part of the biographical details cf. *Ep.* iii. 318 A ff.

EPISTLE VII

Dionysius on the following day—" I have decided to remain. I request you, however," I said, " not to regard me as Dion's master, but to join with me yourself in sending him a letter explaining what we have now decided, and asking him whether it satisfies him; and if not, and if he desires and claims other conditions, let him write them to us immediately; and do you refrain till then from taking any new step in regard to his affairs." This is what was said, and this is what we agreed, pretty nearly in the terms I have now stated.[1]

After this the vessels had put to sea and it was no longer possible for me to sail; and then it was that Dionysius remembered to tell me that one half of the property ought to belong to Dion, the other half to his son; and he said that he would sell it, and when sold he would give me the one half to convey to Dion, and leave the half intended for his son where it was; for that was the most equitable arrangement. I, then, although I was dumbfounded at his statement, deemed that it would be utterly ridiculous to gainsay him any more; I replied, however, that we ought to wait for the letter from Dion, and then send him back this proposal by letter. But immediately after this he proceeded to sell the whole of Dion's property in a very highhanded fashion, where and how and to what purchasers he chose, without ever saying a single word to me about the matter; and verily I, in like manner, forbore to talk to him at all any longer about Dion's affairs; for I thought that there was no longer any profit in so doing.

Now up to this time I had been assisting in this way philosophy and my friends; but after this, the

PLATO

348 ἐζῶμεν ἐγὼ καὶ Διονύσιος, ἐγὼ μὲν βλέπων ἔξω, καθάπερ ὄρνις ποθῶν ποθὲν ἀναπτέσθαι, ὁ δὲ διαμηχανώμενος τίνα τρόπον ἀνασοβήσοι με μηδὲν ἀποδοὺς τῶν Δίωνος. ὅμως δὲ ἐφάνημεν[1] ἑταῖροί γε εἶναι πρὸς πᾶσαν Σικελίαν.

Τῶν δὴ μισθοφόρων τοὺς πρεσβυτέρους Διονύσιος ἐπεχείρησεν ὀλιγομισθοτέρους ποιεῖν παρὰ τὰ τοῦ πατρὸς ἔθη, θυμωθέντες δὲ οἱ στρατιῶται ξυνελέγησαν ἀθρόοι καὶ οὐκ ἔφασαν ἐπιτρέψειν. ὁ δ' ἐπεχείρει βιάζεσθαι κλείσας τὰς τῆς ἀκροπόλεως
B πύλας, οἱ δ' ἐφέροντο εὐθὺς πρὸς τὰ τείχη, παιωνά τινα ἀναβοήσαντες βάρβαρον καὶ πολεμικόν· οὗ δὴ περιδεὴς Διονύσιος γενόμενος ἅπαντα συνεχώρησε καὶ ἔτι πλείω τοῖς τότε συλλεχθεῖσι τῶν πελταστῶν.

Λόγος δή τις ταχὺ διῆλθεν ὡς Ἡρακλείδης αἴτιος εἴη γεγονὼς πάντων τούτων. ὃν ἀκούσας ὁ μὲν Ἡρακλείδης ἐκποδὼν αὐτὸν ἔσχεν ἀφανῆ, Διονύσιος δὲ ἐζήτει λαβεῖν. ἀπορῶν δέ, Θεοδότην
C μεταπεμψάμενος εἰς τὸν κῆπον· ἔτυχον δ' ἐν τῷ κήπῳ καὶ ἐγὼ τότε περιπατῶν· τὰ μὲν οὖν ἄλλα οὔτ' οἶδα οὔτ' ἤκουον διαλεγομένων, ἃ δὲ ἐναντίον εἶπε Θεοδότης ἐμοῦ πρὸς Διονύσιον, οἶδά τε καὶ μέμνημαι. Πλάτων γάρ, ἔφη, Διονύσιον ἐγὼ πείθω τουτονί, ἐὰν ἐγὼ γένωμαι δεῦρο Ἡρακλείδην κομίσαι δυνατὸς ἡμῖν εἰς λόγους περὶ τῶν ἐγκλημάτων αὐτῷ τῶν νῦν γεγονότων, ἂν ἄρα μὴ δόξῃ δεῖν αὐτὸν οἰκεῖν ἐν Σικελίᾳ, τόν τε υἱὸν λαβόντα

---

[1] ἐφάνημεν Apelt: ἔφαμεν mss., Zur.

---

[1] *Cf. Phaedr.* 249 D.
[2] The mercenaries lived in the island of Ortygia, but

EPISTLE VII

kind of life we lived, Dionysius and I, was this—
I was gazing out of my cage, like a bird[1] that is
longing to fly off and away, while he was scheming how
he might shoo me back without paying away any of
Dion's money; nevertheless, to the whole of Sicily
we appeared to be comrades.

Now Dionysius attempted, contrary to his father's
practice, to reduce the pay of the older members of
his mercenary force, and the soldiers, being infuriated,
assembled together and refused to permit it. And
when he kept trying to force them by closing the
gates of the citadel,[2] they immediately rushed up to
the walls shouting out a kind of barbaric war-chant;
whereupon Dionysius became terribly alarmed and
conceded all and even more than all to those of the
peltasts that were then assembled.

Then a report quickly got abroad that Heracleides[3]
was to blame for all this trouble; and Heracleides,
on hearing this, took himself off and vanished. Then
Dionysius was seeking to capture him, and finding
himself at a loss he summoned Theodotes to his
garden; and it happened that at the time I too was
walking in the garden. Now the rest of their conversation I neither know nor heard, but I both know
and remember what Theodotes said to Dionysius in
my presence. "Plato," he said, "I am urging this
course on our friend Dionysius: if I prove able to
fetch Heracleides here to answer the charges now
made against him, in case it is decided that he must
not reside in Sicily, I claim that he should have a

beyond the walls of the Acropolis; so when Plato had to
quit the Acropolis he was surrounded by them in his new
lodgings.

[3] *Cf. Ep.* iii. 318 c for Heracleides, Theodotes, and Eurybius.

D καὶ τὴν γυναῖκα ἀξιῶ εἰς Πελοπόννησον ἀποπλεῖν, οἰκεῖν τε βλάπτοντα μηδὲν Διονύσιον ἐκεῖ, καρπούμενον δὲ τὰ ἑαυτοῦ. μετεπεμψάμην μὲν οὖν καὶ πρότερον αὐτόν, μεταπέμψομαι δὲ καὶ νῦν, ἄν τ' οὖν ἀπὸ τῆς προτέρας μεταπομπῆς ἄν τε καὶ ἀπὸ τῆς νῦν ὑπακούσῃ μοι. Διονύσιον δὲ ἀξιῶ καὶ δέομαι, ἄν τις ἐντυγχάνῃ Ἡρακλείδῃ ἐάν τ' ἐν ἀγρῷ ἐάν τ' ἐνθάδε, μηδὲν ἄλλο αὐτῷ φλαῦρον
E γίγνεσθαι, μεταστῆναι δ' ἐκ τῆς χώρας, ἕως ἂν ἄλλο τι Διονυσίῳ δόξῃ. ταῦτα, ἔφη, συγχωρεῖς; λέγων πρὸς τὸν Διονύσιον. Συγχωρῶ μηδ' ἂν πρὸς τῇ σῇ, ἔφη, φανῇ οἰκίᾳ πείσεσθαι φλαῦρον μηδὲν παρὰ τὰ νῦν εἰρημένα.

Τῇ δὴ μετὰ ταύτην τὴν ἡμέραν δείλης Εὐρύβιος καὶ Θεοδότης προσηλθέτην μοι σπουδῇ τεθορυβημένω θαυμαστῶς, καὶ ὁ Θεοδότης λέγει, Πλάτων, ἔφη, παρῆσθα χθὲς οἷς περὶ Ἡρακλείδου Διονύσιος ὡμολόγει πρὸς ἐμὲ καὶ σέ; Πῶς δὲ οὔκ; ἔφην. Νῦν τοίνυν, ἦ δ' ὅς, περιθέουσι πελτασταὶ λαβεῖν Ἡρακλείδην ζητοῦντες, ὁ δὲ εἶναί πῃ ταύτῃ κινδυνεύει. ἀλλ' ἡμῖν, ἔφη, συνακολού-
349 θησον πρὸς Διονύσιον ἁπάσῃ μηχανῇ. ᾠχόμεθα οὖν καὶ εἰσήλθομεν παρ' αὐτόν, καὶ τὼ μὲν ἐστάτην σιγῇ δακρύοντε, ἐγὼ δὲ εἶπον Οἵδε πεφόβηνται μή τι σὺ παρὰ τὰ χθὲς ὡμολογημένα ποιήσῃς περὶ Ἡρακλείδην νεώτερον· δοκεῖ γάρ μοι ταύτῃ πῃ γεγονέναι φανερὸς ἀποτετραμμένος. ὁ δὲ ἀκούσας ἀνεφλέχθη τε καὶ παντοδαπὰ χρώματα ἧκεν, οἷα

EPISTLE VII

passage to the Peloponnese, taking his son and his wife, and reside there without doing injury to Dionysius, and enjoying the income from his property. In fact I have already sent to fetch him, and I will now send again, in case he should obey either my former summons or the present one. And I request and beseech Dionysius that, should anyone meet with Heracleides, whether in the country or here in the city, no harm should be inflicted on him beyond his removal out of the country until Dionysius has come to some further decision." And addressing Dionysius he said, " Do you agree to this ? " " I agree," he replied, " that even if he be seen at your house he shall suffer no harm beyond what has now been mentioned."

Now on the next day, at evening, Eurybius and Theodotes came to me hurriedly, in an extraordinary state of perturbation ; and Theodotes said—" Plato, were you present yesterday at the agreement which Dionysius made with us both concerning Heracleides ? " " Of course I was," I replied. " But now," he said, " peltasts [1] are running about seeking to capture Heracleides, and he is probably somewhere about here. But do you now by all means accompany us to Dionysius." So we set off and went in to where he was ; and while they two stood in silence, weeping, I said to him—" My friends here are alarmed lest you should take any fresh step regarding Heracleides, contrary to our agreement of yesterday ; for I believe it is known that he has taken refuge somewhere hereabouts." On hearing this, Dionysius fired up and went all colours, just as an angry man

---

[1] *i.e.* light-armed soldiers, so called from the kind of light shield they carried.

ἂν θυμούμενος ἀφείη· προσπεσὼν δ' αὐτῷ ὁ
Θεοδότης, λαβόμενος τῆς χειρός, ἐδάκρυσέ τε
καὶ ἱκέτευε μηδὲν τοιοῦτον ποιεῖν. ὑπολαβὼν δ'
ἐγὼ παραμυθούμενος, Θάρρει, Θεοδότα, ἔφην·
οὐ γὰρ τολμήσει Διονύσιος παρὰ τὰ χθὲς ὡμολογημένα ἄλλα ποτὲ δρᾶν. καὶ ὃς ἐμβλέψας μοι
καὶ μάλα τυραννικῶς, Σοί, ἔφη, ἐγὼ οὔ τέ τι
σμικρὸν οὔτε μέγα ὡμολόγησα. Νὴ τοὺς θεούς,
ἦν δ' ἐγώ, σύ γε ταῦτα ἃ σοῦ νῦν οὗτος δεῖται μὴ
ποιεῖν. καὶ εἰπὼν ταῦτα ἀποστρεφόμενος ᾠχόμην
ἔξω. τὸ μετὰ ταῦτα ὁ μὲν ἐκυνήγει τὸν Ἡρακλείδην, Θεοδότης δὲ ἀγγέλους πέμπων Ἡρακλείδῃ
φεύγειν διεκελεύετο. ὁ δὲ ἐκπέμψας Τισίαν καὶ
πελταστὰς διώκειν ἐκέλευε· φθάνει δέ, ὡς ἐλέγετο,
Ἡρακλείδης εἰς τὴν Καρχηδονίων ἐπικράτειαν
ἐκφυγὼν ἡμέρας σμικρῷ τινι μέρει.

Τὸ δὴ μετὰ τοῦτο ἡ πάλαι ἐπιβουλὴ Διονυσίῳ
τοῦ μὴ ἀποδοῦναι τὰ Δίωνος χρήματα ἔδοξεν
ἔχθρας λόγον ἔχειν ἂν πρός με πιθανόν, καὶ πρῶτον
μὲν ἐκ τῆς ἀκροπόλεως ἐκπέμπει με, εὑρὼν πρόφασιν ὡς τὰς γυναῖκας ἐν τῷ κήπῳ ἐν ᾧ κατῴκουν
ἐγὼ δέοι θῦσαι θυσίαν τινὰ δεχήμερον· ἔξω δή
με παρ' Ἀρχεδήμῳ προσέταττε τὸν χρόνον τοῦτον
μεῖναι. ὄντος δ' ἐμοῦ ἐκεῖ Θεοδότης μεταπεμ-
ψάμενός με πολλὰ περὶ τῶν τότε πραχθέντων
ἠγανάκτει καὶ ἐμέμφετο Διονυσίῳ· ὁ δ' ἀκούσας
ὅτι παρὰ Θεοδότην εἴην εἰσεληλυθώς, πρόφασιν
αὖ ταύτην ἄλλην τῆς πρὸς ἐμὲ διαφορᾶς ποιούμενος, ἀδελφὴν τῆς πρόσθεν, πέμψας τινὰ ἠρώτα
με εἰ ξυγγενοίμην ὄντως μεταπεμψαμένου με

EPISTLE VII

would do; and Theodotes fell at his knees and grasping his hand besought him with tears to do no such thing. And I interposed and said by way of encouragement—" Cheer up, Theodotes ; for Dionysius will never dare to act otherwise contrary to yesterday's agreement." Then Dionysius, with a highly tyrannical glare at me, said—" With you I made no agreement, great or small." " Heaven is witness," I replied, " that you did,—not to do what this man is now begging you not to do." And when I had said this I turned away and went out. After this Dionysius kept on hunting after Heracleides, while Theodotes kept sending messengers to Heracleides bidding him to flee. And Dionysius sent out Tisias and his peltasts with orders to pursue him; but Heracleides, as it was reported, forestalled them by a fraction of a day and made his escape into the Carthaginians' province.

Now after this Dionysius decided that his previous plot of refusing to pay over Dion's money would furnish him with a plausible ground for a quarrel with me ; and, as a first step, he sent me out of the citadel, inventing the excuse that the women had to perform a sacrifice of ten days' duration in the garden where I was lodging ; so during this period he gave orders that I should stay outside with Archedemus. And while I was there Theodotes sent for me and was loud in his indignation at what had then taken place and in his blame of Dionysius ; but the latter, when he heard that I had gone to the house of Theodotes, by way of making this a new pretext, akin to the old, for his quarrel against me, sent a man to ask me whether I had really visited Theodotes when he invited me. "Cer-

PLATO

349 Θεοδότου. κἀγὼ Παντάπασιν ἔφην· ὁ δέ, Ἐκέλευε τοίνυν, ἔφη, σοὶ φράζειν ὅτι καλῶς οὐδαμῇ ποιεῖς Δίωνα καὶ τοὺς Δίωνος φίλους ἀεὶ περὶ πλείονος αὑτοῦ ποιούμενος. ταῦτ' ἐρρήθη, καὶ οὐκέτι μετεπέμψατό με εἰς τὴν οἴκησιν πάλιν, ὡς ἤδη σαφῶς Θεοδότου μὲν ὄντος μου καὶ Ἡρακλείδου φίλου, αὑτοῦ δ' ἐχθροῦ, καὶ οὐκ εὐνοεῖν ᾤετό με, ὅτι Δίωνι τὰ χρήματα ἔρρει παντελῶς.

Ὤικουν δὴ τὸ μετὰ τοῦτο ἔξω τῆς ἀκροπόλεως ἐν
350 τοῖς μισθοφόροις· προσιόντες δέ μοι ἄλλοι τε καὶ οἱ τῶν ὑπηρεσιῶν ὄντες Ἀθήνηθεν, ἐμοὶ πολῖται, ἀπήγγελλον ὅτι διαβεβλημένος εἴην ἐν τοῖς πελτασταῖς καί μοί τινες ἀπειλοῖεν, εἴ που λήψονταί με, διαφθερεῖν. μηχανῶμαι δή τινα τοιάνδε σωτηρίαν· πέμπω παρ' Ἀρχύτην καὶ τοὺς ἄλλους φίλους εἰς Τάραντα, φράζων ἐν οἷς ὢν τυγχάνω· οἱ δὲ πρόφασίν τινα πρεσβείας πορισάμενοι παρὰ τῆς
B πόλεως πέμπουσι τριακόντορόν τε καὶ Λαμίσκον αὐτῶν ἕνα, ὃς ἐλθὼν ἐδεῖτο Διονυσίου περὶ ἐμοῦ, λέγων ὅτι βουλοίμην ἀπιέναι, καὶ μηδαμῶς ἄλλως ποιεῖν. ὁ δὲ ξυνωμολόγησε καὶ ἀπέπεμψεν ἐφόδια δούς· τῶν Δίωνος δὲ χρημάτων οὔτ' ἐγώ τι ἀπῄτουν οὔτε τις ἀπέδωκεν.

Ἐλθὼν δὲ εἰς Πελοπόννησον εἰς Ὀλυμπίαν,¹ Δίωνα καταλαβὼν θεωροῦντα, ἤγγελλον τὰ γεγονότα. ὁ δὲ τὸν Δία ἐπιμαρτυράμενος εὐθὺς

---

¹ i.e. for the festival of 360 B.C.

EPISTLE VII

tainly," I replied; and he said—"Well then, he ordered me to tell you that you are not acting at all honourably in always preferring Dion and Dion's friends to him." Such were his words; and after this he did not summon me again to his house, as though it was now quite clear that I was friendly towards Theodotes and Heracleides but hostile to him; and he supposed that I bore him no goodwill because of the clean sweep he was making of Dion's moneys.

Thereafter I was residing outside the citadel among the mercenaries; and amongst others some of the servants who were from Athens, fellow-citizens of my own, came to me and reported that I had been slanderously spoken of amongst the peltasts; and that some of them were threatening that if they could catch me they would make away with me. So I devised the following plan to save myself: I sent to Archytas and my other friends in Tarentum stating the position in which I found myself: and they, having found some pretext for an Embassy from the State, dispatched a thirty-oared vessel, and with it one of themselves, called Lamiscus; and he, when he came, made request to Dionysius concerning me, saying that I was desirous to depart, and begging him by all means to give his consent. To this he agreed, and he sent me forth after giving me supplies for the journey; but as to Dion's money, neither did I ask for any of it nor did anyone pay me any.

On arriving at Olympia,[1] in the Peloponnese, I came upon Dion, who was attending the Games; and I reported what had taken place. And he, calling Zeus to witness, was invoking me and my

PLATO

350

παρήγγελλεν ἐμοὶ καὶ τοῖς ἐμοῖς οἰκείοις καὶ
C φίλοις παρασκευάζεσθαι τιμωρεῖσθαι Διονύσιον,
ἡμᾶς μὲν ξεναπατίας χάριν, οὕτω γὰρ ἔλεγέ τε
καὶ ἐνόει, αὐτὸν δ' ἐκβολῆς ἀδίκου καὶ φυγῆς.
ἀκούσας δ' ἐγὼ τοὺς μὲν φίλους παρακαλεῖν
ἐκέλευον, εἰ βούλοιντο· Ἐμὲ δ', εἶπον ὅτι, σὺ μετὰ
τῶν ἄλλων βίᾳ τινὰ τρόπον σύσσιτον καὶ συνέστιον
καὶ κοινωνὸν ἱερῶν Διονυσίῳ ἐποίησας, ὃς ἴσως
ἡγεῖτο διαβαλλόντων πολλῶν ἐπιβουλεύειν ἐμὲ
μετὰ σοῦ ἑαυτῷ καὶ τῇ τυραννίδι, καὶ ὅμως οὐκ
D ἀπέκτεινεν, ᾐδέσθη δέ. οὔτ' οὖν ἡλικίαν ἔχω
συμπολεμεῖν ἔτι σχεδὸν οὐδενί, κοινός τε ὑμῖν
εἰμί, ἄν ποτέ τι πρὸς ἀλλήλους δεηθέντες φιλίας
ἀγαθόν τι ποιεῖν βουληθῆτε· κακὰ δὲ ἕως ἂν
ἐπιθυμῆτε, ἄλλους παρακαλεῖτε. ταῦτα εἶπον με-
μισηκὼς τὴν περὶ Σικελίαν πλάνην καὶ ἀτυχίαν.
ἀπειθοῦντες δὲ καὶ οὐ πειθόμενοι ταῖς ὑπ' ἐμοῦ
διαλλάξεσι[1] πάντων τῶν νῦν γεγονότων κακῶν
αὐτοὶ αἴτιοι ἐγένοντο αὑτοῖς, ὧν, εἰ Διονύσιος
E ἀπέδωκε τὰ χρήματα Δίωνι ἢ καὶ παντάπασι
κατηλλάγη, οὐκ ἄν ποτε ἐγένετο οὐδέν, ὅσα γε δὴ
τἀνθρώπινα· Δίωνα γὰρ ἐγὼ καὶ τῷ βούλεσθαι
καὶ τῷ δύνασθαι κατεῖχον ἂν ῥᾳδίως· νῦν δὲ
ὁρμήσαντες ἐπ' ἀλλήλους κακῶν πάντα ἐμπεπλή-
κασι.

351 Καί τοι τὴν γε αὐτὴν Δίων εἶχε βούλησιν ἥνπερ
ἂν ἐγὼ φαίην δεῖν ἐμὲ καὶ ἄλλον, ὅστις μετρίως[2]
περί τε τῆς αὐτοῦ δυνάμεως καὶ φίλων καὶ περὶ

---

[1] διαλλάξεσι some mss.: διαλέξεσι best mss., Zur.
[2] μετρίως] μέτριος mss., Zur.

---

[1] Perhaps an allusion to "the wanderings of Ulysses"; cf.
345 E.

EPISTLE VII

relatives and friends to prepare at once to take vengeance on Dionysius,—we on account of his treachery to guests (for that was what Dion said and meant), and he himself on account of his wrongful expulsion and banishment. And I, when I heard this, bade him summon my friends to his aid, should they be willing—" But as for me," I said, " it was you yourself, with the others, who by main force, so to say, made me an associate of Dionysius at table and at hearth and a partaker in his holy rites ; and he, though he probably believed that I, as many slanderers asserted, was conspiring with you against himself and his throne, yet refrained from killing me, and showed compunction. Thus, not only am I no longer, as I may say, of an age to assist anyone in war, but I also have ties in common with you both, in case you should ever come to crave at all for mutual friendship and wish to do one another good ; but so long as you desire to do evil, summon others." This I said because I loathed my Sicilian wandering [1] and its ill-success. They, however, by their disobedience and their refusal to heed my attempts at conciliation have themselves to blame for all the evils which have now happened ; for, in all human probability, none of these would ever have occurred if Dionysius had paid over the money to Dion or had even become wholly reconciled to him, for both my will and my power were such that I could have easily restrained Dion. But, as things are, by rushing the one against the other they have flooded the world with woes.

And yet Dion had the same designs as I myself should have had (for so I would maintain) or anyone else whose purpose regarding his own power and his

PLATO

351

πόλεως τῆς αὑτοῦ διανοοῖτ᾽ ἂν εὐεργετῶν ἐν δυνάμει καὶ τιμαῖσι γενέσθαι τὰ μέγιστα ἐν ταῖς μεγίσταις. ἔστι δὲ οὐκ ἄν τις πλούσιον ἑαυτὸν ποιήσῃ καὶ ἑταίρους καὶ πόλιν ἐπιβουλεύσας καὶ ξυνωμότας συναγαγών, πένης ὢν καὶ ἑαυτοῦ μὴ κρατῶν, ὑπὸ δειλίας τῆς πρὸς τὰς ἡδονὰς ἡττη-

B μένος, εἶτα τοὺς τὰς οὐσίας κεκτημένους ἀποκτείνας, ἐχθροὺς καλῶν τούτους, διαφορῇ τὰ τούτων χρήματα καὶ τοῖς συνεργοῖς τε καὶ ἑταίροις παρακελεύηται ὅπως μηδεὶς αὐτῷ ἐγκαλεῖ πένης φάσκων εἶναι· ταὐτὸν δὲ καὶ τὴν πόλιν ἂν οὕτω τις εὐεργετῶν τιμᾶται ὑπ᾽ αὐτῆς, τοῖς πολλοῖς τὰ τῶν ὀλίγων ὑπὸ ψηφισμάτων διανέμων, ἢ μεγάλης προεστὼς πόλεως καὶ πολλῶν ἀρχούσης ἐλαττόνων τῇ ἑαυτοῦ πόλει τὰ τῶν σμικροτέρων

C χρήματα διανέμῃ μὴ κατὰ δίκην. οὕτω μὲν γὰρ οὔτε Δίων οὔτε ἄλλος ποτὲ οὐδεὶς ἐπὶ δύναμιν ἑκὼν εἶσιν ἀλιτηριώδη ἑαυτῷ τε καὶ γένει εἰς τὸν ἀεὶ χρόνον, ἐπὶ πολιτείαν δὲ καὶ νόμων κατασκευὴν τῶν δικαιοτάτων τε καὶ ἀρίστων, ὅ τι[1] δι᾽ ὀλιγίστων θανάτων καὶ φυγῶν γιγνομένην.

Ἃ νῦν δὴ Δίων πράττων, προτιμήσας τὸ πάσχειν ἀνόσια τοῦ δρᾶσαι πρότερον, διευλαβούμενος δὲ μὴ παθεῖν, ὅμως ἔπταισεν ἐπ᾽ ἄκρον ἐλθὼν τοῦ περιγενέσθαι τῶν ἐχθρῶν, θαυμαστὸν παθὼν οὐδέν.

D ὅσιος γὰρ ἄνθρωπος ἀνοσίων πέρι, σώφρων τε καὶ

---

[1] ὅ τι I emend (so too Howald): οὔ τι mss., Zur.

---

[1] According to the Socratic dictum, "No one sins voluntarily."

EPISTLE VII

friends and his city was the reasonable one of achieving the greatest height of power and privilege by conferring the greatest benefits. But a man does not do this if he enriches himself, his comrades, and his city by means of plotting and collecting conspirators, while in reality he himself is poor and not his own master but the cowardly slave of pleasures; nor does he do so if he proceeds next to slay the owners of property, dubbing them "enemies," and to dissipate their goods, and to charge his accomplices and comrades not to blame him if any of them complains of poverty. So likewise if a man receives honour from a city for conferring on it such benefits as distributing the goods of the few to the many by means of decrees; or if, when he is at the head of a large city which holds sway over many smaller ones, he distributes the funds of the smaller cities to his own, contrary to what is just. For neither Dion nor any other will ever voluntarily [1] aim thus at a power that would bring upon himself and his race an everlasting curse, but rather at a moderate government and the establishment of the justest and best of laws by means of the fewest possible exiles and executions.

Yet when Dion was now pursuing this course, resolved to suffer rather than to do unholy deeds—although guarding himself against so suffering [2]—none the less when he had attained the highest pitch of superiority over his foes he stumbled. And therein he suffered no surprising fate. For while, in dealing with the unrighteous, a righteous man who is sober and sound of mind will never be wholly

[2] For "suffering" wrong as a bar to complete happiness *cf. Laws* 829 A.

351
ἔμφρων, τὸ μὲν ὅλον οὐκ ἄν ποτε διαψευσθείη τῆς
ψυχῆς τῶν τοιούτων πέρι, κυβερνήτου· δὲ ἀγαθοῦ
πάθος ἂν ἴσως οὐ θαυμαστὸν εἰ πάθοι, ὃν χειμὼν
μὲν ἐσόμενος οὐκ ἂν πάνυ λάθοι, χειμώνων δὲ
ἐξαίσιον καὶ ἀπροσδόκητον μέγεθος λάθοι τ' ἂν
καὶ λαθὸν κατακλύσειε βίᾳ. ταὐτὸν δὴ καὶ Δίωνα
ἔσφηλε [δι' ὀλιγίστων]¹· κακοὶ μὲν γὰρ ὄντες
αὐτὸν σφόδρα οὐκ ἔλαθον οἱ σφήλαντες, ὅσον δὲ
E ὕψος ἀμαθίας εἶχον καὶ τῆς ἄλλης μοχθηρίας τε
καὶ λαιμαργίας, ἔλαθον, ᾧ δὴ σφαλεὶς κεῖται,
Σικελίαν πένθει περιβαλὼν μυρίῳ.

Τὰ δὴ μετὰ τὰ νῦν ῥηθέντα ἃ ξυμβουλεύω,
352 σχεδὸν εἴρηταί τέ μοι καὶ εἰρήσθω· ὧν δ' ἐπαν-
έλαβον ἕνεκα τὴν εἰς Σικελίαν ἄφιξιν τὴν δευτέραν,
ἀναγκαῖον εἶναι ἔδοξέ μοι ῥηθῆναι δεῖν διὰ τὴν
ἀτοπίαν καὶ ἀλογίαν τῶν λεγομένων.² εἰ δ' ἄρα
τινὶ τὰ νῦν ῥηθέντα εὐλογώτερα ἐφάνη καὶ προ-
φάσεις πρὸς τὰ γενόμενα ἱκανὰς ἔχειν ἔδοξέ τῳ,
μετρίως ἂν ἡμῖν καὶ ἱκανῶς εἴη τὰ νῦν εἰρημένα.

¹ δι' ὀλιγίστων omitted by best MSS.
² λεγομένων] γενομένων MSS., Zur.

EPISTLE VII

deceived concerning the souls of such men; yet it would not, perhaps, be surprising if he were to share the fate of a good pilot, who, though he certainly would not fail to notice the oncoming of a storm, yet might fail to realize its extraordinary and unexpected violence, and in consequence of that failure might be forcibly overwhelmed. And Dion's downfall was, in fact, due to the same cause; for while he most certainly did not fail to notice that those who brought him down were evil men, yet he did fail to realize to what a pitch of folly they had come, and of depravity also and voracious greed; and thereby he was brought down and lies fallen, enveloping Sicily in immeasurable woe.

What counsel I have to offer, after this narrative of events, has been given already, and so let it suffice. But I deemed it necessary to explain the reasons why I undertook my second journey to Sicily [1] because absurd and irrational stories are being told about it. If, therefore, the account I have now given appears to anyone more rational, and if anyone believes that it supplies sufficient excuses for what took place, then I shall regard that account as both reasonable and sufficient.

---

[1] *i.e.* Plato's *third* Sicilian visit (as he does not count the first), *cf.* 330 c, 337 E.

EPISTLE VIII

Prefatory Note.—This letter—assuming it to be Plato's—appears to have been written some months after the seventh letter, *i.e.* in 353 B.C., shortly before Callippus, the murderer of Dion, had been driven out in turn by Hipparinus, the son of Dionysius the Elder and the nephew of Dion.

The argument of the letter is briefly this: The advice which Plato will now give is intended to benefit all parties alike—that of Dion, that of Dionysius, and the democrats. But these parties are now in conflict over one point only—the restoration or abolition of the tyranny. The history of the recent turmoils plainly shows that the continuation of this conflict can only end in general ruin. Compromise therefore is necessary, though it may be vain to suggest it—as " a prayer " rather than a possibility. In recommending some compromise as the only escape from the present *impasse*, Plato begins (353 A) by bidding the other parties recollect how much Sicily owed in the past to the royal house of Dionysius, which had saved her from the domination of the barbarian. They should consider, too, that neither of the opposing parties, democrats or tyrants, is ever likely to defeat the other completely, so that the inevitable result of continued civil war will be the re-enslavement of Sicily by the Carthaginians.

EPISTLE VIII

All Greeks should unite to avert that catastrophe. The rulers, avoiding the rôle of tyrant, should don the mantle of constitutional monarchy; and as a check on the kingly power other magistrates and authorities should be established, after the wise example of Lycurgus (354 A-D). The popular party, likewise, should reflect on the dangers of extreme and unlimited freedom, how it leads to anarchy, and anarchy in turn breeds tyranny.

And herein Plato is speaking for Dion as well as for himself; and from this point (355 A) on to the end of the letter he acts as the mouthpiece of the lost leader of the party he is addressing. Dion would bid the men of Syracuse seek first laws that are just and justly make money-making subservient to the well-being of body and of soul, instead of idolizing riches. He would also reiterate the advice already given by Plato that the warring parties should effect a compromise—the monarchical party accepting a power limited by such laws, and the democratic party accepting a liberty limited and controlled by the same laws; whereby a government would be established which would constitute a mean between the extremes of anarchy and despotism, the supreme power being in the hands of Justice and of Law. On this basis let three members of the royal house unite to share the kingship—the son of Dion, the son of Dionysius the Elder, and Dionysius the Younger (355 E ff.). Then, with these kings acting at least as the national chief-priests, let representatives of all parties associate with them in the government various magistrates, judges, and asemblies, so as to secure a balance of power. To establish thus a constitutional government (Dion would say) would have

been my first care, had I lived (357 A). And I should have tried next to re-colonize the Greek cities in Sicily. This double task now falls upon you; and if Hipparinus and my own son combine to undertake it, it is by no means impossible. Strive then by all means to realize this vision of the future which I have set before you, and may your efforts be crowned with success.

It is hardly to be supposed that Plato thought that he could influence the course of affairs in Sicily by this effusion. If it is really his we must regard it as a philosophic manifesto, under the guise of a letter, rather than as a serious contribution to practical politics seriously addressed to an actual Sicilian faction. What we have here, as the gist of this letter, is simply a reiteration of the theme so familiar to readers of the *Republic* and the *Laws* that the secret of the successful and happy State is ordered liberty, a balance of power, and the reign of Justice and of Law. The maxim "Be temperate in all things," or "Nothing too much," is once more enforced and illustrated in the sphere of politics by reference to the long history of misrule in Sicily with its unending succession of tyrants and tyrannicides, all equally vicious and vile. Dion alone stands out as the champion of moderation; and here again, as in the seventh letter, Plato represents Dion as sharing his own political convictions and advocating the very policy which he himself would advise. The suggestion, put into Dion's mouth, that there should be three kings of Sicily may sound strange enough, and Plato, by putting it into Dion's mouth, may have wished to avoid responsibility for it. But none the less it is logical enough if we may suppose that the

three represent conflicting interests amongst the monarchical party. And, moreover, Plato is careful to suggest that these so-called " kings " should have no control except over matters of religion.

In connexion with this proposal for a triple kingship, there is considerable controversy regarding the identity of the first of the kings suggested—Dion's son (355 E). For, according to Plutarch and Cornelius Nepos, Dion's son Hipparinus died before his father, and therefore cannot be the person alluded to here. Some scholars accept this testimony, and, relying on another statement of Plutarch, suppose that the allusion is to a posthumous son of Dion, born in prison. Others regard the stories about Hipparinus's death as unreliable and believe that he is really the person intended by Plato. Both these views, however, seem open to grave objection. Even if we suspect the details of the death of Hipparinus as reported by Plutarch and Nepos, it would be rash indeed to suppose that there is no ground for the assertion that he did die before his father; in fact, apart from the references in this letter and the seventh, there is every reason to suppose that he did. And, on the other hand, granting that there was a posthumous son, after Callippus had imprisoned Dion's wife Arete, it is quite incredible that Plato could have alluded to an infant in arms in the terms here employed, and seriously proposed him as a colleague of grown men, and fit to share the responsibilities of kingship. The most plausible solution of this crux is to suppose that the reference is really to Hipparinus, and that Plato remained in ignorance of his death until long after its occurrence. Nor is this so improbable as it might seem at first sight, when we

PLATO

realize that under the rule of Callippus communication between Athens and Sicily was largely suspended, and if we assume, further, that this eighth letter was composed before the final expulsion of Callippus from the citadel of Syracuse.

If we accept this view [1] it allows us also to identify the Hipparinus mentioned in *Ep.* vii. 324 A with the same son of Dion; and it certainly is preferable, if possible, to suppose that the allusion there is to the son rather than the nephew, since the context all seems to point that way. And, conversely, the fact that Dion's son is the person most likely to be indicated in 324 A goes to confirm the view that Plato was for long in ignorance of his death.

Further, the view that this letter was written during, rather than after, the rule of Callippus is borne out by several other indications. Thus, in speaking of Hipparinus (Dion's nephew) as a candidate for the kingship, he mentions " his present assistance," and how he " is in the act of freeing Sicily "; which language would hardly be natural if Hipparinus had already ousted Callippus and taken his place as master of Syracuse. And it would be still more strange for Plato to advocate his scheme for a threefold kingship if Hipparinus was already in possession of supreme power. The fitting time to propose such a scheme was rather while the monarchical party was as yet some way from achieving full success, and while it was advisable to hold all three sections of it together—that of Dion, and of his nephew Hipparinus, and of the exiled tyrant Dionysius—by promising all a share in the future

---

[1] This is the view taken by Mr. J. Harward: see his discussion in *Classical Quarterly* xxii. 3-4.

EPISTLE VIII

government, so that all might co-operate the more ardently in the common assault that was then being made against their common foe Callippus.

Failing the acceptance of this view as to the date of this letter and the identity of "Dion's son," we should be constrained to fall back on the view of many of the earlier scholars that this, like so many of the other letters, is not really by Plato but the product of a later age.

H

352

Πλάτων τοῖς Δίωνος οἰκείοις τε καὶ ἑταίροις
εὖ πράττειν.

B "Ἃ δ' ἂν διανοηθέντες μάλιστα εὖ πράττοιτε
ὄντως, πειράσομαι ταῦθ' ὑμῖν κατὰ δύναμιν διεξελθεῖν. ἐλπίζω δὲ οὐχ ὑμῖν μόνοις ξυμβουλεύσειν
τὰ ξυμφέροντα, μάλιστά γε μὴν ὑμῖν, καὶ δευτέροις
C πᾶσι τοῖς ἐν Συρακούσαις, τρίτοις δὲ ὑμῶν καὶ
τοῖς ἐχθροῖς καὶ πολεμίοις, πλὴν εἴ τις αὐτῶν
ἀνοσιουργὸς γέγονε· ταῦτα γὰρ ἀνίατα καὶ οὐκ ἄν
ποτέ τις αὐτὰ ἐκνίψειε. νοήσατε δὲ ἃ λέγω νῦν.

Ἔσθ' ὑμῖν κατὰ Σικελίαν πᾶσαν λελυμένης τῆς
τυραννίδος πᾶσα μάχη περὶ αὐτῶν τούτων, τῶν
μὲν βουλομένων ἀναλαβεῖν πάλιν τὴν ἀρχήν, τῶν
δὲ τῇ τῆς τυραννίδος ἀποφυγῇ τέλος ἐπιθεῖναι.
ξυμβουλὴ δὴ περὶ τῶν τοιούτων ὀρθὴ δοκεῖ ἑκά-
D στοτε τοῖς πολλοῖς εἶναι ταῦτα ξυμβουλεύειν δεῖν ἃ
τοὺς μὲν πολεμίους ὡς πλεῖστα κακὰ ἐξεργάσεται,
τοὺς δὲ φίλους ὡς πλεῖστα ἀγαθά. τὸ δὲ οὐδαμῶς
ῥᾴδιον, πολλὰ κακὰ δρῶντα τοὺς ἄλλους μὴ οὐ καὶ
πάσχειν αὐτὸν πολλὰ ἕτερα. δεῖ δὲ οὐ μακρὰν

---

[1] For this reference to the phrasing of the opening saluta-
tion *cf. Ep.* iii. *ad init.*
[2] Alluding to Callippus, the murderer of Dion.

EPISTLE VIII

PLATO TO THE RELATIVES AND COMPANIONS OF DION
WISHES WELL-DOING

THE policy which would best serve to secure your real " well-doing "[1] is that which I shall now endeavour as best I can to describe to you. And I hope that my advice will not only be salutary to you (though to you in special), but also to all the Syracusans, in the second place, and, in the third, to your enemies and your foes, unless any of them be a doer of impious deeds[2]; for such deeds are irremediable and none could ever wash out their stain.[3] Mark, then, what I now say.

Now that the tyranny is broken down over the whole of Sicily all your fighting rages round this one subject of dispute, the one party desiring to recover the headship, and the other to put the finishing touch to the expulsion of the tyrants. Now the majority of men always believe that the right advice about these matters is the advising of such action as will do the greatest possible harm to one's enemies and the greatest possible good to one's friends; whereas it is by no means easy to do much harm to others without also suffering in turn much harm oneself. And without going far afield one may

[3] *Cf. Gorg.* 525 c.

PLATO

352
ἐλθόντας ποι τὰ τοιαῦτα ἐναργῶς ἰδεῖν ἀλλ' ὅσα
νῦν γέγονε τῇδε αὐτοῦ περὶ Σικελίαν, τῶν μὲν
ἐπιχειρούντων δρᾶν, τῶν δὲ ἀμύνασθαι τοὺς
δρῶντας· ἃ κἂν ἄλλοις μυθολογοῦντες ἱκανοὶ
E γίγνοισθ' ἂν ἑκάστοτε διδάσκαλοι. τούτων μὲν δὴ
σχεδὸν οὐκ ἀπορία· τῶν δὲ ὅσα γένοιτ' ἂν ἢ πᾶσι
συμφέροντα ἐχθροῖς τε καὶ φίλοις ἢ ὅ τι σμικρότατα κακὰ ἀμφοῖν, ταῦτα οὔτε ῥᾴδιον ὁρᾶν οὔτε
ἰδόντα ἐπιτελεῖν, εὐχῇ δὲ προσέοικεν ἡ τοιαύτη
ξυμβουλή τε καὶ ἐπιχείρησις τοῦ λόγου. ἔστω δὴ
παντάπασι μὲν εὐχή τις, ἀπὸ γὰρ θεῶν χρὴ πάντα
353 ἀρχόμενον ἀεὶ λέγειν τε καὶ νοεῖν, ἐπιτελὴς δ' εἴη
σημαίνουσα ἡμῖν τοιόνδε τινὰ λόγον· Νῦν ὑμῖν καὶ
τοῖς πολεμίοις σχεδὸν ἐξ οὗπερ γέγονεν ὁ πόλεμος
συγγένεια ἄρχει μία διὰ τέλους, ἥν ποτε κατέστησαν οἱ πατέρες ὑμῶν ἐς ἀπορίαν ἐλθόντες τὴν
ἅπασαν, τόθ' ὅτε κίνδυνος ἐγένετο ἔσχατος Σικελίᾳ
τῇ τῶν Ἑλλήνων ὑπὸ Καρχηδονίων ἀνάστατον
ὅλην ἐκβαρβαρωθεῖσαν γενέσθαι. τότε γὰρ εἵλοντο
Διονύσιον μὲν ὡς νέον καὶ πολεμικὸν ἐπὶ τὰς τοῦ
B πολέμου πρεπούσας αὐτῷ πράξεις, σύμβουλον δὲ
καὶ πρεσβύτερον Ἱππαρῖνον, ἐπὶ σωτηρίᾳ τῆς
Σικελίας αὐτοκράτορας, ὥς φασι, τυράννους ἐπονο-
μάζοντες. καὶ εἴτε δὴ θείαν τις ἡγεῖσθαι βούλεται
τύχην καὶ θεὸν εἴτε τὴν τῶν ἀρχόντων ἀρετὴν εἴτε
καὶ τὸ ξυναμφότερον μετὰ τῶν τότε πολιτῶν τῆς
σωτηρίας αἰτίαν ξυμβῆναι γενομένην, ἔστω ταύτῃ
ὅπῃ τις ὑπολαμβάνει· σωτηρία δ' οὖν οὕτω συνέβη

---

[1] " Prayer " in the sense of a " pious wish " unlikely to
be fulfilled, or a " last resort."

[2] The struggle against the Carthaginians, which had lasted
with hardly a break, since 409 B.C.

EPISTLE VIII

see such consequences clearly in the recent events in Sicily itself, where the one faction is trying to inflict injury and the other to ward off the injurers; and the tale thereof, if ever you told it to others, would inevitably prove a most impressive lesson. Of such policies, one may say, there is no lack; but as for a policy which would prove beneficial to all alike, foes as well as friends, or at least as little detrimental as possible to either, such a policy is neither easy to discern, nor, when discerned, easy to carry out; and to advise such a policy or attempt to describe it is much like saying a prayer.[1] Be it so, then, that this is nothing but a prayer (and in truth every man ought always to begin his speaking and his thinking with the gods); yet may it attain fulfilment in indicating some such counsel as this:—Now and almost ever since the war [2] began both you and your enemies have been ruled continuously by that one family which your fathers set on the throne in the hour of their greatest distress, when Greek Sicily was in the utmost danger of being entirely overrun by the Carthaginians and barbarized. On that occasion they chose Dionysius because of his youth and warlike prowess to take charge of the military operations for which he was suited, with Hipparinus, who was older, as his fellow-counsellor, appointing them dictators for the safeguarding of Sicily, with the title, as men say, of "tyrants." But whether one prefers to suppose that the cause which ultimately brought about their salvation was divine Fortune and the Deity, or the virtue of the rulers, or possibly the combination of both assisted by the citizens of that age—as to this let everyone form his own notion; in any case this was the way in which

353 τοῖς τότε γενομένοις. τοιούτων οὖν αὐτῶν γεγονότων δίκαιόν που τοῖς σώσασι πάντας χάριν ἔχειν. εἰ δέ τι τὸν μετέπειτα χρόνον ἡ τυραννὶς οὐκ ὀρθῶς τῇ τῆς πόλεως δωρεᾷ κατακέχρηται, τούτων δίκας τὰς μὲν ἔχει, τὰς δὲ τινέτω. τίνες οὖν δὴ δίκαι ἀναγκαίως ὀρθαὶ γίγνοιντ' ἂν ἐκ τῶν ὑπαρχόντων αὐτοῖς; εἰ μὲν ῥᾳδίως ὑμεῖς ἀποφυγεῖν οἷοί τ' ἦτε αὐτοὺς καὶ ἄνευ μεγάλων κινδύνων καὶ πόνων, ἢ 'κεῖνοι ἑλεῖν εὐπετῶς πάλιν τὴν ἀρχήν, οὐδ' ἂν συμβουλεύειν οἷόν τ' ἦν τὰ μέλλοντα ῥηθήσεσθαι· νῦν δ' ἐννοεῖν ὑμᾶς ἀμφοτέρους χρεὼν καὶ ἀναμιμνήσκεσθαι ποσάκις ἐν ἐλπίδι ἑκάτεροι γεγόνατε τοῦ [νῦν][1] οἴεσθαι σχεδὸν ἀεί τινος σμικροῦ ἐπιδεεῖς εἶναι τὸ μὴ πάντα κατὰ νοῦν πράττειν, καὶ δὴ καὶ ὅτι τὸ σμικρὸν τοῦτο μεγάλων καὶ μυρίων κακῶν αἴτιον ἑκάστοτε ξυμβαίνει γιγνόμενον, καὶ πέρας οὐδέν ποτε τελεῖται, ξυνάπτει δὲ ἀεὶ παλαιὰ τελευτὴ δοκοῦσα ἀρχῇ φυομένῃ νέᾳ, διολέσθαι δ' ὑπὸ τοῦ κύκλου τούτου κινδυνεύσει καὶ τὸ τυραννικὸν ἅπαν καὶ τὸ δημοτικὸν γένος, ἥξει δέ, ἐάνπερ τῶν εἰκότων γίγνηταί τι καὶ ἀπευκτῶν, σχεδὸν εἰς ἐρημίαν τῆς Ἑλληνικῆς φωνῆς Σικελία πᾶσα, Φοινίκων ἢ Ὀπικῶν μεταβαλοῦσα εἴς τινα δυναστείαν καὶ κράτος. τούτων δὴ χρὴ πάσῃ προθυμίᾳ πάντας τοὺς Ἕλληνας τέμνειν φάρμακον. εἰ μὲν δή τις

[1] νῦν bracketed by W.-Möllendorff.

---

[1] Alluding to the expulsion of Dionysius from Sicily; he retired to Locri in Italy.

EPISTLE VIII

salvation for the men of that generation came about. Seeing, then, that they proved themselves men of such a quality, it is surely right that they should be repaid with gratitude by all those whom they saved. But if in after times the tyrant's house has wrongly abused the bounty of the city, the penalty for this it has suffered in part,[1] and in part it will have to pay. What, then, is the penalty rightly to be exacted from them under existing circumstances? If you were able to get quit of them easily, without serious dangers and trouble, or if they were able to regain the empire without difficulty, then, in either case, it would not have been possible for me so much as to offer the advice which I am now about to utter; but as it is, both of you ought to bear in mind and remember how many times each party has hopefully imagined that it lacked but a little of achieving complete success almost every time; and, what is more, that it is precisely this little deficiency which is always turning out to be the cause of great and numberless evils. And of these evils no limit is ever reached, but what seems to be the end of the old is always being linked on to the beginning of a new brood; and because of this endless chain of evil the whole tribe of tyrants and democrats alike will be in danger of destruction. But should any of these consequences—likely as they are though lamentable —come to pass, hardly a trace of the Greek tongue will remain in all Sicily, since it will have been transformed into a province or dependency of Phoenicians or Opicians.[2] Against this all the Greeks must with all zeal provide a remedy. If, therefore, any man

[2] Probably some tribes of central Italy, Samnites or Campanians.

PLATO

353
354 ὀρθότερον ἄμεινόν τ' ἔχει τοῦ ὑπ' ἐμοῦ ῥηθησομένου, ἐνεγκὼν εἰς τὸ μέσον ὀρθότατα φιλέλλην ἂν λεχθείη· ὃ δέ μοι φαίνεταί πῃ τὰ νῦν, ἐγὼ πειράσομαι πάσῃ παρρησίᾳ καὶ κοινῷ τινι δικαίῳ λόγῳ χρώμενος δηλοῦν. λέγω γὰρ δὴ διαιτητοῦ τινὰ τρόπον διαλεγόμενος [ὡς]¹ δυοῖν τυραννεύσαντί τε καὶ τυραννευθέντι, ὡς ἑνὶ ἑκατέρῳ, παλαιὰν ἐμὴν ξυμβουλήν. καὶ νῦν δ' ὅ γ' ἐμὸς λόγος ἂν εἴη ξύμβουλος τυράννῳ παντὶ φεύγειν μὲν τοὔνομά τε καὶ τοὔργον τοῦτο, εἰς βασιλείαν δέ, εἰ δυνατὸν
B εἴη, μεταβαλεῖν. δυνατὸν δέ, ὡς ἔδειξεν ἔργῳ σοφὸς ἀνὴρ καὶ ἀγαθὸς Λυκοῦργος, ὃς ἰδὼν τὸ τῶν οἰκείων γένος ἐν Ἄργει καὶ Μεσσήνῃ ἐκ βασιλέων εἰς τυράννων δύναμιν ἀφικομένους καὶ διαφθείραντας ἑαυτούς τε καὶ τὴν πόλιν ἑκατέρους ἑκατέραν, δείσας περὶ τῆς αὑτοῦ πόλεως ἅμα καὶ γένους, φάρμακον ἐπήνεγκε τὴν τῶν γερόντων ἀρχὴν καὶ τὴν² τῶν ἐφόρων, δεσμὸν τῆς βασιλικῆς ἀρχῆς σωτήριον, ὥστε γενεὰς τοσαύτας ἤδη μετ'
C εὐκλείας σώζεσθαι, νόμος ἐπειδὴ κύριος ἐγένετο βασιλεὺς τῶν ἀνθρώπων, ἀλλ' οὐκ ἄνθρωποι τύραννοι νόμων.

Ὃ δὴ καὶ νῦν οὑμὸς λόγος πᾶσι παρακελεύεται, τοῖς μὲν τυραννίδος ἐφιεμένοις ἀποτρέπεσθαι καὶ φεύγειν φυγῇ ἀπλήστως πεινώντων εὐδαιμόνισμα ἀνθρώπων καὶ ἀνοήτων, εἰς βασιλέως δ' εἶδος πειρᾶσθαι μεταβάλλειν καὶ δουλεῦσαι νόμοις βασιλικοῖς, τὰς μεγίστας τιμὰς κεκτημένους παρ' ἑκόν-

---

¹ ὡς I bracket.
² τὴν W.-Möllendorff: τὸν mss., Zur.

---

¹ *Cf. Ep.* iv. 320 D.   ² *Cf. Laws* 692 A.

EPISTLE VIII

knows of a remedy that is truer and better than that which I am now about to propose, and puts it openly before us, he shall have the best right to the title " Friend of Greece." The remedy, however, which commends itself to me I shall now endeavour to explain, using the utmost freedom of speech and a tone of impartial justice. For indeed I am speaking somewhat like an arbitrator, and addressing to the two parties, the former despot and his subjects, as though each were a single person, the counsel I gave of old. And now also my word of advice to every despot would be that he should shun the despot's title and his task, and change his despotism for kingship. That this is possible has been actually proved by that wise and good man Lycurgus [1]; for when he saw that the family of his kinsmen in Argos and in Messene had in both cases destroyed both themselves and their city by advancing from kingship to despotic power, he was alarmed about his own city as well as his own family, and as a remedy he introduced the authority of the Elders and of the Ephors to serve as a bond of safety for the kingly power [2]; and because of this they have already been kept safe and glorious all these generations since Law became with them supreme king over men instead of men being despots over the laws.

And now also I urgently admonish you all to do the same. Those of you who are rushing after despotic power I exhort to change their course and to flee betimes from what is counted as " bliss " by men of insatiable cravings and empty heads, and to try to transform themselves into the semblance of a king, and to become subject to kingly laws, owing their possession of the highest honours to the voluntary

PLATO

τῶν τε ἀνθρώπων καὶ τῶν νόμων· τοῖς δὲ δὴ
ἐλεύθερα διώκουσιν ἤθη καὶ φεύγουσι τὸν δούλειον
ζυγὸν ὡς ὂν κακόν, εὐλαβεῖσθαι ξυμβουλεύοιμ' ἂν
μή ποτε ἀπληστίᾳ ἐλευθερίας ἀκαίρου τινὸς εἰς
τὸ τῶν προγόνων νόσημα ἐμπέσωσιν, ὃ διὰ τὴν
ἄγαν ἀναρχίαν οἱ τότε ἔπαθον, ἀμέτρῳ ἐλευθερίας
χρώμενοι ἔρωτι· οἱ γὰρ πρὸ Διονυσίου καὶ Ἱππαρίνου ἄρξαντες Σικελιῶται τότε ὡς ᾤοντο εὐ-
δαιμόνως ἔζων, τρυφῶντές τε καὶ ἅμα ἀρχόντων
ἄρχοντες· οἳ καὶ τοὺς δέκα στρατηγοὺς κατέλευσαν
βάλλοντες τοὺς πρὸ Διονυσίου, κατὰ νόμον οὐδένα
κρίναντες, ἵνα δὴ δουλεύοιεν μηδενὶ μήτε σὺν δίκῃ
μήτε νόμῳ δεσπότῃ, ἐλεύθεροι δ' εἶεν πάντῃ
πάντως· ὅθεν αἱ τυραννίδες ἐγένοντο αὐτοῖς. δουλεία γὰρ καὶ ἐλευθερία ὑπερβάλλουσα μὲν ἑκατέρα
πάγκακον, ἔμμετρος δὲ οὖσα πανάγαθον· μετρία
δὲ ἡ θεῷ δουλεία, ἄμετρος δὲ ἡ τοῖς ἀνθρώποις·
θεὸς δὲ ἀνθρώποις σώφροσι νόμος, ἄφροσι δὲ
ἡδονή.

Τούτων δὴ ταύτῃ πεφυκότων, ἃ ξυμβουλεύω
Συρακοσίοις πᾶσι φράζειν παρακελεύομαι τοῖς
Δίωνος φίλοις ἐκείνου καὶ ἐμὴν κοινὴν ξυμβουλήν·
ἐγὼ δὲ ἑρμηνεύσω ἃ ἐκεῖνος ἔμπνους ὢν καὶ
δυνάμενος εἶπεν ⟨ἂν⟩[1] νῦν πρὸς ὑμᾶς. τίν' οὖν
δή τις ἂν εἴποι λόγον ἀποφαίνεται ἡμῖν περὶ τῶν
νῦν παρόντων ἡ Δίωνος ξυμβουλή; τόνδε.

[1] ἂν added by Bekker.

---

[1] Plato is here in error, apparently: the stoning took place at an earlier date at Agrigentum.
[2] Law is divine as "the dispensation of Reason" (νόμος being derived from νοῦς), *cf. Laws* 762 E. For evils of excessive freedom *cf. Rep.* 564 A.

EPISTLE VIII

goodwill of the citizens and to the laws. And I should counsel those who follow after the ways of freedom, and shun as a really evil thing the yoke of bondage, to beware lest by their insatiable craving for an immoderate freedom they should ever fall sick of their forefathers' disease, which the men of that time suffered because of their excessive anarchy, through indulging an unmeasured love of freedom. For the Siceliots of the age before Dionysius and Hipparinus began to rule were living blissfully, as they supposed, being in luxury and ruling also over their rulers; and they even stoned to death the ten generals who preceded Dionysius, without any legal trial,[1] to show that they were no slaves of any rightful master, nor of any law, but were in all ways altogether free. Hence it was that the rule of the despots befell them. For as regards both slavery and freedom, when either is in excess it is wholly evil, but when in moderation wholly good; and moderate slavery consists in being the slave of God, immoderate, in being the slave of men; and men of sound sense have Law for their God,[2] but men without sense Pleasure.

Since these things are naturally ordained thus, I exhort Dion's friends to declare what I am advising to all the Syracusans, as being the joint advice both of Dion and myself; and I will be the interpreter of what he would have said to you now, were he alive and able to speak.[3] " Pray then," someone might say, " what message does the advice of Dion declare to us concerning the present situation ? " It is this:

[3] For this artifice of putting words into the mouth of an absent speaker *cf. Menex.* 246 c ff., *Ep.* vii. 328 D.

PLATO

355

Δέξασθε, ὦ Συρακόσιοι, πάντων πρῶτον νόμους,
B οἵτινες ἂν ὑμῖν φαίνωνται μὴ πρὸς χρηματισμὸν καὶ πλοῦτον τρέψοντες τὰς γνώμας ὑμῶν μετ' ἐπιθυμίας, ἀλλ' ὄντων τριῶν, ψυχῆς καὶ σώματος, ἔτι δὲ χρημάτων, τὴν τῆς ψυχῆς ἀρετὴν ἐντιμοτάτην ποιοῦντες, δευτέραν δὲ τὴν τοῦ σώματος, ὑπὸ τῇ τῆς ψυχῆς κειμένην, τρίτην δὲ καὶ ὑστάτην τὴν τῶν χρημάτων τιμήν, δουλεύουσαν τῷ σώματί τε καὶ τῇ ψυχῇ. καὶ ὁ μὲν ταῦτα ἀπεργαζόμενος
C θεσμὸς νόμος ἂν ὀρθῶς ὑμῖν εἴη κείμενος, ὄντως εὐδαίμονας ἀποτελῶν τοὺς χρωμένους· ὁ δὲ τοὺς πλουσίους εὐδαίμονας ὀνομάζων λόγος αὐτός τε ἄθλιος, γυναικῶν καὶ παίδων ὢν λόγος ἄνους, τοὺς πειθομένους τε ἀπεργάζεται τοιούτους. ὅτι δ' ἀληθῆ ταῦτ' ἐγὼ παρακελεύομαι, ἐὰν γεύσησθε τῶν νῦν λεγομένων περὶ νόμων, ἔργῳ γνώσεσθε· ἡ δὴ βάσανος ἀληθεστάτη δοκεῖ γίγνεσθαι τῶν πάντων πέρι.

Δεξάμενοι δὲ τοὺς τοιούτους νόμους, ἐπειδὴ
D κατέχει κίνδυνος Σικελίαν, καὶ οὔτε κρατεῖτε ἱκανῶς οὔτ' αὖ διαφερόντως κρατεῖσθε, δίκαιον ἂν ἴσως καὶ ξυμφέρον γίγνοιτο ὑμῖν πᾶσι μέσον τεμεῖν, τοῖς τε φεύγουσι τῆς ἀρχῆς τὴν χαλεπότητα ὑμῖν καὶ τοῖς τῆς ἀρχῆς πάλιν ἐρῶσι τυχεῖν ὧν οἱ πρόγονοι τότε, τὸ μέγιστον, ἔσωσαν ἀπὸ βαρβάρων τοὺς Ἕλληνας, ὥστ' ἐξεῖναι περὶ πολιτείας νῦν ποιεῖσθαι λόγους· ἔρρουσι δὲ τότε οὔτε λόγος οὔτ' ἐλπὶς ἐλείπετ' ἂν οὐδαμῇ οὐδαμῶς. νῦν οὖν τοῖς μὲν ἐλευθερία γιγνέσθω μετὰ βασιλικῆς ἀρχῆς,

---

[1] For this classification of "goods" cf. Gorg. 477 c; Laws 697 B, 726 A ff.
[2] Cf. Laws 631 B; also 355 c infra, Ep. vi. 323 D.
[3] Cf. Rep. 408 E ff., 452 D ff.

EPISTLE VIII

"Above all else, O ye Syracusans, accept such laws as do not appear to you likely to turn your minds covetously to money-making and wealth; but rather—since there are three objects, the soul, the body, and money besides,—accept such laws as cause the virtue of the soul to be held first in honour, that of the body second, subordinate to that of the soul, and the honour paid to money to come third and last, in subjection to both the body and the soul.[1] The ordinance which effects this will be truly laid down by you as law, since it really makes those who obey it blessed[2]; whereas the phrase which terms the rich "blessed" is not only a miserable one in itself, being the senseless phrase of women and children, but also renders those who believe it equally miserable. That this exhortation of mine is true you will learn by actual experience if you make trial of what I am now saying concerning laws; for in all matters experience is held to be the truest test.[3]

And when you have accepted laws of this kind, inasmuch as Sicily is beset with dangers, and you are neither complete victors nor utterly vanquished, it will be, no doubt, both just and profitable for you all to pursue a middle course—not only those of you who flee from the harshness of the tyranny, but also those who crave to win back that tyranny—the men whose ancestors in those days performed the mightiest deed in saving the Greeks from the barbarians, with the result that it is possible for us now to talk about constitutions; whereas, if they had then been ruined, no place would have been left at all for either talk or hope. So, then, let the one party of you gain freedom by the aid of kingly rule, and the other

PLATO

τοῖς δὲ ἀρχὴ ὑπεύθυνος βασιλική, δεσποζόντων νόμων τῶν τε ἄλλων πολιτῶν καὶ τῶν βασιλέων αὐτῶν, ἄν τι παράνομον πράττωσιν. ἐπὶ δὲ τούτοις ξύμπασιν ἀδόλῳ γνώμῃ καὶ ὑγιεῖ μετὰ θεῶν βασιλέας στήσασθε, πρῶτον μὲν τὸν ἐμὸν υἱὸν χαρίτων ἕνεκα διττῶν, τῆς τε παρ' ἐμοῦ καὶ τοῦ ἐμοῦ πατρός· ὁ μὲν γὰρ ἀπὸ βαρβάρων ἠλευθέρωσεν ἐν τῷ τότε χρόνῳ τὴν πόλιν, ἐγὼ δὲ ἀπὸ τυράννων νῦν δίς, ὧν αὐτοὶ μάρτυρες ὑμεῖς γεγόνατε· δεύτερον δὲ δὴ ποιεῖσθε βασιλέα τὸν τῷ μὲν ἐμῷ πατρὶ ταὐτὸν κεκτημένον ὄνομα, υἱὸν δὲ Διονυσίου, χάριν τῆς τε δὴ νῦν βοηθείας καὶ ὁσίου τρόπου· ὃς γενόμενος τυράννου πατρὸς ἑκὼν τὴν πόλιν ἐλευθεροῖ, τιμὴν αὑτῷ καὶ γένει ἀείζωον ἀντὶ τυραννίδος ἐφημέρου καὶ ἀδίκου κτώμενος. τρίτον δὲ προκαλεῖσθαι χρὴ βασιλέα γίγνεσθαι Συρακουσῶν, ἑκόντα ἑκούσης τῆς πόλεως, τὸν νῦν τοῦ τῶν πολεμίων ἄρχοντα στρατοπέδου, Διονύσιον τὸν Διονυσίου, ἐὰν ἐθέλῃ ἑκὼν εἰς βασιλέως σχῆμα ἀπαλλάττεσθαι, δεδιὼς μὲν τὰς τύχας, ἐλεῶν δὲ πατρίδα καὶ ἱερῶν ἀθεραπευσίαν καὶ τάφων,[1] μὴ διὰ φιλονεικίαν πάντως πάντα ἀπολέσῃ βαρβάροις ἐπίχαρτος γενόμενος.

Τρεῖς δ' ὄντας βασιλέας, εἴτ' οὖν τὴν Λακωνικὴν δύναμιν αὐτοῖς δόντες εἴτε ἀφελόντες καὶ ξυνομο-

---

[1] τάφων Estienne: τάφους mss., Zur.

---

[1] *i.e.* Hipparinus, who was about twenty years old at this time; *cf.* Prefatory Note, and *Ep.* vii. 324 A.
[2] *Cf. Ep.* vii. 333 B.
[3] *i.e.* Dionysius the Elder: *cf.* 357 c. This Hipparinus Dion's nephew, was now assisting Dion's party in their attacks on Callippus from their base at Leontini.

EPISTLE VIII

gain a form of kingly rule that is not irresponsible, with the laws exercising despotic sway over the kings themselves as well as the rest of the citizens, in case they do anything illegal. On these conditions set up kings for all of you, by the help of the gods and with honest and sound intent,—my own son [1] first in return for twofold favours, namely that conferred by me and that conferred by my father; for he delivered the city from barbarians in his own day, while I, in the present day, have twice delivered it from tyrants,[2] whereof you yourselves are witnesses. And as your second king create the man who possesses the same name as my father and is son to Dionysius,[3] in return for his present assistance and for his pious disposition; for he, though he is sprung from a tyrant's loins, is in act of delivering the city of his own free will, gaining thereby for himself and for his race everlasting honour in place of a transitory and unrighteous tyranny. And, thirdly, you ought to invite to become king of Syracuse—as willing king of a willing city—him who is now commander of your enemies' army, Dionysius, son of Dionysius, if so be that he is willing of his own accord to transform himself into a king, being moved thereto by fear of fortune's changes, and by pity for his country and the untended state of her temples and her tombs, lest because of his ambition he utterly ruin all and become a cause of rejoicing to the barbarians.

And these three,—whether you grant them the power of the Laconian kings [4] or curtail that power by a common agreement,—you should establish as

---

[4] That power was little more than nominal, dealing chiefly with matters of religion.

λογησάμενοι, καταστήσασθε τρόπῳ τινὶ τοιῷδε, ὃς εἴρηται μὲν καὶ πρότερον ὑμῖν, ὅμως δ' ἔτι καὶ νῦν ἀκούετε.

Ἐὰν ἐθέλῃ τὸ γένος ὑμῖν τὸ Διονυσίου τε καὶ Ἱππαρίνου ἐπὶ σωτηρίᾳ Σικελίας παύσασθαι τῶν νῦν παρόντων κακῶν, τιμὰς αὑτοῖς καὶ γένει λαβόντες εἴς τε τὸν ἔπειτα καὶ τὸν νῦν χρόνον, ἐπὶ τούτοις καλεῖτε, ὥσπερ καὶ πρότερον ἐρρήθη, πρέσβεις οὓς ἂν ἐθελήσωσι κυρίους ποιησάμενοι τῶν διαλλαγῶν, εἴτε τινὰς αὐτόθεν εἴτε ἔξωθεν εἴτε ἀμφότερα, καὶ ὁπόσους ἂν συγχωρήσωσι. τούτους δ' ἐλθόντας νόμους μὲν πρῶτον θεῖναι καὶ πολιτείαν τοιαύτην, ἐν ᾗ βασιλέας ἁρμόττει γίγνεσθαι κυρίους ἱερῶν τε καὶ ὅσων ἄλλων πρέπει τοῖς γενομένοις ποτὲ εὐεργέταις· πολέμου δὲ καὶ εἰρήνης ἄρχοντας νομοφύλακας ποιήσασθαι ἀριθμὸν τριάκοντα καὶ πέντε μετά τε δήμου καὶ βουλῆς. δικαστήρια δὲ ἄλλα μὲν ἄλλων, θανάτου δὲ καὶ φυγῆς τούς τε πέντε καὶ τριάκοντα ὑπάρχειν. πρὸς τούτοις τε ἐκλεκτοὺς γίγνεσθαι δικαστὰς ἐκ τῶν [νῦν]¹ ἀεὶ περυσινῶν ἀρχόντων, ἕνα ἀφ' ἑκάστης τῆς ἀρχῆς τὸν ἄριστον δόξαντ' εἶναι καὶ δικαιότατον· τούτους δὲ τὸν ἐπιόντα ἐνιαυτὸν δικάζειν ὅσα θανάτου καὶ δεσμοῦ καὶ μεταστάσεως τῶν πολιτῶν· βασιλέα δὲ τῶν τοιούτων δικῶν μὴ ἐξεῖναι δικαστὴν γίγνεσθαι, καθάπερ ἱερέα φόνου καθαρεύοντα καὶ δεσμοῦ καὶ φυγῆς.

¹ νῦν omitted in some mss.

¹ Cf. *Ep.* vii. 337 B ff.

EPISTLE VIII

kings in some such manner as the following, which indeed has been described to you before,[1] yet listen to it now again.

If you find that the family of Dionysius and Hipparinus is willing to make an end of the evils now occurring in order to secure the salvation of Sicily provided that they receive honours both in the present and for the future for themselves and for their family, then on these terms, as was said before, convoke envoys empowered to negotiate a pact, such men as they may choose, whether they come from Sicily or from abroad or both, and in such numbers as may be mutually agreed. And these men, on their arrival, should first lay down laws and a constitution which is so framed as to permit the kings to be put in control of the temples and of all else that fitly belongs to those who once were benefactors. And as controllers of war and peace they should appoint Law-wardens, thirty-five in number, in conjunction with the People and the Council. And there should be various courts of law for various suits, but in matters involving death or exile the Thirty-five should form the court; and in addition to these there should be judges selected from the magistrates of each preceding year, one from each magistracy—the one, that is, who is approved as the most good and just; and these should decide for the ensuing year all cases which involve the death, imprisonment or transportation of citizens; and it should not be permissible for a king to be a judge of such suits, but he, like a priest, should remain clean from bloodshed and imprisonment and exile.[2]

[2] For the scheme here proposed *cf. Ep.* vii. 337 B ff., *Laws* 752 D ff., 762 c ff., 767 c ff., 855 c.

357

Ταῦθ' ὑμῖν ἐγὼ καὶ ζῶν διενοήθην γίγνεσθαι καὶ νῦν διανοοῦμαι. καὶ τότε κρατήσας τῶν ἐχθρῶν μεθ' ὑμῶν, εἰ μὴ ξενικαὶ ἐρινύες ἐκώλυσαν, κατέστησα ἂν ᾗπερ καὶ διενοούμην, καὶ μετὰ ταῦτα Σικελίαν ἂν τὴν ἄλλην, εἴπερ ἔργα ἐπὶ νῷ ἐγίγνετο, κατῴκισα, τοὺς μὲν βαρβάρους ἣν νῦν ἔχουσιν ἀφελόμενος, ὅσοι μὴ ὑπὲρ τῆς κοινῆς ἐλευθερίας διεπολέμησαν πρὸς τὴν τυραννίδα, τοὺς
B δ' ἔμπροθεν οἰκητὰς τῶν Ἑλληνικῶν τόπων εἰς τὰς ἀρχαίας καὶ πατρῴας οἰκήσεις κατοικίσας. ταὐτὰ δὲ ταῦτα καὶ νῦν πᾶσι συμβουλεύω κοινῇ διανοηθῆναι καὶ πράττειν τε καὶ παρακαλεῖν ἐπὶ ταύτας τὰς πράξεις πάντας, τὸν μὴ θέλοντα δὲ πολέμιον ἡγεῖσθαι κοινῇ. ἔστι δὲ ταῦτα οὐκ ἀδύνατα· ἃ γὰρ ἐν δυοῖν τε ὄντα ψυχαῖν τυγχάνει καὶ λογισαμένοις εὑρεῖν βέλτιστα ἑτοίμως ἔχει, ταῦτα δὴ[1] σχεδὸν ὁ κρίνων ἀδύνατα οὐκ εὖ φρονεῖ. λέγω δὲ τὰς δύο
C τήν τε Ἱππαρίνου τοῦ Διονυσίου υἱέος καὶ τὴν τοῦ ἐμοῦ υἱέος· τούτοιν γὰρ ξυνομολογησάντοιν τοῖς γε ἄλλοις Συρακουσίοις οἶμαι πᾶσιν ὅσοιπερ τῆς πόλεως κήδονται ξυνδοκεῖν.

Ἀλλὰ θεοῖς τε πᾶσι τιμὰς μετ' εὐχῶν δόντες τοῖς τε ἄλλοις ὅσοις μετὰ θεῶν πρέπει, πείθοντες καὶ προκαλούμενοι φίλους καὶ διαφόρους μαλακῶς τε καὶ πάντως μὴ ἀποστῆτε πρὶν ἂν τὰ νῦν ὑφ' ἡμῶν λεχθέντα, οἷον ὀνείρατα θεῖα ἐπιστάντα
D ἐγρηγορόσιν, ἐναργῆ τε ἐξεργάσησθε τελεσθέντα καὶ εὐτυχῆ.

[1] δὴ H. Richards: δὲ Zur. (om. some mss.).

[1] Alluding to Dion's murderers, Callippus and Philostratus; *cf. Ep.* vii. 333 E ff.

EPISTLE VIII

This is what I planned for you when I was alive, and it is still my plan now. With your aid, had not Furies in the guise of guests [1] prevented me, I should then have overcome our foes, and established the State in the way I planned; and after this, had my intentions been realized, I should have re-settled the rest of Sicily by depriving the barbarians of the land they now hold—excepting those who fought in defence of the common liberty against the tyranny—and restoring the former occupiers of the Greek regions to their ancient and ancestral homes. And now likewise I counsel you all with one accord to adopt and execute these same plans, and to summon all to this task, and to count him who refuses as a common enemy. Nor is such a course impossible; for when plans actually exist in two souls, and when they are readily perceived upon reflexion to be the best, he who pronounces such plans impossible is hardly a man of understanding. And by the "two souls" I mean the soul of Hipparinus the son of Dionysius and that of my own son; for should these agree together, I believe that all the rest of the Syracusans who have a care for their city will consent.

Well then, when you have paid due honour, with prayer, to all the gods and all the other powers to whom, along with the gods, it is due, cease not from urging and exhorting both friends and opponents by gentle means and every means, until, like a heaven-sent dream presented to waking eyes,[2] the plan which I have pictured in words be wrought by you into plain deeds and brought to a happy consummation."

[2] *Cf. Soph.* 266 c, *Rep.* 533 c.

EPISTLE IX

Prefatory Note.—This letter, together with the twelfth, which also is addressed to Archytas, is generally recognized as a forgery, in spite of the evidence of Cicero, who quotes it as Plato's twice (*De fin.* ii. 14, and *De offic.* i. 7). It is a colourless and commonplace effusion which we would not willingly ascribe to Plato, and which no correspondent of his would be likely to preserve. Moreover, there are certain peculiarities of diction which point to a later hand.

If Platonic, the letter must be dated after Plato's first voyage to Sicily, when he first met Archytas, *i.e.* after 387 B.C. And at that date the Echecrates mentioned in the *Phaedo* (if he is the person alluded to in 358 B) could not possibly be described as a "youth."

Θ

Πλάτων Ἀρχύτᾳ Ταραντίνῳ εὖ πράττειν.

Ἀφίκοντο πρὸς ἡμᾶς οἱ περὶ Ἄρχιππον καὶ Φιλωνίδην, τήν τε ἐπιστολὴν φέροντες ἣν σὺ αὐτοῖς ἔδωκας, καὶ ἀπαγγέλλοντες τὰ παρὰ σοῦ. τὰ μὲν οὖν πρὸς τὴν πόλιν οὐ χαλεπῶς διεπράξαντο· καὶ γὰρ οὐδὲ παντελῶς ἦν ἐργώδη· τὰ δὲ παρὰ σοῦ διῆλθον ἡμῖν λέγοντες ὑποδυσφορεῖν σε ὅτι οὐ δύνασαι τῆς περὶ τὰ κοινὰ ἀσχολίας ἀπολυθῆναι. ὅτι μὲν οὖν ἥδιστόν ἐστιν ἐν τῷ βίῳ τὸ τὰ αὑτοῦ πράττειν, ἄλλως τε καὶ εἴ τις ἕλοιτο τοιαῦτα πράττειν οἷα καὶ σύ, σχεδὸν παντὶ δῆλον· ἀλλὰ κἀκεῖνο δεῖ σε ἐνθυμεῖσθαι, ὅτι ἕκαστος ἡμῶν οὐχ αὑτῷ μόνον γέγονεν, ἀλλὰ τῆς γενέσεως ἡμῶν τὸ μέν τι ἡ πατρὶς μερίζεται, τὸ δέ τι οἱ γεννήσαντες, τὸ δὲ οἱ λοιποὶ φίλοι, πολλὰ δὲ καὶ τοῖς καιροῖς δίδοται τοῖς τὸν βίον ἡμῶν καταλαμβάνουσι. καλούσης δὲ τῆς πατρίδος αὐτῆς πρὸς τὰ κοινὰ ἄτοπον ἴσως τὸ μὴ ὑπακούειν· ἅμα γὰρ ξυμβαίνει καὶ χώραν καταλιμπάνειν φαύλοις ἀνθρώποις, οἳ οὐκ ἀπὸ τοῦ βελτίστου πρὸς τὰ κοινὰ προσέρχονται.

---

[1] *Cf. Ep.* vii. 338 c, 350 A. Archippus and Philonides were also members of the Pythagorean School, as was Echecrates (in 358 B).

EPISTLE IX

PLATO TO ARCHYTAS [1] OF TARENTUM WISHES WELL-DOING

ARCHIPPUS and Philonides and their party have arrived, bringing us the letter which you gave them, and also reporting your news. Their business with the city they have completed without difficulty—for in truth it was not at all a hard task; and they have given us a full account of you, telling us that you are somewhat distressed at not being able to get free from your public engagements. Now it is plain to almost everyone that the pleasantest thing in life is to attend to one's own business, especially when the business one chooses is such as yours; yet you ought also to bear in mind that no one of us exists for himself alone, but one share of our existence belongs to our country, another to our parents, a third to the rest of our friends, while a great part is given over to those needs of the hour with which our life is beset. And when our country itself calls us to public duties, it were surely improper not to hearken to the call [2]; for to do so will involve the further consequence of leaving room to worthless men who engage in public affairs from motives that are by no means the best.

[1] *Cf. Rep.* 347, 521, 540.

PLATO

358

Περὶ τούτων μὲν οὖν ἱκανῶς, Ἐχεκράτους δὲ καὶ νῦν ἐπιμέλειαν ἔχομεν καὶ εἰς τὸν λοιπὸν χρόνον ἕξομεν καὶ διὰ σὲ καὶ διὰ τὸν πατέρα αὐτοῦ Φρυνίωνα καὶ δι' αὐτὸν τὸν νεανίσκον.

EPISTLE IX

Enough, however, of this subject. We are looking after Echecrates now and we shall do so in the future also, for your sake and that of his father Phrynion, as well as for the sake of the youth himself.

EPISTLE X

Prefatory Note.—Of Aristodorus, to whom this letter is addressed, nothing is known beyond what we learn from the letter itself. In it he is commended for his loyalty to Dion, who was—we may suppose—in exile. What is here said of the nature of "true philosophy" and how it contrasts with mere "expertness" has fairly close parallel in the Dialogues (*e.g. cf. Rep.* 409 D, 499 A ff.; *Theaet.* 176 C); but the blunt way in which "philosophy" is identified with purely moral qualities, with no reference to intellectual endowments, is foreign to Plato's manner. There need be no hesitation, therefore, in rejecting this letter also as a spurious composition.

I

Πλάτων Ἀριστοδώρῳ εὖ πράττειν.

Ἀκούω Δίωνος ἐν τοῖς μάλιστα ἑταῖρον εἶναί τέ σε νῦν καὶ γεγονέναι διὰ παντός, τὸ σοφώτατον ἦθος τῶν εἰς φιλοσοφίαν παρεχόμενον· τὸ γὰρ βέβαιον καὶ πιστὸν καὶ ὑγιές, τοῦτο ἐγώ φημι εἶναι τὴν ἀληθινὴν φιλοσοφίαν, τὰς δὲ ἄλλας τε καὶ εἰς ἄλλα τεινούσας σοφίας τε καὶ δεινότητας κομψότητας οἶμαι προσαγορεύων ὀρθῶς ὀνομάζειν.

Ἀλλ' ἔρρωσό τε καὶ μένε ἐν τοῖς ἤθεσιν οἷσπερ καὶ νῦν μένεις.

EPISTLE X

PLATO TO ARISTODORUS WISHES WELL-DOING

I HEAR that you now are and always have been one of Dion's most intimate companions, since of all who pursue philosophy you exhibit the most philosophic disposition; for steadfastness, trustiness, and sincerity—these I affirm to be the genuine philosophy, but as to all other forms of science and cleverness which tend in other directions, I shall, I believe, be giving them their right names if I dub them "parlour-tricks.[1]"

So farewell, and continue in the same disposition in which you are continuing now.

[1] *Cf. Gorg.* 486 c, 521 D.

EPISTLE XI

Prefatory Note.—This letter is a reply to a request for help in drawing up a code of laws for a new colony. The Laodamas to whom it is addressed may be the mathematician of Thasos who is said to have invented the analytical method in geometry; but here he appears solely as a statesman. It has been conjectured that the colony alluded to may have been Crenidae or Datos—both founded about 360-359 B.C.; and if so, this letter, supposing it to be Plato's, may be dated shortly after his return from his third visit to Sicily.

The "Socrates" referred to (in 358 D-E) is generally supposed to be the "younger Socrates" of the *Politicus*.

The authenticity of the letter might possibly be allowed but for one wholly un-Platonic phrase (about "the illness of Socrates") which seems to disprove it sufficiently.

ΙΑ

Πλάτων Λαοδάμαντι εὖ πράττειν.

Ἐπέστειλα μέν σοι καὶ πρότερον ὅτι πολὺ διαφέρει πρὸς ἅπαντα ἃ λέγεις αὐτὸν ἀφικέσθαι σε Ἀθήναζε· ἐπειδὴ δὲ σὺ φῇς ἀδύνατον εἶναι, μετὰ τοῦτο ἦν δεύτερον, εἰ δυνατὸν ἐμὲ ἀφικέσθαι ἢ Σωκράτη, ὥσπερ ἐπέστειλας. νῦν δὲ Σωκράτης μέν ἐστι περὶ ἀσθένειαν τὴν τῆς στραγγουρίας, ἐμὲ δὲ ἀφικόμενον ἐνταῦθα ἄσχημον ἂν εἴη μὴ διαπράξασθαι ἐφ' ἅπερ σὺ παρακαλεῖς. ἐγὼ δὲ ταῦτα γενέσθαι ἂν οὐ πολλὴν ἐλπίδα ἔχω· δι'. ἃ δέ, μακρᾶς ἑτέρας δέοιτ' ἂν ἐπιστολῆς ἥτις¹ πάντα διεξίοι· καὶ ἅμα οὐδὲ τῷ σώματι διὰ τὴν ἡλικίαν ἱκανῶς ἔχω πλανᾶσθαι καὶ κινδυνεύειν κατά τε γῆν καὶ κατὰ θάλατταν οἷα ἅπαντα· καὶ νῦν πάντα κινδύνων ἐν ταῖς πορείαις ἐστὶ μεστά. συμβουλεῦσαι μέντοι ἔχω σοί τε καὶ τοῖς οἰκισταῖς, ὃ "εἰπόντος μὲν ἐμοῦ," φησὶν Ἡσίοδος, δόξαι ἂν εἶναι "φαῦλον χαλεπὸν δὲ νοῆσαι." εἰ γὰρ οἴονθ' ὑπὸ νόμων θέσεως καὶ ὧν τινῶν εὖ ποτε πολιτείαν² ἂν κατασκευασθῆναι, ἄνευ τοῦ εἶναί τι κύριον

¹ ἥτις MSS.: εἴ τις Zur.
² πολιτείαν MSS. corr.: πόλιν MSS., Zur.

---

¹ Probably an allusion to the prevalence of pirates (such as Alexander of Pherae) in the Aegean Sea.

EPISTLE XI

PLATO TO LAODAMAS WISHES WELL-DOING

I WROTE to you before that in view of all that you say it is of great importance that you yourself should come to Athens. But since you say that this is impossible, the second best course would have been that I, if possible, or Socrates should go to you, as in fact you said in your letter. At present, however, Socrates is laid up with an attack of strangury; while if I were to go there, it would be humiliating if I failed to succeed in the task for which you are inviting me. But I myself have no great hopes of success (as to my reasons for this, another long letter would be required to explain them in full), and moreover, because of my age, I am not physically fit to go wandering about and to run such risks as one encounters both by sea and land; and at present there is nothing but danger for travellers everywhere.[1] I am able, however, to give you and the settlers advice which may seem to be, as Hesiod [2] says, "Trivial when uttered by me, but hard to be understanded." For they are mistaken if they believe that a constitution could ever be well established by any kind of legislation whatsoever without the

---

[2] A fragment (229) of Hesiod, otherwise unknown: *cf.* Hesiod, *Op. et D.* 483-484.

PLATO

359 ἐπιμελούμενον ἐν τῇ πόλει τῆς καθ' ἡμέραν διαίτης, ὅπως ἂν ᾖ σώφρων τε καὶ ἀνδρική, δούλων τε καὶ ἐλευθέρων, οὐκ ὀρθῶς διανοοῦνται. τοῦτο δ' αὖ, εἰ μὲν εἰσὶν ἤδη ἄνδρες ἄξιοι τῆς ἀρχῆς ταύτης, B γένοιτ' ἄν· εἰ δ' ἐπὶ τὸ παιδεῦσαι δεῖ τινός, οὔτε ὁ παιδεύσων οὔτε οἱ παιδευθησόμενοι, ὡς ἐγὼ οἶμαι, εἰσὶν ὑμῖν, ἀλλὰ τὸ λοιπὸν τοῖς θεοῖς εὔχεσθαι. καὶ γὰρ σχεδόν τι καὶ αἱ ἔμπροσθεν πόλεις οὕτω κατεσκευάσθησαν, καὶ ἔπειτα εὖ ᾤκησαν, ὑπὸ ξυμβάσεων πραγμάτων μεγάλων καὶ κατὰ πόλεμον καὶ κατὰ τὰς ἄλλας πράξεις γενομένων, ὅταν ἐν τοιούτοις καιροῖς ἀνὴρ καλός τε καὶ ἀγαθὸς ἐγγένηται μεγάλην δύναμιν ἔχων.

Τὸ δ' ἔμπροσθεν αὐτὰ προθυμεῖσθαι μὲν χρὴ καὶ C ἀνάγκη, διανοεῖσθαι μέντοι αὐτὰ οἷα λέγω, καὶ μὴ ἀνοηταίνειν οἰομένους τι ἑτοίμως διαπράξεσθαι.[1] Εὐτύχει.

[1] διαπράξεσθαι H. Richards: διαπράξασθαι mss., Zur.

---

[1] Cf. *Laws* 962 B, *Ep.* vii. 326 C, D.

EPISTLE XI

existence of some authority[1] in the State which supervises the daily life both of slaves and freemen, to see that it is both temperate and manly. And this condition might be secured if you already possess men who are worthy of such authority. If, however, you require someone to train them, you do not, in my opinion, possess either the trainer or the pupils to be trained; so it only remains for you to pray to the gods.[2] For, in truth, the earlier States also were mostly organized in this way; and they came to have a good constitution at a later date, as a result of their being confronted with grave troubles, either through war or other difficulties, whenever there arose in their midst at such a crisis a man of noble character in possession of great power.

So it is both right and necessary that you should at first be eager for these results, but also that you should conceive of them in the way I suggest, and not be so foolish as to suppose that you will readily accomplish anything. Good-fortune attend you!

[2] For prayer in cases where "with men it is impossible" cf. *Ep.* viii. 352 E, *Rep.* 540 D.

EPISTLE XII

Prefatory Note.—Like the ninth, this letter, addressed to Archytas, is certainly spurious. In it Plato professes to have received certain treatises; he compliments their author; and he informs Archytas that he is sending him certain unfinished treatises of his own.

The treatises here mentioned as coming from Archytas are said by Diogenes Laertius (viii. 80) to have been those of Ocellos of Lucania, a Pythagorean. But, as the writings which bear his name are undoubtedly forgeries (perhaps of the first century B.C.), this letter is probably written by the same forger with the object of stamping his effusions with the authority of Plato.

Another indication that these letters to Archytas do not come from Plato is the fact that they spell the name "Archytas," whereas Plato elsewhere spells it "Archytes."

IB

Πλάτων Ἀρχύτᾳ Ταραντίνῳ εὖ πράττειν.

Τὰ μὲν παρὰ σοῦ ἐλθόνθ' ὑπομνήματα θαυμαστῶς ὡς ἄσμενοί τε ἐλάβομεν καὶ τοῦ γράψαντος αὐτὰ ἠγάσθημεν ὡς ἔνι μάλιστα, καὶ ἔδοξεν ἡμῖν εἶναι ὁ ἀνὴρ ἄξιος ἐκείνων τῶν πάλαι προγόνων· λέγονται γὰρ δὴ οἱ ἄνδρες οὗτοι Μύριοι εἶναι, οὗτοι δ' ἦσαν τῶν ἐπὶ Λαομέδοντος ἐξαναστάντων Τρώων, ἄνδρες ἀγαθοί, ὡς ὁ παραδεδομένος μῦθος δηλοῖ. τὰ δὲ παρ' ἐμοὶ ὑπομνήματα περὶ ὧν ἐπέστειλας, ἱκανῶς μὲν οὔπω ἔχει, ὡς δέ ποτε τυγχάνει ἔχοντα, ἀπέσταλκά σοι· περὶ δὲ τῆς φυλακῆς ἀμφότεροι συμφωνοῦμεν, ὥστ' οὐδὲν δεῖ παρακελεύεσθαι.

(Ἀντιλέγεται ὡς οὐ Πλάτωνος.)

---

[1] Father of Priam, king of Troy. Nothing is told us elsewhere of this Trojan colony in Italy; so we may regard it as an invention of the writer.

EPISTLE XII

PLATO TO ARCHYTAS OF TARENTUM WISHES WELL-
DOING

We have been wonderfully pleased at receiving the treatises which have come from you and felt the utmost possible admiration for their author; indeed we judged the man to be worthy of those ancient ancestors of his. For in truth these men are said to be Myrians; and they were amongst those Trojans who emigrated in the reign of Laomedon [1] —valiant men, as the traditional story declares. As to those treatises of mine about which you wrote, they are not as yet completed, but I have sent them to you just in the state in which they happen to be; as concerns their preservation [2] we are both in accord, so that there is no need to give directions.

(Denied to be Plato's.)

[2] *Cf. Ep.* ii. 314 A, xiii. 363 E.

EPISTLE XIII

Prefatory Note.—This letter to Dionysius is of a private character and—if genuine—would seem to have been written shortly after Plato's return from his second visit to Sicily, *i.e.* in 366-365 B.C. In the letter Plato alludes to the friendly terms on which he had been with the tyrant and expresses a hope that this friendship may continue to the mutual benefit of both. Then he mentions certain treatises he is sending to Dionysius, and he commends to him a scientist called Helicon. From this he passes on to certain purchases Dionysius had asked him to make, and various presents he had bought; and he gives a summary of the domestic expenses he has to meet. This leads on to the subject of the tyrant's financial standing at Athens and the difficulty of obtaining loans; and advice is given as to the expediency of prompt repayment. Next there is a reference to Dion, with a hint of certain proposals affecting him which Dionysius had made. And the epistle ends with a number of disjointed comments on various matters of private interest and on personal acquaintances.

It is obvious at a first reading that the Plato disclosed to us in this letter is quite a different Plato from that presented to us in the Dialogues or in the seventh epistle. He is no longer the sublime philo-

EPISTLE XIII

sopher, contemptuous of all that is sold in the merchant's mart, but a business man engrossed in the financial operations of the money-market, buying and borrowing and advising the Sicilian tyrant how best to maintain his hold on the bankers of Athens. This contrast between all the personal traits we had hitherto imputed to Plato and the portrait which he gives of himself in this letter inevitably suggests grave doubts in our minds as to its authenticity. And these doubts are confirmed by further reflexion. For, apart from the general improbability of its tone, the letter contains several statements or references which are highly suspicious. Thus the emphasis laid on the form of greeting with which the letter begins as a " sign " or " token " of authenticity (360 A) sounds too much like the style of the third letter (*ad init.*) to be easily accepted as Platonic. So too with the " sign " on which so much stress is laid in 363 B as a mark of distinction between " serious " and " non-serious " letters : it looks much more like an invention of a later theologian than a genuine device of the author of the *Timaeus*. And in any case such precautions would be superfluous in regard to this letter seeing that it was to be conveyed by a messenger as trusty and well-informed as Leptines. The concluding sentence of the letter is also suggestive of a provident forger ; for by making " Plato " write " preserve this letter " he tries to forestall any possible objection against it on the ground that such a letter was likely to have been destroyed as soon as the business it deals with had been completed.

Among other stones of stumbling which the reader encounters are the following. The story told at the beginning (360 A-B) which exhibits the vanity of

PLATO

Plato and his appetite for flattery. The vague description of certain treatises (360 B) as "Pythagoreans and Divisions," which some scholars wish to identify with the *Timaeus* and the *Sophist* and *Politicus*, although it is more than doubtful whether these works had been written at the supposed date of this letter (366-365 B.C.). The mention of Isocrates and Polyxenus as amongst the teachers of Helicon, in the commendation of that scientist. The implication that Plato had been ill at Syracuse and nursed by Dionysius's wife (361 A). The way in which Plato claims to dispose of other people's money as if it were his own (361 C). The fact that several nieces died at the same time; and the strange incident of Plato refusing to "wear a crown" (what sort of crown?) as an indication of the time (361 D). The niggardly spirit exhibited in connexion with the nieces' "portions," together with the callous brevity of the reference to the prospective death of his mother. The unlikelihood of Plato, at the age of sixty-two, having a mother alive at all. The inherent improbability of the position here ascribed to Plato as the agent in full charge of the Athenian business of the Syracusan court, acting as a sort of Sicilian consul, and also of the description given of the financial standing of that court (362 A ff.). The suggestion that loans should be promptly repaid not because honesty requires this but because it makes loans easy to get and profits the borrower (362 D, 363 D). The readiness of Plato to act in collusion with Dionysius in the matter of robbing Dion of his wife and handing her over to a favourite of the tyrant—no touch of John the Baptist here! (362 E). The way in which the name of Cebes is dragged in, for the

EPISTLE XIII

sake of bringing up an allusion to the *Phaedo* (363 A). The complacent reference to the praise of "the ambassadors," and the device by which "what Philaedes said" is conveniently omitted (363 B-C). The number of details about persons and things which are quite unverifiable; and the studiously vague indications of date by means of the word "then" or "at that time."

When we consider all these points we seem driven to the conclusion that this letter is spurious, the work of an artist in epistolary fabrication who specialized in the accumulation of private and personal details with the object of producing an impression of verisimilitude, gratifying the appetite of the public for biographical "gossip," and imposing (as he has imposed) on a credulous posterity.

ΙΓ

Πλάτων Διονυσίῳ τυράννῳ Συρακουσῶν εὖ
πράττειν.

Ἀρχή σοι τῆς ἐπιστολῆς ἔστω καὶ ἅμα ξύμβολον ὅτι παρ' ἐμοῦ ἐστί. τοὺς Λοκροὺς ποθ' ἑστιῶν
νεανίσκους, πόρρω κατακείμενος ἀπ' ἐμοῦ, ἀνέστης
παρ' ἐμὲ καὶ φιλοφρονούμενος εἶπες εὖ τι ῥῆμα
ἔχον, ὡς ἐμοί τε ἐδόκει καὶ τῷ παρακατακειμένῳ·
ἦν δ' οὗτος τῶν καλῶν τις· ὃς τότε εἶπεν Ἦ που
πολλά, ὦ Διονύσιε, εἰς σοφίαν ὠφελεῖ ὑπὸ Πλάτωνος. σὺ δ' εἶπες Καὶ εἰς ἄλλα πολλά, ἐπεὶ κα
ἀπ' αὐτῆς τῆς μεταπέμψεως, ὅτι μετεπεμψάμη
αὐτόν, δι' αὐτὸ τοῦτο εὐθὺς ὠφελήθην. τοῦτ' οὖ
διασωστέον, ὅπως ἂν αὐξάνηται ἀεὶ ἡμῖν ἡ ἀπ
ἀλλήλων ὠφέλεια. καὶ ἐγὼ νῦν τοῦτ' αὐτὸ παρα
σκευάζων τῶν τε Πυθαγορείων πέμπω σοι καὶ τῶ
διαιρέσεων, καὶ ἄνδρα, ὥσπερ ἐδόκει ἡμῖν τότε, ᾧ
γε σὺ καὶ Ἀρχύτης, εἴπερ ἥκει παρὰ σὲ Ἀρχύτης
χρῆσθαι δύναισθ' ἄν. ἔστι δὲ ὄνομα μὲν Ἑλίκων
τὸ δὲ γένος ἐκ Κυζίκου, μαθητὴς δὲ Εὐδόξου κα
περὶ πάντα τὰ ἐκείνου πάνυ χαριέντως ἔχων· ἔτ

---

[1] For the significance of the greeting "well-doing" se
Ep. iii. *ad init.*; *cf.* 363 B *infra*.
[2] A famous astronomer.

EPISTLE XIII

PLATO TO DIONYSIUS, TYRANT OF SYRACUSE, WISHES WELL-DOING

LET this greeting not only commence my letter but serve at the same time as a token that it is from me.[1] Once when you were feasting the Locrian youths and were seated at a distance from me, you got up and came over to me and in a friendly spirit made some remark which I thought excellent, as also did my neighbour at the table, who was one of the beautiful youths. And he then said—"No doubt, Dionysius, you find Plato of great benefit as regards philosophy!" And you replied—"Yes, and in regard to much else; since from the very moment of my inviting him I derived benefit at once from the very fact that I had invited him." This tone, then, should be carefully preserved, in order that the mutual benefit we derive from one another may always go on increasing. So by way of helping towards this end I am now sending you some of the Pythagorean works and of the "Divisions," and also, as we arranged at that time, a man of whom you and Archytas—if Archytas has come to your court—may be able to make use. His name is Helicon, he is a native of Cyzicus, and he is a pupil of Eudoxus[2] and exceedingly well versed in all his doctrine.

360
δὲ καὶ τῶν Ἰσοκράτους μαθητῶν τῳ ξυγγέγονε
καὶ Πολυξένῳ τῶν Βρύσωνός τινι ἑταίρων. ὃ δὲ
σπάνιον ἐπὶ τούτοις, οὔτε ἄχαρίς ἐστιν ἐντυχεῖν
οὔτε κακοήθει ἔοικεν, ἀλλὰ μᾶλλον ἐλαφρὸς καὶ
D εὐήθης δόξειεν ἂν εἶναι. δεδιὼς δὲ λέγω ταῦτα,
ὅτι ὑπὲρ ἀνθρώπου δόξαν ἀποφαίνομαι, οὐ φαύλου
ζῴου ἀλλ' εὐμεταβόλου, πλὴν πάνυ ὀλίγων τινῶν
καὶ εἰς ὀλίγα· ἐπεὶ καὶ περὶ τούτου φοβούμενος καὶ
ἀπιστῶν ἐσκόπουν αὐτός τε ἐντυγχάνων καὶ ἐπυν-
θανόμην τῶν πολιτῶν αὐτοῦ, καὶ οὐδεὶς οὐδὲν
φλαῦρον ἔλεγε τὸν ἄνδρα. σκόπει δὲ καὶ αὐτὸς
καὶ εὐλαβοῦ. μάλιστα μὲν οὖν, ἂν καὶ ὁπωστιοῦν
E σχολάζῃς, μάνθανε παρ' αὐτοῦ καὶ τἆλλα φιλο-
σόφει· εἰ δὲ μή, ἐκδίδαξαί τινα, ἵνα κατὰ σχολὴν
μανθάνων βελτίων γίγνῃ καὶ εὐδοξῇς, ὅπως τὸ δι'
ἐμὲ ὠφελεῖσθαί σε μὴ ἀνιῇ. καὶ ταῦτα μὲν δὴ
ταύτῃ.

361 Περὶ δὲ ὧν ἐπέστελλές μοι ἀποπέμπειν σοι, τὸν
μὲν Ἀπόλλω ἐποιησάμην τε καὶ ἄγει σοι Λεπτίνης,
νέου καὶ ἀγαθοῦ δημιουργοῦ· ὄνομα δ' ἔστιν αὐτῷ
Λεωχάρης. ἕτερον δὲ παρ' αὐτῷ ἔργον ἦν πάνυ
κομψόν, ὡς ἐδόκει· ἐπριάμην οὖν αὐτὸ βουλόμενός
σου τῇ γυναικὶ δοῦναι, ὅτι μου ἐπεμελεῖτο καὶ
ὑγιαίνοντος καὶ ἀσθενοῦντος ἀξίως ἐμοῦ τε καὶ
σοῦ· δὸς οὖν αὐτῇ, ἂν μή τι σοὶ ἄλλο δόξῃ. πέμπω

---

[1] *Cf. Ep.* ii. 310 c. Bryson "the Sophist" was a mathematician who claimed, it is said, to have "squared the circle" (*cf.* Aristot. *An. Post.* i. 9, *Rhet.* iii. 2).

[2] *Cf. Ep.* vi. 323 B, vii. 335 E.

[3] A Pythagorean of this name is said to have murdered Callippus at Rhegium.

EPISTLE XIII

Moreover, he has associated with one of the pupils of Isocrates and with Polyxenus,[1] one of Bryson's companions; and, what is rare in these cases, he is not without charm of address nor is he of a churlish disposition; rather he would seem to be gay and good-tempered. This, however, I say with trepidation, since I am uttering an opinion about a man, and man though not a worthless is an inconstant creature,[2] save in very few instances and in few respects. For even in this man's case my fears and suspicions were such that, when I met him, I observed him carefully myself and I made inquiry also from his fellow-citizens, and no one had anything bad to say of the man. But do you yourself also keep him under observation and be cautious. It were best, then, if you have any leisure at all, to take lessons from him in addition to your other studies in philosophy; but if not, get someone else thoroughly taught so that you may learn from him when you have leisure, and thereby make progress and gain glory,—that so the benefit you gain from me may still continue. So much, then, for this subject.

As regards the things you wrote to me to send you, I have had the Apollo made and Leptines[3] is bringing it to you. It is by a young and good craftsman named Leochares.[4] He had at his shop another piece which was, as I thought, very artistic; so I bought it with the intention of presenting it to your wife,[5] because she tended me both in health and sickness in a manner which did credit both to you and to me. So will you give it to her, unless you

[4] A sculptor of some eminence, pupil of Scopas.
[5] Sophrosyne ("Prudence"), daughter of Dionysius the Elder and niece of Dion.

361 δὲ καὶ οἴνου γλυκέος δώδεκα σταμνία τοῖς παισ.
B καὶ μέλιτος δύο. ἰσχάδων δὲ ὕστερον ἤλθομεν τῆς
ἀποθέσεως, τὰ δὲ μύρτα ἀποτεθέντα κατεσάπη·
ἀλλ' αὖθις βέλτιον ἐπιμελησόμεθα. περὶ δὲ φυτῶν
Λεπτίνης σοι ἐρεῖ.

Ἀργύριον δ' εἰς ταῦτα ἕνεκά τε τούτων καὶ
εἰσφορῶν τινῶν εἰς τὴν πόλιν ἔλαβον παρὰ Λεπτίνου, λέγων ἅ μοι ἐδόκει εὐσχημονέστατα ἡμῖν
εἶναι καὶ ἀληθῆ λέγειν, ὅτι ἡμέτερον εἴη ὃ εἰς
τὴν ναῦν ἀνηλώσαμεν τὴν Λευκαδίαν, σχεδὸν ἑκ-
C καίδεκα μναῖ. τοῦτ' οὖν ἔλαβον, καὶ λαβὼν αὐτὸς
τε ἐχρησάμην καὶ ὑμῖν ταῦτα ἀπέπεμψα. τὸ δὴ
μετὰ τοῦτο περὶ χρημάτων ἄκουε ὥς σοι ἔχει,
περί τε τὰ σὰ τὰ Ἀθήνησι καὶ περὶ τὰ ἐμά. ἐγὼ
τοῖς σοῖς χρήμασιν, ὥσπερ τότε σοι ἔλεγον,
χρήσομαι καθάπερ τοῖς τῶν ἄλλων ἐπιτηδείων,
χρῶμαι δὲ ὡς ἂν δύνωμαι ὀλιγίστοις, ὅσα ἀναγκαῖα
ἢ δίκαια ἢ εὐσχήμονα ἐμοί τε δοκεῖ καὶ παρ' οὗ
ἂν λαμβάνω. ἐμοὶ δὴ τοιοῦτον νῦν ξυμβέβηκεν.
εἰσί μοι ἀδελφιδῶν θυγατέρες τῶν ἀποθανουσῶν
D τότε ὅτ' ἐγὼ οὐκ ἐστεφανούμην, σὺ δ' ἐκέλευες,
τέτταρες, ἡ μὲν νῦν ἐπίγαμος, ἡ δὲ ὀκταέτις, ἡ δὲ
σμικρὸν πρὸς τρισὶν ἔτεσιν, ἡ δὲ οὔπω ἐνιαυσία.
ταύτας ἐκδοτέον ἐμοί ἐστι καὶ τοῖς ἐμοῖς ἐπιτη-
δείοις, αἷς ἂν ἐγὼ ἐπιβιῶ· αἷς δ' ἂν μή, χαιρόντων.
καὶ ὧν ἂν γένωνται οἱ πατέρες αὐτῶν ἐμοῦ πλου-
σιώτεροι, οὐκ ἐκδοτέον· τὰ δὲ νῦν ἐγὼ αὐτῶν
εὐπορώτατος,[1] καὶ τὰς μητέρας δὲ αὐτῶν ἐγὼ
E ἐξέδωκα καὶ μετ' ἄλλων καὶ μετὰ Δίωνος. ἡ μὲν

[1] εὐπορώτατος MSS.: εὐπορώτερος Zur.

EPISTLE XIII

prefer to do otherwise. I am also sending twelve jars of sweet wine for the children and two of honey. We arrived too late for the storing of the figs, and the myrtle-berries that were stored have rotted; but in future we shall take better care of them. About the plants Leptines will tell you.

The money to meet these expenses—I mean for the purchases mentioned and for certain State taxes—I obtained from Leptines, telling him what I thought it best became us to tell him, it being also true,—that the sum of about sixteen minas which we spent on the Leucadian ship belonged to us; this, then, was the sum I obtained, and on obtaining it I used it myself and sent off these purchases to you.

Next, let me tell you what your position is in regard to money, both what you have at Athens and my own. I shall make use of your money, as I told you previously, just as I do that of all my other friends; I use as little as I possibly can, only just so much as I and the man I get it from agree to be necessary or right or fitting. Now this is how I am situated at present. I have in my charge four daughters of those nieces of mine who died at the time when you bade me to wear a crown, and I refused; and of these one is of marriageable age, one eight years old, one a little over three years, and the fourth not yet a year old. To these girls I and my friends must give portions—to all of them, that is, whom I live to see married; as to the rest, they must look to themselves. Nor should I give portions to any whose fathers may get to be richer than I; though at present I am the wealthiest of them, and it was I who, with the help of Dion and others, gave their mothers their portions. Now the eldest one is

PLATO

οὖν Σπευσίππῳ γαμεῖται, ἀδελφῆς οὖσα αὐτῷ θυγάτηρ· δεῖ δὴ ταύτῃ οὐδὲν πλέον ἢ τριάκοντα μνῶν· μέτριαι γὰρ αὗται ἡμῖν προῖκες. ἔτι δὲ ἐὰν ἡ μήτηρ τελευτήσῃ ἡ ἐμή, οὐδὲν αὖ πλείονος ἢ δέκα μνῶν δέοι ἂν εἰς τὴν οἰκοδομίαν τοῦ τάφου. καὶ περὶ ταῦτα τὰ μὲν ἐμὰ ἀναγκαῖα σχεδόν τι ἐν τῷ νῦν ταῦτά ἐστιν· ἐὰν δέ τι ἄλλο γίγνηται ἴδιον ἢ δημόσιον ἀνάλωμα διὰ τὴν παρὰ σὲ ἄφιξιν, ὥσπερ τότε ἔλεγον δεῖ ποιεῖν, ἐμὲ μὲν διαμάχεσθαι ὅπως ὡς ὀλίγιστον γένηται τὸ ἀνάλωμα, ὃ δ' ἂν μὴ δύνωμαι, σὴν εἶναι τὴν δαπάνην.

Τὸ δὴ μετὰ ταῦτα λέγω περὶ τῶν σῶν αὖ χρημάτων τῶν Ἀθήνησι τῆς ἀναλώσεως, ὅτι πρῶτον μὲν ἐάν τι δέῃ ἐμὲ ἀναλίσκειν εἰς χορηγίαν ἢ τι τοιοῦτον, οὐκ ἔστι σοι ξένος οὐδεὶς ὅστις δώσει, ὡς ᾠόμεθα, ἔπειτα[1] καὶ ἄν τι σοὶ αὐτῷ διαφέρῃ μέγα, ὥστε ἀναλωθὲν μὲν ἤδη ὀνῆσαι, μὴ ἀναλωθὲν δὲ ἀλλ' ἐγχρονισθέν, ἕως ἄν τις παρὰ σοῦ ἔλθῃ, βλάψαι, πρὸς τῷ χαλεπῷ τὸ τοιοῦτόν σοί ἐστι καὶ αἰσχρόν. ἐγὼ γὰρ δὴ ταῦτά γε ἐξήτασα, παρ' Ἀνδρομήδῃ τὸν Αἰγινήτην πέμψας Ἔραστον, παρ' οὗ ἐκέλευες τοῦ ὑμετέρου ξένου, εἴ τι δεοίμην, λαμβάνειν, βουλόμενος καὶ ἄλλα μείζονα ἃ ἐπέστελλες πέμπειν. ὁ δὲ εἶπεν εἰκότα καὶ ἀνθρώπινα, ὅτι καὶ πρότερον ἀναλώσας τῷ πατρί σου μόλις κομίσαιτο, καὶ νῦν σμικρὰ μὲν δοίη ἄν, πλείω δὲ οὔ. οὕτω δὴ παρὰ Λεπτίνου ἔλαβον· καὶ τοῦτό γε ἄξιον ἐπαινέσαι Λεπτίνην, οὐχ ὅτι ἔδωκεν, ἀλλ'

---

[1] ἔπειτα Schneider : ἐπεὶ MSS., Zur.

---

[1] Cf. Ep. ii. 314 E.
[2] Cf. Ep. vi. 322 D (if this is the same person).

EPISTLE XIII

marrying Speusippus,[1] she being his sister's daughter. So for her I require no more than thirty minas, that being for us a reasonable dowry. Moreover, in case my own mother should die, no more than ten minas would be required for the building of her tomb. For such purposes, then, these are pretty well all my necessary requirements at the present time. And should any further expense, private or public, be incurred owing to my visit to your court, we must do as I said before : I must strive hard to keep the expense as low as possible, and if ever that is beyond my power, the charge must fall upon you.

In the next place, as regards the spending of your own money at Athens, I have to tell you, first of all, that, contrary to what we supposed, you have not a single friend who will advance money in case I am required to spend something on furnishing a chorus or the like ; and further, if you yourself have some urgent affair on hand in which prompt expenditure is to your advantage, whereas it is to your disadvantage to have the expenditure deferred until the arrival of a messenger from you, such a state of affairs is not only awkward but reflects also on your honour. And in fact I discovered this myself when I sent Erastus [2] to Andromedes the Aeginetan—from whom, as a friend of yours, you told me to borrow what I needed ; as I wished to send you also some other valuable items which you had written for. He replied— naturally enough, as any man might—that when, on a previous occasion, he had advanced money on your father's account he had had difficulty in recovering it, and that he would now loan a small amount but no more. That was how I came to borrow from Leptines ; and for this Leptines is deserving of

PLATO

ὅτι προθύμως, καὶ τὰ ἄλλα περὶ σὲ καὶ λέγων καὶ πράττων, ὅ τι οἷός τ' ἦν ἐπιτήδειος, φανερὸς ἦν. χρὴ γὰρ δὴ καὶ τὰ τοιαῦτα καὶ τἀναντία τούτων ἐμὲ ἀπαγγέλλειν, ὁποῖός τις ἂν ἕκαστος ἐμοὶ φαίνηται περὶ σέ.

Τὸ δ' οὖν περὶ τῶν χρημάτων ἐγώ σοι παρρησιάσομαι· δίκαιον γάρ, καὶ ἅμα ἐμπείρως ἔχων τῶν παρὰ σοὶ λέγοιμ' ἄν. οἱ προσαγγέλλοντες ἑκάστοτέ σοι, ὅ τι ἂν οἴωνται ἀνάλωμα εἰσαγγέλλειν, οὐκ ἐθέλουσι προσαγγέλλειν, ὡς δὴ ἀπεχθησόμενοι. ἔθιζε οὖν αὐτοὺς καὶ ἀνάγκαζε φράζειν καὶ ταῦτα καὶ τὰ ἄλλα· σὲ γὰρ δεῖ εἰδέναι τε τὰ πάντα κατὰ δύναμιν καὶ κριτὴν εἶναι καὶ μὴ φεύγειν τὸ εἰδέναι. πάντων γὰρ ἄριστόν σοι ἔσται πρὸς τὴν ἀρχήν· τὰ γὰρ ἀναλώματα ὀρθῶς ἀναλισκόμενα καὶ ὀρθῶς ἀποδιδόμενα πρός τε τἆλλα καὶ πρὸς αὐτὴν τὴν τῶν χρημάτων κτῆσιν καὶ σὺ δὴ φὴς ἀγαθὸν εἶναι καὶ φήσεις. μὴ οὖν σε διαβαλλόντων πρὸς τοὺς ἀνθρώπους οἱ κήδεσθαί σου φάσκοντες· τοῦτο γὰρ οὔτε ἀγαθὸν οὔτε καλὸν πρὸς δόξαν σοι, δοκεῖν δυσσύμβολον[1] εἶναι.

Τὰ μετὰ ταῦτα περὶ Δίωνος λέγοιμ' ἄν. τὰ μὲν οὖν ἄλλ' οὔπω ἔχω λέγειν, πρὶν ἂν παρὰ σοῦ ἔλθωσιν αἱ ἐπιστολαί, ὥσπερ ἔφης· περὶ μέντοι ἐκείνων ὧν οὐκ εἴας μεμνῆσθαι πρὸς αὐτόν, οὔτε ἐμνήσθην οὔτε διελέχθην, ἐξεπειρώμην δὲ εἴτε χαλεπῶς εἴτε ῥᾳδίως οἴσει γιγνομένων, καί μοι ἐδόκει οὐκ ἠρέμα ἂν ἄχθεσθαι εἰ γίγνοιτο. τὰ δὲ

---

[1] δοκεῖν δυσσύμβολον Schneider: δοκεῖ· ξύμβολον MSS., Zur.

[1] This may be a reference to Dionysius's plan for giving Dion's wife Arete to a favourite of his own (*cf.* Plutarch, *Dion* 21).

EPISTLE XIII

praise, not that he gave it, but that he did so readily, and plainly showed his friendship and its quality in all else that he did or said regarding you. For it is surely right that I should report such actions, as well as the opposite kind, to show what I believe to be each man's attitude towards you.

However, I will tell you candidly the position with regard to money matters; for it is right to do so, and, moreover, I shall be speaking from experience of your court. The agents who bring you the reports every time are unwilling to report anything which they think entails an expense, as being likely to bring them odium. Do you therefore accustom them and compel them to declare these matters as well as the rest; for it is right that you should know the whole state of affairs so far as you can and act as the judge, and not avoid this knowledge. For such a course will best serve to enhance your authority. For expenditure that is rightly laid out and rightly paid back is a good thing—as you yourself maintain and will maintain—not only for other purposes but also for the acquisition of money itself. Therefore, do not let those who profess to be devoted to you slander you before the world; for to have the reputation of being ill to deal with is neither good for your reputation nor honourable.

In the next place I shall speak about Dion. Other matters I cannot speak of as yet, until the letters from you arrive, as you said; with regard, however, to those matters which you forbade me to mention to him,[1] I neither mentioned nor discussed them, but I did try to discover whether he would take their occurrence hardly or calmly, and it seemed to me that if they occurred it would cause him no small

PLATO

362 ἄλλα περί σε καὶ λόγῳ καὶ ἔργῳ μέτριός μοι δοκεῖ εἶναι Δίων.

363 Κρατίνῳ τῷ Τιμοθέου μὲν ἀδελφῷ, ἐμῷ δ' ἑταίρῳ, θώρακα δωρησώμεθα ὁπλιτικὸν τῶν μαλακῶν[1] τῶν πεζῶν, καὶ ταῖς Κέβητος θυγατράσι χιτώνια τρία ἑπταπήχη, μὴ τῶν πολυτελῶν τῶν Ἀμοργίνων, ἀλλὰ τῶν Σικελικῶν τῶν λινῶν. ἐπιεικῶς δὲ γιγνώσκεις τοὔνομα Κέβητος· γεγραμμένος γάρ ἐστιν ἐν τοῖς Σωκρατείοις λόγοις μετὰ Σιμμίου Σωκράτει διαλεγόμενος ἐν τῷ περὶ ψυχῆς λόγῳ, ἀνὴρ πᾶσιν ἡμῖν οἰκεῖός τε καὶ εὔνους.

B Περὶ δὲ δὴ τοῦ ξυμβόλου τοῦ περὶ τὰς ἐπιστολάς, ὅσας τε ἂν ἐπιστέλλω σπουδῇ καὶ ὅσας ἂν μή, οἶμαι μέν σε μεμνῆσθαι, ὅμως δ' ἐννόει καὶ πάνυ πρόσεχε τὸν νοῦν· πολλοὶ γὰρ οἱ κελεύοντες γράφειν, οὓς οὐ ῥᾴδιον φανερῶς διωθεῖσθαι. τῆς μὲν γὰρ σπουδαίας ἐπιστολῆς θεὸς ἄρχει, θεοὶ δὲ τῆς ἧττον.

Οἱ πρέσβεις καὶ ἐδέοντο ἐπιστέλλειν σοι καὶ εἰκός· πάνυ γὰρ προθύμως σὲ πανταχοῦ καὶ ἐμὲ ἐγκωμιάζουσι, καὶ οὐχ ἥκιστα Φίλαγρος, ὃς τότε τὴν χεῖρα ἠσθένει. καὶ Φιλαΐδης [ὁ][2] παρὰ
C βασιλέως ἥκων τοῦ μεγάλου ἔλεγε περὶ σοῦ· εἰ δὲ μὴ πάνυ μακρᾶς ἐπιστολῆς ἦν, ἔγραψα ἂν ἃ ἔλεγε, νῦν δὲ Λεπτίνου πυνθάνου.

Ἂν τὸν θώρακα ἢ ἄλλο τι ὧν ἐπιστέλλω πέμπῃς, ἂν μὲν αὐτός τῳ βούλῃ, εἰ δὲ μή, Τηρίλλῳ δός· ἔστι δὲ τῶν ἀεὶ πλεόντων, ἡμέτερος ἐπιτήδειος καὶ τὰ ἄλλα καὶ περὶ φιλοσοφίαν χαρίεις. Τείσωνος δ'

[1] μαλακῶν mss.: μάλα καλῶν mss. corr., Zur.
[2] [ὁ] om. best mss.

---

[1] *i.e.* the *Phaedo*.   [2] *Cf.* 360 A *supra*.

EPISTLE XIII

vexation. As to all else Dion's attitude towards you seems to me to be reasonable both in word and deed.

To Cratinus the brother of Timotheus, and my own companion, let us present a hoplite's corslet, one of the soft kind for foot-soldiers; and to the daughters of Cebes three tunics of seven cubits, not made of the costly Amorgos stuff but of the Sicilian linen. The name of Cebes you probably know; for he is mentioned in writing in the Socratic discourses as conversing with Socrates, in company with Simmias, in the discourse concerning the Soul,[1] he being an intimate and kindly friend of us all.

Concerning the sign [2] which indicates which of my letters are seriously written and which not, I suppose that you remember it, but none the less bear it in mind and pay the utmost attention; for there are many bidding me to write, whom it is not easy to repulse openly. "God," then, is at the head of the serious letter, but "gods" of the less serious.

The ambassadors requested me to write to you, and naturally so; for they are everywhere lauding both you and me with the utmost zeal; and not least Philagrus, who was then suffering with his hand. Philaides also, on his arrival from the Great King, was talking about you; and if it had not required a very long letter I would have told you in writing what he said; but as it is, ask Leptines to tell you.

If you are sending the corslet or any of the other things I have written about, in case you have anyone you prefer yourself, give it to him, but if not, give it to Terillus; he is one of those who are constantly making the voyage, and he is a friend of ours who is skilled in philosophy as well as in other things. He

ἔστι κηδεστής, ὃς τότε ὅθ' ἡμεῖς ἀπεπλέομεν
ἐπολιανόμει.

Ἔρρωσο καὶ φιλοσόφει καὶ τοὺς ἄλλους προ-
τρέπου τοὺς νεωτέρους, καὶ τοὺς συσφαιριστὰς
ἀσπάζου ὑπὲρ ἐμοῦ, καὶ πρόσταττε τοῖς τε ἄλλοις
καὶ Ἀριστοκρίτῳ, ἐάν τις παρ' ἐμοῦ λόγος ἢ
ἐπιστολὴ ἴῃ παρὰ σέ, ἐπιμελεῖσθαι ὅπως ὡς τάχιστα
σὺ αἴσθῃ, καὶ ὑπομιμνῄσκειν σε ἵνα ἐπιμελῇ τῶν
ἐπισταλέντων. καὶ νῦν Λεπτίνῃ τῆς ἀποδόσεως
τοῦ ἀργυρίου μὴ ἀμελήσῃς, ἀλλ' ὡς τάχιστα ἀπό-
δος, ἵνα καὶ οἱ ἄλλοι πρὸς τοῦτον ὁρῶντες προ-
θυμότεροι ὦσιν ἡμῖν ὑπηρετεῖν.

Ἰατροκλῆς ὁ μετὰ Μυρωνίδου τότε ἐλεύθερος
ἀφεθεὶς ὑπ' ἐμοῦ πλεῖ νῦν μετὰ τῶν πεμπομένων
παρ' ἐμοῦ· ἔμμισθον οὖν που αὐτὸν κατάστησον ὡς
ὄντα σοι εὔνουν, καὶ ἄν τι βούλῃ, αὐτῷ χρῶ. καὶ
τὴν ἐπιστολὴν ἢ αὐτὴν ἢ [εἰ] ὑπόμνημα αὐτῆς
σῷζέ τε[1] καὶ ὁ αὐτὸς[2] ἴσθι.

---

[1] σῷζέ τε Schneider: σώζεται mss., Zur.
[2] ὁ αὐτὸς one ms.: αὐτὸς Zur.

---

[1] The Greek word "fellow-spherists" suggests a play on

EPISTLE XIII

is also a son-in-law of Teison who was city-steward at the time when we sailed away.

Keep well and study philosophy and exhort thereto all the other young men; and greet for me your comrades at the game of ball [1]; and charge Aristocritus, as well as the rest, that if any message or letter from me should come to your palace, he must take care that you are informed of it as soon as possible; and bid him remind you not to neglect the contents of my letters. So too now, do not neglect to repay Leptines his money, but pay it back as promptly as possible, in order that the others also, seeing how you deal with him, may be the more ready to assist us.

Iatrocles, the man whom I released on that occasion, along with Myronides, is now sailing with the things that I am sending: I ask you, then, to give him some paid post, as he is well-disposed towards you, and employ him for whatever you wish. Preserve also this letter, either itself or a précis of it, and continue as you are.

the double meaning of "sphere" as "ball" and "globe" (*cf. Ep.* ii. 312 D); so that the real meaning may be "fellow-astronomers."

INDEX

Abdomen, 191
Above (oppd. to Below), 157 ff.
Academy, 386 f., 415 n.
Acid, 169, 195
Acropolis, 275, 277, 289, 293, 495, 553 n.
Adamant, 147; (Diamond), 397
Address (Epistolary form of), 425 (cf. 615)
Aegina, 447
Aeginetan, 621
Aegospotami, 365 n.
Aegyptus, 369
Aeschines, 488 n.
Aether, 145
Affections (bodily), 163 ff.
Agamemnon, 405
Air (element), 59 f., 113 ff., 135 ff., 139 ff., 161, 207 ff., 229
Alcmaeon, 14
Alexander, of Pherae, 602 n.
Alkalies, 169
Allotment, 23, 279 f., 289, 299
Amasis, 31
Amazons, 349
Amorgos, 625
Ampheres, 283
Amynander, 31
Anarchy, 531
Anaxagoras, 5, 405
Ancient (traditions), 27 ff., 33 ff., 269, 511
Andromedes, 621
Animals, 251 f.
Antalcidas (Peace of), 331, 363 n.
Antiphon, 339
Anytus, 480 n.
Apaturia, feast of, 28 n., 29
Apollo, 617 (cf. 425)
Archedemus, 403, 413 f., 437, 523, 557

Archippus, 592 f.
Archytas, 387 f., 402 n., 465, **468**, 521 ff., 559, 591 ff., 607 ff., 615
Arete, 622 n.
Arginusae, 361 n.
Argives, 349, 365 f.
Argos, 579
Aristippus, 488 n.
Aristocracy, 347
Aristocritus, 437, 627
Aristodorus, 597 ff.
Ariston, 417
Aristonymus, 313
Artemisium (battle of), 353
Asia, 39, 41, 265, 279, 349 ff.
Asopus, 271
Aspasia, 330 f., 336 n., 337 f., 381
Ast, 392
Astringent, 167
Atarneus, 454 ff.
Athena, 31, 207, 277, 342 n.
Athens, 256, 267, 465, 475
Atlantic, 41, 283
Atlantis, 3, 41, 256, 265, 279 ff.
Atlas, 283
Autochthon, 283
Azaes, 283

Baccheius, 397
Barbarians, 349, 353 ff., 503, 585 f.
Becoming (oppd. to Being), 49, 53, 75, 113, 117, 125, 147
Bees, 273
Being (essential), 53, 65 f., 73, 125, 398
Belief (oppd. to Truth), 53, 75
Belly, 207 ff., 215
Below (oppd. to Above), 157 ff.
Bile, 223 ff., 229 ff.
Birds, 251; (caged), 553
Birth (second), 93, 249

629

INDEX

Bitter (flavour), 169, 175
Black (colour), 175
Bladder, 249
Bliss, Blissful Life, 488, 487; 579, 581 f.
Blood, 211, 217, 231
Blue (colour), 177
Body (oppd. to Soul), 65, 99 ff., 179, 241, 817, 535
Body-guard, 293
Boeotians, 365 f.
Boiling, 169
Bone (nature of), 165, 191 ff., 227 ff.
Brain, 193
Brass, 289
Brilliant (colour), 175
Bronze (metal), 147
Bryson, 403 n., 616 f.
Bubbles, 169, 225, 229
Bulls (sacred), 301

Cadmeians, 349
Cadmus, 369
Callippus, 389, 463, 473, 506 n., 588 n.
Cambyses, 351 n.
Carthaginians, 505, 514 n., 557, 575
Catarrhs (cause of), 231
Cause (two kinds of), 49, 51, 55, 105 ff., 111, 155, 177 f., 237, 461
Cebes, 542 n., 612, 615
Cecrops, 269
Cereals, 215
Chain (of evil), 577
Chariots, 299
Charmides, 478 n.
Charybdis, 545
Chestnut (colour), 177
Cimon, 356 n.
Circle, 65, 71 ff., 287 ff., 533 ff.
Cithaeron, 271
Cleito, 281, 289
Cleitophon, 311 ff.
Cold (oppd. to Heat), 157, 231
Colours (classified), 173 ff., 289
Compression (centripetal), 143
Conjunctions (of stars), 87
Connus, 337
Conon, 366 n.
Copy (oppd. to Model), 58, 113, 117 ff.
Corinth, 369, 433
Corinthians, 365 f.

Coriscus, 454 ff.
Corslet, 625
Cosmos, 51 ff., 55, 95
Council, 587
Counsel, 449 ff., 463 ff., 469, 497
Cratinus, 625
Cratistolus, 403
Crenides, 601
Creon, 405
Critias, 3 f., 27 f., 43, 256 ff., 478 n.
Critias (grandfather of above), 29 f., 43 (cf. 279)
Croesus, 405
Cronos, 87
Crown, 612, 619
Ctesippus, 331
Cubic (number), 59; (form), 133 n.
Cupping-glass, 213
Cureotis (feast-day), 29
Cyprus, 357
Cyrus, 351, 398, 405, 440, 445
Cyzicus, 360 n., 615

Daemon, 245 f. (cf. Deity, 515)
Danaus, 369
Darius, 351, 501
Datis, 351 f.
Datos, 601
Day (oppd. to Night), 81, 107
Dazzling (colour), 175
Dead, the, 377, 407
Death (causes of), 219
Definition (logical), 466, 532 ff.
Delphi, 425
Demiurge, 7
Democracy, 449 ff., 485
Democritus, 6
Demos, 453 (cf. People, 587)
Desires, 181 ff.
Despot, Despotism, 464, 579
Deucalion, 33, 275
Diaprepes, 283
Diarrhoea, 233
Dictator, 393, 575
Diogenes Laertius, 607
Dion, 385 ff., 463 ff.
Dionysius, the Elder, 386, 463, 487, 501 f., 567, 575, 585
Dionysius, the Younger, 385 ff., 463 ff., 610 ff.
Diseases (classified), 219 ff., 233 ff.
Divination, Diviners, 187
Divisions (treatise on), 612 f.
Docks (naval), 287 f.

630

INDEX

Doctor, 464, 497 (*cf.* Physician, 239, 445)
Dorian (life), 515
Doris, 411
Dropides, 29
Drugs, 243 f.
Dysentery, 233

Earnestness (oppd. to Jest), 455, 461 (*cf.* Serious)
Earth, 79, 85; (kinds of), 151 ff.
Earthenware, 151
Echecrates, 591 ff.
Ecliptic, 72 n.
Education, 315
Egypt, 283
Egyptian (priests), 269, 279
Elasippus, 283
Elders, 579
Electron, 215
Elements (material), 11 ff., 111 ff., 219
Elephants, 285
Eleusis, 363
Empedocles, 14
Entrails, 191
Ephors, 579
Epicurus, 390
Equator, 72 n.
Erastus, 454 ff., 621
Erechtheus, 36 n., 269
Eretrians, 351 f.
Erichthonius, 269
Eridanus, 275
Erysichthon, 269
Essence, 466, 532 n., 535 ff. (*cf.* Being)
Eternity, 77
Euboea, 450 n.
Eudoxus, 13, 615
Eumelus, 283
Eumolpus, 349
Eunuch, the, 501
Euphraeus, 449 ff.
Euripides, 397, 405 n.
Europe, 41 f., 279, 349
Eurybius, 435, 553 ff.
Eurymedon, 357
Eurystheus, 349 n.
Evaemon, 283
Evenor, 281
Excellence, 443 f.
Existent (oppd. to Becoming), 49
Expenditure (advice on), 621 f.

Experience, 457 f., 583
Extreme (terms), 67 ff.
Eyes (use of), 101, 107

Faction (civil), 517
Fear, 91, 517
Fermenting (process of), 169
Fevers (classified), 233
Fibrine (of blood), 221 ff., 231 f.
Ficinus, 393
Fire (element), 101; (kinds of), 145 ff.
First (principle), 398, 411
Fish (nature of), 253
Fish-weel, 207 ff.
Flame, 145
Flavours (classified), 167
Flesh (nature of), 191, 195 f., 221 ff.
Flood, the, 33, 275
Fluidity, 145
Folly, 233 (*cf.* Ignorance)
Forms (animate), 83; (principles), 113; (ideal), 105, 121
Frankincense, 155
Freedom, 581
"Friend of Greece," 579
Friendship, 323 f., 363, 457 ff., 503, 507
Fruit, 215, 285
Furies, 589
Fusible (water), 145 f.

Gadeira, 283
Gadeirus, 283
Galen, 391
Gê, 37, 87
Gelon, 505
"Giver of Titles," 209
Glass (nature of), 155
Globe (celestial), 86 n., 411, 627 n.; (brain), 193
God, Gods, 55, 79, 83, 87, 93, 127 f., 135, 177 ff., 187, 267, 279, 315, 343 f., 425, 454, 461, 625
Gold (nature of), 147; 277, 289 f., 397, 457
Goods (classified), 582 n.
Government (forms of), 484 n.
Greek (names), 279
Green (colour), 177
Grey (colour), 177
Guardians, 19, 271
Gums, 285
Gymnasia, 277

INDEX

Gymnastics, 241 f., 321

Hades, 99
Hail (nature of), 149
Hair (nature of), 165, 201 f., 251
Harbours, 287, 295
Hard (oppd. to Soft), 157
Harmony, 109, 179, 215, 247, 315
Harsh (flavour), 167, 173
Head (structure of), 99, 181, 193, 197 ff., 247, 251
Hearing, 109, 173
Heart (function of), 181 f.
Heat (oppd. to Cold), 155 f.
Heaven, 51, 84 ff. (*cf.* Cosmos).
Heavy (oppd. to Light), 157, 161
Hegesippus, 417
Helen (of Troy), 439
Helicon, 610 ff.
Helios, 33
Hellas, 277
Hellespont, 361
Hephaestus, 37, 267, 277
Hera, 87
Heraclean (stone), 215
Heracleidae, 349, 435 ff.
Heracleides, 388 f., 445 f., 468, 553 ff.
Heracles (pillars of), 41 f., 265, 283
Hermeias, 454 ff.
Hermes (star of), 79
Hermocrates, 3 f., 27
Hesiod, 31; (quoted), 603
Hicetas, 389
Hiero, 405, 515
Himera, 505 n.
Hipparinus (father of Dion), 575, 581, 587
Hipparinus (son of Dion), 476 ff., 584 n.
Hipparinus (nephew of Dion), 389, 489 n., 518 n., 584 n., 589
Hippocrates, 504 n.
Hoar-frost, 149
Homer, 31; (quoted) 87 n., 405 n., 540 n., 544 n.
Honey (nature of), 151
Humours (bodily), 235

Iatrocles, 627
Ice, 149
Idea (of Good, etc.), 7, 12, 400, 457, 472 ff., 530 n.
Ideal State, 465, 470, 518 n.
Ignorance, 233, 515

Ilissus, 275
Image (oppd. to Real Object), 77, 466 (*cf.* Copy)
Immortality, 247, 511
Incontinence, 235
Inflammation, 231 f.
Injustice, 317
Interval (in number-series), 67 ff., 97
Involuntary (evil), 235 f., 317, 562 n.
Isocrates, 390, 441, 474, 612 ff., 617
Isthmus, 271
Italy, 386, 427, 431, 525

Joinery (art of), 323
Justice, 315 ff., 373, 459

Kidneys (nature of the), 249
Kinds (Four, *i.e.* elements), 125 ff.
King (Archon), 347
King (of All), 400, 411
King (the Great, *i.e.* Persian, etc.), 257, 293 ff., 299 ff., 365 ff., 625
Kingship (oppd. to Despotism), 579 ff. (*cf.* 439)
Knowledge (nature of), 75, 466, 473, 533 ff.

Lacedaemon, 447
Lacedaemonians, 353 ff.
Laconian (kings), 585
Lamiscus, 559
Lamprus, 337
Laodamas, 601 ff.
Laomedon, 609
Law (reign of, etc.), 465, 518 n., 579 ff.
Laws (of Nature), 213, 225
Law-wardens, 587
Lechaeum, 369
Leochares, 617
Leontini, 359, 389, 473, 518 n.
Leptines, 500 n., 611, 617 ff.
Letters ("open" oppd. to "private"), 390 ff., 422, 474; (of alphabet), 110 n.
Leucadian (ship), 619
Leucippe, 281
Libya, 41, 265, 351
Light (nature of), 157, 161
Likely, the, 53, 111, 137, 263 (*cf.* Probable)
Likeness (in art), 261 ff.
Liver (nature of the), 185 ff.

INDEX

Living Creature, 55 f., 61, 75, 81 ff., 179, 203 ff., 287, 253, 585
Locrian (youths), 615
Love (Goddess of), 571
Love (sex-instinct of), 249
Lucanians, 387, 430 n., 607
Lungs (nature of), 183, 209 ff., 229, 249
Luxury (Sicilian) 474, 487
Lycabettus, 275
Lycophron, 417
Lycurgus, 440, 445, 579
Lye, 151
Lynceus, 539
Lysias, 313
Lysicleides, 419
Lysis, 331

Macedon, 449 f.
Man (nature of), 91, 177, 197, 249, 511, 617 (cf. 459)
Marathon, 353 ff., 367
Mardonius, 257
Marriage (regulations), 21
Marrow (nature of), 191 f., 205, 217 f., 221 ff., 227, 233 f., 249
Matter (Platonic), 10
Mean (oppd. to Extreme terms), 67 ff., 97
Measure (due), 237
Medes, 351, 501
Medicine (art of), 321
Mediterranean, 283
Megara, 491
Meletus, 480 n.
Memory, 539 f.
Menexenus, 330 ff.
Messene, 579
Mestor, 283
Metals, 144 n., 285
Midriff, 181, 205, 229
Minos, 405
Mirror (action of), 103 f., 185
Mnemosyne, 265
Mnaseus, 283
Mobile (substances), 145, 163
Model (oppd. to Copy), 53, 75, 83, 87, 113
Month (origin of), 81, 107
Moon, 79 f.
Moorlands, 273
Morning Star, 79
Mother (Plato's), 621
Mother (of the World), 119

Mother (Country), 343 f.
Motion (kinds of), 9, 55, 63, 71, 95, 141 ff., 241 ff.
Mountains, 273, 295 ff.
Mouth (nature of), 199, 209 f.
Murder (of Dion), 507
Muses, 109, 265
Music, 19, 241
Myrians, 609
Myronides, 627
Mysteries, 506 n.
Mytilene, 361

Nails (of fingers and toes), 203
Name (in logic), 466, 532 ff.
Name (ancient), 269, 279
Navel, 183, 205
Necessity (as Cosmic factor), 91, 109, 127
Neïth, 31
Nereids, 291
Nestor, 405
Nieces (Plato's), 619
Night (origin of), 81, 107
Nile, 35
Niobe, 33
Nostrils (nature of), 171, 211
Number, 59, 107, 127
Nurse (of Becoming), 113 241; (Earth), 85
Nutriment, 205 f., 215 ff.
Nysaeus, 389

Ocellos, 607
Ochre, 177
Odysseus, 405
Oenophyte, 357
Oil (nature of), 149
Old age, 219
Olympia, 388, 399, 403, 468, 559
"Open" Letters, see Letters
Opicians, 577
Opinion (oppd. to Knowledge), 75, 121 f.
Opisthotonus, 229
Orichalcum, 285, 289 f., 301
Oropia, 271
Ortygia, 388, 552 n.
Other, the (oppd. to "the Same"), 8, 65 f., 71 ff., 79 f., 97

Pain, 163 ff., 181, 233 f., 425
Painters, 261 ff.
Paion, 265

633

INDEX

Palace, 287 f.
Palamedes, 405
Palinode, 422, 439
Panathenaea (feast), 28 n.
Parmenides, 6
Parnes, 271
Passion, 180 ff.
Pattern (oppd. to Copy), 57
Pausanias, 405
Peiraeus, 363
Peloponnesus, 505, 515, 545 f., 559
Pelops, 369
Peltasts, 555 f.
Perdiccas, 449 f.
Periander, 415
Pericles, 337 f., 405
Persia, 256
Persians, 349 ff., 503
Phaedo, the, 612, 624 n.
Phaethon, 33
Philaedes, 612, 625
Philagrus, 625
Philip, 390
Philistides, 427
Philistion, 417
Philistus, 488 n., 494 n.
Philonides, 592 f.
Philosopher-king, 386, 437 n., 463, 465, 469
Philosophy, Philosopher, 25, 107, 251, 407 ff., 459, 465 ff., 472 ff., 483, 491 ff., 527 ff.
Philostratus, 506 n., 588 n.
Phlegm, 223 ff., 229 ff.
Phoenicians, 577
Phorkys, 87
Phoroneus, 33
Phrynion, 595
Piety (filial), 498 n.
Piraeus, 479
Pirates, 602 n.
Plain (of Atlantis), 295 ff.
Planets, 72, 79 ff.
Plants, 203
Plataea, 355, 367
Pleasure, 163 ff., 181, 233 f., 425
Pnyx, 275
Poets, 25, 31
Politics, 319, 345 f.
Polyeidus, 405
Polyxenus, 403, 417, 488 n., 612 ff
Portents, 87
Poseidon, 279 f., 289 f., 301, 342 n.
Power, 454, 457, 513

Prayer, 259, 605 (*cf.* 49)
Preludes (of laws), 427
Priam, 608
Priests, 33 ff., 269, 277
Probable (oppd. to True), 93 127, 137, 147, 249 (*cf.* Likely)
Projectiles, 213
Prometheus, 405
Prophets, 187
Proportion, 59, 61, 179
Pseudo-Smerdis, 501 n.
Pungent (flavour), 169
Purple (colour), 177
Purpose (divine), 203
Pyramid (form), 133 n., 137
Pyrrha, 33
Pythagorean (treatises), 612, 614

Quality, 398, 411, 466, 537 (*cf.*115 ff.)

Race-course, 293
Realities (*i.e.* Ideas), 121, 147, 398, 466
Reason, 7, 9 ff., 83, 107 f., 121 f.
Receptacle, Recipient (*i.e.* material substrate), 119, 125, 141
Red (colour), 175, 217
Republic (ref. to the), 19, 401, 482 n.
Reservoirs, 297
Respiration (nature or), 207 ff.
Rest (oppd. to Motion), 141
Reverence, 517
Revolution (political), 479
Revolution (of Circles, etc.), 71 ff., 95 ff., 109 ff., 247
Rhea, 87
Rhythm, 109
Riddling (exposition), 411
Rites (funeral), 379
Roughness, 163
Rust (nature of), 147

Sacred (Disease), 231
Sacrifice, 301 f., 557
Salamis, 353 ff., 367
Saline (flavour), 169
Salt (nature of), 153, 195
Same, Sameness (oppd. to Other), 8, 65 f., 71 ff., 81, 85, 97
Sap (vegetable), 149
Saviour (God), 111, 509, 527
Scabs, 229
Scepsis, 454 ff.
Science, 457

INDEX

Scopas, 617 n.
Scylla, 545
Scythians, 351
Seed (nature of), 235, 249
Sensation, Sense, Sensible, 49 f., 75, 91, 95 f., 119 ff., 155 f.
Serious (oppd. to Play), 467, 541, 611 (*cf.* Earnest)
Serum, 220 ff.
Ship-yards, 293
Shivering (cause of), 157, 231
Sicily, 359, 386 ff., 419 ff., 515
Silver, 277
Simmias, 625
Simonides, 405
Sinew (nature of), 195 f., 221 ff., 227 f.
Skin (nature of), 201 ff.
Slavery, 339, 581
Sleep (nature of), 103
Smells (classified), 171
Smoothness, 163
Snow (nature of), 149
Socrates, 479 ff., 601 ff., 625
Socrates (young), 417, 603
Soft (to touch), 157, 173
Solid (body), 59, 127, 132 f.; (angle), 132 f.
Solon, 29 ff., 265, 269, 279, 398
Sophists, 25
Sophrosyne, 617 n.
Soul (nature of), 8 ff., 65, 91, 179 ff., 241, 245 ff., 317
Sound (classified), 109, 173, 213 ff.
Space (as Matter), 10 ff.
Speech (purpose of), 107
Speusippus, 417, 546 n., 621
Sphagia, 359
Sphere, 63
Spherical (form), 99, 159, 535
Spleen (nature of), 189
Springs (water), 281, 291
Square (numbers), 59
Stars, 79 ff., 85
Stesichorus, 489
Stone (nature of), 151, 289
Stone-quarries, 417
Strabo, 454
Stream (visual), 101 ff., 165, 173 ff.
Substrate (material), 10 ff.
Such-like (oppd. to This), 115 ff.
Sun (planet), 79, 81, 107
Sweat, 225, 229
Sweet (flavour), 171

Syllables (oppd. to Letters), 111
Symmetry, 237 f.
Syracuse, 386 ff., 419 ff., 483

Tanagra, 357
Tarentum, 387, 521 ff., 559
Tears, 175, 225
Teison, 627
Temples (of Atlantis), 287 f., 291 f.; 585 f.
Terillus, 625
Test (of sincerity), 527 f.
Tetanus, 229
Tethys, 87
Tetters, 229
Thales, 405
Thasos, 601
Theaetetus, 13
Thearidas, 500 n.
Theatres, 445
Theban (dialect), 543
Theodotes, 435, 445 f., 468, 553 ff.
Theramenes, 480 n.
Theseus, 269, 348 n.
Thirty, the, 478 f.
Thorax, 181
Thrasybulus, 362 n., 480 n.
Thrasyllus, 385, 480 n.
Thrasymachus, 311
Thunderbolts (cause of), 215
Timaeus (of Locri), 3, 25, 47, 256 ff.
Time (creation of), 77 ff., 107
Timotheus, 625
Tin (metal), 289
Tiresias, 405
Tisias, 557
Tissaphernes, 360 n.
Token (of authenticity), 615 (*cf.* Sign, 625)
Tongue (nature of), 167, 199
Transformation (of animals), 251 f.
Transparent (forms), 173
Trench (of Atlantis), 297 f.
Triangles (basic), 12, 127 ff., 141, 145, 191, 217 f., 243
Trojans, 609
Truth (oppd. to Belief), 33, 53, 97, 247, 443, 495, 533 ff.
Tuscany, 43, 283
Tyrants, Tyranny, 395, 493 ff., 573 ff.

Unanimity, 325
Uniformity (physical), 143 ff.

635

INDEX

Universe, 3, 49, 57 ff. (*passim* in *Tim.*, *cf.* Cosmos, Heaven, All, World-all)
Uranus, 87

Veins (nature of), 167, 205
Verjuice, 151
Violet (colour), 177
Vision (benefit of), 107
Voices (of polities), 449
Void (space), 143, 151, 211, 215
Voluntary (evil), 235, 562 n.

Water (kinds of), 145 ff.
Wax (nature of), 155
Wealth, 433
White (colour), 175
Windpipe, 183, 209
Wine, 149

Wisdom, 454
Wolf-love, 437
Woman (nature of, etc.), 21 f., 93, 249 f., 271
Womb (nature of the), 249
World-Soul, 8, 91
World-Year, 82 n.
Writing (dangers of), 417, 531 ff., 541

Xanthippus, 337
Xenocrates, 546

Year (origin of), 81 f., 107
Yellow (colour), 175, 195

Zeller, 392
Zeus, 87, 275, 405, 493, 559, 609